Delano's Domain
A History of Warren Delano's Mining Towns of Vintondale, Wehrum and Claghorn
Volume I
1789-1930

by

Denise Dusza Weber

The A. G. Halldin Publishing Company, Inc.
Indiana, Pennsylvania 15701-0667

ISBN 0-935648-33-X

Copyright ©, 1991 Denise Dusza Weber

All rights reserved. No part of the material protected by this copyright notice may be reproduced or utilized in any form or by any means, electronic or mechanial, including photocopying, recording, or by any informational storage and retrieval system, without written permission from the copyright owner. Printed in the United States of America by The A. G. Halldin Publishing Company Inc., Indiana, Pennsylvania 15701-0667.

FOREWARD

Well, Aggie and Todd, it's finally done. Sure hope it meets your approval, not to mention "Eddie" Thompson's, Malcolm Cowley's and Emmet Averi's. Tood, you wrote and said you'd be watching over me, nudging me on. Well, I think your poking finally did some good. The book is certainly long enough and has enough photographs, but it still is not what I naively expected to publish way back in 1982. After two sabbatical leaves and numerous summers to finish the writing and the research, I thought that I owed everyone something concrete. So **Delano's Domain** is the result. It is not footnoted as planned, but then I can rationalize that my audience will not be an academic one. There is no index, but I can number the pages in Volume II consecutively and have time later to complete an index. I know it is over eight years overdue; but look at the bright side, this book is much better researched than the first one would have been. When I re-read my first rough drafts, I could not believe that I accepted some of that information as accurate.

This book is the culmination of ten years of research and writing. My only purpose, then and now, is to preserve, in writing and through photographs, the history of a unique mining town before it is lost forever. The chapters on Wehrum, Claghorn, and Bracken were an afterthought, perhaps an even more urgent afterthought as they disappeared over sixty years ago. Those who have never lived in Vintondale have difficulty understanding what draws her native sons and daughters back every year to the Homecoming. Perhaps this book will help the reader understand that Vintondale magnetism a little better.

Pleasant Reading!
Denise Weber

ABBREVIATIONS

The following abbreviations appear throughout the book:

BL&YC RR	Black Lick and Yellow Creek Railroad
BL&IC	Blacklick Land and Improvement Company
BR&P RR	Buffalo, Rochester and Pittsburgh Railroad
C&I RR	Cambria and Indiana Railroad
GC&C	Graceton Coal and Coke Company
LC&C	Lackawanna Coal and Coke Company
LI&SC	Lackawanna Iron and Steel Company
PRR	Pennsylvania Railroad Company
UMWA	United Miner Workers of America
VCC	Vinton Coal and Coke Company
VLC	Vinton Lumber Company

The following abbreviations appear in the appendix:

BV Twp.	Brush Valley Township
CCDB	Cambria County Deed Book
EW Twp.	East Wheatfield Township
ICDB	Indiana County Deed Book
WW Twp.	West Wheatfield Township

TABLE OF CONTENTS

Chapter I.
B.D.-Before Delano 1
   A. Land Grants 1
   B. **Eliza** 4
   C. Ritter's and Rodgers' Land Transactions 8
   D. Between the Iron and the Coal 10
   E. Blacklick Land and Improvement Company 14

Chapter II.
Lumber, Coal, and Railroads 15
   A. Lumbering in Vintondale 15
   B. Cambria and Indiana Railroad 21
   C. Malcolm Cowley 24
   D. Rexis: Another Mystery in History 26
   E. Railroads in the Valley 28
   F. Roads and Bridges 32

Chapter III.
Biographical Sketches of Coal and Lumbering Entrepreneurs 35
   A. Warren Delano 35
   B. Clarence Claghorn 45
   C. Augustine Vinton Barker 49
   D. Webster Griffith 50
   E. Samuel Lemon Reed 51
   F. James Mitchell 52

Chapter IV.
Mining in Vintondale, 1894-1930 53
   A. Vintondale Mine Operations 53
   B. Cambridge Bituminous Coal Company 57
   C. Algernon Taylor Burr 58
   D. Schwerins 60
   E. Otto Hoffman 61
   F. Who's Minding the Office? 64

Chapter V.
Union Activity: 1894-1930 88
   A. Vinton Colliery vs. The United Mine Workers of America: 1894-1921 88
   B. Strike of 1922 90
   C. Labor Activity in 1923 96
   D. Strike of 1924 96

Chapter VI.
Vintondale Starts to Grow 99
   A. Vintondale Before World War I 99
   B. First Families 101
   C. Town Improvements 107
   D. Memories of Vintondale 113
   E. Vintondale and Vinton Colliery Go to War 127
   F. Ethnic Heritage 129
   G. Vintondale's Ethnic Minorities 133
   H. Town Tidbits 137

Chapter VII.
Public Services 139
   A. Water Companies in Vintondale 139
   B. Vintondale Borough Council 139
   C. Medical Care in Vintondale 141
   D. Education in Vintondale 146
   E. Vintondale Volunteer Fire Company 150
   F. Post Office 151

Chapter VIII.
Businesses 152
   A. Vintondale Supply Company 152
   B. Vintondale Inn 154
   C. Vintondale's Jewish Merchants 157
   D. S.R. Williams 158
   E. Billie's Confectionery 160
   F. Nevy Brothers 160
   G. Freemans 162
   H. Cupp Stores 162
   I. Vintondale State Bank 162
   J. Vintondale Amusement Company 164

Chapter IX.
Churches 166
   A. Vintondale Baptist Church 166
   B. Catholic Churches in the Valley 167
   C. Church of God 169
   D. Sts. Peter and Paul Russian Orthodox Greek Catholic Church 169
   E. Sts. Peter and Paul Cemeteries 172
   F. Presbyterian Church 173
   G. Hungarian Reformed Church 173
   H. Christian and Missionary Alliance Church 177

Chapter X.
Potpourri 178
   A. Automobiles 178
   B. Brief Biographies 178
   C. Dance Your Troubles Away 181
   D. Sports 181
   E. Town Tragedies 184
   F. Vintondale Hunting and Fishing Club 186
   G. Bracken 187

Chapter XI.
The Notorious Side 190
   A. Rough, Rowdy, and Ready to Fight 190
   B. Black Hand 194
   C. Ku Klux Klan 196

Chapter XII.
Wehrum 197
   A. Lackawanna Coal and Coke Company, 1901-1904 197
   B. Henry Wehrum 202
   C. John A. Scott 203
   D. Wehrum's Second Chance 203
   E. The Wehrum Mine Explosion 204
   F. Lackawanna After the Explosion 208

G. Labor Activity 211
H. Wehrum: The Town, 1901-1904 212
I. Roads and Bridges in Wehrum 219
J. Wehrum Schools 220
K. Wehrum Churches 221
L. Leisure Time in Wehrum 223
M. Wehrum Anecdotes 225
N. Wehrum Town Tragedies 225
O. Ya Done Good Boy! 226
P. And the Bomb Dropped! 227

Chapter XIII.
Delano's Other Towns 234
  A. Claghorn 234
  B. Graceton 245

APPENDIX
  A. Birth Certificates: 1906-1921 258
  B. Birth Certificates: Bracken 294
  C. Death Certificates: 1906-1921 297
  D. Death Certificates: Bracken 325
  E. Vinton Colliery Company: 327
    1. Annual Report 327
    2. Financial Records 330
    3. Fatal Accidents 336
    4. Non-Fatal Accidents 337
    5. Mine Operations - 1912, 1913, 1914, 1917, and 1923 341
    6. Heshbon Lands 357
    7. Lands Subject to 1916 Lease 359
  F. Lackawanna Coal and Coke Company 361
    1. Annual Report 361
    2. Fatal Accidents 363
    3. Non-Fatal Accidents 364
    4. Daily Time Report: March 9, 1923 366
    5. Bethlehem Mines Corporation: Police Department Reports, 1926-29 ... 367
  G. Graceton Coal and Coke Company: Annual Report 369
  H. Commercial Coal Mining Company 371
    1. Annual Report 371
    2. Fatal Accidents 372
    3. Non-Fatal Accidents 372
  I. Mine Examinations 373
  J. Prosecutions for Violations of Mining Laws 375
  K. Coal Miners' Pocketbook 376
  L. Evictions 377
    1. Vinton Colliery Company 377
    2. Claghorn 379
    3. Graceton Coal and Coke Company 379
    4. Lackawanna Coal and Coke Company 380
  M. Receipts From the 1909 Vintondale Strike 381
  N. Fires in Vintondale 390
  O. Privately Owned Lots in Vintondale 391

- P. Tax Assessor's Records ... 392
  - 1. Claghorn ... 392
  - 2. Rexis ... 398
  - 3. Wehrum ... 402
  - 4. Foreign Miners: Wehrum ... 410
- Q. Obituaries ... 418
- R. Wehrum Cemetery ... 427
  - 1. Translation of Tombstones ... 427
  - 2. Cemetery List ... 430
  - 3. Russian Church Deeds ... 432
- S. Sales of A.V. Barker Lots ... 433

BIBLIOGRAPHIES
- A. Primary Sources ... 434
- B. Secondary Sources ... 442

Chapter I
B.D.- Before Delano

LAND GRANTS

In 1681, King Charles II of England granted the territory between New York and Maryland to William Penn, a noted Quaker. The land, which was named Pennsylvania (Penn's Woods) by Charles II, was in lieu of payment of a debt owed to Penn's father, Admiral Penn. A charter was signed that year, and as early as 1682, William Penn set up a land office to deal with the problems of charting the boundaries of the various parcels of land which he sold in his new colony. Between 1682 and 1684, Penn sold 600,000 acres of land to immigrants who had been attracted to Pennsylvania through Penn's promotional efforts. Penn's selling price was five pounds per one hundred acres.

Upon arriving in his proprietorship in 1682, Penn purchased the land from the Indians. His heirs, however, did not adhere to the same principles of fairness and cheated the Indians of their land in the Walking Purchase of 1736. Through an act of the General Assembly in 1779, the Penns lost the proprietorship of Pennsylvania, and the land office became a state function.

The geographical focus of this book is not the entire state of Pennsylvania, but one of her less important streams: the north and south branches of the Blacklick Creek, both of which rise in Cambria County near Ebensburg in the hills west of the Allegheny Front. The historical focus involves three towns built along that Blacklick Creek; the purpose of these towns was to house miners and lumbermen who would exploit the natural resources of the Blacklick Valley.

Blacklick Valley Indians

Before the arrival of the white man, the Indians who roamed this valley were members of the Shawnee and Delaware tribes, migrating freely throughout Western Pennsylvania as traders, hunters and trappers. The Kittanning Path, a major east-west trail, was only five miles to the north of present-day Vintondale via the north branch of the Blacklick. Today, Route 422 basically follows that trail.

The fork of the Blacklick would have been a prime location for an Indian trading center, if not a permanent village. The Blacklick could have been used for fishing and transportation. The forests abounded with game. The large flood plain could have been used for agriculture. It is not known if there were any Indian settlements in the vicinity of Vintondale, but arrowheads found at the ball diamond, especially after the scraping and leveling for games, indicate that the Indians at least passed through the area. Unfortunately, this land today is probably too scarred from lumbering and mining to conduct a proper archeological dig for Indian remains.

Arrival of the White Man

In 1764, George Findley was one of the first white men to settle in what became West Wheatfield Township. The date of his arrival coincided with the victory of the British in the French and Indian War (1756-1763). The territory west of the Allegheny Front was now open for settlers, who continued to face problems with the Indians until the 1780's. The Beards, Brackens, Ragers, and Shumans were early pioneers in the Rager Hollow and upper Maple Street areas of Chickaree Mountain. The small streams passing through what became Vintondale were named after these early pioneers, i.e., Shuman's Run and Bracken's Run. After the defeat of the British in the American Revolution, the land west of the Allegheny Front was waiting to be claimed.

Selling the Western Lands

Both in the proprietory era and afterward, in order to claim any piece of land, the purchaser had to apply to the land office for a warrant, which was permission to have the tract in question surveyed. When the survey was returned to the land office, the applicant was then expected to pay for the land and apply for a patent, which granted a clear title for the land to the applicant. Many would-be purchasers thought that the warrant gave them possession of the land and never bothered to apply for the patent. As late as 1945, Cambria County had 210 tracts of 63,525 acres without titles from the state. Indiana County had 387 tracts with 136,600 acres untitled.

In 1768, the Penn heirs were represented at the negotiations for the Treaty of Fort Stanwix, held in New York. By terms of the treaty, the Indians ceded all land south of the Ohio and east of the Allegheny to the Penns. This concluded the Penn's final purchase from the Indians. Land south of the Purchase Line was open for settlement. There were so many applicants for this open land that the land office tried a lottery that year. This was quickly abandoned, and applications were then taken on a first come-first served basis. This time, the purchaser was limited to one application of 300 acres. Land speculators took out applications under aliases or had others apply for warrants for them. The price of this land under the Penns was 15 pounds, 10 shillings per 100 acres in the east and 3 pounds, 10 shillings for land in the west. This was considered high by many settlers and helped discourage purchase of lands in the west. Between 1792 and 1814, the Pennsylvania government raised the acreage limit to 400 acres. Unimproved land sold for 50 shillings per 100 acres. There had to be proof of settlement on the land in order to get the warrant; grain was to be grown on this land as a sign of improvement. The easiest way to claim improvement was to girdle (removing a band of bark on the tree trunk) an acre of trees on the tract.

After 1790, the wilderness bordering the north and south branches of the Blacklick Creek was surveyed and claimed; surveys were made in 1794; the patents were awarded in a majority of cases. Since the counties of Cambria and Indiana had not yet been created, these tracts were located within the boundaries of Westmoreland, Bedford, Huntingdon and Somerset Counties. In less than a year after the warrants had been issued, the tracts, which had been given stately names reminiscent of English manors, were transferred to James Fisher, a Philadelphia businessman.

The tracts of land with the names of the applicants, title of estate, acreage, date of warrant, and date of transfer to James Fisher are listed below. (Please check the map for location of the tracts.) The price paid by the applicants was 5 shillings, 9 pence.

Sources of estate names are unknown, as is the elusive James Fisher. Research indicates that he was a Philadelphia resident, but judging from how quickly Fisher's purchases followed the granting of the warrants of these lands, one assumes that he was a land speculator. These tracts passed through several hands, such as David Ritter, the Rex Family and the Cambria Land Company until 1892 when Judge Augustine Vinton Barker began to purchase these various tracts for the Blacklick Land and Improvement Company.

NAME	ESTATE NAME	ACREAGE	WARRANT	TRANSFER
Daniel Levy	"Milford"		5/13/1793	1/4/1794
John Simpson	"Castlemartyr"		5/3/1793	1/11/1794
Alexander Hunt		433a.	12/21/1793	
Paul Lebo	"Castleblakeney"	405.105	12/20/1793	5/3/1794
Jasper Ewing		433.153	12/21/1792	
Daniel Hurley		405	3/3/1793	
Jacob Pressinger	"Leinster"	405.105	5/3/1793	1/2/1794
James Reese	"Santry"	405.105	5/3/1793	3/10/1796
William Morris		405	5/3/1793	
Adam Rand	"Darby"	405.105	5/3/1793	11/5/1793
Isaac Bonser	"Emyvale"	405.105	5/3/1793	1/20/1794
Henry Shoemaker			5/3/1793	
John Hambright	two parcels 50 and 148a.		12/19/1792	
Jacob Messersmith		433	12./21/1792	
Jeremiah Moshier	"Trim"	405.103	5/3/1793	12/20/1793
Mathias Young		405	5/3/1793	
Jacob Rupp	"Millford"	405	5/3/1793	
John Hubley	"Drewstown"		5/3/1793	7/4/1793
Christian Stoner		405.105	5/3/1793	
John Stoner	"Christiana"	405.105	5/3/1793	1/18/1794
Mathias Slough (Slow)	"Algiers"	405.105	5/3/1793	7/3/1793

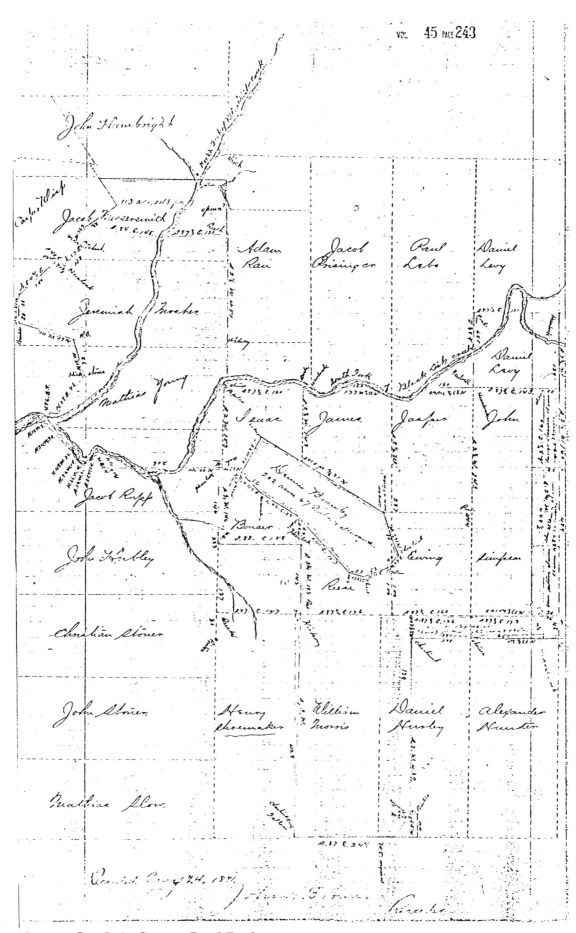

Source: *Cambria County Deed Book*

ELIZA FURNACE
"THE OLD IRON FURNACE"

Coming down the straggling road into Vintondale from the west, the first thing which attracts one's attention is the immense, ragged structure of brown stone that is Ritter's Iron Furnace. It is really Vintondale's only claim to fame since according to the "Johnstown Democrat", the furnace is of grand proportions, and from its shape and the rough-hewn stones fitted beautifully together without mortar, makes one think immediately of the Egyptian pyramids. It rears itself about a hundred feet from the road from a clump of delicate elder and locust bushes. Topping the furnace and behind the dead black coils of blast pipes, the feathery tops of little trees peep out, their roots bedded in the decay of the ancient stones.

Time has been kind to the old furnace. It has removed all semblance of industrialism and has left it clothed in softened hues, great quiet, the comings and goings of no more than small birds, the winds, the sun and the rain. In fact, to the inhabitants of the community, it has ceased to be Ritter's Furnace, and is simply the "Old Iron Furnace". The years have passed into decades and the decades have rolled past the century mark, but the old furnace remains aloof taking toll of it all. Man in his greedy search for coal has changed the mountains, which the poet thought were changeless: he has gashed them of their timber and so subjected them to the erosion of countless rains until they no longer resemble the mountains of a century ago. But the Old Iron Furnace stands there watching it all.

Some time about the year 1820, Ritter, an Englishman, built three charcoal furnaces in the Black Lick Valley. He secured his ore from the mountains along the Black Lick Creek. The making of charcoal was a simple matter since the timber in Western Pennsylvania was then uncut. The first of these furnaces was the one which concerns us now; the second was built near Wheatfield in Indiana County and is now completely gone; the third, the Buena Vista Furnace, was built near Dias in Indiana County and is still in fair condition. The ore, smelted in the furnaces, was hauled across the mountains to Conemaugh.

As so frequently occurs about things that pass beyond our ken of time, legends have grown up about the Old Iron Furnace. The tales have been colored by local superstition and imagination. An old, old woman whose grandfather was a contemporary of Ritter, tells as fact what everyone guesses at. The whole of Ritter's venture was a miserable failure and Ritter is said to have hanged himself one cold winter night at the front of the furnace. Some wise people say, however, that he hanged himself not because of finances but because his six-year old son fell in the furnace and burned to death.

And of course there are those thrill-loving inhabitants that say on a cold winter night Ritter may be seen hanging from the front on the furnace. I have never been fortunate enough to see him, and somehow I cannot help feeling the persons who utter such inanities are poor indeed. Can they perhaps live all their life there in the shadow of that calm, ageless pile of simple beauty and learn no more from it than a foolish tale?

Bruce Lybarger
Vintondale, Pa.

Eliza Furnace, c. 1985.
Courtesy of Diane Dusza.

The above essay was written in the 1930's by Bruce Lybarger, a teacher at the Vintondale High School. This essay appeared on a Christmas card sent by Milton Brandon, Superintendent of the Vinton Colliery Company. Mr. Lybarger's writings conger up some mysterious events at the furnace, but contrary to popular thought, these events have proven to be nothing more than legends. This essay was reprinted with the verbal permission of the late Katherine Morris Lybarger Uzo.

Industry Comes to the Blacklick Valley

Coal was not the first natural resource which drew speculators to the Blacklick Valley. Even lumber was preceded by another resource to which few, if any, pay attention today. Investors were attracted by a band of carbonate iron ore at a depth which ranged from six inches to two feet thick resting in a bed of shale. The 1877 **Pennsylvania Geological Survey** reported that near Vintondale the ore was worked at the forks of the Blacklick Creek. The ore outcropped on both sides of the North Branch of the Blacklick, and a road, later nicknamed the "China Road", led from the furnace to the ore beds on the same side of the hill. Traces of iron workings, along with an old apple orchard, can be found above Rexis on the opposite hill. It is believed that houses for the employees were once located near the diggings. In addition to the iron ore, the **Geological Survey** reported several layers of limestone and three coal beds in the valley.

With the necessary ingredients available to manufacture iron: limestone, iron ore, ample timber for charcoal, and water power for the bellows, the fork of the Blacklick seemed an ideal site for construction. It was here in 1845-46 that David Ritter and George Rodgers of Ebensburg purchased large tracts of timber for the manufacture of charcoal and then constructed the Eliza Furnace.

Iron Production at the Eliza Furnace

With all the ingredients readily available, Ritter and Rodgers contracted a stone mason to construct the furnace against a bank about one hundred feet from the North Branch and three hundred feet from the fork. The furnace stack of cut stone was reportedly built by Tommy Devereaux, a well-known local stone mason who lived outside of Ebensburg. John and William Gillin, Irish immigrants who settled near Vinco around 1830, also laid claim to being the stone masons who built the furnace. Constructed of cut stone fitted without mortar, Eliza rose thirty feet above the flood plain. Her bosh, the interior opening, was nine feet in diameter. The iron ore, charcoal and limestone were dumped into the bosh from an opening at the top of the furnace. A bridge and cast house connected the furnace to the hillside.

The charcoal, the carbonaceous substance produced by heating wood without oxygen, was produced locally in large charcoal pits scattered around the area. There were supposedly traces of charcoal pits along the Dinkey Tract near Sandy Beach (across the Blacklick from Lackawanna's #3 mine). Local woodsmen split hardwoods, especially oak, into three or four foot lengths and stacked them in a circle, leaving an open space for the fire. The wood was piled in a cylindrical shape ten to twelve feet high with a vent on top. The sides of the pile were sealed with layers of leaves and dirt. Once lit, the fire inside the pile was controlled by the woodsmen who kept several small openings in the pile to observe the burning process. After the pile had burned down and cooled, the charcoal was sent to the furnace where it served a dual purpose in the iron-making process. When lit, charcoal provided the necessary heat to smelt the iron ore. Also the gasses given off by the burning charcoal united with the oxides in the iron ore and formed carbon dioxide, leaving behind pure molten iron.

The limestone acted as a flux, a substance which helped the iron and its impurities melt at the same temperature. The impurities then floated to the top of the molten iron and were drawn off.

During the smelting process, a blast of air was injected into the furnace by means of water-driven bellows. **Eliza** was one of the first area furnaces to attempt the hot blast method. Hot air was gathered at the base of the furnace and directed by pipe to the coils at the top. This hot air circulated around the coils and heated the cold blast which was produced by the bellows. A second pipe carried the cold blast from the bellows to the coils. The warmed air then entered the furnace near the top of the bosh. However, this method did not work well, and the air temperature was only raised by about ten degrees.

At the base of the furnace was a flat area covered with sand. Molds in the shape of bricks or simple utensils were carved into the sand. When the smelting process was finished, the furnace was "tapped", and the molten iron ran from the base of the furnace into the molds. After these blocks of iron, called "pigs", had cooled, they were removed from the molds and sold as is or further refined and fashioned into wrought iron. Production of iron at **Eliza** in 1848 was listed as 1080 tons with a capacity of making 1800 tons.

Such an operation, from the making of the charcoal to the shipping of the "pigs", required many laborers. It was estimated that as many as ninety men were employed at the furnace. Since the nearest water transportation was on the Pennsylvania Canal at Ninevah, the iron had to be carted there by horsae and wagon, a distance of at least ten miles. Consequently, forty-five horses and mules were used to haul raw materials and finished iron. There is no concrete evidence of any auxiliary furnace structure surviving into the 20th century. Possible surviving buildings were the white building behind the iron furnace and the house owned by Edward Findley and later Bill Clarkson. Wilbur Misner said that the Findley/Clarkson house, adjacent to the iron bridge, was moved across the road when the railroad was extended to Wehrum about 1901.

Failure of the Furnace

The Eliza Furnace was never a financial success for several reasons. 1) The local iron ore was of poor grade, and in 1844, the Mesabi Range in Minnesota was discovered, making high grade ore readily available. 2) In 1846, the US government lowered the tariffs on imported iron; local producers no longer had an advantage over imported iron, especially cheap iron rails from England. 3) Because the Blacklick Creek was unsuited for transportation, the iron had to be hauled overland to the Pennsylvania Canal at Ninevah, an expensive venture. 4) Many thought that the Pennsylvania Railroad was going to be built through the Blacklick Valley instead of the Conemaugh Valley. This would have greatly cut transportation costs for Ritter. When the Eliza Furnace was constructed in 1845, Ritter and Rodgers may have been gambling on an almost sure thing. After all, the Blacklick Valley was one of several possible routes surveyed by Colonel Charles Schlatter in 1840 and the one recommended to the Pennsylvania State Legislature. Local groups lobbied hard to influence the railroad's decision; a group of those favoring the Blacklick route held a meeting in Blairsville in November, 1845. The Pennsylvania Railroad itself was not chartered until April, 1846. As of March, 1848, the decision of exactly where to build the mainline was still not made; an Indiana County interest group, in a last ditch effort, named a committee to purchase stock in the Pennsylvania Railroad as a means of pressuring it to proceed with the Blacklick Valley route. Two of the committeemen involved in this vain effort were Elias McClelland of **Buena Vista Furnace** and David Stewart of **Blacklick Furnace**. It is not known if David Ritter was part of any lobbying group, but by 1848, his furnace had failed. Ritter disappeared from the area, and Lot Irvin, Ritter's new partner, took over management of the Greenville Furnace. However, even if the railroad had come through the Blacklick Valley, it is doubtful that the Eliza Furnace could have been profitable because advances in technology, such as the Bessemer process, in the 1850's and 1860's soon made that type of furnace obsolete. Local historian, Louise Bem, mused on the fate of the **Eliza**, "Who knows, if the PRR had been built through Vintondale instead of the Conemaugh Valley, maybe the Eliza Furnace instead of the Cambria Furnace would have become Bethlehem Steel." An interesting thought!

The Legends Eliza Holds

In September, 1845, George Rodgers sold his interests in the furnace and timberlands to David Ritter, who then was able to attract a new partner, Lot Irwin or Lot Irvin, supposedly a wealthy Huntingdon County farmer. (See following section of Ritter-Rodgers-Irvin land transactions.) However, the financial situation at the furnace did not improve. This led to the rise of several legends about the furnace and its owners. One said that Ritter went to Pittsburgh to get financial assistance from Dr. Peter Shoenberger, wealthy ironmaker and one of the founders of the Cambria Iron Company of Johnstown in the 1850's. When Ritter returned with the money, he used it for himself instead of covering the furnace's debts and the workers' back wages. According to another version, his partner, Irwin, in despair, committed suicide on his farm in Huntingdon County.

Still another legend said that Ritter hung himself because his wife ran away with his partner. Another version said he hung himself because his small son fell into the furnace and burned to death. In either case, Ritter did not commit suicide at the furnace. He died on February 12, 1858 at the age of 63 in Cattawissa, Columbia County, Pennsylvania.

Even the source of the name Eliza presents a mystery. Some have said the furnace was named after Dr. Shoenberger's daughter, Elizabeth. A more logical explanation was discovered in the early 1980's in Lloyd Cemetery in Ebensburg by Blair Tarr. A single monument marks the graves of Eliza Ritter, one of Ritter's wives, her son David M. and his wife, Catherine Jones Ritter. Eliza Ritter, the most likely source of the furnace's name, died on November 23, 1873 at the age of 57. By calculating the dates on the obituary notice and the tombstone, David Ritter was twenty-one years older than Eliza. David M. Ritter was definitely a son of furnace builder, David Ritter.

Over the years, the Eliza Furnace has been called various names. The 1871 **Indiana County Atlas** listed it as the Ritter and Nevins Furnace and placed in entirely in Indiana County on the west side of the Blacklick Creek. Emma Jane Daly Mathews said as a child she heard it called the Baker Furnace. No evidence supporting either name has been located.

Another misconception about Ritter is that he constructed three furnaces: **Eliza, Blacklick Furnace** at Wheatfield, and the **Buena Vista Furnace** on Route 56. Research done by Clarence Stephenson, Indiana County Historian, showed that **Buena Vista** was built mainly by Dr. Stephen Johnston of Armagh. This furnace eventually came into the hands of the Vinton Colliery Company in 1916. The **Blacklick Furnace**, which was across the Wheatfield Bridge, was erected by David Stewart and George King in the 1840's. The furnace was advertised for sale or rent in 1849 by Edward Shoemaker, an Ebensburg attorney. In the Cambria County tax records, King and Peter Shoenberger are listed as owners of Stewart's timberlands from 1847 to 1855, when their Cambria Iron Company assumed the tax responsibility. Cambria Iron held these coal and timber rights in Cambria and Indiana Counties until the development of Wehrum in 1901. Today no traces of the Blacklick Furnace remain.

Will the Real Lot Irvin Please Stand Up?

Another secret which the furnace retains is the true identity of Ritter's second partner. Was his name Lot Irvin or Lot Irwin? Did he really commit suicide? Letters to historical societies and searches of deed books and other documents added to the mystery. Deedbooks in Indiana and Cambria Counties have listed him as Lot Irwin. Documents in the Indiana County Prothonotary's Office list the name as Lot Irvin. None of these deeds list an address, which was supposedly a farm in Huntingdon County. An 1847 power of attorney document, filed in the Blair County Courthouse in which Irvin turned over his interest in a piece of Hollidaysburg property, is the only reference found for Lot Irwin (Irvin) in Blair and Huntingdon County Courthouses; Blair County was created from Huntingdon County in 1846. An 1867 court order in the Indiana County Orphan's Court Docket listed Irwin's heirs as three minor children, but there is no death date listed for Lot on the court order.

There definitely was a Lot Irvin who was involved in iron-making and who did commit suicide. This Lot Irvin originally came from Centre County and managed a furnace in Greenville, Mercer County. He got into the family business through his brother, General James Irvin, owner of numerous iron furnaces and donor of two hundred acres to the new school which eventually became Penn State. Lot Irvin managed the Greenville Furnace of the Mercer Iron Works for his brother and proceeded to build himself a mansion worthy of an ironmaster. The Greenville Furnace failed partly because of poor business conditions, but also because of the expense of the ironmaster's house. In despair, Lot Irvin, aged 31 and listed as single, hanged himself in the bridge house of the furnace.

The author urges extreme caution when depending on old county histories for accurate factual background. Because she took a Mercer County history on its word that its Lot Irvin was a bachelor, the author concluded that there were two Lot Irvins and would have published this chapter incorrectly. Thanks to Beryl Sternagle of Hollidaysburg, a guest on the America's Industrial Heritage Project bus trip through Vintondale and Wehrum in June, 1990, the mystery of the two Lot Irvins was solved. Mrs. Sternagle discovered the name Lot Irvin while researching the history of her farm outside of Hollidaysburg. Lot Irvin of Eliza Furnace and Lot Irvin of Greenville were one and the same. Lot W. Irvin was born at Linden Hall, Penns Valley in Centre County around 1820, son of John and Ann Watson Irvin. John Irvin had been a stone mason and storekeeper and became one of the largest landowners in Penns Valley. Lot's brother was that same locally famous General James Irvin, ironmaster of ten furnaces, member of Congress in 1841, candidate for governor of Pennsylvania, and donor of the land on which present-day Pennsylvania State University sits.

Lot Irvin married Charlotte Henderson Moore on March 27, 1844 in Hollidaysburg; Charlotte was the daughter of Silas and Lucretia Henderson Moore. Silas was a real estate speculator as well as a merchant and stage operator. As the local partner of two Phiadelphia businessmen, Moore laid out and sold lots in Hollidaysburg. Lot and Charlotte had three children: Anna M. Irvin, 1845-1912; Lucretia Moore Irvin, 1846-1904, married to Rev. Howard King; and William W. Irvin, 1848-1915.

In 1858, Charlotte moved from Bellefonte to Hollidaysburg with her children. In the 1860 census, Charlotte's age was listed as 35; Annie was 14 and William was 12. In 1863, Charlotte purchased a two-story brick house on Allegheny Street in Hollidaysburg for $2,500. Charlotte Irvin died on June 10, 1886 at the age of 63. In her will, Charlotte labeled herself as a widow and left the house and portraits of her and Lot to Annie with the stipulation that she make a home for William as long as she remained single. Annie never married and was still living in the family home at the time of her death on July 31, 1912. By that time, she spelled her last name Irvine. Ann bequeathed her share in the home to her brother and her nephews, William and Howard King. William King also received the portraits of her mother and father which hung in the parlor of the Allegheny Street house. The family **Bible** went to Howard King.

The confusion concerning Lot Irvin stemmed from the Mercer County histories which listed Irvin as single and the Vintondale stories which said that he was a farmer from Huntingdon County. No property transfers for Lot Irvin (Irvin) were listed in Huntingdon or Blair Counties, nor were any divorce records found in Blair County. If there had been a divorce between Lot and Charlotte Irvin, it probably would have been recorded in Centre County. Yet, after discovering that the Lot Irvins were one and the same person, the records add to the mystique of Lot Irvin. Why did he assign the power of attorney in 1847? Why would he not have taken his wife and family to Greenville? Did he desert his family for another woman? **Eliza** knows the answers, but is not about to reveal them.

Present Situation

The Eliza Furnace passed through several owners and eventually became the property of the Blacklick Land and Improvement Company. It was then transferred to the Vinton Colliery Company and its successors. In August, 1959, the Cambria County Historical Society attempted to purchase the site from the Collins Fuel Company to use as a tourist attraction. Manor Realty, a subsidiary of the Pennsylvania Railroad, acquired the furnace in November, 1959 and offered to lease the furnace site to the Historical Society for $20 per year beginning in 1960. The matter stalled for three years while legal problems were worked out. Among the legalities to be solved was the exact location of the Indiana/Cambria County line which supposedly ran through the stack. The area needed to be surveyed again; and a problem arose about a right-of-way for the Driscoll Mining Company, successor to Collins Fuel. Driscoll refused to allow Manor Realty to convey the property without receiving a satisfactory price because they felt they would need to build another access road to some of their properties. The furnace was finally in the hands of the Historical Society in July, 1965. However, due to a lack of funds, the restoration plans have never materialized.

The site was improved in 1975 by the AFL-CIO's Labor Committee for participation in the Bicentennial. The furnace badly needed a clean-up at that time. Trees were growing from the top of the stack and brush covered the ground around the base. In cooperation with the Labor Committee, the 876th Engineering Battalion graded the site. A drainage system was put in; a small parking lot and access road constructed; and a flag pole erected. The furnace was also commemorated during the Bicentennial by the Vintondale Homecoming Committee which sold T-shirts and license plates imprinted with a picture of **Eliza**.

Occasional cleanups have been held since 1975. One took place in August, 1981 in time for the annual homecoming. The September 2 issue of the **Nanty Glo Journal** carried a series of photographs of the operation. Brush was cleared from the top, and the vent pipe was reattached. Equipment for the job was furnished by the fire company, and the Fire Ladies Auxiliary served lunch to the volunteers. Another cleanup was organized by Tom Columbus, Jr. in 1983. Work was done by Blacklick Township Boy Scout Troop 85. In 1989, funds for basic maintenance were provided by the Amercia's Industrial Heritage Project. Plans for future development of the site are tentative. However, vandalism this past year has been hard on the furnace, forcing the sealing off of the openings to the bosh to prevent people from climbing inside the furnace. Shamefully, spray paint graffiti has also appeared on her stones.

Today, **Eliza** which initiated the industrial revolution in the Blacklick Valley, has withstood the test of time and has silently watched the rise and fall of the timber and coal industries and all the good times and bad times in Vintondale.

RITTER'S AND RODGERS' LAND TRANSACTIONS

The Cambria County Deed Books provided an interesting, but confusing succession of deeds on the acquisition and subsequent sale of the Eliza Furnace and the surrounding timber lands. George Rodgers' first purchase in the Blacklick Valley was finalized on September 29, 1837. It included three tracts which were originally warranted in 1793 to Jacob Pressing, Paul Lebo and Daniel Levy. The total acreage was 989 acres, 27 perches. (A perch was one rod or five and one-half yards.) This land, which borders the north side of the South Branch of the Blacklick Creek toward Twin Rocks, had been conveyed to James Fisher of Philadelphia in 1800. Fisher sold it in 1837 to Rodgers and Richard Lewis, Esquire. In an article of agreement dated November 15, 1844, Lewis sold his undivided one-half interest in the above tract to David Ritter. Rodgers sold his share to Ritter on September 8, 1845 for $2,500.

In May of 1845, Ritter and Rodgers made a large land purchase from John Murray, Esquire, sheriff and later judge of Cambria County. The acquisition for $1,611 included parts of tracts warranted in 1793 to Mathias Young (231 acres), Jacob Messersmith (807 acres) and Jeremiah Moshier (528 acres). The total acreage purchased was 528 acres, 22 perches. This parcel bordered the North Branch of the Blacklick. These properties had also passed to James Fisher in 1800, and their heirs sold the land to John Murray in 1841. When Murray sold the land to Ritter and Rodgers, he reserved one acre for himself for a salt well and as much coal as was necessary to manufacture the salt, a venture which was never successful.

Ritter and Rodgers also purchased 1,922 acres from Edward Shoemaker, Ebensburg attorney for James Fisher's heir, Ann. For $418, Ritter and Rodgers also bought 159 acres of the Daniel Levy tract. (Shoemaker was also involved in the advertising for the 1849 sale or lease of the Blacklick Furnace at Wheatfield.)

The tax lists for the late 1840's do not indicate the number and location of buildings owned by nor any other improvements to properties made by Ritter and Rodgers or Ritter and Irwin. The records do show they paid taxes on a sawmill, in 1845, and that they were assessed $500 in Cambria Township for stone. This may support the contention that Deveraux was the stone mason for Eliza Furnace.

For some unexplained reason, George Rodgers gave up all of his interest in the 1922 Shoemaker acres in September 1845; Ritter agreed to pay for the land and keep Rodgers' name harmless of any liabilities. Ritter also agreed to pay Edward Shoemaker the purchase money for the 159 Levy acres. $45.75 was due in January, 1846, and $50 was due January, 1847. In another clause of the agreement, Ritter consented to pay all the debts of the Eliza Furnace and to pay for all goods bought in Philadelphia and Pittsburgh for the furnace. Both men agreed to pay the debts owed an unnamed Ebensburg store.

At the same time, Rodgers sold Ritter 100 acres in Jackson Township for the selling price of $2,500. Rodgers had purchased it for back taxes on January 1, 1845 from Sheriff Murray. James Fisher had originally sold it to John Luke.

Rodgers granted Ritter a mortgage for $2,564.07. The first payment of $700 was due on November 1, 1845, the second for $932.03 on November 1, 1846 and the third for $923.03 on November 11, 1847. This mortgage was never marked as satisfied in the deed book.

Ritter did not pay off all the debts as agreed. In July, 1847, Charles Koons and Amos Keilman of Ebensburg went to court to sue for a debt of $1,177.40 owed by Ritter. On July 12, 1847, the court ordered Sheriff Jessee Patterson to sell the "goods and chattels" of Ritter and Irwin and present the money to the court by October 1, 1847.

Because the sheriff could find no buyers, he seized the property and advertised for a sheriff's sale on January 3-6, 1848. At this sale, George King, future Cambria Iron partner, puchased the 250 acres originally warranted to Christian Stoner for $103. The 250 acre John Stoner tract was purchased by King for $150. Sheriff Patterson also sold the 159 acre Daniel

Levy tract for $150. This land had been partially developed. Thirty-five acres had been cleared and seven houses, a coal shed and stable had been constructed on the land. (The Stoner Tracts are near the top of Chickaree Hill bordering either side of the Rager Hollow road.)

An additional sheriff's sale was held July 3-5, 1848. At this sale, Solomon Alter and John Replier, both of Philadelphia, were the highest bidders. They bought the 807 acre Jacob Messersmith tract for $5.00. The twenty-three acre section of Mathias Young tract sold for $1,900. Included was the furnace stack, a bridge house, a carting house, wheel and bellows, two-two story frame houses, twenty-one hewed log dwellings, a frame stove house, an office, a smith shop, a wagon-maker shop, a smokehouse and a log stable. The Jeremiah Moshier tract, now reduced to 284 acres, sold for $520.

Other tracts purchased at this sale by Alter and Replier were:

1) Daniel Farley   181 acres, 77 perches $1
2) Isaac Bonser   208 acres, 40 perches $1
3) John Simpson   413.96 acres
                   with 3 log houses $5
4) James Reese   89 acres, 148 perches $1
                   Unimproved

The actual acreage varied from deed to deed.

George Rodgers had given up his interest in the furnace in 1845, but he continued working with Richard Lewis in buying and selling property in the Blacklick Valley. In February 1846, they sold 100 acres, parts of the Lebo and Pressing tracts to William Cameron for $100. In October 1853, Rodgers sold 1,048 acres, 27 perches (the Pressing, Lebo and Levy lands) to William Piper and Thomas Moore, for $2,000. $1,000 was to be in cash and two equal payments were to be made without interest. Rodgers and Lewis had reobtained this land at a sheriff's sale. Lewis had died by 1853, and Rodgers had to petition the Orphan's Court for permission to settle a balance due on an article of agreement which Lewis and Rodgers made in November, 1844.

Rodgers was listed in the deed books as selling the same property in February, 1866 for $2,000 to J. Blair Moore. The deed lists the 1853 sale to Thomas Moore but does not explain how Rodgers would be able to resell the property. In 1873, J. Blair Moore conveyed the same tract to Johnston Moore for $1. This is the Moore Syndicate Tract near Twin Rocks. Eventually Webster Griffith and A.V. Barker, Ebensburg businessmen, controlled most of the shares in the Syndicate.

Other property sales made by Rodgers were to William and John Duncan in 1853 for 97 acres in Pine and Blacklick Townships for $225. Rodgers had purchased it in 1849 from William Davis. Another transaction was to Hezekiah Rager in 1856 who bought 209.5 acres in Jackson Township for $700. Rodgers had purchased the land from a Morris of Philadelphia in 1846.

George Rodgers' last property transaction was on a February 26, 1873 deed which listed him as an insolvent debtor. George Huntley was the court-appointed assignee in the bankruptcy; he sold five acres in Cambria Township to A.A. Barker for $262.

Ritter's Indiana County Purchases

Before he was involved with the Eliza Furnace, David Ritter bought several pieces of property in Indiana County. On December 28, 1842, Sheriff Ralston deeded Ritter lots #2 and #4 in Armagh for $425. Then on June 28, 1843, the sheriff deeded him three-fourths of an acre in Ninevah, along the Conemaugh River. The property, which consisted of two frame houses and a frame stable, bordered both sides of the Pennsylvania Canal. The cost of the transaction was $156.

By 1846, Ritter was in financial trouble. His ex-partner Rodgers brought suit for $350 against Ritter. In June 1848, the sheriff deeded the Armagh property to Rodgers for $350. The next month, Rodgers sold the lots for $400 to Alexander Elder of Armagh. Rodgers received the deed for the Ninevah property in December, 1848 and sold it for $900 in March, 1851 to Thomas Taylor of Ninevah.

Ritter and his new furnace partner, Lot Irvin, were granted as warrant on April 27, 1846 for 253 acres, 131 perches in Brush Valley Township. A patent was granted in 1860 to the heirs of Ritter and Irvin. However, the deed for this transaction was not recorded at the Indiana County Courthouse until June 5, 1893. The author has been unable to locate this piece of property on early maps. The state survey books list the property on both sides of Yellow Creek.

The Orphans Court Docket of December Term 1866 contains a petition to sell the tract of 253 acres, 199 perches in Brush Valley Township. It was filed by William Kittell, guardian of Mary, David M., John, and Charles Ritter, minor children of David Ritter. Kittell was also listed as the attorney in fact for two additional Ritter heirs, Eve and Virginia Ritter, legal representatives of David Ritter. Virginia and Eve of Catawissa directed Kittell in November 1866 to sell their undivided 1/6 part of the undivided half. The Irvin children, Annie, Lucretia, and William were represented by Attorney William Jack of Hollidaysburg. The tract was offered for sale at a public auction on February 16, 1867 at the Cresswell Hotel in Strongstown. For lack of an acceptable price, the property was again offered for sale on June 1, 1867. Although the Blair County had directed Mr. Jack to raise at least $1,000 from the sale, James Simpson of Indiana bought the Ritter interest in the property for $333 and the Irwin's share for $250. This was the last transfer on property in Indiana and Cambria Counties accorded to David Ritter and Lot Irvin.

BETWEEN THE IRON AND THE COAL

Factual information on the development of the middle Blacklick between 1850 and 1890 is scanty. Deeds, tax, and census records provide a sketchy puzzle to piece together. Most of Ritter's land along the Blacklick was obtained by Philadelphia businessmen at sheriff's sales. A few of these properties were partially cleared and farmed. In some cases, a farmer thought he owned the land because he cleared it and paid taxes on it, and then lost it to some entrepreneur who had applied for the previously ungranted warrant.

A Pringle farm was drawn on early Vintondale mine maps, but did not appear on the 1890 county atlas. That piece of property contained 127 acres, 8 perches and was purchased by David and Agnes Pringle of Summerhill on March 2, 1871. The acreage was sold to them for $350 by Ephriam McKelvey of East Wheatfield Township. McKelvey had purchased the land in 1867 from F.A. Shoemaker of Ebensburg, who had bought it the same year. Records indicate that the Pringles operated some type of mill, either grist or lumber. According to a Cambria County Road Docket, there was a private road which connected the mill with the Rager's Hollow road. The Pringles sold their land to lumberman Thomas Griffith of Ebensburg in January, 1884 for $300. The Griffith heirs then transferred that parcel to the Blacklick Land and Improvement Company in 1892. Supposedly the farmhouse on the Pringle farm is now the home of Jennie Pioli on Second Street.

The myriad of deed transactions which took place between 1848-1892 are worth deciphering because so much of Vintondale's history has been determined by outside investors from Philadelphia and New York. In 1848, Solomon Alter and John Replier had purchased the Ritter Furnace and properties at a sheriff's sale, but as far as is known did not produce any iron. Alter and Replier sold 2,355.64 acres to Dr. Abraham Rex of Philadelphia in June, 1865. The selling price was to be $5 per acre. According to the deed, Rex was to grant the sellers 1,000 shares of Pittsburgh and Petrona Oil Company stock. Alter and Replier were to hold the stock for four months, and receive $3,500 plus intrest.

George and Abraham Rex

This sale may have been nullified because a deed dated January 1, 1868 sold some of the same acreage to George Rex, Philadelphia physician. It is believed that George Rex was the brother of Abraham Rex. This was a three party transaction. The sellers were Alter and Replier, first party; John Mulford and Joseph Rank, Philadelphia, second party; and George Rex, third party. The first two parties sold 1,534 acres to Rex for $11,777. This transaction included 350 acres obtained by Alter and Replier in 1853 from the James Fisher heirs; 284 acres of the Moshier Tract and 231 acres of the Matthias Young Tract.

On December 31, 1880, George Rex made an agreement with Thomas Griffith, one of the leading lumbermen of Cambria County, which granted Griffith all the poplar, cherry and ash trees on Rex's acreage. The price agreed upon was $5.50 per thousand scale measure feet of lumber. Griffith agreed to pay Rex $6,000 per year for the duration of the agreement, 1881-1883. On December 31, 1880 Rex received $2,500 with $1,000 due March 1, 1881. However, on February 14, 1881, Rex transferred the lumber agreement to the newly organized Cambria Land Company, incorporated January 25, 1881. For $10,000, Rex transferred his 3,000 acres to the Cambria Land Company on February 14. First president of Cambria Land Company was William D. Leslie; treasurer was Abraham Rex.

Around the same time, George Rex borrowed $9,000 from John Krause, a Philadelphia merchant. When the debt went unpaid, Krause sued. At a shareholders' meeting on July 25, 1890, the Cambria Land Company agreed to sell their 3,000 acres for $20,000 to Philadelphian Isaac Grubb; $9,500 of that amount was applied to Rex's debt to Krause. The clerk and treasurer of the Cambria Land Company in 1890 was Cyrus Rex, who was also administrator of the George Rex Estate. According to the deed books, both Abraham and George Rex had died by 1890. Cambria County Sheriff Stineman issued a sheriff's deed to Grubb.

On November 4, 1890, Isaac Grubb sold the 3,000 acres to John Krause, Theodore Bechtel, and Augustine V. Barker. Purchase price was $25,500. These men were some of the original investors in the Blacklick Land and Improvement Company which was formed to exploit the resources in the valley.

Mr. Krause, who owned one third of the Rex land, died on March 3, 1891. His executors, Newton Keim and the Philadelphia First Safe Deposit and Insurance Company granted Krause's share to Warren Delano on March 29, 1892 for $20,000. When the Rex properties passed into Barker's and Delano's hands, the future of Vintondale was sealed. There were, however, other Rex properties in the area. In the Indiana County Tax Records, R. Rex was assessed for two parcels; one 180 acres and the other 16 acres, from 1870-74. The next year, George Rex was taxed for the 196 acres of unseated (undeveloped) lands. In the late 1870's, he was also assessed for 19 acres and 12 acres of lots. Are these the same acres developed twenty years earlier by Ritter? Regardless, these 12 acres of cleared land were sold in 1893 to the Blacklick Land and Improvement Company.

The nineteen acres also landed into BLIC hands after a very confusing exchange of owners. These nineteen acres were part of the Mathias Young Tract, but not part of the 231 acres purchased by Ritter and sold to Alter and Replier. George Rex purchased the 19 acres from William Cameron in April 1877. Within the same year G. Rex sold it to John Brock of Philadelphia.

Brock's heirs conveyed it for $1.00 to Abraham Rex in January, 1884. Abraham Rex died January 31, 1890; his will was dated 1864. His personal estate was worth $2,390, and his debts were $12,932. According to the will, the executor of the estate was to be Dr. George Rex, but he had died several years before Abraham. On April 2, 1892, the Real Estate Title & Insurance & Trust Company filed a petition in Indiana County Orphan's Court for permission to sell the 19 acres to help pay Abraham Rex's debts. Abraham's two parcels in West Mahoning Township were given to the Brock Heirs to clear a debt to their father's estate. The 19 acres were to be sold at a private sale on June 1, 1891. For lack of an acceptable bid, a second sale was ordered on February 7, 1892. The purchaser was William R. George of Cambria County with a high bid of $275. This same property was sold to BLIC on May 19, 1892 for $320. This location and ownership was very hard to trace but has significance for the history of Rexis. The BLIC sold 10.98 of the tract to the Vinton Lumber Company, and about one half of the village of Rexis is built on this land.

The 1890 **Cambria County Atlas** shows little growth and/or industrial development in Jackson Township, whose entire population in the 1890 census was 1,104. The only town of any size was Fairview (Vinco), population 71. A road led from the pike (future Rt. 22) through Rager's Hollow to Buffington (Wheatfield Bridge area). The only marked properties along this road were J. Shuman, 30 acres; D. Shuman, 158 acres; W. Shuman, 30 acres; M. Rager, 500 acres; and C. Rager, 50 acres. The only local merchant/farmer listed for Rager's Hollow area of Jackson Township was Daniel Shuman, address Buffington, farmer and stock raiser. (This may be a case of promising to puchase a copy of the book if your name was listed in it. This happened frequently with biographical sections of county histories.) The timber in the immediate Vintondale area was marked Thomas Griffith and Martin & Griffith. Also in 1890, no railroad had been built in Jackson Township. The boom was yet to come.

On the Blacklick Township map, the iron furnace was indicated, as was a gas well beside it. A road led from the furnace up over #6 hill and connected with the road leading from Twin Rocks (Expedit) to Belsano. Several unidentified structures are drawn on the Adam Rand Tract. Most of the timber lands surrounding Vintondale were owned by the Cambria Land Company, but the timbering rights had been conveyed to Thomas Griffith.

The Early Philadelphia Investors

Here is an excellent opportunity to break with the chronology of the development of Vintondale to insert biographical information on Theodore Bechtel, John Krause and George Rex. Bechtel's biography is very brief. All that the author could research on him was that he was a member of the firm of Jones, Hoar and Company of 1748 Parker Avenue, Philadelphia.

When John Krause died on March 3, 1891, his will was filed and available for examination. He was a retired merchant who lived at 539 North Sixth Street in Philadelphia. His grocery firm, John H. Krause and Company was located at North Fourth and Vine. In his will, he left $10,000 of Newton Reading Railroad bonds and his house and lot to Isaac Grubb's wife and discharged Grubb from the $4,475 debt that he owed Krause. In addition to leaving money and/or property to relatives and employees, he granted in his will endowments to a church in Douglasville for upkeep of the family plot, to the Baptist Home, and to the Baptist Orphans in Philadelphia. The final inventory of his estate was valued at $346,819.52. Would it not be fascinating to be able to read between the lines of the will? Was Grubb a Krause employee and acting as a front for him on these land purchases? Was he a relative? Why would he leave his house to Grubb's wife and money to others named Grubb?

The author located an excellent biography of George A. Rex, but the deeds show that the George Rex in question had died before 1890. Perhaps the George listed below is the son of George V., also a physician and debtor. George Abraham Rex, physician of 2023 Pine Street and nephew of Abraham Rex, was born on April 28, 1845 in Chestnut Hill, son of George Valentine Rex and Mary Catherine Lentz Rex. During the Civil War, Rex served as a highly acclaimed engineer in the Navy and then entered the University of Pennsylvania in 1866. From 1870-1873 he was assistant demonstrator of anatomy at Penn. Rex was a member of the College of Physicians, Academy of Natural Sciences and its Council of Management. While Dr. Rex was vaccinating a patient, his brother, Rev. Henry Rex, heard a gasping noise and calling to Dr. Rex received no response. Dr. George Rex, age 50, was dead of a heart attack. He was survived by a sister Louise and three brothers, Alfred; Rev. Henry, ex-Registrar of Wills for Philadelphia; and Dr. Thomas A., Pittsburgh. The funeral was held at his late residence; a private burial took place at Ivy Hill Cemetery. His executors were Louise Rex and Henry Rex. Cash value of the estate from accounts and the bank was $4,265. In either case, both George Rexes were hopefully better physicians than investors.

All of these property exchanges indicate that a great deal of money was invested in the middle Blacklick Valley, but with little return on the investment. Not until there was easy transportation into and especially out of the valley could a profit be achieved. Providing the acreage for development and a means of transporting the area's resources for the Eastern capitalists was the role accepted by Augustine Vinton Barker of Ebensburg.

J.A. Caldwell.
"Illustrated Historical Combination Atlas of Cambria County, 1890."

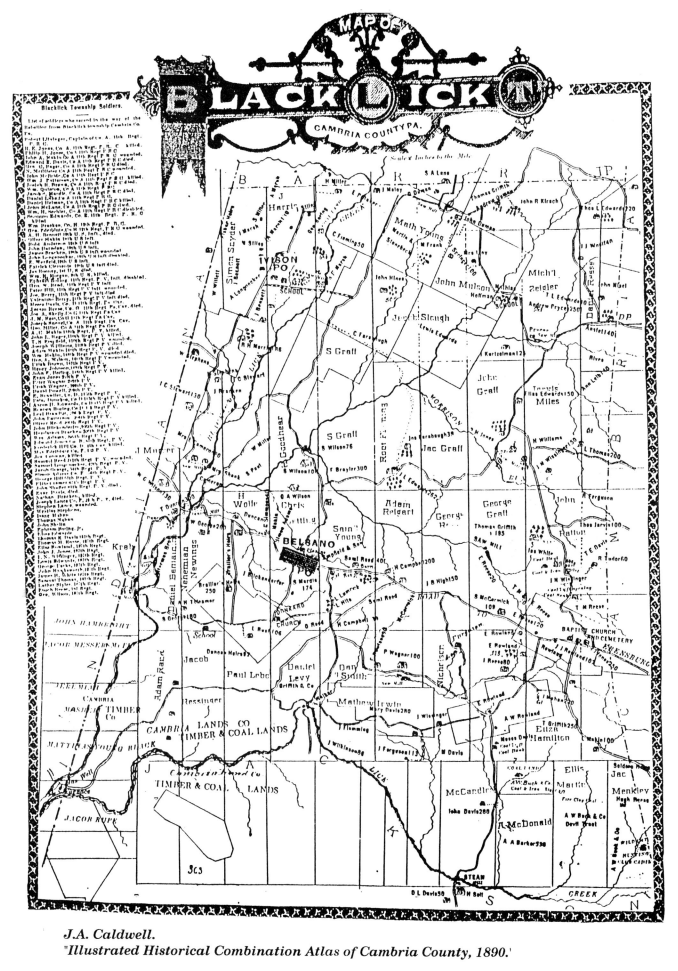

J.A. Caldwell.
"Illustrated Historical Combination Atlas of Cambria County, 1890."

BLACKLICK LAND AND IMPROVEMENT COMPANY

In the late 1880's and early 1890's, local Ebensburg businessmen saw the financial opportunities in the Blacklick Valley knocking at their doors. Judge Augustine Vinton Barker stands out in particular as a businessman who "done good" in his own back yard. Barker purchased lands and mineral rights in Jackson and Blacklick Townships for unpaid taxes. Other tracts were obtained in Buffington, Brush Valley, East Wheatfield and West Wheatfield Townships in Indiana County. Many of the earlier sales of these lands, in particular, the Simon and Alter purchase of Ritter's land in 1848, were not recorded until 1892 or 1893 when Barker and others were attempting to obtain clear titles to the land.

In 1890 or 1891, Barker was able to attract Eastern capitalists to invest in the Blacklick Valley. His acquaintance with these men may have stemmed from college or from the Barker commercial ties with Philadelphia. This merger of Ebensburg and Philadelphia entrepreneurs was chartered as the Blacklick Land and Improvement Company. Barker's fellow investors were: Theodore Bechtel and John Krause of Philadelphia and Warren Delano of East Orange, New Jersey. When Mr. Krause died on March 3, 1891, his shares were conveyed to Warren Delano.

The purpose of the company was to consolidate the land which Barker had bought, open a mine or mines to exploit the coal, and plan a suitable community for the employees. The first purchase recorded in Cambria County in the name of the Blacklick Land and Improvement Company was on May 19, 1892. The next year, Judge Barker conveyed to BL&IC his interest in nine parcels in Blacklick and Jackson Townships which he purchased on July 7, 1890 for non-payment of taxes. These tracts were assessed in the names of the Cambria Land Company (the Rexes), Isaac Bonser-320 acres, Daniel Levy-63 acres, Daniel Farley-188 acres, Jasper Ewing-430 acres, and James Reese. These tracts had belonged to David Ritter in the 1840's and formed the coal reserves for the future #3 mine.

From the Thomas Griffith heirs, the company also purchased the mineral rights for what became #1 mine and the surface rights for the new planned community. These tracts were found in Jackson and East Wheatfield Townships. Parcels included the Wigton Survey, Pringle Land, John Hubley Tract, Adam Shuman Tract and William Shuman Tract. Sale price to the Griffiths was $25 of BL&IC stock per acre for coal and $35 worth of stock per acre for the timber. The Griffiths reserved the timber rights for fifteen years.

Within two years, a town plan was made and laid out accordingly. The chosen site for this new community was in Jackson and Blacklick Townships in Cambria County near the fork of the Blacklick Creek. The majority of the lots were laid out in Jackson Township between Shuman Run and the Ritter Furnace. Lots were planned on the Blacklick Township side of the creek where Vinton Colliery's #6 mine was eventually built.

While all the site preparations were taking place, other related projects were moving ahead. The Ebensburg and Blacklick Railroad, subsidiary of the Pennsylvania Railroad, edged its way toward the new town. Without a rail line for shipping coal, the plans would have stalled. To supply water to the mine and to residents, the Vintondale Water Company was formed. In March, 1894, the **Johnstown Tribune** reported that telephone lines were being strung between Ebensburg and Vinton. Eight men worked on the crew and walked seven miles to and from work. The first coal from the new #1 mine was shipped in October, 1894. Presumably, the Vinton Colliery Company leased #1 from the BL&IC, just as Madiera Hill leased #2, which opened in 1895. Warren Delano's hand was prominent in the development of the Blacklick Land and Improvement Company and the Vinton Colliery Company; his connection with Madiera Hill may have been through its board of directors.

The new community suffered through several name changes. The first was Barker City; then it was called Vinton, Bark City, and Bark. Its first post office was opened under the name of Vintonville in June, 1894 with Clarence Claghorn as its first postmaster. The exact date that the new town received its permanent name of Vintondale is unknown.

When lots were offered to the public, Barker Brothers owned the majority of the parcels on what became Plank Road and Maple Street, where they operated a sawmill. (See appendix for the sales of their lots.) The lots on Maple Street and the western side of Plank Road were not part of the town plan. On Maple Street there was a large boarding house which may be the oldest house in town.

In 1894, it appeared that Vintondale was going to be a unique mining town. It would not only have company-owned houses for rent, but Vintondale would also have lots which were privately owned. George Blewitt, who transferred from J.R. Claghorn's Bernice mine in Sullivan County, purchased two lots, 9 and 11 in Block J for $300. Francis Farabaugh purchased the corner lot in Block B for $350. This building, which functioned as a wholesale liquor store, was resold to the company in 1912. Other purchasers included Warren Delano, Clarence Claghorn, Dr. Abner Griffith, Annie Griffith Lyte and Valentine Barker. Many of these owners later sold their properties to the Vinton Colliery Company or the Vinton Land Company. The Blacklick Land and Improvement Company continued to sell lots until 1900 when it disbanded and sold its property in Vintondale to the Lackawanna Iron and Steel Company. (See related chapter.)

Chapter II
Lumber, Coal, and Railroads

LUMBERING IN VINTONDALE

About forty years after the decline of the iron industry, a second industrial revolution created a boom town atmosphere in the middle Blacklick Valley. The new coal, lumber and railroad industries complemented each other just as the procurement of limestone, iron ore and the charcoal had earlier. The successful marketing of coal hinged on construction of a railroad, and both the mines and the railroad demanded a ready supply of lumber. Lured by coal, the Ebensburg and Blacklick Branch of the Pennyslvania Railroad was completed in 1894; the south branch of the Blacklick Creek was soon dotted with sawmills and coal mines. The development of the north branch followed in less than ten years.

The lumber companies were seeking hardwoods, hemlock and tanning bark. They offered area farmers quick money for dissipating their woodlands with no offer of reforestation. Left standing were the birch and beech trees, not commercially useful to them. Also left behind was the brush from the logging operations.

Known Lumbering Enterprises

A large, but almost forgotten, lumbering operation called the Vintondale Lumber Company was located between Vintondale and Twin Rocks; it was a subsidiary of the William Whitmer and Sons Lumber Company of Sunbury, Pennsylvania. This operation, opened around 1896, was believed to be a "wildcatter", which meant that the logs were brought to the mill on a wooden tramway, often powered by gravity. As logging technology improved, the wooden rails were replaced by steel rails, and the special logging locomotives, like the Climax and the Porter, replaced the horses. According to the few surviving records, the tramroad consisted of 36" gauge steel rails and covered ten miles. Its locomotive was a sixteen ton Climax-Type A built in 1893 and sent to Vintondale from the Reitz and Whitmer operation in Tusseyville, Centre County. The mill and yard were located in Twin Rocks, and its circular saw had the capacity of cutting 25,000-30,000 board feet per day. The company owned timber rights in Blacklick Township. Known Whitmer employees who transferred from the Union County operation were Ed Jolly and Henry Shoulter. By 1904, the Whitmer operation was shut down and almost forgotten except for the efforts of three railroad historians, Thomas Taber, Benjamin Kline, and Walter Casler who were determined to preserve a history of Pennsylvania's logging railroads. They were responsible for researching and publishing a twelve book series called **Logging Railroad Era of Lumbering in Pennsylvania**.

Local Cambria County entrepreneurs who had a hand in developing the lumber industry near Vintondale were Ebensburgers Thomas Griffith, Webster Griffith and Augustine Vinton Barker (see biographies). These men bought large tracts of timber and coal, often at sheriff's sales. Barker Brothers owned an estimated 1 and 1/2 million feet of hemlock and other timber in the Blacklick Valley. In 1894, the Barker mill in Vintondale was located near Plank Road. John Griffith, mill superintendent, had his office in the house which was later occupied by the Andy Jacobs family.

Thomas Griffith had obtained most of the timber located on the #1 hill. When his heirs sold their coal lands to the Blacklick Land and Improvement Company in 1892, they retained the timber rights for fifteen years. Their mill was located along the stream near Sixth Street. In 1905, Webster Griffith sued the Lackawanna Coal and Coke Company, successor to BL&IC, on a trespass action. Griffith wanted to continue to have the right to timber the hill, use the water from the stream, and to have rights to cross Lackawanna's narrow gauge railroad (Dinkey Track). An agreement was worked out which granted Griffith an additional five years to complete his logging operation. One known Griffith employee was Charles Cameron who left Vintondale in 1905 to start timbering on the Isaac Mahan farm at Pindleton. (Near the Bethel Baptist Church between Belsano and Ebensburg.)

Making a more permanent impact on the local scene was Blair Shaffer, who moved to Vintondale from Strongstown. Mr. Shaffer purchased land in Indiana County from the Cambria Iron Company, who retained mineral rights. Some of this land was divided into lots and became part of Rexis. In Vintondale, he constructed a planing mill at the lower end of town called the Vintondale Planing Mill. His letterhead listed him as a "Contractor and Builder and Dealer in all kinds of worked lumber, sashes, frames, etc." In 1895, Mrs. Emma Shaffer purchased the lot in Vintondale on which the family home was built. Mrs. Emma Shaffer lived there until her death in 1925. Her son Homer (Jack) and daughter Mary lived there until their deaths; it is now occupied by Shaffer's granddaughter, Lynetta Daly Tarr. Mr. Shaffer also built the company house on Third and Griffith which had been occupied by Grandpa Blewitt and owned by the Pisaneschi family for over forty years. Blair Shaffer also won the contract to build the houses in the newly planned town of Wehrum. However, this contract was never fulfilled because Shaffer died unexpectedly of paralysis on March 19, 1902; he was fifty-two years old. Eventually his daughter Ella and son-in-law Herb Daly built a house on one of the lots where the planing mill had been located.

Other sawmills in the Vintondale area were the Johnson & Company, which suffered a fire loss of $14,000 in September, 1899 and the Weber mill. Exact locations of these mills are not known. Also working in the area was James George, who represented the Roaring Spring Pulp Wood Company.

In 1898, the local newspapers reported a boom taking place in the Vintondale lumber industry. Ground bordering the North Branch of the Blacklick was being cleared for a big mill for Crandle and Son of Crawford County. That land had belonged to the Blacklick Land and Improvement Company, who retained the mineral rights. A wooden tram road was constructed to Nipton (Red Mill), site of the late J.W. Duncan's grist mill, sawmill and shingle mill. This property had been sold in June, 1898 to Evan Morgan and W.J. Donnelly for $10,500. Local papers hinted that these men represented Eastern capitalists. A pine tree, felled on the Duncan tract by John Davis, a logging contractor from Grisemore, produced 6,835 board feet in lumber, giving an indication of the value of the timber in the North Branch.

The contract for constructing Crandall's mill went to E.J. Blakely. In 1898, Crandle also purchased Lot 12 in Block AD in Vintondale from Thomas and Mathilda Wray. (In 1925 that property was sold to Bernard Metz by A.W. Lee, liquidator of the Vinton Lumber Company.)

Even before Crandle started constructing his mill, representatives of the Clearfield Lumber Company, namely George Dimeling and E.J. Hoover, were in the north branch area purchasing stands of timber. As early as 1895, these men purchashed timber from Webster Griffith. In September, 1898, the Clearfield Lumber Company purchased Crandle's sawmill and timber rights for $37,000. Merrick Crandle's only son had just died of typhoid fever. Because of the loss of his son and his advanced age, Crandle decided to retire from the lumbering business. Hoover and Dimeling added to the Crandle holdings by purchasing the timber on the Davis Tract from Donnelly. Also in October 1898, the Clearfield Lumber Company enlarged their new mill by forty feet.

Organization of the Vinton Lumber Company

On January 3, 1899, stockholders of the Clearfield Lumber Company created a new organization known as the Vinton Lumber Company. This limited partnership of joint stock association had capital stock worth $100,000, consisting of 1,000 shares of par value of $100 each. $50,000 was to remain in the company treasury in cash, and the remaining $50,000 was to be used to purchase timber, set up sawmills and cut timber to market requirements. The sawmill was located in Buffington Township with the nearest post office in Vintondale. The Vinton Lumber Company's business office was in Clearfield.

The officers and investors were Clearfield and Osceola Mills businessmen. The chairman was Ashbury W. Lee; W.B. Townsend was the secretary-treasurer. each officer served for a one year term. Other managers and officers were George Dimeling, W.A. Crist, D. McGaughey and John Dimeling. The names of the subscribers and their shares are as follows:

E.J. Hoover	100 shares	$10,000
Dr. C.S. Stewart	50 shares	5,000
John Dimeling	100 shares	10,000
W.A. Crist	125 shares	12,500
George Dimeling	125 shares	12,500
A.W. Lee	125 shares	12,500
D. McGaughey	125 shares	12,500
D. McGaughey	125 shares	12,500

Foundation for Vinton Lumber Company mill - May 13, 1984.

Courtesy of Diane Dusza.

Mr. Lee was the leading lumberman in Clearfield County in the 1890's, in addition to being president of the Clearfield Bank and Trust Company. Some of his direct decendents continue to live in Clearfield; however, attempts by the author to contact his grandson, A.W. Lee III, were fruitless.

The Lee lumbering operations in Clearfield County were found at Belsano Mills, Houtzdale, Bear Run, Kerrmoor, LaJose and Faunce. As the timber at one area was exhausted, the company moved its operations to their other holdings. As their Clearfield County holdings were becoming exchausted, the company began eyeing the timber in Cambria and Indiana Counties, manily those tracts bordering the north branch of the Blacklick Creek.

In addition to purchasing timber rights along the north branch (see appendix), the Vinton Lumber Company in October, 1899 purchased 10.98 acres of surface from the BL&IC for $1,647. This land bordered their log pond on the east and Blair Shaffer's property on the west. The BL&IC reserved the right to repurchase the property when the lumber company had completed the timbering of the valley.

Remains of the log pond of Vinton Lumber Company - May 13, 1984.
Courtesy of Diane Dusza.

Vinton Lumber Company Mill Operations

The sawmill was located in the Indiana County side of the North Branch, across the creek from the Eliza Furnace. Acquired from Crandle and Company, it was a large operation, capable of cutting 100,000 to 200,000 board feet of lumber per day. The timber cut was mainly hemlock, white pine, and hardwoods. Some of the logs were up to sixty-two feet long. A shotgun carriage and trailer carriage allowed the logs that length to be cut. Logs which were too large had to be blasted with black powder to fit into the saw carriages. The mill used a large band saw fifty feet long and eight to ten inches wide. It ran on two large pulleys, six feet in diameter, one above the cutting floor, one below. When the saw was cutting, it gave a constant downward motion. The work was done rapidly, and repairs to the saw were made easily and inexpensively. The initial slab, if suitable, was cut into plaster lath by crosscut circular saws. Any refuse was carried down a chute to a "sawmill hog", ground up and burned in the sawmill's boiler. The better boards were sent to the "edger", who used graded circular rip saws. Then the boards were cut into ten to sixteen foot lengths. Large amounts of hardwood were cut into squares two inches by two inches for chair rungs. In February 1899, the engine installed by Crandle was replaced with a 250 horsepower engine.

Vinton Lumber Company - 1905. Mary Lynch Clarkson on left and Bill Clarkson in center.
Courtesy of Leona Clarkson Dusza.

In July 1899, the Vinton Lumber Company suffered a major fire and quickly rebuilt. By October framework for the new 160' x 55' mill was completed, and George Dimeling, general manager, was impatiently waiting for the delivery of machinery.

The Vinton Lumber Company erected a three story boarding house, measuring 28' x 40' with a 16' x 16' attached kitchen. Eventually the VLC purchased or constructed nineteen single houses, two double houses, one boarding house and an office in addition to the mill. Total assessed value of all their property in 1905 was $7,390 with county taxes totaling $99.95.

A large lot storage dam was constructed across the Blacklick where logs remained until needed. It is not known if any were floated down the creek to the dam. A chain on pulleys, studded with spikes, ran under the water, caught the logs on the spikes and drew the logs endwise onto a logway. A machine called a "steam-nigger" positioned the logs on the carriage.

The log pond was cleared once a year on the Fourth of July. Since the North Fork was not yet polluted with mine water, the workmen were able to wade into the shallow water and gather fish by the tubful. The dam also provided a good fishing spot for neighborhood boys. In 1901, the **Indiana Progress** reported that a Vintondale boy caught a 2-1/2 pound mountain trout in the pond. The remains of the dam's breastwork can still be seen in Rexis.

Breastwork for Vinton Lumber Company dam - May 13, 1984.

Courtesy of Diane Dusza.

Breastwork for Vinton Lumber Company dam - May 13, 1984.

Courtesy of Diane Dusza.

Forest fires also posed a major threat to any lumber company. A serious set of fires hit the north branch in May, 1900. All VLC hands were enlisted in fighting the fires which spread rapidly because of the branches left behind as a result of the timbering. The farm buildings of Jacob Kuckenbrod and Robert Skiles were in danger, but saved. On the Mack, McFeaters, Bracken, and Duncan tracts, wood for paper and thousands of split rails were destroyed. Some blamed the fires on an arsonist; others credited the sparks from the dinkey trains.

In addition to providing ready money for timber to struggling farmers and jobs in the woods to their sons, the advent of lumber camps and mining towns was a boon to the local farmer. Lumberjacks and miners had to be fed, and many local farmers benefitted from sell ing produce and livestock to the companies. E.W. George of Nipton butchered forty-two hogs and sold the meat to lumber camps in his neighborhood. Mules were scarce, and Thomas Graham got a good price selling his pair to George Dimeling.

Company Employees

In a very useful 1975 work called **Logging Railroads of West-Central Pennsylvania**, the late John Misner of Rexis was listed as a surviving employee of the Vinton Lumber Company. By interviewing Mr. Misner, author Benjamine Kline was able to gain much first-hand information about the company. According to Mr. Misner, the mill superintendent was Frank McGuire; sawyer, Bill Downs; assistant superintendent, Lloyd David; mill engineer, Tom Beard; setter, John Woods; dogger, Ed Walker; edger, William Misner; in the yard, Mike Misner, Henry Misner, Will Findley. Other employees were: Frank David, Tom Brinnon, Laush Dodson, Charles Gill, Sid Conrad and Silas Conrad. Oscar Geulock of Clearfield County was the lumber estimator. (See the Rexis tax list in the appendix.)

There was only one known fatality; William Misner died after being thrown forty feet when a board that he was trying to free suddenly came loose. Injuries were very common as logging was a dangerous occupation. Within the span of February and March, 1900, five accidents were reported in the Indiana paper. A man named Murphy, on night shift at the VLC, fell from a scaffold and was badly bruised. The foreman broke a leg, and J.P. Phillips of Nolo, another VLC employee, smashed his thumb. Even children, who lived in the lumber camps with their families, were not accident-free. Tom Davis, small son of Mr. and Mrs. John Davis of the Davis Camp, located one mile north of Rexis, fell and struck his right hand on an ax while cutting kindling. Andrew Nicewonger was stuck by a log while loading a car and suffered a broken arm and leg. David Cowley, uncle of Malcolm Cowley, was maimed for life in a logging accident timbering the Shaffer land for John McCullough, who had the contract with the VLC. The late Rev. Clarence Bennett said Cowley's life was saved by Dr. Comerer, Vintondale's company doctor. Cowley was treated at the house the Bennett's were renting in Rexis, second to the last house on the left. Lumbering was also dangerous for the horses; a log ran over Harvey Sheesley's team, killing one horse and injuring another. (Many of these names are also found on the Rexis tax list.)

The Blacklick and Yellow Creek Railroad

In addition to getting the logs to the mill, shipping the finished lumber also presented problems to the Vinton Lumber Company in 1899-1900. The Ebensburg and Blacklick branch of the Pennsylvania extended only to Vintondale at that time. To get to the freight station, lumber had to be hauled by horse and wagon, fording the Blacklick at a low spot. To facilitate shipping, the VLC made an agreement with the Ebensburg and Blacklick Railroad for the construction of a railroad to the mill. A third party in the agreement, the Pennsylvania Railroad was to furnish the rails when the roadbed was done. At the same time, rights of way were purchased by VLC to construct a logging railroad along the north branch of the Blacklick. Construction of a permanent railroad rather than a tramway probably was done to exploit the coal deposits in addition to the timber.

To complete the connection of the railroads, Blair Shaffer sold the Ebensburg and Blacklick Railroad two strips of land totaling 1.824 acres in December, 1901 for $500. The mineral rights on the two strips remained in the hands of the Cambria Iron Company. The lumber company also agreed to grant a release to its timbering rights on the strip to the railroad for $1. In addition, Shaffer agreed to sell eight acres to the Blacklick Land and Improvement Company. Before these transactions were recorded, Mr. Shaffer died of paralysis on March 19, 1902. His widow had to petition the Orphan's Court on August 5, 1902 for permission to convey the eight acres to the Lackawanna Coal and Coke Company, the successor to the BL&IC.

On June 15, 1904 the Blacklick and Yellow Creek Railroad was chartered in Harrisburg and was owned by A.W. Lee and Associates as a subsidiary of the VLC. L.H. Davis was the superintendent of the B&LC RR; he was also one of the first automobile owners in Vintondale, purchasing a 1909 Model 10 Buick Runabout from the newly formed Indiana Motor Company.

Railroad Equipment

Actual number of locomotives and cars owned by the railroad are unknown. John Misner said that the BL&YC owned a 0-4-2 saddle tank engine and a 4-6-0 Porter. One verified locomotive was PRR #835, Class M, Type 0-6-0, which was built at the Altoona Machine Shop in November 1882. it was sold to the Vinton Lumber Company for $2,500 on February 17, 1902. Its disposition is unknown. The H.K. Porter, #2001, was built in 1899 and purchased new by the VLC. It was sold to the Morehead and North Fork Railroad (1906-1973) of Clearfield, Kentucky. This railroad's predecessor was the Morehead and West Liberty Railroad which operated from 1901-1906. Two other engines used were an 0-4-2, #1751 Porter built in 1897, which may have come from Clearfield Lumber Company's #2 mill in Faunce and a type 4-6-0 possibly obtained from the PRR. Along this railroad were several logging camps including the Davis camp. Also there were several branch lines to the camps. These branches were removed when the VLC was disbanded.

Vinton Lumber Company - Blacklick and Yellow Creek Railroad - 1906.
Courtesy of Leona Clarkson Dusza.

According to its charter, the life of the Vinton Lumber Company was not to exceed fifteen years. By 1906, the company was completing its operations in Rexis and was making plans to open sawmills near Morehead, Kentucky. On a postcard of the lumber company, Verna Findly wrote to her sister that the VLC would be gone by February, 1907. Today, there is a town of Clearfield located near Morehead. Short of a trip to Kentucky, the author's attempts to learn more about the Kentucky operation have been fruitless. In a letter, Harry Caudill, author of **Night Comes to the Cumberlands**, provided useful information of Warren Delano's Kentucky holdings but did not have any information on the VLC. A letter to A.W. Lee III was returned; a letter to the tourist bureau of Morehead was not answered.

Some VLC employees went to Kentucky to set up operations there, but many of the families remained in Vintondale and sought emloyment in the mines or on the railroads. John Misner, Charles Gill, and William Findley were some of the employees who remained behind.

At a stockholders' meeting on May 22, 1909,

Thomas Fisher, W. Clark Millen and John W. Wrigley were elected as the liquidating trustees of the Vinton Lumber Comany. In November 1910, the Vinton Colliery Company, short of housing, signed an agreement to pay rent for $6.00 per month all of the VLC's houses not already rented by VLC employees. The year before, against the wishes of the coal company, many of these same houses had been rented to United Mine Workers for Vinton Colliery employees who were on strike in May and June of 1909. In the 1910 agreement, the lumber company agreed to put the houses in rentable condition and maintain them. In particular, spigots were to be repaired, and the Vinton Colliery agreed to continue to supply water to the houses. The disrepair of the houses is born out in a note in the 1911 tax assessor's books describing several houses as dilapidated. In particular, the boarding house was "useless for habitation."

In 1914, the LVC trustees sold their houses, the mill site and railroad tracks leading to the juncture with the Ebensburg and Blacklick to the Pennsylvania Railroad for $3,750. The mineral rights reamined with the Vinton Colliery Company.

Except for a few foundations, all traces of the Vinton Lumber Company have disappeared from Vintondale. However, the trustees carried out one lasting transaction; they sold the railroad rather than abandoning it. In 1910, the Blacklick and Yellow Creek Railroad was conveyed to coal operators, J. Heil Weaver and B. Dawson Coleman.

Blacklick and Yellow Creek Railroad - 1907. Standing on the platform is Charlie Gill, identified by daughter Ella Gill Lantzy.
  Courtesy of Leona Clarkson Dusza.

Blacklick and Yellow Creek Railroad - 1907.
  Courtesy of Leona Clarkson Dusza.

Blacklick and Yellow Creek Railroad - 1907.
  Courtesy of Leona Clarkson Dusza.

THE CAMBRIA AND INDIANA RAILROAD

A typical logging line was often a wooden tramway or a temporary system which was torn up when the company ceased operation and moved on. The Blacklick and Yellow Creek Railroad was obviously built to be more than just a logging line. In the planning and construction stages, the Vinton Lumber Company was part of a three party agreement with the Ebensburg and Blacklick Railroad and the Pennsylvania Railroad, whereby the VLC obtained the rails at little or no cost from the PRR and the E&B completed the roadbed. It is not known who exactly constructed the B&YC, but the VLC obtained the rights of way and definitely was in charge of the grading of the W.R. George farm just north of White Mill. Using the name Blacklick and Yellow Creek Railroad, the line had a short seven year life span, but remained a lasting reminder of the lumber industry in the valley because of its conversion to a coal carrier.

When the Vinton Lumber Company closed operations and moved to Kentucky, the line offered for sale. On June 30, 1910, S.H. Jencks, Ebensburg engineer representing Weaver and Coleman of Dixon Coal Company, made an inspection of the B&YC with Lloyd Duncan, engineer and caretaker of the line from Rexis to Burns Station, ten miles north of Rexis. In July 1910, the trustees of the VLC sold the Blacklick and Yellow Creek Railroad to J. Heil Weaver and B. Dawson Coleman of Dixon Coal Company, Idamar, Indiana County. Weaver and Coleman were also the co-owners of the Ebensburg Coal Company which was developing the mines and town of Colver. (The name Colver was a combination of the first three letters of Coleman's name and the last three of Weaver's. Later, when they organized the Monroe Coal Company, they reversed the letters of Colver and named the new town Revloc. The Heisley mine in Nanty-Glo was taken from Mr. Weaver's middle name.)

Coleman and Weaver paid $10,000 cash for the option of running the railroad for sixty days and then paid $100,000 cash for the line. Its new president was Mr. Coleman of Lebanon; other officials were: General Manager, J. Edgar Long; Secretary, Earle Long; General Auditor, W.E. Dobson; Superintendent, T.E. Dunn; Chief Engineer, S.H. Jencks. (Mr. Jencks wrote a chronology of his experiences with the railroads. This is available for study at the Cambria County Historical Society in Ebensburg.) The tie inspector was Quince Brickley, and the locomotive engineer was Buckie Mentch. L.E. Summers became resident engineer from Rexis to Stile's Station and from the junction to Colver; his office was a room at the Village Hotel in Vintondale. Summer's corps included: Bowden, Ben Mahaffey, Ralph Rodgers and Frank Sabatto. Andy Little was engineer from Stiles Station to Pine Flats and had an office in a building on the Williams' farm near Williams' Summit. On Little's crew were John Miller, Ed Clark, Ralph Rodgers, Barney Scanlon, Phil Bender and John Harrison. Sam Carlson, track foreman and later supervisor, was in charge of laying track from White Mill to Colver. Clair Bearer, after working four years for PRR as an extra agent, went to work for the C&I because wages and conditions were better.

Near William's Summit in Pine Township, the C&I worked out an agreement with William Sides to enlarge his lake so that the C&I engines would have an ample water supply for their locomotives. In return, Mr. Sides maintained his right to use the lake for fish propagation and for cutting ice in the winter. Cost of the transaction was $1.00.

On April 20, 1911, the Blacklick and Yellow Creek Railroad was officially incorporated as the Cambria and Indiana Railroad. It is the only common carrier in Pennsylvania which can directly trace its origin directly to a logging railroad.

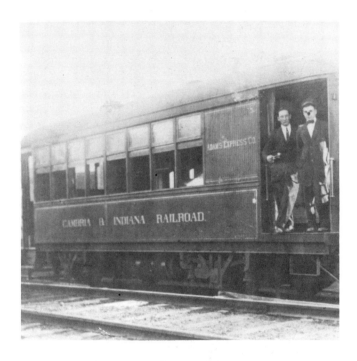

Cambria and Indiana County Railroad- 1925 - Hootlebug.

Courtesy of George Lantzy.

The C&I's first listing in **Poor's Manual**, a railroad directory, was in 1911. The line's length was 18.8 miles, from Rexis to Manver. (Again, the name is taken from Weaver and Coleman.) At that time, the C&I owned four locomotives, three passenger cars, and 612 freight cars. It connected with the PRR at Rexis and the New York Central at Manver. Main offices were in Philadelphia.

The 1923 volume of **Moody's Railroad Investments** shows that the railroad expanded to Colver, Nanty Glo, and Revloc, all Weaver-Coleman enterprises. Total trackage, including sidings, was 59.15 miles. In 1923, the line had a large debt encumbered to pay for the purchase of steel hoppers. Previous equipment purchases included 300 cars in 1912, 393 in 1913, 500 in 1914, 100 in 1916, and 1000 in 1917. In 1922, net revenue was -$221,785; this loss could possibly be attributed to the 1922 UMWA strike. The next year, the net revenue was $56,827. The company paid dividends of 20% in 1916, 25% in 1918, 8% in 1920 and 10% in 1922. Their rolling stock in the early 1920's included nine locomotives, three passenger cars, 2,497 freight cars and two work cars. A repair shop and brick duplexes for employees were built for the workers between Nanty Glo and Nettleton (Cardiff). This area is still called The C&I Houses.

Single coach passenger service, called the "Stump Dodger" by locals, was provided from Rexis to Colver, including any requested stops along the route. There were two morning departures from Rexis at 8:25 and 10:30 and one evening departure at 7:35 PM. Trains arrived at 8:13 AM, 10:35 AM and 3:50 PM. Not too long after the 1924 train robbery, the steam engine was replaced with first a battery-operated and then a gasoline-operated "Hootlebug". Passenger service was also provided between Nanty Glo and Colver. On December 1, 1931, due to the loss of a US Mail contract in Nanty Glo and Rexis, passenger service was ended. Freight service from Rexis to Colver continued in spite of the loss of passenger service.

C&I Rexis Marker - May 13, 1984.

Before Prohibition, several boxcars a day made the trip to Colver; included were usually two carloads of beer, but no whiskey. Two houses half-way up the hill in Colver handled a lot of that beer. One Colver house even had a sidewalk made of beer caps.

A dinkey engine belonging to either the lumber company or the C&I sat abandoned for years in the railyard near Debona's house in Rexis. This engine was eventually scrapped for the war effort in the 1940's.

The C&I almost obtained the contract to ship Vintondale coal in 1911. Warren Delano, unhappy with the PRR for undisclosed reasons, but perhaps because of a constant lack of hoppers, met with S.R. Jencks on April 11 to go over the property to see if the railroad could be connected with Vinton Colliery's #6 mine. Apparently Delano's problem with the PRR was worked out because the mine and the C&I were never "connected". However, in addition to the coal shipped from Colver, Commercial Coal Company's #16 mine had a portal just south of Red Mill. Also nearby was a sand quarry with several houses built for the laborers.

The Great Train Robbery

The Wild West returns to Western Pennsylvania! The ghosts of Billy the Kid and Jesse James would have enjoyed the events unfolding on the Stump Dodger on Saturday, October 11, 1924. Joseph Davis and James Garman were on board; at their feet was a small safe containing $33,054.64, the payroll for the Colver mine. Two well-dressed foreigners, who spoke good English, also boarded the train at Rexis, making a total of eight passengers in the car. At 9:20 AM, engineer Bucky Mentch stopped the train at Chrysler's Crossing near the bridge at White Mill to pick up another passenger. At that moment, the two passengers who boarded at Rexis shot and killed Garman, age 65, and disarmed Davis. They also fired two shots at the other four passengers, one of whom was Maggie Bracken, seated directly behind Garman. Immediately, four men rushed the train and covered the engineer, allowing the bandits to drag the safe to a getaway car.

The train proceeded to Colver where the crime was reported, allowing time for the robbers to get away. According to the late Clair Bearer, C&I supply clerk and storekeeper, the engineer had no way of reporting the crime until he reached Colver. Mr. Bearer said he met the train and assisted Mr. Davis, who was in shock, up the hill to the community hospital.

A posse of 100 to 200 men was formed to search for clues. In the meantime, an armed convoy from Johnstown delivered a replacement payroll. A number of suspects were taken into custody, and a combined reward of $2,500 was offered by the county and the express company. In addition, the state police sent circulars to other states describing the suspects.

A break came in the case within two weeks; two troopers on patrol in Terre Haute, Indiana, spotted two men fitting the descriptions on the circulars. The suspects said they were coal miners from northern Indiana, but the troopers were suspecious because their hands were uncalloused. Michelo Bassi and Anthony

Pezzi were taken into custody for further questioning. Each man had $3,000 on them, and two revolvers were found in their hotel room.

A Spangler man, Peter Antionucci, was arrested and charged with complicity in the crime. It was believed that his car was used as the getaway car. Bassi's trial began on December 11, 1924. He was positively identified as Garman's murderer by trainmen and passengers. During the testimony, it was revealed that he had stayed at a Twin Rocks hotel, moving out just before the robbery. On December 12, the jury, after deliberating only 56 minutes, found Bassi guilty of first degree murder. Pezzi was also found guilty of first degree murder after a one day trial. Antionucci's trial started March 9, 1925. The prosecution tried to prove that it was his automobile which served as the getaway car. The defense produced witnesses who showed Antionucci did not have his Dodge automobile at the time of the robbery and that he was not in the area when the robbery took place. Antionucci was found not guilty.

The two convicted men were housed in the Cambria County Jail. After trying to escape, they were moved to Rockview State Penitentiary near Bellefonte. There, they went on a twenty-nine day hunger strike to proclaim their innocence. However, both men were electrocuted on February 23, 1925. The safe and the remainder of the money have never been found.

The C&I Today

Eventually, the partnership of Weaver and Coleman broke up. Mr. Coleman, whom Clair Bearer described as a real gentleman, retained the Ebensburg Coal Company and Colver for sentimental reasons. Mr. Weaver, a hardheaded businessman, obtained the Heisley Mines and the C&I Railroad. In Mr. Bearer's opinion, Mr. Weaver got the better part of the bargain.

Mr. Coleman died in Philadelphia at the age of 68 in March, 1933. At the time of his death, he was president of the First National Bank, Lebanon; a director of: Coleman Coal Company, Girard Trust, Penn Mutual Life Insurance, and Baldwin Locomotive. Mr. Bearer knew both men personally and attended Weaver's funeral in Williamsport with C&I superintendent Hooper and engineer Jencks.

Eventually, the Bethlehem Mines Corporation acquired the Heisley mine and operated it until the late 1970's. Revloc and the C&I also became Bethlehem enterprises. The railroad is still operating, but at a much smaller capacity than in the past. The damages caused by the 1977 flood, which included a washed-out bridge, and the closing of the Colver mine in the mid-1970's led to the abandonment of the Colver-Rexis line. Today, the line services mainly Bethlehem's Cambria Slope mine outside of Ebensburg.

In December 1989, officials of the C&I held a press conference in Nanty Glo at the site of its two-unit bridge which crossed Beulah Road and the Blacklick to the Heisley mine. The bridges were built in 1916 at a cost of $26,000 for the highway span and $31,000 for the unit which crossed the creek. At the conference, officials announced that the two spans would be demolished in the spring of 1990. Salvage rights were awarded to Jack Calandra of Cresson. Before deciding on demolition, the railroad contacted the Pennsylvania Historical Society (Pennsylvania Historical and Museum Commission?) to see if the bridges were of any historical value. However, the structures were of a very common architectural style and too recent to be considered for historic preservation. In addition to the bridges, the C&I wanted the stone abuttments removed. Work was held up temporarily in the spring when an owl's nest was found on one of the abuttments. The work stoppage lasted until the chicks were able to leave the nest.

Gone forever are the evenings of driving to Rexis to watch the C&I switch the coal cars on to the sidings or being allowed to hitch a short ride on the caboose. Gone are the days of finding your car sandwiched between the iron bridge and the C&I tracks with no where to go while waiting for the PRR to shuttle the Colver coal cars onto its tracks. Or remember sitting on the other side of the bridge near Charlie Lynch's house and watching the hoppers bounce on the loose rails, hoping all the time that none would derail?

Cambria and Indiana Railroad before removal of rails and ties - May 13, 1984.
Courtesy of Diane Dusza.

In 1984, the ties and rails were removed by Kovalchick Salvage. Perhaps these rights of way can be used for recreational purposes, especially cross-country skiing. Today the area is slowly returning to its pre-Vinton Lumber Company days. Maybe in our own lifetimes, the north branch of the Blacklick Creek will again produce 2 1/2 pound trout.

MALCOLM COWLEY

Who is Malcolm Cowley? What is his role in the history of Vintondale? Technically, nothing, but thanks to Cowley's interest in the Vinton Lumber Company and the efforts of the late Iona "Eddie" Thompson and Clair Bearer, the author had the privilege of playing chauffeur to a world-renowned author, poet, and literary critic. Malcolm Cowley was born on his grandparent's farm near White Mill on August 24, 1898. His father, Dr. William Cowley, was a Pittsburgh physician who lived in East Liberty; his mother was the former Josephine Hutmacher. Malcolm attended Peabody High School, Bryn Athyn Academy and Harvard University. However, as a youngster, his summers and early fall were spent in carefree roaming in the area around Belsano. He attended school there until it was too cold to go barefoot and then reluctantly returned to Pittsburgh for the remainder of the academic year.

Malcolm Cowley, White Mill Hotel, May 10, 1985.

When the United States entered World War I, Cowley was at Harvard; he tested for and was accepted by the American Field Service Volunteer Corps, a noncombat medical organization. Serving as an ambulance and truck driver from May to October, 1917, Malcolm returned to Pittsburgh where he published his first piece of writing, an article about his experiences in France. Cowley then volunteered for the balloon corps, but failed the physical because of a minor eye problem. He returned to Harvard in February, 1918. After enlisting in the army in September, he was sent to Camp Taylor, Kentucky, but the Armistice intervened. Malcolm never made it to the front as a doughboy. Since the de-mobilization of the army kept him out of Harvard for the 1919 spring semester, Cowley spent that semester in Greenwich Village, married artist Peggy Baird and lived in poverty as did most of the writers and artists of the time. With the help of a scholarship, Cowley returned to Harvard and graduated. He returned to Greenwich Village to write and live in poverty. Applying to the American Field Service for a graduate fellowship in Europe, Cowley received one to study at the University of Montpellier in France.

In July, 1921, he returned to Paris and joined that elite group of US exiles who were disillusioned with American society. There he spent two years studying, traveling on a shoestring, and writing poetry, essays and book reviews. Upon his return in August, 1923, Cowley regained his old copywriter's job at Sweet's Catalog, but was disillusioned at the lack of success of his personal writing. In the spring of 1926, he moved to Sherman, Connecticut to be close to nature, but still close to the literary world of New York City, seventy miles away. In October, 1929, he became an editorial assistant at **The New Republic** and soon advanced to editor in charge of the book department. Cowley is credited as one of the creators of modern literary criticism. He was accused in the 1930's and 1950's as having leftist leanings. However, he survived these accusations and along the way befriended and supported F. Scott Fitzgerald, John Cheever, and Harte Crane. He promoted the Lost Generation so they would not have '"fallen into neglect"'. His greatest literary triumph was introducing William Faulkner to the reading public in 1946 by editing an anthology of his works called **The Portable Faulkner**. Cowley remained in Sherman, Connecticut until his death in March, 1989. He was survived by his second wife Muriel and son Rob.

However, Malcolm Cowley never lost his love of the hills of Pennsylvania. One of his first published collections of poetry was called **Blue Juniata**. He chose the Juniata River as part of the title because Blue Blacklick would not have been appropriate. Malcolm detested the destruction of the landscape by the logging and mining. Perhaps his love of the area can be traced to his mother, who tried to protect the remaining trees on her property when the cement bridge was built at White Mill. The late Rev. Clarence Bennett, Blacklick Township teacher and Methodist minister, drove bulldozer at the bridge site. Before Mrs. Cowley would give the trucks the right of way to use the old road to haul out the dirt, she made them tie wire and boards around the trees. She also wanted all the rotten logs, which had been left behind by the Vinton Lumber Company, hauled to her house as kindling.

However, this man who walked among the giants of 20th century literature received few accolades at home. Most Blacklick Valley residents respond with "Malcolm who?" when his name is mentioned. The only local newspaper to give him any coverage was the

Ebensburg Mountaineer-Herald, owned by the Thompson Family. That was because "Eddie" Thompson, columnist for her husband David's paper, made the acquaintance of the Cowleys in the 1960's and maintained an on-going friendship until their deaths in 1989.

Local recognition was finally granted Malcolm in 1985 when Indiana University of Pennsylvania awarded him an honorary doctorate. This was due to the efforts of Philip Zorich, special collections librarian. To get an opportunity to meet Malcolm personally, the author accomplished a major feat of wheeling and dealing. By volunteering to chauffeur Malcolm around the back roads of White Mill and Red Mill, she obtained tickets to graduation, luncheon in the Blue Room, and an invitation to a reception for Malcolm in the president's apartment in Sutton Hall.

In return, the author had the pleasure of driving Malcolm Cowley through the byways of Cambria and Indiana Counties. When she first proposed stopping at the White Mill Hotel, Malcolm's birthplace, he was reluctant to stop, but then changed his mind. He ordered a beer at the bar, which was the room in which he was born. From there, we took the back roads to Red Mill and Belsano. As we drove down the dusty roads, Malcolm's eyes lit up as he remembered the families who lived along the road so many years ago.

Malcolm Cowley, May 10, 1985. Taken by author at the White Mill Hotel near Belsano, birthplace of the world-renowned poet and essayist.

On the way to Ebensburg, we drove through Revloc so that he could see the changes, or lack of changes, over the past fifty plus years. His opinions about coal operators had not changed much since his youth. In Ebensburg, we made a short visit with the Thompsons who had not been able to attend the graduation ceremonies. On the return route to Indiana, the entourage, consisting of Malcolm, the author, Craig Swauger, Jim DeGeorge, Bill Betts, Margaret Binney Smith and Jim Kempf and a few others took the long route. (Kempf had written a book and Smith a master's dissertation to Cowley.) To let Malcolm see the changes in the mining towns, we drove through Nanty Glo, Vintondale and Wehrum. Little was left to remind him of the days he rode the Mountain Goat from Pittsburgh to Vintondale, debarking at the station and walking a mile to Rexis to catch the Stump Dodger to White Mill.

The two-bit tour exhausted him, but Malcolm was gracious enough to invite the group to join him for supper. A return trip was planned for July; this time his son Rob was going to bring him. Malcolm was not sure that he could make the trip the second time, explaining on a May 19th post card,

. . . May see you 13 July; can't make promises. When I got home after more than 900 miles on the road, my legs wouldn't bear me. Yours, as ever - Malcolm Cowley.

The same group: Smith, Betts, Weber, DeGeorge, and Swauger had the pleasure of dinner with the Cowleys at the Greenery. On the return trip, Malcolm wanted to show Rob the beauty of Cook Forest in Clarion County, the scene of past pleasant memories.

Correspondence continued occasionally between the author and the Cowley, who continued to chide her about completing this book.

. . . Get back to work on your book & to hell with teaching duties for this year. Vintondale needs the book. Yours, Malcolm Cowley.

Of course, the advice was easily given, but not easy to follow. Malcolm's health continued to deteriorate; he also suffered from partial deafness since the 1940's. On March 27, 1989, he died of a heart attack at the age of ninety. Our mutual friend "Eddie" Thompson died September 13, 1989.

In his letters, Cowley made some blunt, but lively statements about Vintondale and the surrounding areas. These are quoted below.

November 17, 1967, MC to "Eddie" Thompson:
Let's face up to it, Vintondale was a hellhole. It was owned by the coal company, lock stock and constables, and it was about as ravaged and grimy as a town can be. I first saw it in the days before you were born, when the Vinton Lumber Company had its big mill beside the North Branch."

July 29, 1982, MC to Denise Weber:
I've about given up on the Vinton Lumber Co. All I know is that it ravaged the countryside, cutting all the first-growth hemlock and hardwood (the pine had gone fifth years before), then setting a fire accidentally--was that 1904?--that destroyed everything else. Then it dismantled its mill in Rexis and moved away--to Kentucky, so I heard.*

Undated letter, Summer, 1982, MC to Denise Weber:
And the local landowners, who were mostly poor, except John L. Edwards, who amassed something like 1,100 acres of land rich in coal and timber. He got a pretty good price for it in the end, more than a million--no I'm exaggerating; perhaps $200,000. The great buyer of coal rights was Manor Real Estate and Trust Co., of Philadelphia, which paid usually $100 an acre. Some of the farmers who sold their coal earlier were poor innocents, for example, John Bracken, who traded off his 74 acres of coal to Reed and Griffith (Judge Lemon Reed, I think) for a team of mules. (Nobody cheated Judge Lemon Reed.)

August 28, 1982, MC to Denise Weber:
It isn't your job to be original. Your job is to put the book together with a firm structure, a beginning and an end; to make it informative and useful. Good luck with it.

*Cowley's parentheses.

REXIS: ANOTHER MYSTERY IN HISTORY

By using first-hand evidence, the historian attempts to answer the questions of who, what, when, where, and why. When it comes to compiling an accurate account of the development of Rexis, the evidence is limited and often contradictory. Who, what, where, when, and why are used here to attempt to piece together all the confusing facts about Rexis.

WHO: This looks like a simple question, but after which Rex was Rexis named? Is it Abraham Rex? George Valentine Rex? or George Abraham Rex, son? Unfortunately, the deed books only list George Rex as a Philadelphia physician and do not include a middle initial in the name. A. Rex and one of the G. Rexes owned a nineteen acre tract on which half of Rexis was built. George Rex, physician, purchased the tract from William Cameron in 1877. This had been part of the Mathias Young tract, but not a part owned by David Ritter. Cameron had purchased it from Thomas Bracken. A.V. Barker obtained the property in 1892 from the Indiana County Treasurer for back taxes owed by the Rexes.

A. Rex was treasurer of the Cambria Land Company and died around 1890. An 1890 treasurer's deed to Isaac Grubb for other properties in Jackson Township listed George Rex as deceased; George A. Rex, physician of Philadelphia, died of a heart attack in 1895 at the age of 50. So, to whom do we attribute the name? The Rex Family as a whole? That's the simplest guess.

WHAT: What is Rexis? A village? Yes, but a unique village. Half of it was owned by the Vinton Lumber Company and represented a company town, and the other half consisted of privately-owned houses. Blair Shaffer had purchased the land west of the road from the Cambria Iron Company. In 1899, he sold the timber on the acreage to the Vinton Lumber Company and agreed not to construct any more houses in Rexis except the one he was building. (First house on the left.) For the next two years, he sold lots on the upper side of Rexis. Because he sold land to private owners, the village has survived.

WHEN: Was the village called Rexis before 1901-1902? Deed and early maps show no development on the site before the 1890's. The **1896 Indiana County Directory** listed the addresses of known Rexis area residents as Vintondale. In the 1899 charter of the Vinton Lumber Company, its mill was listed as being in Buffington Township with the nearest post office in Vintondale. The Indiana County tax assessor's books do not list a Rexis address until 1902, which co-incides with the opening of the Rexis post office on August 1, 1902. G.W. Campbell was the first postmaster. John Krumbine assumed the position of postmaster until he moved his furniture and undertaking business to Vintondale. The Rexis pst office was in operation until August 14, 1909. The position of postmaster was declared open in 1909, and a civil service test was offered, but no one wanted the job. Service in Rexis was re-absorbed into the Vintondale rural postal system. (By 1909, the Vinton Lumber Company had moved its operation to Kentucky.) The 1911 county directory again listed the mailing addresses of Rexis residents as Vintondale. To whom can credit by given for naming the village Rexis? To the Vinton Lumber Company or Blair Shaffer? Probably the lumber company. Some residents thought the name Rexis came from a man named Rex who owned the lumber company. The author found no evidence to support that assumption.

WHERE: This is one of the easier questions to answer. Rexis is on the eastern border of Indiana County, in Buffington Township, across the north branch of the Blacklick Creek from the Eliza Furance.

WHY: Agan, a very easy question. Rexis' **raison d'etre** was lumbering. The village provided some of the housing for the millworkers and immediate supervisors. Several of the top officials lived in privately owned houses in Vintondale. (The author's family home was one of these houses.) With the departure of the Vinton Lumber Company, Rexis continued to serve as a railroad departure site for the Cambria and Indiana Railroad and railyard for the Pennsylvania Railroad. Some of the former lumber company houses were rented to employees of the Pennsylvania Railroad.

Rexis: The Village

If there are buildings in Rexis which originate from **Eliza's** heyday, the tax records, county histories and county atlases have refused to reveal the facts. However, there was an apple orchard in Rexis which predates the development of the sawmill. Located where Hess' and Mottin's houses are today, the orchard may have been planted during the furnace period or by Mr. Cameron before he sold the land to the Rexes.

The Rexis we speak of is of a much more recent vintage, 1898-1910. Blair Shaffer had purchased thirty-nine acres from Cambria Iron in 1895; the price was $239.15. Mineral rights were retained by Cambria Iron. In June, 1899, he bought another tract for $601.31. In April, 1899, he sold 140 acres of timber to the Vinton Lumber Company for $1,250 upon signing with an addition $1,000 paid on April 6, 1900.

Shaffer had a town plan which unfortunately was never filed with the county. Lot 9 was sold to Marlin Cameron for $80; other puchasers were: Elizabeth Gill, $100; James Hagens, $100; Samuel Lemon Reed, $125; and Annie Campbell, $75. Frank and Mary George purchased eight acres and eight perches from Shaffer in July, 1900; their house, now occupied by Mrs. John Rabel, was the last on the left as one rounded the curve on Rexis Hill. The Shaffer lots were on the left side of the township road, except for the Gill lot, which was the last property on the right side.

Charles Gill worked for the lumber company stacking the lumber at the yard. When offered an opportunity to go to Kentucky, he refused because his wife desired to remain near to the Mahaffey Cemetery, where their son was buried. Not only did Mr. Gill purchase the land from Shaffer, but also the lumber for the house. He used ten inch boards when building the house, but nailed them vertically instead of horizontally. Ella Gill Lantzy also said that many of the Rexis houses were papered with flour sacks.

Surprisingly, the small village had several businesses. In 1899, the Shaffers built a storeroom near their house and called it the Commissary. The store catered to the lumbermen and the railroaders. For part of the time, it also served as the post office for Mr. Krumbine. Ella Lantzy said that she usually picked up the family's mail. The mailroom was at the back of the storeroom, and Krumbine had the front room filled with furniture and paint, etc. One day, when she started to the back to retrieve the mail, she saw a body covered with a sheet. After that, she refused to enter the store, and her sister became the family mail carrier. Cameons ran a candy store, and Frank George operated a grocery store near the first house on the right (DeBona's). In addition to owning a house and eight acres of hillside, Mr. George also had a horse and buggy, two work horses, and a cow.

In 1902, many families began constructing houses on their lots and moved into Rexis when completed. Wesley Detwiler purchased Lot #2 and built a house on it; in May, 1904, Emma Shaffer conveyed the lot for $300 to Sylvester Detwiler by agreement of Indiana County Court. Sylvester Detwiler sold the house to Bernard McCloskey two months later; in turn, McCloskey sold it to his brother, James "Frank" McCloskey, for $400 in October, 1911. Present owner, Mrs. Albert McKeel and her late husband bought the house in 1920 for $300. James Hagens purchased his lot for $100 and started on his house in March, 1902; John Misner later owned the property. The Austin Mervine family also moved to Rexis in March, 1902.

Mert Misner, Mary McCloskey, Mary Misner Rexis.
Courtesy of Merle & Winnie Hunter.

Families migrated into and out of Rexis on a regular basis depending on the season of the year and the logging contracts obtained. Many of the laborers were local men who had a family farm nearby to which they could retreat in lean times. The late Reverend Clarence Bennett said that his family lived in Rexis two different times. When he was four, Bennett's father helped cut the Shaffer timber above Rexis. They rented the house presently owned by McKeel's. In 1910, they returned while Mr. Bennett worked on the railroad extension. The Bennett's used a unique mode of transportation to move their household goods to Rexis. They took their furniture and household goods to the edge of the tracks, loaded them on the train, and then unloaded again when the train pulled up in front of their new house.

Rexis also had a school house which was close to Lackawanna's #3 mine. The building burned in 1910, and a building behind the iron furnace was converted into a school. This may possibly have been a Ritter building. Another school was located on Duncan land on the hill above Rexis. A path connected Rexis and the school. Although Vintondale had a grade school just a mile away, Rexis' students attended school in

Wehrum because Rexis was in Indiana County. Students during the 1920's rode the train to school, using a weekly ticket. After train sercice terminated, a bus system was set up. Until the mid-1930's, high school aged students were permitted to attend the Vintondale High School.

Early in Rexis history there were also at least two houses on a large island in the Blacklick, north of the log pond. Ella Gill Lantzy said that her family lived in one of the houses until her father was able to build theirs, and an Italian family resided in the other. During one period of high water, her father was unable to reach the house and had to sit on the tracks all night. One of the sawmill employees lassoed the house and anchored it to a tree so that it would not wash away. These houses were later destroyed when an unusually large ice jam broke loose.

Mary Bernadette McCloskey and William Merle Hunter were married February 15, 1906 at the Vintondale church rectory by Father W.F. Davies. Sponsors were James Frank McCloskey and Blanche Kerr.
Courtesy of Merle & Winnie Hunter.

Rexis 1913. Front Row: James Hunter 6, William Merle Hunter 29, Velma Hunter 3, Mary McCloskey Hunter 26, holding Roy Merle Hunter 1. Back Row: Emma McCloskey 23, James Francis McCloskey 28. Bernard McCloskey 58, William (Boots) McCloskey 19.
Courtesy of Winnie & Merle McCloskey.

The actual number of houses, private and company, are hard to estimate. Below the Gill house was a street of four singles, of which two remain. Five houses, now gone, lined the street which presently leads from Bill Mazey's house. All traces of the three-story boarding house, opeated by Mrs. Lundy, are gone, as is the company office. Vinton Lumber Company's double house has been remodeled and reduced in size by Peter Shilling. In 1914, the lumber company sold their houses, mill site, and rail yard to the Pennsylania Railroad; Vinton Colliery retained the coal rights for the 10.98 acres.

In 1943, the railroad sold some of its houses to Tom Hoffman, Bill Toth, Raymond Hugar, and Albert Mazey. Joe Pluchinsky and Jim DeBona purchased theirs in 1945. Most of the families in Rexis today, such as the Hunters, Misners, Mazeys, DeBonas, McKeels are children and grandchildren of employees of the Vinton Lumber Company, Vinton Colliery Company, the Cambria and Indiana Railroad, or the Pennsylvania Railroad. Until her death in March 1990, Ella Lantzy called Rexis home for nearly ninety years.

RAILROADS IN THE VALLEY

In 1854, the Pennsylvania Railroad successfully completed its climb over the treacherous Allegheny Front between Altoona and Cresson. Part of that trek included the Horseshoe Curve, an engineering masterpiece even yet today. The completion of the railroad spelled doom for the Pennsylvania Canal and its Portage Railroad in Cambria County. From Harrisburg to Pittsburgh, the PRR followed the canal through southern Cambria County, bypassing the proposed route through the Blacklick Valley. Becase of cheap transportation, first by canal and then railroad, along with abundant natural resources and the best technology of the time, Johnstown grew into a major industrial center.

The resources of central and northern Cambria County remained basically untouched for another thirty years. In the 1850's, an eleven mile line connecting Cresson and Ebensburg was initiated and completed in July, 1862. This roadbed hugged the crests of the hills and had steep grades and deep cuts (near Loretto). Called the Ebensburg and Cresson Railroad, it was leased to the PRR. Branching off this line was the sixty-six mile Cambria and Clearfield Railroad, chartered January 13, 1887. The Ebensburg and Cresson was sold under a foreclosure in 1891 and became part of the Cambria and Clearfield in 1893. Another section of this new division was the Ebensburg and Blacklick Railroad, chartered January 18, 1893. This new line was initiated through the efforts of Augustine Vinton Barker and his Eastern "connections." The Ebensburg and Blacklick was twelve miles long and in 1894, stopped east of Vintondale where there was a "Y" switch and water tower to service the trains. In addition to exploiting the resources, the line was also pushed through to prevent the Baltimore and Ohio, the New York Central, and the Beech Creek Railroads from developing the south branch first.

Railroad engineers arrived in Ebensburg in mid-April, 1892; their surveying instruments soon followed. Two routes were being surveyed: Bradley Junction to the headwaters of the Blacklick, missing Ebensburg by two miles or routing the line through Ebensburg. The second plan won out. Also in April, 1892, Charles McFadden, future developer of the Big Bend Mine of Twin Rocks, sent a representative into the region to review the quality of the resources; his reply was positive. As early as December 1892, PRR engineers were studying a route through the Blacklick Valley to Blairsville.

Ebensburg and Blacklick Railroad

The new Ebensburg and Blacklick Railroad was chartered in 1893 with capital stock listed as $500,000. Judge A.V. Barker was one of the parties encouraging the development of the railroad through the Blacklick Valley. Work on the roadbed started in February, 1893 after rights of way had been obtained. In July 1893, newspapers reported that the PRR was also surveying the North Branch of the Blacklick with the intention of building lines from the Ritter Furnace to Yellow Creek, and from there to Homer City. PRR's purpose was to stop development of resources in the Yellow Creek area by the Beech Creek Railroad who also had engineers surveying the North Branch in 1893. The Ebensburg and Blacklick also wanted to prevent any development in the north branch area by the Baltimore and Ohio and the New York Central Railroads.

The early grading of the E&BL was done too hastily, and the inspector said several thousand dollars was needed to repair the sinking roadbed. For a while in January, 1894, work on the E&BL halted because of rumors that the Beech Creek was heading for Johnstown through Hinkston Run and the North Branch. Work resumed when PRR was assured that BCRR was not planning any expansion. Flooding in May caused damages of nearly $10,000 near the big fill at Twin Rocks. Grading was completed by July, and the roadbed was finished in October, 1894. After repairs, the roadbed was described as the best that could be put down; the best stone was used for culverts and the best iron and/or steel for the bridges. That could partially explain the cost per mile of $20,000. The first coal from Vinton Colliery's #1 mine was shipped on October 24, 1894.

Because traversing the Blacklick was a shorter route between Cresson and Blairsville, local newspapers were speculating this was going to be the new main line and urged that the line be completed from Vintondale to Blacklick Station.

Further industrial development west of Vintondale hinged on the extension of the E&BL. In 1900, with the friendly persuasion of the board of the Lackawanna Iron and Steel Company, the line was extended six miles to Dilltown. Civil engineers, headquartered in Vintondale, were scouting the lower Blacklick for a route to complete the railroad from Cresson to Blairsville.

By August 1900, surveying crews, under the leadership of John S. Davis of Greismore, were slashing the right-of-way between Vintondale and Buffington. Eighty foreigners were recruited and put to work immediately in Buffington. Grading was completed by early September. The new six-mile addition between Vintondale and Dilltown was completed in late October 1903, and in 1904, the final leg of the E&BL through the rugged Chestnut Ridge area was terminated at Blacklick Station.

Buffalo, Rochester and Pittsburgh Railroad

The Buffalo, Rochester and Pittsburgh Railroad was originally the Rochester and State Line Railroad, sold in 1881 to Walston Brown, a representative of a New York snydicate. The line, renamed the Buffalo and Rochester, went bankrupt in 1885. Its rescuer was Adrian Iselin, one of the owners of the Rochester and Pittsburgh Coal Company and a director of the Lackawanna Iron and Steel Company. The line was renamed the Buffalo, Rochester and Pittsburgh Railroad; Arthur Yates was appointed as its president and served in that position until his death in 1909. Yates' successor was William Noonan, who had advanced from superintendent to general manager to president in 1910. Under Noonan's leadership, the line became known as one of the best-run small railroads in America. At the same time, the Buffalo, Rochester and Pittsburgh Railroad was pushing soutward towards Pittsburgh. Lackawanna Iron and Steel Company was in the process of transferring its steel mills from Scranton to Buffalo and invested heavily in Wehrum as its coking center. The BR&P was also surveying possible lines to Wehrum, but instead came to an agreement with the E&BL to use its lines as far as Vintondale. After 1906, most of VCC's coke and Wehrum's coal was shipped northward via the BR&P.

The first freight shipped from Vintondale by the BR&P was on June 1, 1904. Twenty cars made up the first shipment, part of which was destined for the Lackawanna mills in Buffalo. Passenger service was initiated on July 18, 1904 with the 7:00 AM westbound train from Vintondale arriving in Indiana at 8:45, in time to connect with the 9:30 which left for Punxsutawney. The train crew remained in Indiana and departed after meeting the 5:30 PM from Punxy. Stops were made in Homer, Coral, Heshbon, Claghorn, Dilltown and Wehrum. At first patronage was sparse. Departure hours were later changed.

In 1906, BR&P's Indiana Branch extended from Indiana to Josephine; a large railyard capable of holding 216 cars was constructed at Josephine. On this branch, including the trackage rights to Vintondale, the BR&P served eighty mines. In its heyday, it hauled 200 cars daily on the Indiana Division. Empty cars were dropped off at the mines on the trip down, and filled loads were picked up on the return trip.

There was one known major mishap on the Vintondale branch of the BR&P. In March, 1920, the bridge over Yellow Creek at Coral collapsed and dropped twenty-eight cars of Wehrum coal into the creek. A train, consisting of forty-three cars, two engines and a pusher, was crossing the bridge. A hopper went off the rails at the south end of the bridge and ran into a pile of railroad ties. The derailment caused the middle span of the three span bridge to give way. The engineer of the pusher engine was not aware of the derailment and forced all the cars between him and the break into the creek. Luckily, there were no injuries. Railroad employees worked day and night for three weeks to repair the bridge.

The BR&P had an excellent efficiency record. The main reason for this was the periodic inspection trips made by President Noonan, who arrived unannounced in his private rail car. He had a "spit and polish" reputation. These impromptu inspection trips were often recorded in Otto Hoffman's mine diaries.

Today the BR&P is part of the CSX System. In 1929, because of economic problems caused by the stock market crash, the BR&P was sold to the Baltimore and Ohio Railroad. The Wehrum shutdown in 1929 and the fire in the Vintondale washery in 1945 ended the profitability of running trains on the Blacklick. Today the line, with very few mines or factories to feed it, runs a train once to twice a week.

Train Service in Vintondale
Once the Ebensburg and Blacklick tracks were completed in 1894, the new planned community of Vintondale (or whatever it was called at the time) had immediate contact with the outside world. A Vintondale resident could board a train for almost any destination in the country; however, most were satisfied with Nanty Glo, Johnstown or Indiana as a final destination. In fact, the Johnstown trip was tedious because one had to take the train to Cresson and wait for another one leaving for Johnstown. This problem was alleviated in 1914 when the Southern Cambria Railways Company initiated trolley service to Johnstown via Conemaugh. However, this was not always the safest route considering the company's safety record, which included a crash near Conemaugh on August 12, 1916 which killed twenty-seven passengers. The trolley service ended in 1928.

Vintondale in 1894 soon had two stations, a passenger and a separate freight station. These were located near the big rock on the lower end of the town where the Blacklick split the village into two sections. To provide comfort to visitors, salesmen and railroaders alike, two privately owned hotels were soon built across from the station. One was the Village Inn, built by Dr. Merritt Shaffer and eventually owned by Fred Bitterdorf; the other was the Red Hotel or Morris Hotel, built by Thomas Morris. Uptown was the imposing Vintondale Inn.

**Vintondale passenger station - 1909.
Courtesy of Lovell Mitchell Esaias.**

The train which wound through the rugged Blacklick between Blairsville and Ebensburg acquired an appropriate nickname, "The Mountain Goat". It stopped in Vintondale at 7:30 AM and 4:00 PM heading east and at 10:30 AM and 7:30 PM heading west. The evening train went as far west as Blairsville and returned to Wehrum where it sat all night. In the morning, the train left Wehrum in time for the 7:30 arrival in Vintondale. On the westbound train, one could go to Indiana in the morning and return at 4:00 PM. Of course, the train was always met by the company policeman who surveyed the passengers. At times, some passengers were told to keep on moving.

Pennsylvania Railroad's "Mountain Goat". Post card photo taken at Wehrum in 1914 by Clair Bearer, Ebensburg.
Courtesy of Clair Bearer.

Passenger station, Bill Clarkson is the man on the left. Date unknown.
Courtesy of Leona Clarkson Dusza.

In addition to delivering mail to the towns, the conductor also delivered newspapers along the line. At the Best farm near Wheatfield, the conductor dropped the paper along side the tracks. Helen Berlitz Neumeyer said that her mother or aunt often sent her down to the tracks for the paper, but she was reluctant to fetch it because of the swampy ground along the tracks and because of the snakes.

Railroad Employees

The freight agent for years was William Lockard. (See biography.) Bill Clarkson acted as ticket agent until illness forced him to retire in 1930. He and his wife acquired the Findley house (last house in Vintondale, actually in Indiana County) and operated a restaurant there for many years. Working for the Pennsy was often unpredictable. Charles McGuire of Cresson was an extra agent for the PRR for ten years; he left to form a confectionery company and later made sales trips into Vintondale to sell his wares. Clair Bearer of Ebensburg was an extra block agent who spent about a month in Vintondale before World War I. Following the war, he returned to the PRR, but left it to work for the C&I for better wages and working conditions. William Lybarger was a crew boss for the PRR and lived in Rexis for many years. He was mudered in 1931 presumably by a member of his crew. This muder was never solved.

Del George was another employee at the Vintondale station. However, his twenty-five years of service were interrupted with periods of unemployment and temporary assignments in Hastings, Coalport, Buffalo and Harrisburg. He eventually purchased the store owned by his father-in-law, S.R. Williams and operated it until illness forced him to retire in the early 1950's.

Wehrum: Passenger Station on left, freight station of right, #4 mine in background. Postcard photograph was taken in 1914 by Clair Bearer, Ebensburg.
Courtesy of Clair Bearer.

Loss of Passenger Service

The automobile made a drastic impact on ridership on the rails, especially after the introduction of Henry Ford's Tin Lizzie. Now families of a lower income base, even a miner's family, could afford an automobile. The Stock Market Crash of 1929 accompanied by the loss of the mail contracts tolled a death knell for passenger service. In 1931, both the PRR and the C&I applied to the Public Utility Commission for permission to abandon passenger service throughout the Blacklick Valley. The last passenger train stopped in Vintondale on August 29, 1931. The mail was afterward to be delivered by star route from Johnstown. Bids went out for the job, and Frank Kovach of Twin Rocks was awarded a two year contract for the route which included Nanty Glo, Twin Rocks, Nettleton (Cardiff) and Vintondale.

Demise of the Coal Trains

Freight traffic on the PRR and the C&I continued to diminish with the closings of the mines in the valley: first Claghorn, then Wehrum, Dilltown, Twin Rocks, Vintondale in 1968, Colver, and finally the mine with longest lifespan, Bethlehem's Heisley in Nanty Glo. Even though the mini-coal boom of the early 1970's, caused in part by the oil embargo, led to the opening of several mines near Dias, these also fell on hard times for the same reasons as Claghorn. (See chapter on Claghorn.)

Mother Nature's awesome might drove the final nail into the railroading coffin on the night of July 20-21, 1977 when the valley endured an all night cloudburst. Flood damages to the roadbed, bridges, etc. were, in the railroad's eyes, not worth repairing. A bridge on the C&I near Red Mill was washed out forcing the abandonment of the Colver to Rexis branch. The PRR abandoned its roadbed from Nanty Glo to Dias. The section between Dias to Blacklick was repaired because the Oneida cleaning plant, using mainly stripped coal, was still in operation. In 1984, nine miles of rights of way for the C&I and Conrail were sold to Kovalchick Salvage of Indiana. Ninety years after the initial construction of the line, the rails and ties were removed.

Vintondale has witnessed major changes in the railroad industry firsthand with wooden coalcars, replaced by steel hoppers, and then 100 ton cars. She lived through the excitement and the dirt created by the steam engine; its successor, the diesel locomotive was more efficient and cleaner, but some of the glamor was lost. The days of watching a "black snake" wind wind its way up the grade, past #6, across the black bridge and around the bend to Twin Rocks are gone forever. Has this scene been preserved on film?

ROADS AND BRIDGES

Vintondale, nice place to visit if you can get to it. Sandwiched in between four hills carved out by the Blacklick and its tributaries, access to the town has been, needless-to-say, difficult. Perhaps this lack of accessibility hindered earlier successful development of the area. One of the reasons for the failure of the Ritter Furnace in the 1940's was the lack of good transportation. Early roads were dusty and full of ruts in the summer, quagmires in the fall and spring, and almost impassible in the winter. Hauling lumber, especially railroad ties, damaged the roads every year. In 1909, according to local newspaper reports, Buffington and Jackson Townships were carrying out their annual much-needed road repairs. Until the 1920's, all roads leading into Vintondale were dirt; Governor Pinchot's administration (1923-1927) is given credit for initiating the program to pave the rural roads of Pennsylvania.

Chickaree Hill

The 1870 **Beers Atlas of Indiana County** showed no public roads leading directly into Vintondale or Rexis. A private road existed when Ritter's Furnace was operating, but it did not appear on the Beers map. Cambria County's 1890 atlas also had no public roads into Vintondale in Jackson Township; the nearest public road was Rager's Hollow Road. According to the Cambria County Road Dockets, Rager's Hollow Road was first laid out in October, 1869; it was to connect the Blacklick Furnace Road, beginning at Beard's Lane in Indiana County, to the upper Johnstown road (Benshoff Hill road) at the Stone Pike (Route 22).

To have a road maintained by the county, a petition requesting a road was first filed with the county court. The judge then appointed three unbiased viewers to study the need for the road, look over the proposed route, and render a decision. Then the judge made a decision based on the viewer's report.

Township Road on Chickaree Hill. Taken by John Huth - 1924.

Rager's Hollow Road was made public in 1882. There was a private road to the Pringle Farm in what is today Vintondale; it connected with Rager's Hollow Road at William Shuman's property. In 1894, a petition was filed in Cambria County Court to vacate part of a public supply road which led from Rager's Hollow Road to the "Old Pringle Mill site" or from the Chickory School to the Ritter Furnace. Court records do not show when this road became public. The part which was vacated was within the new town plan of "Vintonville." Viewers for this project were John Rager, William Shuman; (who left his mark (X)), and Clarence Claghorn. One might have a bit of difficulty labeling Clarence Claghorn as a disinterested viewer, but he and the Vinton Colliery Company received a positive response to the petition.

Ice storm on Chickaree, c. 1917.

In Blacklick Township, there was a road which led from the Ritter Furnace over #6 hill to Belsano, but the author found no record of when that road was officially vacated.

Bridges

Unlike Indiana County, Cambria County does not have a Bridge Book, so little information is available at the courthouse concerning the bridges across the Blacklick between Main Street and the train station. The first bridge may have been built by the company or the county. Around 1916, a concrete single arch bridge was built in the same style as those soon constructed in Wehrum and Claghorn. The strength of the bridge has been attested to many times, but particularly in the 1977 flood when it held back debris and forced the flood waters to find an alternate route on its destructive path. Its second test came when it was being demolished for the new bridge. The contractor was forced to dynamite the foundations in order to remove it.

Approaching From the Indiana County Side

Area farmers in Indiana County realized as early as 1892 that this new coal town of Vintondale meant possible prosperity for them through access to the railroad. New markets were opening up, and the farmers needed a public road to reach Vintondale. Subsequently, two rival citizen groups arose, each promoting its own route into the new town. One of the proposed routes would have been laid out from the road at Duncan's Mills (Red Mill), over the hill and back down through what became Rexis and ended at the Ritter Furnace. There had been a private road there during the heyday of the furnace.

The second group favored a road which skirted the top of the hill and descended to Vintondale via Laurel Run near Three Springs. A petition to reopen the old private road and make it public was filed by the Duncan's Mills group, as were exceptions to confirmation of the petition by the Laurel Run group. Opinions were heard in the Indiana County Court in June and September, 1893; June, 1894; and June, 1895.

In June 1893, the group who wished the county to start the road at the lands of James Altemus and James Cauffield and proceed to the furnace via Laurel Run posted bond of $500. Signing the petition to open the Laurel Run road were: E.J. Blackley, S.E. Allison, W.B. McFeaters, J.T. Bracken, John E. Blackley, Sr., John E. Blackley, Jr., Wm. Davis, Alvis Davis, George Davis, John P. George, Samuel Graham, Harry Bennett, Milton Hoffman, T.J. Davis, Frank Altemus, John Dodson, Ervin Engle, D.C. Syster, and Wellington Cameron. Nineteen other unnamed signers favored the route from Ducan's Mills (Red Mill) through land of Filmore Duncan and George Kerr to the furnace, the eastern route. Judge Harry White appointed J.W. Botsford, Elias Buterbaugh and Jess Keith to view the eastern site. Filmore Duncan had no objections to the road going through his property, but would not sign a release of damages. The owners of the land near the furnace, the Blacklick Land and Improvement Company, did sign a release. The controversy extended to the September term. Judge White again appointed three disinterested viewers to study the proposed sites. D.L. Morehead, Bill Barkley, and John Graham viewed the advantages and disadvantages for both routes and decided to allow for damages. The east route over what would become Rexis was laid out after September, 1893 term, but apparently not opened. The western route was planned after the June, 1894 term. At the June 1895 term of court, the route from Duncan's Mills through Filmore Duncan's and George Kerr's land to the Ritter Furnace was viewed and ordered to be opened. Width of the new county road was to be thirty-three feet.

The Iron Bridge

In 1896, the iron bridge was completed across the north branch of the Blacklick. Before that, people approaching the Ritter furnace or Vintondale had to ford the Blacklick. The single-span bridge was built by Pittsburgh Bridge Company at a cost of $3,847.89. Later repairs were made in 1924 for $35.99 by P.S. Allen; supplies for the repairs came from the J.M. Stewart Company of Indiana. Mr. Allen also painted the bridge for $200. In 1926, the J.S. McIlvaine Bridge Company made repairs for $4,720.

Roads leading into Vintondale were open, but remained in deplorable condition. For instance, Andrew Nicewonger of Mitchell's Mills made a trip to Vintondale in March, 1897. Because of the condition of the roads, he almost lost one of his horses and requested damages from the township supervisors.

The Twin Rocks Road

In Cambria County, a public road which would parallel the Blacklick from Ebensburg to Vintondale was discussed in 1902. Encouraging the construction of this route was the Ebensburg Improvement Society. A road was constructed between Expedit (Twin Rocks) and Vintondale between 1890 and 1906. Part of this road was vacated and relocated by petition in 1907. This relocation may have possibly entailed making the cut in the hill known as the Big Curve. The county commissioners adopted a resolution to take over the Twin Rocks road in May, 1925, but was unable to begin paving until the grand jury approved all the required steps. The length of the road was 13,801 feet, and the cost was estimated to be $145,150.70. Bids were put out on August 13, 1925. Ten companies made bids with Fort Pitt Construction winning with a low bid of $136,553. Work was finished in early December. The completion was marked with a monument and brass plate. County Commissioners at the time of the paving job were J.D. Walker, H.C. George, and W.J. Cavenaugh. County engineer was L.R. Owen. The contractor was Fort Pitt Construction of Pittsburgh.

Monument dedicated in 1926 to mark the paving of the Twin Rocks/Vintondale Road. Taken May 13, 1984.

Courtesy of Diane Dusza.

The new concrete Twin Rocks-Vintondale road abruptly stopped at the township/borough line, and the state did not take over the responsibility of paving Plank Road, Main Street, and Chickaree Hill until the 1930's. Because of the new paved road, Vintondale now had bus service. Four trips daily stopped in Vintondale in 1926. In December 1927, the Cambria Bus Company purchased the Southern Cambria Bus Company. President of the Cambria Bus was H.E. Englehart. The use of the bus along with the cheaper automobile assisted in the loss of rail passenger service to Vintondale in 1931.

Borough Streets

In Vintondale, the borough streets remained in deplorable condition and subject to washouts during heavy downpours. Holes and ruts were packed with ashes from the powerhouse and red dog from the rock dump. In the winter, few drivers, even with a running start, were able to get their automobiles to the top of Second and Third Streets. As more and more townspeople were able to afford the automobile, the demands for better roads were being heard and answered in Ebensburg and Harrisburg, but in Vintondale, until the early 1950's, paved streets were only on the wish list.

Eliza Furnace and Old Schoolhouse, c. 1905. Courtesy of Pauline Bostik Smith & Betty Bostick Biss.

Vinton Lumber Company, Rexis c. 1906. Courtesy of Pauline Bostick Smith & Betty Bostick Biss.

Chapter III
Biographical Sketches of Coal and Lumbering Entrepreneurs

WARREN DELANO

Pennsylvania's history has often centered on the achievements of various entrepreneurs who seized upon opportunities and made themselves into household names. Industrial giants like Asa Packer, Charles Schwab, Henry Clay Frick, Andrew Mellon, and Andrew Carnegie, immediately come to mind. However, Pennsylvania also attracted dozens of lesser entrepreneurs, whose careers have been overshadowed by the giants, but who have made their marks on Pennsylvania history just the same.

Warren Delano III (1852-1920) was one of those risk takers who staked much of his personal fortune on the unpredictable coal industry. Unlike most of the entrepreneurs previously mentioned, Warren Delano was not a self-made man. He was born into a wealthy New York family whose fortunes in the mid-1800's fluctuated with the times. In Pennsylvania, Warren Delano III left his mark in Schuylkill County at the town of New Boston; in Indiana County at Wehrum, Claghorn, and Graceton; and especially in Cambria County in his own community of Vintondale. The hard-coal venture was a result of investments made by his father, Warren Delano, II; but Vintondale, was Warren III's personal project.

**Warren Delano, III.
Courtesy of Clarence Stephenson.**

Family Background

The Delano Family, whose American roots can be traced to 1621, has played a prominent role in the United States history. The first Delano in America, Phillipe de Lannoy, arrived in Plymouth in 1621 on the **Fortune**. Much of the written history of Philippe and his ancestors is based on a mixture of fact and myth. A major perpetrator of these stores was Daniel Delano, author of **Franklin Delano Roosevelt and the Delano Influence**. In his book, Delano wrote that Philippe's grandfather, Gsybert, had been disowned at the age of eleven by his noble family for embracing Protestantism. Gsybert's son Jean migrated to Leiden, a Protestant stronghold in the Netherlands. There Jean and his wife established twelve houses of refuge for exiled English Puritans. One of these Puritan daughters was Priscilla Mullens, who was sent to America because she was attracted to Philippe's older brother, Jacques. Philippe, who was also in love with Priscilla, took the next boat to Massachusetts, but arrived too late; she had already married John Alden. Wonderful story, but full of literary license and little fact.

Frederick Adams, grandson of Warren Delano III, resides in Paris; in 1985, Mr. Adams and his wife traveled to Leiden to trace the history of Philippe de Lannoy. With the assistance of some very helpful archivists, Mr. Adams discovered that his ancestor was not of the noble family of de Lannoys, but were woolworkers originally from the French village of Lannoy, near Lille in the province of Artois. The real Philippe's parents were Huguenots, (French Protestants) who were received into the Walloon Church (French-speaking Dutch Reformed Church) in 1591. Philippe was baptized in that church on November 6, 1603. Mr. and Mrs. Adams also discovered that there were more than 300 children baptized in Leyden who had surnames of De Lanoy, De Lano, de la No, de la Noy. Most of these families had originally been textile workers in Artois. As for the older brother Jacques, the Adamses could find no evidence confirming his existence. The houses of refuge also were not built by Philippe's parents, but were built in 1683 by funds donated by Jean Pesyn and his wife, Marie de Lannoy.

Mr. Adams believes that Philippe de Lannoy sailed to America because his godfather, Francois Cooke, had sailed on the **Mayflower**, and Philippe, a 17 year-old orphan, followed him, not Priscilla Mullens, to Plymouth Colony. Philippe's son and Francois' son, John Cooke, settled Fairhaven, Massachusetts; John was the last surviving male passenger from the **Mayflower**. In Fairhaven, Philippe de Lannoy acquired an 800 tract. Eventually, Warren I, whose last name had been anglicized to Delano, built a thirty-two room house in Fairhaven, which was often used as a summer home by Warren Delano II and his family. According to the will of Warren Delano I, the house was to be held in trust for the entire family for two generations. (In 1942, the house was auctioned off because there were only two surviving grandchildren, one of whom was Frederic Delano.)

The early Delanos turned to the sea for their livelihood. Warren I was engaged in the whaling business. Warren II (1809-1898), a shipper, entered the firm of Russell and Company, which traded in China. He spent most of the years between 1833-1846 in China, participating like most other companies in the illegal, but very profitable opium trade. Upon his return, he invested in New York real estate, Tennessee copper, Maryland fire brick, and Pennsylvania coal.

In 1854, Warren Delano II ventured into the Pennsylvania anthracite fields when he and a group of fellow investors purchased thirteen tracts totaling 5,200 acres in Schulykill County from Chirstopher Loeser of Pottsville. Loeser had purchased the land at a public auction in Oswigsburg in 1851. These lands had been part of the Girard Estate, and after a court battle over ownership, the judge ordered the public auction. Eventually these thirteen tracts of coal lands came under the ownership of the Delano Land Company.

Because of debts incurred during the Panic of 1857, Warren Delano II, signed over his interest in the above properties to his brother, Franklin, and his brother-in-law, Joseph Lyman. This included a 1,500 tract known as the New Boston Lands and 90/100 of the thirteen tracts in Rush Township. At this time, Judge Asa Packer of Mauch Chunk also became an investor. A town of Delano was built in 1861 and eventually became the railroad yard for Packer's Lehigh Valley Railroad. In 1864, the New Boston Coal Company, a Delano concern, opened a mine on that acreage. Connecting the mine with the Lehigh Valley Railroad was an eight mile spur called the Quakake Railroad. The New Boston Land Company, another Delano Company, built the village of New Boston which included thirty blocks of buildings, a school, a company store and a mansion.

In the meantime, to recoup his losses from the Panic of 1857, Warren Delano II sailed again for China. Daniel Delano in his book **FDR and the Delano Influence** claimed that Warren Delano II was President Lincoln's special agent in China, obtaining the opium for medical use in the Union army. However, Geoffry Ward in his work, **Before The Trumpet**, refutes this special agent claim, stating there is no evidence for it. The Delano family itself has not admitted to any stories about Warren II's career in the opium business. In 1862, Warren II sent for his wife, Catherine Lyman Delano, and their seven children. Mrs. Delano and the children set sail on a four month voyage to China in July, 1862 on the sailing ship, **Surprise**. Frederic A. Delano was born in Hong Kong in 1863, and Laura was born there in 1864. (In 1884, she tragically died of burns caused by the explosion of an alcohol lamp which she was using to heat hair curlers.) Four of the children, including Annie and twelve year old Warren, were sent home in 1864 and were cared for by Mr. Delano's bachelor brother, Ned, and maiden sister, Sarah. Warren III was then sent to Colonel Miles' military school in Brattleboro, Vermont.

Warren Delano III

Warren III, a sickly youth, was the second child in the family to be named Warren, and his poor health caused the family some concern. While at military school, Warren acquired traits that remained with him the rest of his life: neatness and orderliness in everything. At the end of the Civil War, he was sent to schools in Celle and Hanover in Germany, and Paris, France. In 1869, he and two sisters returned from Europe, and Warren went to Cambridge to be tutored for entrance to the Harvard (Lawrence) Scientific School. In his brother Frederic's youthful eyes, Warren was a "demi-god" who could do anything, yet was not vain or conceited.

While at Harvard, Warren III was bombarded by letters from his father, which gave practical advice such as: do not borrow money, keep the windows open four to twelve inches, do not put on airs, do not dress better than the schoolmasters, etc. Warren II even recommended that his son visit Henry Wadsworth Longfellow, a longtime friend of his wife's family. Warren III's trips to China and Europe could serve as topics of conversation if necessary.

Upon graduation from Harvard in June, 1874, Delano planned to move to Nevada, but instead became a manager at the Union Mining Company and the Mount Savage Fire Brick Company in Mount Savage, Maryland (near Cumberland). These were family concerns which had been established as early as 1841. Warren Delano II had become involved in the Mt. Savage companies when he helped organize the Cumberland and Pennsylvania Railroad, which serviced the mines around George's Creek. James Roosevelt, father of FDR, was general manager of the railroad. In 1860, the coal mines in the Mount Savage area merged to become Consolidated Coal Corporation. Because William Henry Aspinwall, uncle of Roosevelt, was one of the founders, James Roosevelt was given a seat on the board of directors. Warren Delano also joined its board. In 1875, Roosevelt and Delano lost their seats on the Consolidated Board in a reorganization made in response to the Panic of 1873.

Marriage

On July 11, 1876, Warren Delano III married Jennie Walters, daughter of William Walters of Baltimore. Her father and his brother, Henry, were organizers of the Atlantic Coastline Railroad. William Walters, reportedly a tightwad and eccentric, was a collector of French art and also introduced the Pecheron horse to America. Harry Walters, a Harvard classmate, introduced Warren to his sister; also a Harvard student, Jennie Walters boarded with several other young women at the home of Professor James Bradstreet Greenough, the founder of Radcliffe. William Walters refused to give his blessing to the marriage and said that Jenny was needed at home as household manager and hostess. He ordered her not to write or see Warren for two years. However, the engagement was announced in May, 1876, and the wedding took place at a church near the family's Maryland summer home. Mr. Walters refused to invite the Delano family to the wedding until the last minute. Following the ceremony, he followed the couple down the aisle, but there were no demonstrations of congratulations. After their marriage, he reportedly had a light installed on the front porch which burned day and night to light her way home. Mr. Walters remained bitter until his death and left no provision for Jennie in his will. However, her brother shared some of his inheritance with her. In contrast, the Delano family was delighted with their new daughter-in-law. Her sound ideas on homemaking were even copied by sister-in-law, Sarah, FDR's mother.

Warren and his bride set up housekeeping in "Bruce House" at Mount Savage. A son, Warren, was born to them in 1877. However, the child died in January, 1882, two weeks before the birth of Franklin Delano Roosevelt. Sara Roosevelt wanted to name her son Warren Delano Roosevelt after her father, but Warren III and Jennie felt that they could not bear to hear another child with that name. Instead, Sara's son was named after her favorite uncle, Franklin Delano.

Warren and Jennie had six other children:
Lyman - 1883-1944
Ellen Walters - 1884-?, married Frederick Adams.
Laura Franklin - 1885-?, unmarried, was at Warm Springs when FDR died.
Jean Walters - 1890-?, married George Edgell.
William Walters - 1892-1892
Sara - 1894-?, married Roland Redmond.

Seven grandchildren eventually completed the picture. Lyman had six children, and Ellen Adams had one son, Frederick Adams, who retired from the Morgan Library and presently resides in Paris. (The author has been in contact with Mr. Adams for several years. During the summer of 1989, while in Amsterdam, the author unsuccessfully tried to secure an interview with Mr. Adams. Unfortunately, the French Revolution Bicentennial celebrations intervened.)

Warren III remained at Mt. Savage until the spring of 1882. During his tenure, he made some major improvements and reorganizations. In a letter dated March 12, 1882, brother Frederic wrote:

I hope all things connected with Mount Savage are going smoothly and that an efficient man may be soon got to take your place. At least you can be proud in giving up the position that it is in very different condition from what you found it when you took it over more than seven years ago.

Delano's next venture was in manufacturing, and he moved to Orange, New Jersey. In this particular area, Delano failed. In a letter dated, February 22, 1889, Frederic related:

I was sorry to hear that at last you felt it necessary to give up the phosphate-glue business and appreciate your goodness of advising me of the fact. You have had an uncommonly hard up-hill struggle of it.

Frederic went on to recommend that he and Warren always try to keep their interests together as much as possible. Over the years Frederic, who was mainly involved in railroad management, was a major source of investment money and advice for Warren.

From 1886 until his father's death in 1898, Warren III took on more and more responsiblity of managing his father's interests. Surviving letters from father to son show that Warren II sought his son's advice on all kinds of business matters, such as buying of stocks, settling disputes on coal leases and raising employees salaries. For instance, Warren III wished to increase the salary of Mr. Jones, superintendent of the Mill Creek Coal Company, from $3,000 per year to $6,000 because other collieries were paying that amount. In a letter dated December 18, 1888, the elder Delano wrote "*. . . if other colliers pay more than Mill Creek Coal Company now pay, we must not hesitate to pay the same and I withdraw any objections. . .*" In the same letter he also withdrew any objections to paying a Mill Creek dividend on January 1st.

Under Warren III's leadership, the Mill Creek Coal Company flourished. Dividends were paid on a regular basis to various family stockholders. Among his personal letters were numerous thank yous for Mill Creek dividend checks. In 1901, Sara Delano Roosevelt received $1,968.75, her portion of the distribution of $15,000 from the New Boston Land Account. In 1904, she received checks for $22.50 from the Dodson Coal Company (hard coal), $575.00 from Union Mining Company (Mt. Savage) and $6,847.50 from the Mill Creek Coal Company. Following reception of a check for $6,250 in February, 1905, she wrote her brother that, "*Mill Creek Coal has certainly done splendidly.*"

In addition to investing family money in family concerns, Warren Delano facilitated other purchased for them, such as stocks, bonds and property, such as railroad stocks and properties in West Philadelphia and in the wharf area along the Delaware. He also handled James Roosevelt's estate after his death in 1900.

Delano even took care of ordering coal for other

family members. Even though the mines were family-owned, the coal was not free. In a 1908 letter, Sara Roosevelt requested that Warren order her two carloads of coal: a forty-ton car of furnace coal and a thirty-one ton car of range coal. On the edge of that letter, Warren made notations about the coal. Twenty tons of egg size and twelve tons of stove coal were delivered to Sarah and to FDR. For himself, Warren ordered twenty-five tons of egg and twelve tons of stove. In 1909, the total cost per ton of furnishing coal was $6.60, with the coal costing $3 per ton, freight $.85 per ton and cartage $1.75 per ton.

For handling business matters, the Delanos had a very capable assistant in Algernon Taylor Burr, company accountant and treasurer. His name was frequently mentioned in personal family correspondence. Often, he was asked to check the value of a stock or bond to see if it was a good investment. He even made travel arrangements for the family, booking a yacht for Frederic in 1895. Burr also kept track of the income of most of the family, advising them if they had money to buy stocks or if they should sell. A.T. Burr's grandson, Julian Burr (author's cousin), said that his grandfather handled FDR's allowance and that he was frequently broke. The Delanos seemed to rely on their accountants and set great store by them. Burr's successor, Charles Miller of Mahanoy City, was left a legacy in 1951 in Frederic Delano's will. (Following his retirement from the company, Mr. Miller moved back to Mahanoy City in Schuykill County. He died there in 1985.)

Warren Delano III's early addresses included Mt. Savage, Maryland and East Orange, New Jersey. He later had a house on Park Place and 36th Street in New York City, but in 1894, Warren Delano III became a "country gentleman." His uncle, Franklin Delano, bequeathed his country estate, "Steen Valetje," to Warren. The estate had been given to Franklin and his wife, Laura Astor, by her father, William Blackhouse Astor. Laura Delano had admired the lobby of the Hotel Beau Rivage on Lake Geneva and remodeled the lobby of "Steen Valetje" to resemble it. (Once, she was asked to give up her Beau Rivage suite to the Empress of Austria; Laura refused, saying she had paid for the rooms. The empress stayed in other rooms.)

Warren III took a keen interest in his estate. Numerous letters survive from stock breeders concerning the purchase of new breeding rams and horses; from the American Association of Creamery and Butter Manufactures about a problem of producing butter without an odor; and from various agricultural colleges, such as Cornell, about growing alfalfa on the estate. In 1902, he even sent T.D. Jones, superintendent of the Mill Creek Coal Company, on several trips into the Amish Country of Lancaster and Berks Counties to locate a team of mules. Also, he served as vice-president of the Edgewood Club of Tivoli, New York.

Lyman Delano's widow lived at "Steen Valetje" until her death in the 1950's. Her children then sold the estate; it resold within the last decade for about four million dollars. The house is reportedly in excellent condition, but the long Victorian veranda and the porte cochere hae been removed. In a January 10, 1986 letter, Frederick Adams described that carriage entrance as "disfiguring."

Delano Discovers the Blacklick

As a true entrepreneur, Warren Delano III made his move into the undeveloped coal fields of Cambria County in 1892. A land company called the Blacklick Land and Improvement Company was created to purchase the necessary coal lands. Other known investors in the land company were Theodore Bechtel and John Krause of Philadelphia. The coal on this land was to be mined by the Vinton Colliery Company, chartered in 1894. Site for this venture was in Blacklick and Jackson Townships near the Ritter Iron Furnace at the fork of the North and South Branches of the Blacklick Creek. The prime mover in this deal was Judge Augustine Vinton Barker of Ebensburg. The Barker family was involved in the mercantile business, lumbering and coal mining. Judge Barker had contacts with Philadelphia merchants and financiers through the family lumber business and was able to attract investors to the area. Barker and Webster Griffith, also of Ebensburg, began purchasing properties and/or coal rights in the Blacklick Valley, often for non-payment of taxes. The Griffiths reserved the timber rights on much of the land they conveyed to the new development company.

As Barker was purchasing land, he also was negotiating with the Pennsylvania Railroad to extend its subsidiary, the Ebensburg and Blacklick Railroad, from Ebensburg to the newly planned community which was first named Barker City. Some time in 1894, the town became known as Vintondale. after Judge Barker's middle name. (No where has the author found any evidence to support the popular assumption that the town was named after a Vinton Family.) The spur line from Ebensburg was finished in the fall of 1894, and the first coal from the Vinton Colliery Company's #1 mine was shipped in October, 1894.

Between 1894 and 1900, three mines were opened in Vintondale. Under the supervision of superintendent Clarence R. Claghorn, formerly of Philadelphia and Bernice, a modified longwall mining system was being tried in the new #3 mine, which opened in 1899. Delano was also interested in coking the coal. Eight experimental coke ovens were built at the #3 mine.

As early as August, 1895, Frederic Delano wrote to his brother offering to invest in the coke business because of his confidence in Warren's judgement. He advised:

I should want to be reasonably sure if I were you that some large coke manufacturer, say Frick and

Comapny c'ld not adapt your plans with sufficient modifications to evade your patents or at least those which have not expired and so steal your thunder.

Frederic offered $5,000 to invest in the coke project and offered to distribute a coke prospectus around Chicago. However, due to downturn in business, Frederic could only send him $2,000 in January. Additional amounts were sent in mid-January and February. In an undated letter, he acknowledged receipt of twenty shares each in the Bee Hive Oven By-Products Company. If this was a Delano company, no surviving evidence has been discovered which confirms its existence.

In 1897, the soft-coal industry was in a slump. Vintondale mines were "running full on lean orders." Frederic suggested that Warren find a market for his coal in the Upper Great Lakes region. In spite of poor market conditions, the Mill Creek Collieries continued to make a good showing, and Delano considered using compressed air haulage there.

Also in the 1890's, Warren Delano began some type of attachment with the Lackawanna Iron and Steel Company of Scranton. In a letter dated September 7, 1893, Frederic Delano discussed the conditions of a job offer that his brother had given him and wondered whether Warren could actually afford the "very handsome", but unannounced price he offered. Frederic wrote, *"Do I understand that my accepting your offer is the only condition of your accepting the Lackawanna Iron and Steel Company?"* Judge Barker and Warren Delano, presumably a director on the Lackawanna board, were convincing salesmen to the Lackawanna concern. While Lackawanna Iron and Steel was building a new plant in Lackawanna, New York and dismantling the Scranton works, Judge Barker was also buying additional coal lands in Indiana County. Barker sold around 12,000 acres in 1902 to the newly organized Lackawanna Coal and Coke Company for a personal profit of $141,717. The new subsidiary was formed in 1901 with Delano, Walter Scranton, E.A.S. Clarke and Cornelius Vanderbilt as principal investors.

Work began in 1900 on the new town of Wehrum, named after Henry Wehrum (1843-1906), general manager of the Scranton works. An extension of the Ebensburg and Blacklick Railroad from Vintondale to Dilltown and then on to Blacklick Station was completed. Shafts for two mines, #3 and #4 were sunk in 1901. A sawmill was erected near the site to provide lumber for the mine and houses.

In the 1901 plans was a coke works, probably at the #3 mine, halfway between Vintondale and Wehrum. A narrow gauge (dinky) railroad was built to connect Vintondale and Wehrum. Also that year, the Vinton Colliery Company sold its #1 and #2 mines to Lackawanna for $175,000. Blacklick Land and Improvement Company sold its houses and property to them for $325,000. Clarence Claghorn became superintendent of Wehrum and proceeded to build an English-style mansion on the hill overlooking the town. A company brickyard was also constructed and in 1901 was producing 20,00 bricks per day. Another town, named Claghorn, was being planned further west on the Blacklick near Heshbon.

By late 1901, the coking plans were delayed, and the Lackawanna board decided that the coking was going to done in Buffalo at the new steel plant. The ovens were never built, and in 1904, because of poor market conditions, the entire operations of Wehrum and Claghorn closed. The #3 mine, which had experienced an explosion in May, 1904, remained closed until 1914. Work on Claghorn was abandoned; Delano purchased it in 1917. Vinton Colliery Company operated the Claghorn mines until 1924.

The coal in this part of the Blacklick was, in Lackawanna's eyes, unsuited for coking even after being washed. The sulphur content was too high. Warren Delano disagreed and decided to strike out on his own and prove that this was good coking coal. In 1901, the Vinton Colliery Company had retained the #3 mine in Vintondale; Delano decided to lease the #1 mine in 1904 and then initiated negotiations to repurchase it. Looking for investment money, Warren Delano suggested to his brother that he put $10,000 into houses for the miners. Frederic's reply of November 1, 1905 was:

I am not sure that I want to become the landlord of miners' houses exacting an arbitrarily high rent. There are aspects of making money which I don't care about not that I wish to make pretences at scruples which you would not feel. I simply mean that high return does not tempt me to squeeze a working man. I leave it to your judgment and sense of right and if you think it is good clean business I shall be glad to send the $10,000 on demand.

In a letter dated November 12, 1905, after receiving more explanation from his brother, Frederic declined the offer, saying:

While my scruples are without foundation, I think prefer not to become an "absentee landlord." I would rather buy Vintondale bonds from you or some other owner to the amount of $10,000 worth and let the seller of these bonds become the landlord. Doesn't it seem better in every way to have your houses and lots owned by a man who would be in touch with the needs of the place?

In 1905, Warren III started formal negotiations with Lackawanna to repurchase the #1 mine and coal rights in Blacklick Township. A special stockholders' meeting was held on March 6, 1906 at Wehrum to act on the proposed sale of real estate, mining and other rights. After successfully negotiating the sale, Delano began executing plans to build a new mine (#6) with a coal washery and a battery of 152 coke ovens. (A photographic record of the construction of the ovens and other buildings have been preserved on glass negatives. These have been donated to the Pennsylvania Historical and Museum Commission. Prints of these negatives are available for study at the Archives.)

In October 1906, Frederic was again giving advice and also notifying Warren that he had no spare cash to invest. He questioned the logic of putting everything into the expansion of Vintondale.

In other words if you are making a large investment and going into a coking proposition to prove to the Lackawanna people that they are wrong and you were right, you are doing a thankless task which might cost a lot of money.-I have no doubt that you were right but does it pay for an individual to try to do single handed what a big corporation has found impossible?

Warren did not heed his brother's advice and proceeded to take out a mortgage for $500,000 with Pennsylvania Company for Insurance on Lives and Granting Annuities. Effective date of the mortgage was June 8, 1906. With the addition of this mortgage, the indebtedness of the Vinton Colliery Company rose to $675,000.

Things were not looking up for Warren Delano in 1907. A financial recession set in, and Delano struggled to keep his new mine going. In a 1920 tribute to his brother, Frederic Delano wrote:

Because the thought or even the possibility of failure was not in his scheme of things, he did not follow the rules which most prudent men of business follow, and he inevitably paid dearly at times for his intrepidity. In a great majority of cases he won out, where a less vigorous and energetic man would have failed. The panic of 1907 found him unprepared, and for a time the future looked very somber for him, but he never lost heart. All he wanted was the chance to make good, and given that he worked like a tiger to get results. It was a hard uphill fight for several years, but at last things began to come his way, and finally his confidence was justified and his endeavors richly rewarded.

Grandson, Frederick Adams echoed the same opinion of his grandfather, but also was the only source who acknowledged that Warren Delano received financial assistance from Jennie Delano's brother.

My grandfather was venturesome and forward-looking, but his timing terrible, through no fault of his own. Building Vintondale #6 just before the 1907 depression was alas typical. Fortunately, his brother-in-law Henry Walters was ready to help in emergencies.

In spite of the financial panic, the mines in Vintondale were working a full schedule and were short of help. Due to a shortage of cash in November, 1907, the Vinton Colliery Company began issuing scrip or "cheques" which were redeemable at the Vinton Supply Company, the large company store built in 1906. "Foreign" miners received the scrip without protest, and Vintondale merchants accepted the scrip at face value.

Throughout this period of financial uneasiness, Frederic continued to offer support, both written and financial (when he had it). In 1907, the Vintondale operation was evaluated in the Invilliers Report was found in Frederic Delano's files at the FDR Library

There was an E.V. Invilliers who owned one share in the Vinton Colliery Company, and a C.S. Invilliers owned Lot B1 in Vintondale. Perhaps one of these persons may haven been asked to assess the future of the company based on its resources.) Frederic encouraged his brother to stay with it and see it through rather than sell out his interest which would have meant a heavy scaling down of his equity. Frederic felt that with keeping in touch with actual conditions, watching the sales and collections, the weekly and monthly cost sheets, and dropping employees, Warren could see it through. Part of this cost cutting included shutting down the longwall mining method of the #3 mine. In the same letter, Frederic alluded to the longwall mining, *"Too bad it took so long to find out the error of the "long wall" method."* This error was never elaborated upon, but the Lackawanna Coal and Coke Company had also abandoned the longwall method in 1906 after Clarence Claghorn left their employment.

Delano in Kentucky

While Delano was investing large amounts of money in his Vintondale operation, he was also speculating in the rich coal lands of Eastern Kentucky. As early as 1904, he traveled there to look at coal leases. The lands that he purchased were on Puckett's Creek and Catron's Creek in Lecher County. In 1907, he created the Kentenia Corporation (for Kentucky, Tennessee, and Virginia) to exploit the coal in the region. Stockholders in this new corporation were Delano, Charles Davis, some Roosevelts and local politicians. Its capital stock was valued at $10 million. $4.25 million of it was issued to Davis and Delano. Davis also owned $2 million in stock himself.

Warren Delano made frequent trips to Kentucky to secure the coal leases. Franklin Roosevelt aided him in conducting the title searches. In some cases, it meant traveling by donkey over trails to talk to the "mountaineers." Both men were successful in charming the hill people into selling their coal rights to the Kentenia Corporation. In the opinion of Harry Caudill, author of **And Night Came to the Cumberlands**, Delano's towns were considered some of the better constructed "patches" in that area.

In addition to his Kentenia Corporation, Delano also invested in Charles Edward Hallier's Elk Horn Coal Company, which owned 70,000 acres of prime coal. Delano became a member of the board of directors of the Louisville and Nashville Railroad. It came under control of J.P. Morgan in 1902 and became a subsidiary of the Atlantic Coastline Railroad.

Delano may also have been instrumental in encouraging the Vinton Lumber Company to move to Kentucky in 1907. This company, which cut the timber along the north branch of the Blacklick Creek, closed its Vintondale operation and moved to Morehead, Kentucky. Efforts to connect this move with Delano's Kentucky holdings have proved fruitless.

The Kentucky and Mill Creek properties may have carried the financial burden for Delano. In 1908, coal stripping was initiated at Mill Creek, and in 1909, Frederic was looking ahead at the prospects for the family coal holdings. The Dodson Coal Company "looked pretty hard on the rocks" and needed a miracle to save it. He estimated that they could operate at Mill Creek for eight more years with their present organization.

In addition to financial problems with the Vintondale operation, finding a suitable mine superintendent was a recurring problem. Clarence Claghorn departed from Tacoma, Washington in 1904 when the Wehrum operation closed temporarily. Charles Hower served as Vinton Colliery Company superintendent from 1906 to 1909 until he was dismissed or resigned because of a miners' strike. Frederic aided his brother by recommending several candidates, Stewart Kennedy and S. Kedzie Smith. The major task facing the new superintendent was to make coke at a profit. Smith did take the job for one year. Another, T.W. Hamilton remained from 1912 to 1914. He brought with him Otto Hoffman, an ore boat engineer, who became chief engineer in Vintondale. With Hamilton's departure in 1914, Mr. Hoffman, nicknamed "King Otto" or "Pappy", became superintendent and proceeded to oversee the town as his feudal estate until his death in 1930.

In 1908, Delano also hired Clarence Schwerin I, mining engineer and graduate of the Columbia School of Mines, to manage the coke department of the Vinton Colliery Company. Through his expertise, Vinton coke became a salable product, much of which was shipped to Buffalo via the Buffalo, Rochester and Pittsburgh Railroad. The major customers were American Cynamid and Union Carbide. Delano's confidence in Schwerin manifested itself when Schwerin became president of the Vinton Colliery Company in 1914 and the Mill Creek Coal Company in 1918. Nick Gronland, coal salesman for the marketing division's Delano Coal Company, credited the success of the Vinton Colliery before World War I to Mr. Schwerin. This rapid advancement of Schwerin was indicitive of Delano's attitude toward his employees. Frederic wrote that his brother was:

... liberal and generous with those about him, and men liked to work for him and with him because he was direct, straightforward and is ready to reward and approve a good piece of work as he was to find fault with neglect of duty or disobedience. He put great trust in his men and he liked to see them grow under that treatment. He left some remarkable instances of faith in his associates. His idea was to encourage men by giving them a handsome share in the fruits of success.

By 1914, the financial condition of the Vinton Colliery Company was on an upturn. Frederic received 2,000 VCC certificates and wrote:

It is splendid that you have at last gotten things going so that they won't worry you - what has been accomplished at Vintondale is certainly incredible.

In 1916, Delano's entrepreneurial drive was again at work. He took out another mortgage on the Vintondale properties with The Pennsylvania Company For Insurances On Lives And Granting Annuities for $225,000 to build much needed housing in Vintondale and to develop the town of Claghorn. The Vinton Land Company was formed to handle the rental of houses in Vintondale and to purchase the coal rights in Claghorn from the Lackawanna Coal and Coke Company.

Delano continued to invest heavily in the soft coal business and put faith in his ability to produce coke. In March, 1920 he negotiated the purchase of another mine and coking operation, the Graceton Coke Company, which was renamed the Graceton Coal and Coke Company. The facility, located in Indiana County, remained in the hands of the Vinton Colliery Company until it went into receivership in 1936.

In May of 1920, to show off his towns, Delano brought his family, including several nieces and nephews, to visit his Pennsylvania mines. Initially Franklin and Eleanor Roosevelt had accepted the invitation, but had to back out on April 27, 1920 because FDR expected to be called before a congressional committee.

When Delano did visit Vintondale, he often came in his own private rail car which was parked on a siding near #1. If he stayed at the Vintondale Inn, his room was #7. He frequently brought horses with him and kept them at Charlie Bennett's livery stable. Delano enjoyed surveying his "domain" on horseback.

Prize-winning horses were raised on Delano's Hudson Valley estate, and he entered them annually at the Duchess County Fair. His stable in 1920 included sixty-five saddle, driving, and draft horses. On September 9, 1920, Delano, serving as Superintendent of the Horse Show, had gone to the fairgrounds. That afternoon, he left the fair to go to the Barrytown train station to either pick up some guests or a guest's trunk. He took his favorite horse, Belle, who needed some exercise. However, because the horse was easily frightened by locomotives, Delano had planned to depart before the next train arrived. The job took too long; when Delano heard the train whistle, he climbed into the carriage to calm the horse. Belle seemed to quite down and suddenly bolted and raced around the station into the path of the train. The buggy was carried 150 feet and the horse 100 feet. Mr. Delano, who died instantly from a broken neck, was found still sitting in the buggy seat; the rest of the buggy was in splinters. The buggy wheels were not found.

The private funeral was held on September 13 at "Steen Valetje." Only immediate family and estate workers attended. FDR, who was the 1920 Democratic Vice-Presidential candidate, arrived from Eastport, Maine. A special train carried Delano's casket to Fairhaven for burial in the family plot. Floral tributes were received from Levi Morton, Vincent Astor, Col. Archibald Rogers, Frederick Vanderbilt, Roosevelts, W.D. Dinsmores and the Huntingtons.

After Warren Delano's untimely death, management of the family businesses fell mainly to Clarence Schwerin and to the main office in New York City. Frederic served on the board of directors and usually only attended those meetings. However, in 1940, Frederic was asked by UMWA Local 621 for help in reopening the mines in Vintondale, which were closed without notice in March, 1940. Union consensus is that Frederic Delano was able to arrange a loan to get the company operating again under Chapter 11 Bankruptcy; it was renamed the Vinton Coal and Coke Company.

Warren's son, Lyman, played little or no role in the Delano coal operations. He eventually became chairman of the board of the Seaboard Coast Line Railroad, successor to his grandfather Walters Atlantic Coastline Railroad. Grandson, Frederic Adams, served on the board until the Mill Creek Coal Company, the Delano Coal Company and the Vinton Coal and Coke company were dissolved in 1957 following the premature death of Clarence Schwerin II.

How major was Warren Delano's role in Pennsylvania? In comparison with the captains of industry, one would have to say that his role was miniscule. However, he was in part responsible for the creation of three mining towns: Vintondale, Wehrum and Claghorn. He was also landlord of two older towns: New Boston and Graceton. Wehrum and Claghorn both disappeared within ten years of his death.

Survival of Vintondale perhaps could be called his greatest achievement. But was it? His brother Frederic's unheeded advice was not to get in over his head. Nick Gronland, who admired Delano very much, said that he had purchased the wrong coal properties, and that there was much better grades of coal to be found in nearby Nanty Glo, Ebensburg and Johnstown. Delano took great pride in the development of his towns, visited them often and sought improvements for them, yet he was in the business to make a profit, a true entrepreneur. This can be seen in both Delano's and Schwerin's anti-union stance. They went to great lengths to keep the UMWA out of their towns and welcomed strikes at other mines so that they would have the increased business.

At the same time, Delano was well-respected by those who had contact with him. Ruth Roberts Anderson said that when Delano was in Vintondale, he frequently visited her father, John Roberts. She said that Delano looked like a southern colonel, carrying a cane and wearing a homburg. Nick Gronland left the employment of the Vinton Colliery Company in 1917 to enlist in the service. Mr. Delano asked him to write to him while he was overseas. When Gronland returned, he discovered that Delano had made copies of his letters and shared them with others in the office. John Brugan, shipping clerk for the Vinton Colliery until 1929, stated in a letter to the Cambria County Historical Society that Delano was one of the greatest men he had ever known.

His importance as a family leader was undeniable. All looked to him for leadeship and advice. There was a strong sense of affection and loyalty between Warren and his brothers and sisters, as can be seen from their letters. He took great interest in keeping family photographs up to date. These were framed and hung in his office and in his homes. After the death of his father, Warren III became the patriarch of the clan, presiding over the family gathering at Fairhaven for Thanksgiving, carving the turkey and offering the traditional family toast. With the untimely death of Warren Delano III, the idea of a patriarch of the Delano family and even the idea of a Delano empire seemed to die out.

Charley Miller, Treasurer of Vinton Colliery Company after retirement of A.T. Burr, 1919 following his discharge from the service. Taken at Oceans Grove, New Jersey.
Courtesy of Mrs. Ruth Miller, Mahony City, PA.

Delano Family Gathering in Newburgh, NY, July 13, 1889. Birthday of Warren Delano, II. Pictured: Warren II and wife Catherine (Lyman) in the center, Franklin Delano Roosevelt in back row second from left, Sara Delano Roosevelt next to FDR, Warren, III on right.

Courtesy of FDR Library.

Algonac - Delano Family Home, Newburgh, NY, 1872.
Courtesy of Delano Roosevelt Library.

Stein Valetje - home of Warren Delano III, Barrytown, NY.
Franklin Delano Roosevelt Library.

CLARENCE CLAGHORN

If there was any one person to be credited with shaping the destiny of early Vintondale and Wehrum, that individual was Clarence Raymond Claghorn, first superintendent of the Vinton Colliery Company and Lackawanna Coal and Coke Company. Claghorn supervised the development of Vintondale and Wehrum from the clearing of the land to the construction of suitable houses for himself. An innovator in mining techniques, Claghorn introduced the longwall method in Vintondale and Wehrum and presented papers on the subject at annual meetings of mining engineers. In spite of his early contributions, he also left a legacy of mystery when he departed the area a second time in 1917.

Family Background

Clarence R. Claghorn was born in Philadelphia on August 10, 1864, son of James Raymond and Elizabeth Rice Claghorn. His paternal grandfather was James Logan Claghorn, of the seventh generation of Cleghornes who hailed from Lanarkshire, Scotland. James L. Claghorn, born in 1817, was a well-known and highly-respected Philadelphia merchant and businessman. His father, John W., was an auctioneer; James L. also began his career as an auctioneer in the firm of Claghorn and Hill. Later, he became a clerk in the dry goods firm of Myers and Claghorn. Claghorn's rise in the financial world and his prominence on the social scene in Philadelphia can be traced through the city directories and the U.S. Census Reports. Those who could afford to do so, moved westward into the newer and more prestigious sections of the city.

In the 1860 census, James L. listed property worth $20,000 and personal property of $140,000; the household consisted of Claghorn, his wife, a son and three maids. In 1870, the household included a son, J. Raymond, his six-year old grandson Clarence, and three Irish maids. As Philadelphia grew, Claghorn moved his household westward in the city from 383 Mulberry in 1849 to 1009 Arch in 1858 to 1729 Arch in 1868 to his final home at 222 North 19th Street, bordering West Logan Square.

James Logan Claghorn was a strong supporter of the Union cause in the Civil War, having enrolled in Company K, Gray Reserves, in 1861. After the war he traveled in Europe, Africa, and Asia, returning to the United States in March, 1868. He then became president of the Commercial National Bank and was also elected to the Academy of Fine Arts. He was also a director of the Board of City Trusts, Saving Fund Society, Girard National Bank, and the Northern Home for Friendless Children. The orphans at the Home called him "Father Claghorn." Because James L. was also an avid art collector, it was fitting that, after his death on August 25, 1884, his body be laid out in the art gallery of his town house on West Logan Square. At the private funeral held at the Grace Protestant Episcopal Church, invited guests included bankers, judges, ex-governors, and the orphans of the Northern Home in "full uniform." James L. Claghorn was survived by his widow, the former Julia Raymond of Machias, Maine, whom he married in 1841, and a son, James Raymond, who was born on October 5, 1842.

First superintendent of Vinton Colliery Company and Lackawanna Coal and Coke Company - Clarence Raymond Claghorn, United States Naval Reserve.
Courtesy of Robert Cresswell.

James Raymond Claghorn

J. Raymond's first wife was Elizabeth Rice, whom he married on November 11, 1863. Their children were Clarence Raymond, born August 10, 1864, and Herbert Rice, who was born August 12, 1869, and died at age one. Following the death of his first wife, J. Raymond married Anne Lockwood on April 10, 1872. Their children were Mabel, born August 5, 1874, and Julia, born January 10, 1885.

In his early business career, J. Raymond clerked in the family store, worked in an iron foundry and finally entered real estate and insurance. Some of his later interests included the State Line and Sullivan County Railroad, the Barkley Coal Company and the Long Valley Coal Company of Bradford County. He was also involved with the operation of the Connell Breaker in Bernice, Sullivan County.

J. Raymond Claghorn's disappearance from the Philadelphia business and social scene can be traced to 1898 when his name no longer appeared in the city directories. Following his father's death in 1884, Claghorn continued to live at the Logan Square address until 1896, when the directory gave his address as **The Dresden**, an apartment building at 800 Pine Street. This address did not carry the same prestige as West Logan Square. His funeral notice in 1906 was the first reappearance of his name in Philadelphia papers. J. Ramond Claghorn died on September 9, 1906 at the age of 63 in Hartford, Connecticut. (His daughter Mable had married John Bulkeley of New Haven in 1896.) His funeral was held in Philadelphia. It is the author's opinion that Claghorn had undergone a serious financial failure and was forced to move in with his daughter. Basis for this opinion is the disappearance of his name from the directories, his move from Logan Square to an apartment, a general economic downturn caused by the Panic of 1893, and the sale of the Bernice mine.

James Raymond Claghorn of Philadelphia. "The Barony of Cleghorne."

Education and Early Career of Clarence Claghorn

Clarence R. Claghorn, only surviving son of J. Raymond, was educated at Lauderback's Academy and entered the University of Pennsylvania in 1880, receiving a degree in Mine Engineering from the Towne School of Engineering at Penn in 1884. While at the University, he was a member of the Beta Theta Pi Fraternity. In 1884-85, Claghorn assisted with the Second Geological Survey of Pennsylvania, served superintendent of the State Line and Sullivan County Railroad in Bernice from 1885-88 and was superintendent and general manager of Coal City Coal Company of Birmingham, Alabama, and as assistant geologist and special agent to the Alabama Geological Survey. In 1892 and 1893, he traveled to Berlin and attended the university there, but received no degree.

Clarence R. Claghorn - graduation photograph - University of Pennsylvania Archives.

Claghorn's ties with Vintondale began when he was hired as the first superintendent of the Vinton Colliery Company. How the Claghorns and the Delanos became acquainted is unknown, but many of the early investors of the Blacklick Land and Improvement Company, like Theodore Bechtel and John Krause, hailed from Philadelphia and may have supplied the link.

Clarence Claghorn was thirty when he arrived in Vintondale to supervise the building of the town and the opening of the #1, #2, and #3 mines. At #3, he introduced the modified longwall mining system. He also served as the first postmaster of both Vintondale and Wehrum. At the corner of Lovell and Second Streets, he built a large English-style house. Later this house became the company doctor's house and was purchased by Roy Roberts in the 1940's. In 1984, the house was purchased by Paul and Betsy Shandor from Dick Roberts. Present owner is Betsy Shandor Helsel.

When the Lackawanna Coal and Coke Company was created in 1900, Claghorn was an investor and the logical candidate for the position of general superintendent of all the Lackawanna mines. Claghorn moved to Wehrum and built a large mansion overlooking the town. According to local stories, the house included twenty-nine rooms; Claghorn had a coach-in-four complete with coach dogs; and he was upset because the town was named Wehrum instead of Claghorn and that is why he left Wehrum. However, the facts do not support the hearsay. In 1903, Lackawanna started to build the town of Claghorn, located between Dias and Heshbon on the Blakclick Creek. Several house were built, and one or two drifts were opened. However, due to poor economic conditions, all of the Lackawanna operations in the Blacklick were closed in 1904, and Wehrum did not reopen until a year late. Lackawanna #1 and #2 were sold back to Delano, and Lackawanna's #3 and the town of Claghorn remained closed for at least a decade.

When Wehrum closed, Claghorn moved to Tacoma, Washington and became superintendent of coal operations for the Northern Pacific Railroad. While there, he joined the Naval Reserve and served as a lieutenant from 1909-28.

Lieut. Clarence R. Claghorn, of Tacoma.

In 1916, Delano decided to purchase the abandoned town of Claghorn and its coal lands from the Lackawanna Coal and Coke and recalled Clarence Claghorn to be the superintendent. Claghorn returned in the summer of 1917 and spent part of that time staying with the Blewitts. He built another English-style house in Claghorn, but stayed only about one year or less. Where he went and what he did between 1917 and 1931 are unknown, but his Penn alumnus address in 1931 was Baltimore, Maryland. His last known address, for the year 1935, was Chicago, Illinois.

Claghorn was married twice. His first wife was

Margaret Montgomery of Muncy, daughter of Colonel David Montgomery, United States Army. They were married on December 30, 1886, and had three sons: James Lawrence (6/10/1889), David Montgomery (3/5/1891) and Richard Raymond (9/26/1895), who attended the Naval Academy. A daughter was born on March 21, 1906. Margaret Claghorn suffered from glaucoma and lost her sight; to relieve the pressure on her eyes, he eyeballs had been removed. A former resident who remembered Claghorns thought that they were "uppity." Claghorn's second wife's name was Jeanne Lemaitre.

Efforts to locate more information on Clarence Clghorn's later career and date of death have been fruitless. A cousin who had been a minister in Johnstown reported that the family had lost all contact with Clarence's family. Another cousin, descended from John W. Claghorn did not reply to inquiries. So Clarence Claghorn disappeared into the pages of history. His town is a ghost town, but one of his three houses in the valley still stands stately at the top of the hill overlooking Vintondale.

**James Lawrence Claghorn, of Tacoma.
"The Barony of Cleghorne."**

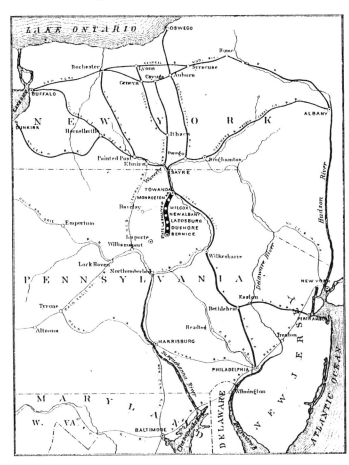

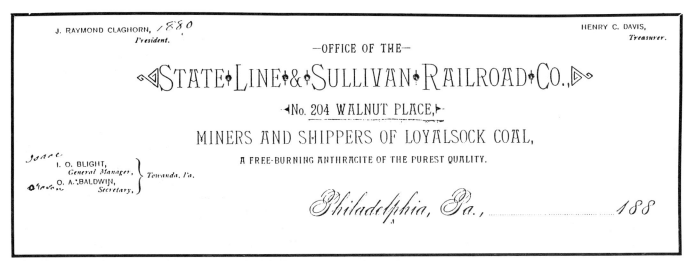

Courtesy of Nedra Snyder - Sullivan County Historical Society.

AUGUSTINE VINTON BARKER

Towering above his fellow Cambria Countians in the field of land speculation was Judge Augustine Vinton Barker of Ebensburg who served as the intermediary with the "Eastern industrialists." Through his salesmanship, the Ebensburg and Blacklick Railroad was inititated; the Blacklick Land and Improvement Company incorporated; and the Lackawanna Iron and Steel Corporation drawn to the Blacklick Valley. All of these companies have faded from the scene, but there is one lasting achievement that can be credited to A.V. Barker, the town of Vintondale. Approaching its centennial year, Vintondale took its label from the middle name of Judge Barker.

Judge A.V. Barker. Courtesy of Biographical & Portrait Cyclopedia of Cambria County, Pennsylvania, Union Publishing Co., Philadelphia, 1896, p. 436.

A.V. Barker, an Ebensburg lawyer, was the son of Abraham A. Barker (1816-1898). A native of Lovell, Maine, A.A. Barker married Losina Little on 6/24/1842. Their union produced four sons: Valentine S., 8/15/43-1906; Florentine, 2/8/1847; Augustine Vinton, 6/20/1849-8/20/1928; and Constantine, 9/20/1854-3/8/1924. A.A. Barker migrated to Carrolltown in 1854, and two years later moved to Ebensburg, taking up the lumbering business and adding a mercantile interest in 1858. Florentine Baker joined his father in business in 1866 as Barker and Son. Abraham Barker was credited as one of the Ebensburg businessmen instrumental in facilitating the construction of the Ebensburg and Cresson Railroad.

As an ardent abolitionist, A.A. Barker reportedly ran a station on the underground railroad. He was also one of the area's first Republicans and served as a delegate to the 1860 Republican Convention. In 1864, he won a seat in the House of Representatives for the 17th District. Later, he left the Republican Party to join the Prohibition Party and delivered many speeches in the surrounding counties on the evils of alcohol. His prohibition tendencies had a major impact on his sons. When A.V. Barker or Barker Brothers sold a lot in Vintondale, the deed usually included a provision that "no spiritous or malt intoxicating liquors shall be sold or kept for sale on the premises." If that stipulation was violated, the Barkers reserved the right to confiscate the property.

In 1880, the firm Barker and Son was renamed Barker Brothers. The company owned the Ivory Hill mine in Nanty Glo and the family mercantile store, an imposing structure at the corner of High and Center Streets in Ebensburg. Later, Murtha Furniture occupied the building; it was razed in the mid-1980's and replaced with a convenience store and parking lot.

Augustine Vinton Barker

Augustine Vinton Barker was born on June 20, 1849 in Lovell, Maine, the third son of Abraham and Losina Little Barker. At the age of fourteen, he served in the Civil War. Following discharge as a private from the army, he attended Darthmouth College, graduating with a B.A. in 1872 (1868?). Barker studied law under Shoemaker and Sechler of Ebensburg and also Judge E. Evans of Chicago. In August, 1874, Barker was admitted to the Cambria County Bar.

Barker married Katherine Zahn on June 1, 1875. They had three children: Fred, 5/6/1876-1917; Lovell, 12/12/1884; and Helen, 8/18/1890. (Lovell Street in Vintondale is named after Barker's daughter.)

Augustine Vinton Barker's judicial career was initiated when Governor James Beaver appointed him to the bench to fill the vacancy of R. Johnson in 1890. Election to a ten-year term as Judge of the Pennsylvania 47th Judicial District followed in November, 1891. In his re-election bid in 1901, Barker lost by 75 votes to Attorney Francis O'Connor of Johnstown. Judge George Griffith, Barker's nephew, attributed the failure of his uncle's re-election bid to his campaign manager, Florentine Barker, who "rubbed a lot of people the wrong way." Webster Griffith, brother-in-law by marriage to both men, volunteered to assist with the campaign, but was told by Florentine that it was under control. In December, 1901, Judge Barker filed with Attorney General Elkin in Harrisburg an application to contest the election. Barker either dropped the appeal or lost it because he retired from the bench in 1902.

A good example of how Florentine affected the Democratic voters of Cambria County was conveyed to the author by Judge Griffith. In September 1915, a Civil War monument was erected in Ebensburg in the park near the county jail. Judge Harry White of Indiana was the guest speaker. In his address, Cambria County Judge John Kephart complimented Florentine Barker, Grand Army of the Republic Post 556 Commander, for his efforts in securing the monument. As the stars and stripes were being raised, someone shouted, "Hoorah for Old Glory!" J.A.C. Beers, a prominent Democrat from Hastings, thought the person said, "Hoorah for Old Florie!" He indignantly rose, faced the crowd and countered with, "The hell with Old Florrie!"

During his twelve year tenure on the bench, Barker was industriously purchasing coal lands in the Blacklick Valley. In most cases, he purchased the coal rights and often the surface rights for non-payment of taxes and then sold them to the land or mining companies. (See appendix for his sales to Delano and Lackawanna.) He also was one of the initial investors in the Blacklick Land and Improvement Company and the Lackawanna Coal and Coke Company. After the lands were secured for Wehrum, he continued purchasing coal lands between Dilltown and Heshbon. By comparing his selling prices to Lackawanna in 1902 with the price paid to farmers for coal rights, the author calculated Barker's profits to be $141,717.

After his retirement from the bench in 1902, he practiced law for seven years with his son, Fred D. Barker. In 1909, he retired to Bradenton, Florida each winter and returned when able in the summer. Webster Griffith puchased Barker's house south of the courthouse. Today, the site is a parking lot.

When World War I was declared, his only son, Fred, was too old at 44 to enlist in the army; patriotically, he joined the Red Cross Overseas Service. While serving in France, Fred Barker was killed in 1917. Surviving him were his widow, the former Mary Griffith, and three children.

A.V. Barker began disposing of his Vintondale properties as early as 1906; a few houses sold as late as 1917. Both of these dates are significant in the expansion of the Vinton Colliery Company. A list of A.V. Barker properties and their purchasers are listed in the appendix. Numerous lots were also owned by his brothers, operating as Barker Brothers.

Mrs. Katherine Barker died in Bradenton on January 17, 1915 following a four year illness. In October, 1925, testimonial dinner honored of Judge Barker, the only living ex-judge in Cambria County. A.V. Barker died in Florida on August 20, 1928. His funeral was held in Ebensburg; the courthouse and local businesses closed as a sign of respect. He and his wife were buried in Lloyd Cemetery. His heirs included daughters, Lovell Gates of Williamsport; Helen Land of Florida; and the two sons of the late Fred Barker, William G. and Fred V. Barker.

One-third of the estate went to each daughter, and one-sixth belonged to the two grandsons. However, the $1,200 yearly amount granted to his daughter-in-law, Mary, was to be taken from the shares of the grandsons. Executor of the estate was Attorney M.D. Kittell. The value of his estate was $66,215.66 in real estate and $237,089.04 in stocks, bonds, savings, etc. Deductions from the estate for taxes, funeral expenses, executor's expenses, etc. was $31,849.53. Total value of the estate was $271,455.17. Oddly enough, the inventory listed no stock in Vinton Colliery Company or Vinton Land Company or property in Vintondale. It is strange that an entrepreneur would have no monetary interests in the town which was named for him.

Judge Barker was a director of the First National Bank of Ebensburg, as was Florentine Barker and Webster Griffith; bank president was A.B. Buck. Technically, the bank could have been renamed the Zahn Bank as all of these men were married to Zahn sisters. At the time of his death, Barker's interest in the bank amounted to 120 shares at $309.89 per share with a value of $37,184.

A.V. Barker and his brothers had a major impact on the development on the central Cambria County region, yet little has been preserved in Ebensburg that can be traced to the Barker family.

WEBSTER GRIFFITH

The Griffith's have been one of the most prominent Welsh families in the Ebensburg area for over 150 years. The scion of the family was Thomas Griffith, born on October 20, 1818; he was the son of Thomas Griffith who arrived from Wales around 1806. As a young man, the younger Thomas and his brother purchased the family farm, a carding mill and a water-powered sawmill located outside of Ebensburg. Eventually Thomas Griffith had sawmills and large tracts of timber in Cambria, McKean and Elk Counties. His Vintondale holdings included the Wigton Survey, Pringle Land, the Adam Rand Tract and 5/12 of the Moore Syndicate near Twin Rocks. In McKean County, several thousand acres of timber were purchased from General Thomas Kane. Several of the Griffith children settled permanently in the town of Kane and managed the Griffith interests there.

Thomas Griffith was the largest dealer of cherry lumber in Pennsylvania, with a large retail yard in Philadelphia at Market and 20th. Griffith also shipped a lot of lumber to a furniture manufacturer in Williamsport. At the time of his death in 1890, Thomas Griffith owned about 4,000 acres of coal and timber in the Blacklick Valley at the site of the future town of Vintondale. The coal rights on this land were sold by his heirs to the Blacklick Land and Improvement Company, but the timber rights were reserved for the heirs for fifteen years.

In 1844, Thomas Griffith married Mary Davis; the couple had five children:
1. William W., 8/15/1845-1877, managed the lumber interests in Kane.
2. John T., 6/19/1853-6/40/1892, managed the Kane interests after William's death.

3. Dr. Abner, b. 8/5/1855, graduated 1879 from University of Pennsylvania, abandoned medical profession in 1888 because of poor health, wintered in Florida. Married to Elizabeth Evans. S.L. Lemon was legal guardian for Abner's minor children.

4. Annie, b. 12/3/1858, lived in Kane, married to F.A. Lyte.

5. Webster, 6/5/1860-1928, began supervising his father's mills at age 16. Had a lumbering and oil business in Kane with brother John until 1892 when John died. Executor of his father's estate. Married Alice Zahn on 11/28/1894, two sons, George and Thomas.

Webster Griffith was one of the best know businessmen in Northern Cambria County. Besides gaining first hand experience early in managing a sawmill, Griffith also learned how to obtain good timberlands and coal rights for the best price--back taxes. In addition to the land his father obtained, which had to be divided five ways amoung his heirs, Webster collaborated with S.L. Reed, Attorney Kittell and Alvin Evans to obtain additional mineral rights which they sold to the Vinton Colliery Company in 1916.

Griffith operated a sawmill in Vintondale before 1910, and also owned mills in Ebensburg, Beulah and Cardiff. His mill in Cardiff operated between approximately 1910-1923 and had five mile logging railroad. Superintendent of the mill was Elisha Mahan. The mill area was sold to a coal company around 1923. In Nanty Glo, Griffith's coal rights were leased to Lincoln Mining Company. The Webster mine in Nanty Glo was named after Webster Griffith.

The Griffith family were long time Republicans. The first Republican sheriff of Cambria County was Thomas Griffith, elected in 1880. The returns were slow in coming in that election night, the next day he discovered that his Democratic customers in Barr Township voted for him and helped carry the election. Webster Griffith also served as Cambria County Sheriff, which brought him into conflict from time to time with Otto Hoffman, Vinton Colliery Company superintendent. According to George Griffith, Hoffman called Griffith during a strike and said that he was not going to let anyone into Vintondale. Griffith responded that he had a posse of 100 men and was coming. He met no opposition in Vintondale, but had to endure a cold shoulder from Hoffman for several years. The two men did renew the friendship before their deaths.

Warren Delano was also a friend of Griffith, often stopping in Ebensburg to visit when he was in the area. Once, he even stopped in to see the Griffiths when they were vacationing at Atlantic City.

In addition to his coal, timber and sheriffing interests, Webster Griffith served on the board of directors of the First National Bank. Most of the other board members were his brothers-in-law by marriage. In 1927, Griffith was in ill health. That year, all the Griffiths heirs signed an agreement on the estate of Thomas Griffith because the family was becoming too scattered. On April 2, 1928, former sheriff Webster Griffith died. In his honor, the courthouse and Ebensburg businessmen closed for the funeral.

The executors for the estate were his wife, Alice, and sons, George and Thomas. His real estate holdings included eighty-two parcels and 1/4 of the holdings of the Thomas Griffith Estate. When inventoried, Webster's real estate was worth $137,088.20. The personal property, including a Chrysler automobile and house with furnishings which were left to his wife, were listed at $107,681.93. Unspecified debts owed by the estate amounted to $21,535.35 leaving the clear value of the estate at $123,234.35. At the time of his death, Webster Griffith still operated a sawmill in Jackson Township. On his farm, he raised prize Guernsey cattle; each animal was listed on the inventory. A $500 legacy was listed for Mary Griffith Barker, widow of Fred Barker. Grandsons Webster II and George Jr. also received $500. Similar to Barker's estate, Griffith's inventory was devoid of any Vintondale stock.

Alice Griffith died October 24, 1932. Thomas took care of the office after his father's death and closed things up; George became Chief Judge of the Court of Common Pleas of Cambria County, remaining active until about a year before his death in 1988.

SAMUEL LEMON REED

Another prominent Ebensburger who speculated in the coal rights rush at the turn of the century was Samuel Lemon Reed, Esquire. The son of Samuel and Ella Reed, S.L. Reed was born on March 13, 1864 in Blacklick Township. He attended public schools, the Ebensburg Academy and the Strongstown Academy. For seven years, he taught school. In 1888, he passed a pre-law test and studied law under George Reed (Reade?). S.L. Reed was admitted to the Cambria County Bar in July, 1890 and formed a partnership with Mathiot Reade in 1892. In addition to practicing law, Reed was also an inventor, having obtained with his brother, William S. Reed, a patent on a railway crossing gate. Reed also worked closely with Webster Griffith, Judge Evans, his brother and Attorney M.D. Kittell in purchasing acreage in the Blacklick Valley and in Pine Township.

In 1919, S.L. Reed was appointed judge; in 1934, while serving as a substitute judge for Allegheny County, Judge Reed died at St. Francis Hospital. In his honor, the Cambria County Courthouse adjourned for three days, and Ebensburg businesses closed for the funeral. Mrs. Reed, the former Elizabeth Evans of Ebensburg, preceded her husband in death in 1933. The Reeds had no children.

In addition to serving on the bench, Reed was president of the American National Bank and on the Selective Service Board. Judge Reed was a staunch Republican and a member of the Methodist Episcopal Church.

Judge George Griffith said that his father and

Judge Reed were great friends. Every Sunday afternoon, the two men could be seen strolling the avenues of Ebensburg. Following the death of Webster Griffith in 1928, Judge Reed asked George Griffith to accompany him on the weekly promenade. These walks continued until S.L. Reed's death.

In his will, Reed left legacies for his sisters, half-sisters and/or their heirs. In addition, the Methodist Episcopal Church of Belsano received a gift of $5,000; the Congregational Church of Ebensburg was also a beneficiary of a $5,000 bequest. Judge Reed also stipulated that a masoleum be built for him in Lloyd Cemetery; the cost was to be no less than $5,000 and no more than $8,000. A codicial was added to the will that a masoleum had been built during his lifetime. This expenditure was many times more than Judge Reed paid the average farmer for his coal rights.

His estate included a house on Center Street in Ebensburg, stocks, bonds, real estate, mortgages, notes, judgments and his wife's estate. Total value of the inventory was $415,579.00. His executors were George W. Griffith and Frank Hartman. Reed was the only member of the Ebensburg triumvirate who owned Vintondale bonds at his death. The inventory of his estate listed thirty Vinton Land Company bonds, issued for $1,000 each. In 1934, the value of these bonds was only listed as $4,500. Did he retain these for sentimental reasons, or because he could not unload them? One will never know, but as Malcolm Cowley wrote in a letter to the author, you never got one up on Judge Reed.

JAMES MITCHELL

The mystery man in the race to buy coal rights between 1890 and 1910 was James L. Mitchell. Factual information about him is very sketchy. S.H. Jencks of the C&I Railroad described him as an 1890's Tyrone coal operator. In 1904, his address was Gallitzin; shortly afterward he was living on the Russell Edwards farm near the Ebensburg Fairgrounds on the Carrolltown road. During this period, he was puchasing extensive coal properties in Indiana and Cambria Counties. The majority of these holdings were in Pine, Buffington, Cherryhill, and Green Townships in Indiana County and in Blacklick, Jackson and Barr Townships in Cambria County. By consulting the deed books in both counties, one can count up dozens of transactions for thousands of acres. Deeds filed between 1907 to 1914 list Mitchell's address as Philadelphia. On February 26, 1914, Mitchell granted power of attorney to John S. Fisher of Indiana, future governor of Pennsylvania.

One year later on February 24, 1915, Mitchell was residing in Royalton, Illinois. At this time, he made a deed of conveyance to John E. Evans of Ebensburg for $1.00 for eighty tracts of land in both Pennyslvania and Illinois. Due to "sundry losses and misfortunes," Mitchell was unable to pay his debts and assigned all his property to Evans for the purpose of making a just distribution among his creditors. Evans was at reasonable cost to pay the creditors; he completed this task and filed his account in the Cambria County Court of Common Pleas on March 3, 1924. The account was filed with the auditor on June 7, 1924.

Mitchell died in Royalton in the fall of 1923; executors of the estate were Ralph Mitchell and Ben Link. The executors petitioned the Cambria County Court to re-convey the remaining lands to the Mitchell estate. Remaining claims to the estate belonged to R.A. Whiteside and Elmer Davis of the Real Estate and Mortgage Company of Pittsburgh. George Griffith was appointed auditor to investigate the claims. His report dismissed the claims and recommended a deed of re-conveyance of five tracts in Pennsylvania and thirty-six in Illinois.

Why bother including a biography of James Mitchell in a history of Vintondale? What contributions did he make to the developement of the town? First, to show that not all coal specualtors reaped great fortunes, and second, some of his local coal parcels passed to Griffith and Reed were included in the large land sale to Warren Delano in 1916.

Eliza Furnace c. 1900. Donated to the Vintondale Homecoming Committee by the Feldman Family.

**Will Findley on left. Man on the right unknown. Taken at the side of the Eliza Furnace c. 1905.
Courtesy of Pauline Bostik Smith & Betty Bostik Biss.**

Chapter IV
Mining in Vintondale, 1894-1930

VINTONDALE MINE OPERATIONS

On July 6, 1894, the Vinton Colliery Company applied to the governor of Pennsylvania for a charter. The purpose of this new company was to mine coal and clay and to manufacture coke and fireclay. Shares in the company numbered 300, par value of which was $100. These were held mainly by Warren Delano (197) and Clarence Claghorn (100). The other investors were Theodore Bechtel of Philadelphia (1), E.V. Invilliers of Philadelphia (1), and R.A. Cook of Pine Plains, New York (1). Mr. Bechtel was an investor in the Blacklick Land and Improvement Company, and Cook owned shares in the Vintondale Supply Company. The directors for the first year were Claghorn, Cook, and Delano; treasurer of the Vinton Colliery Company was Algernon T. Burr. Its main office was 1 Broadway, New York City. The reader should have alertly noticed that Judge Barker was not among the Vinton Colliery Company stockholders; the reason for this is not known. Also, this charter was not recorded in the Cambria County Courthouse until December 15, 1900. The author surmises that the late recording date may be associated with the creation of the Lackawanna Coal and Coke Company in 1900, in which Barker was a shareholder.

Vinton Colliery Company developed its first mine in 1894, and the first coal was shipped that October when the Ebensburg and Blacklick Branch of the Pennsylvania Railroad was completed. In the first few years, two mines were in operation: #1, situated between Second Street and Chickaree Hill, and #2, located on the hill behind the Honor Roll and Roberts' Service Center.

Pennsylvania's mine safety laws were years ahead of the federal standards. Each regional inspector in both the anthracite and bituminous regions submitted an annual report to the Pennsylvania Department of Mines. Included in these reports were mine production, coke production, accident and fatality reports, mine improvements, and other vital statistics for every mine in the inspector's district. These volumes, especially up to 1918, provide vital primary statistics needed to supplement a history such as this. In the **1894 Department of Mines Annual Report** Vinton Colliery Company's #1 mine was listed as a new operation with mining to be done by machines driven by electricity. All haulage and tipple work was also to be done by electricity supplied by a new plant. Located at the eastern end of Barker Street, the powerplant was not operational in 1894. Later, it was replaced by the #6 powerplant, and its bricks were used to build brattiches at #6.

For the annual production, days worked, and the superintendents of all mines, consult the appendix in the back of the book. In 1894, #1 mine only worked 43 days and employed 22 men. The first superintendent of the Vinton Colliery Company was Clarence Claghorn. (See biography.)

In the **1895 Annual Report**, the mine inspector made a glowing report of conditions in the Vintondale mines. #1 and #2 had first class ventilation and drainage, and the "very fine" electric plant "lights up the beautiful town of Vintondale." The foremen of these mines were George Blewitt and John Good. Blewitt was a former employee of the Claghorn family mine at Bernice in Sullivan County.

Within two years, #1 had 120 inside and 14 outside employees. A large tipple was constructed adjacent to the railroad siding on the opposite bank of Shuman Run; coal was shuttled to the tipple over a trestle. #1 had three locomotives, two steam boilers, and twelve horses and mules. The animals were housed in a mule barn below the mine, but on Saturday nights Clarence "Socks" Stephens drove the mules down the Dinkey Track to graze in a field across the Blacklick Creek from the ball field. Hence the field became known as Mule Field and was the site for numerous school picnics, weiner roasts, and parties until the late 1950's.

#2 mine had a very short life span. It opened in 1895, and the Blacklick Land and Improvement Company leased it to the Madeira Hill Coal Mining Company. Clarence Claghorn was still listed as its superintendent in the **Annual Report**. In 1895, #2 worked 24 days and employed 20 people. The coal was machine-mined, but was brought out by rope haulage. In 1896, it worked 29 days, and in 1897 and 1898, no production figures were given. The production was probably listed with that of #1. In 1899, the lease was transferred to the Vinton Colliery Company, and in 1901, the mine was included in the sale to Lackawanna Coal and Coke Company.

Between 1901 and 1905 while Lackawanna was operating #1 and #2, the Vinton Colliery Company retained #3, which it opened in 1899. This mine was located between the last house on Plank Road and the "Big Curve." The mine was a drift-type, fan-ventilated and utilized rope haulage and compressed-air mining machines. #3 was unique in that the coal was machine-mined using the longwall method. Clarence Claghorn was responsible for introducing this system and described its application in Vintondale in several professional publications. A special machine mined several sections 200 to 300 feet wide, taking out all the coal and leaving pillars to protect the entries and throwing the roof down behind them. The face of the coal was kept clear with a battery of props. Conveyors brought the coal to the main haulage way, which was low enough so that the conveyors could empty directly into the cars. The ventilation in #3 was rated the best in 1901. William Alexander was the inside foreman when it opened. Harry Hampson became the weighmaster after he arrived from Bernice around 1900. The mine cars came out of #3 on a downhill grade; often there were runway trips which were slowed down by means of a runabout before reaching the tipple. As weighmaster, Hampson was expected to try to prevent a collision or derailment of the runaway cars.

The coal seams in #2 and #3 outcropped above Maple Street near the Goughnour farm. Tracks were built across the farm to connect the drifts. The opening on the #3 side may have been called the Panama Drift. This would have been above the Goat Hill houses.

In 1902, the Vinton Colliery Company used thirteen compressed air mining machines: two Ingersol, six Sullivan, two Harrison and three Jeffrey (longwall). A twelve-foot Stine fan was also installed that year. The steam-powered fan provided 33,500 cubic feet of air per minute at the mine entrance; there were two air splits in the mine. #3 also had eight experimental coke ovens which were located across Plank Road near the #3 tipple. No production for these ovens was ever listed in the **Annual Report**.

In 1904, the **Annual Report** credited the Vinton Colliery Company as owning seven tubular boilers, two electric locomotives, four steam engines, and two electric dynamos. The Stine fan was replaced with a twelve foot Robinson fan, and the air entering the mine was increased to 48,000 feet per minute. A new opening at the main heading was also cut in 1904.

In 1905, Warren Delano repurchased the #1 mine from the Lackawanna Coal and Coke Company. #1 and #3 mines were listed in good condition that year. #2 was ide, and #5, which had been idle, was leased by Vinton Colliery from the Cambridge Bituminous Coal Company and reopened in December, 1905.

The **1906 Annual Report** mentioned a Vinton #4 whose production was listed with #1. Numerous inquiries as to its location led nowhere. Even the late Russell Dodson could not identify the drift. There were other #4 mines in the area: Commercial #4 (Bracken) and Lackawanna #4 (Wehrum). The answer appeared in an unusual way. The author was using a 1906 glass negative of the town plan on an overhead projector in her classroom. By enlarging the map, the #4 mine showed up clearly as a shaft under Fourth Street. (This helps explain why there are frequent collapses in the road.) The #4 drift was in production from 1906 to 1910. When #6 was about mined out in the 1960's, some additional coal was retrived from these old workings.

The finalization of the resale of #1 and #2 to Vinton Colliery Company and the opening of #6 were the big newsmakers in coal operations in 1906. The actual sale is discussed in the chapter on Lackawanna Coal and Coke. The new "boom" in Vintondale became a feature in all the local papers. A photographic record of the actual construction of the coke ovens and mine buildings have been preserved on glass negatives. The majority of these 1906 negatives have been donated in the memory of John Huth to the Pennsylvania Historical and Museum Commission, and copies of the photographs are available for study at the State Archives in Harrisburg. Some of these copies have been reproduced in this book with permission of the PHMC.

To pay for the purchase of Lackawanna lands and to finance the construction of the #6 mine and 152 coke ovens, the Vinton Colliery Company stockholders at a May 16, 1906 meeting voted to increase the company debt from $175,000 to $675,000. They authorized the sale of $500,000 worth of bonds and secured the payment of the same by a mortgage. The bonds were in denominations of $1,000 and were due November, 1935 at 5% interest. This mortgage was not paid off in 1935 and was finally satisfied in 1942 by court order during the bankruptcy and reorganization proceedings.

The new #6 mine with its washery, power plant, tipple and coke ovens were built on land originally laid out in lots in the 1901 Lackawanna town plan. Large numbers of men were drawn to the valley to work as laborers, lumbermen, miners, carpenters, masons, teamsters, etc. The estimates of men working in Vintondale in 1906 runs as high as 2,000. On the coke oven construction alone were 400 men. **The Indiana County Progress** called Vintondale a "coal belt phenomenon," and **The Johnstown Weekly Tribune** said it was "surpassed by none in enterprise or prosperity." The concrete, fire brick, and cut stone ovens were completed in December, 1906 and were lit for the first time in early 1907. A huge washery was built to clean the coal before coking. Vintondale coal was high in BTU's, but was also very high in sulphur, so the washing process helped to remove some of the impurities. Waste water from the washer was emptied into the Blacklick along with the acid mine water, polluting the once pristine stream.

#6 mine was a slope with two air splits in 1906. It employed 22 men and produced 10,491 tons the first year. It also operated on the longwall method during the early years. 1906 was one of the few years in which all six mines of the Vinton Colliery Company were in operation.

All mines were reported in good condition in relation to ventilation and drainage in 1907. Mines 1, 2, 3, and 6 were long-wall mined; 4 and 5 were room and pillar. The methods of ventilation were mainly be fan; however, #4 was ventilated by furnace. #1 had a 5-foot electric Sirocco fan; #2 had a 6-foot electric Disk fan; #3 had a 14-foot steam Robinson fan; #5 had a 14-foot steam Brazil fan; and #6 had a 7-foot Stine fan. The 1907 statistics also list the Vinton Colliery

Company as owning twelve tubular boilers, two steam locomotives, five electric locomotives, twenty-four steam engines, three pumps, six electric dynamos and four air compressors.

In 1908, #2 and #3 were listed as idle, having only worked 13 days. The other mines were in fair to good condition. #5 was abandoned in 1909, and its coal was removed through #6. #2 and #3 both remained idle that year. Two new overcases were built in #1 and #4 to carry the air currents over openings in the main heading. In #6, a new overcast was built to split the air into separate currents.

In 1910, #3 was reopened after being idle almost three years. #2 remained idle. The other mines were in good condition except #6 where there were drainage problems from the #5 workings. Some major improvements were made that year. #1 received a new 8.5-foot Sirocco fan, and the old one was installed in #6. Six brick and steel overcasts were built in #1 in the lower workings, and masonry stoppings were constructed. Seventy-pound rails were relaid on the haulage road. In #6, shelter holes were built along the new slope road. Masonry air stopping were built; overcasts of brick and steel were also erected. A DuBois and Aldrich pump was installed. The main air return to the face of the workings was cleaned, and forty-pound rails were laid on the new main hauling road.

Additional improvements were made in 1912. A new office, motor barn and sand house were built at #1. New machinery, such as one short-wall machine and three puncher machines, were purchased. A masonry overcast and five shelter holes on the main drift were constructed. Six of the old headings were wired for electric haulage. Mother Nature also plagued the Vinton Colliery Company. There was below zero temperatures in January, and water shortages in the summer. Accompanying these was the perennial problem of obtaining enough coal hoppers from the Pennsylvania Railroad and the Buffalo, Rochester & Pittsburgh Railroad. Personnel problems also abounded in 1912.

Fortunately, a closer examination of the Vintondale operations can be made on a day-to-day basis. Several mine superintendent diaries, 1912, 1913, 1914, 1917 and 1923 have survived. These diaries are a treasure trove of mining trivia which show us Vintondale in its heyday. A more detailed description of the day-to-day events and problems can be found in the appendix. The person or persons who wrote the entires in the 1912 and 1914 books did not identify themselves; the author persumes that the writing in 1912-1914 may have been made by Mr. Smillee, general superintendent and T.W. Hamilton, superintendent. The 1917 and 1923 diaries are definitely the work of Otto Hoffman; Lloyd Arbogast, chief clerk, made the entries when Hoffman was out of town.

#3 closed in 1915. Its production for that year was 2,113 tons, and it worked 34 days. No reasons were given for the closings, but by checking the mine maps, both #3 and #2 had large untouched reserves.

1917 brought more improvements in #6. A twelve foot circular shaft was sunk near the face of #2 slope. The shaft was 180 feet deep and was concreted the entire length. 190 brick stoppings and two brick overcasts were also built. In town, the company also built ten houses. 1917 was also the year of the purchase and opening of the "Heshbon Lands" which were bought from Lackawanna Coal and Coke in 1916 for $100. This deal and the history of Claghorn is discussed in separate chapters.

Vintondale's #1 and #6 mines received favorable reports from the mine inspector in 1918. The ventilation was listed as good to very good, and drainage was satisfactory to good. Ventilation at both mines was improved because of the installation of new fans at new shafts. #1 installed a new shaft on the Rummell farm, and a Jeffrey fan was in operation by early 1919. Rager Hollow employees, such as the Edmistons, often used this entrance to go to work. There was a power line strung from the #6 power plant to the shaft and substation. The shaft was 48 feet deep, and there was a ladder to get to the working face. Russell Dodson said that Mr. Monar, Roy Moore, Dan Stutzman and John Misner cut lumber and put in a new liner for the shaft. He said he was afraid to use that shaft because of the rattlesnakes which sunned themselves on the rocks surrounding the entrance. Occasionally they would crawl down the shaft and and lay on beams. One time, the power kept going off in #1. When someone was sent to investigate the problem, he found a rattlesnake had crawled into the switch-blade and had shorted out the power. Needless to say, the snake was well-cooked.

Since the subject of rattlesnakes has surfaced. this is a good diversionary point to add other rattlesnake stories. As a child, the author was always told to stay away from #6 hill because of rattlesnakes. Some did slither down to the flat in the summertime. Russ Dodson said that if you stamped your foot on the wooden floor of the old mine office, you could hear the snakes buzz. Later the office was replaced with a concrete block building with a cement floor. Snake skins were also found behind a shower house that had been installed near the mine entrance where an old fan house had burned down. There were two close encounters with snakes at #6. One had crawled into a cap which was hung on the wall. The men killed it and threw it outside. The next day, they checked to see what kind it was and found it was a rattler. The second episode involved Mario Bianucci, who was sitting on a bench at the mine entrance waiting for the mantrip. Someone said that there was something under the bench. Mario did not believe him, and looked under the bench. Staring back at him was four foot rattlesnake. It was quickly killed, and the men looked around for its mate and found it wrapped around a telephone pole. Fortunately, no one was bitten in the line of duty at the Vinton Colliery Company.

Returning to the subjects of mine shafts and ventilation, 1918 was the last year that the Department of Mines listed the accident report and equipment improvements. #1 had a Sirocco fan whose blades were 8.5 feet in diameter which supplied 67,000 cubic feet of air at the entrance. There were five air splits in #1, and at the working face, there were 50,000 cubic feet of air circulating. #6 had a twelve foot Jeffrey fan which had 108,000 cubic feet at the entrance and 82,600 cubic feet at the face. In 1918, #6 had ten air splits. Russell Dodson said in 1981 that the mines must supply 5,000 cubic feet per man. In earlier years, some mines only provided 5,000 cubic feet for all the men.

The electricity to power these fans, be it #1 or #6, was supplied by the VCC powerhouse at #6. So, in addition to obtaining the property rights for shaft sites, the company also had to purchase rights-of-way for the power lines. In May 1915, Webster Griffith granted VCC the right to build a power line and sink a shaft on his property in Jackson Township. The transfer was achieved for $1 and Griffith's right to cut timber within five feet of the line and use power from the line to cut the timber. In 1916, Henry Rummel of Rager Hollow agreed to allow the construction of an air shaft, fan and machinery on his property. An additional agreement was procured with Webster Griffith to extend the power line 1400 feet to that of Lackawanna Coal and Coke Company. For $1, Griffith granted VCC a 25 foot right-of-way. Griffith received a royalty of two cents/prop, four cents/mine tie, and two dollars for a chestnut pole thirty to thirty-five feet long. All locust trees were to be cut into seven foot fence ties for Griffith. VCC had the right to use the sawmill road which connected with Second Street.

Photographs of Vintondale in the 1930's and the 1940's clearly show the two pole lines, one intersecting #6 hill and one crossing the #1 hill leading to Rager Hollow. The pole line was also a clear route for Vintondale kids to use to cross the hill to Wehrum Dam or to a sawmill pond at the top of the hill. Both sites were dangerous swimming holes. The author presumes that this pond belonged to Griffith because of its location to the pole line.

Supplying air to the working face was a tricky business, especially in #6, which was such a large mine. The fans actually pulled the air out of the mine, drawing the fresh air in from the entrance. One air shaft was located along the north branch about two miles north of Rexis. Access was by train, and when the passenger service ended, by foot. There was a three room house near the tracks for the caretaker, who regulated the fans and the electricity. Mine water was also pumped out of that spot. In the 1930's, Tim Casper lived there for four years with his wife, the former Helen Shestak, and daughter, Elaine. Because of the isolation, grocery orders were called in from the mine and delivered by Mr. Feldman or Charlie Shestak, Helen's father. The Caspers had no refrigerator, making storage of food extremely difficult. To get to town, they rigged up a baby carriage with wide wheels that could ride the rails. Lack of available medical care for Elaine during a bout of whooping cough convinced them to move into town.

As #6 tapped the coal reserves purchased in 1916, an additional air shaft was a must. In 1923,. a shaft was built off the Red Mill Road which became known as the Anderson shaft because Joe and Andy Anderson lived there for years in the houses provided by the company. They had moved there about a year after Claghorn closed.

In February, 1923, the construction bid went ot the S.J. Harry Company whose bid was $3,000 less than Dravo's. The mine diary for 1923 did not reveal the home base of the company, but most think that the shaft was dug by a Pittsburgh contractor. Actual construction began in early March when the engineering crew surveyed the site; a car of coal to be used for sinking the shaft was shipped to Red Mill. At least once a week, Hoffman or another companyman trekked to the site to survey the progress. That weekly visit often turned into a long hike; on March 21st, the roads were so bad that Huth and Hoffman had to park the "machine" at Kerr's farm and walk to the shaft. Total trip took three hours. Also that week, forman Clark found four miners who were willing to double-shift on the second slope so that the heading would be there when the shaft was completed.

In May, the shaft site was threatened by the annual forest fires. Huth and Hoffman visited the site on May 19th to check the site for the fan foundation. Each month, the two men made a site visit to compile the monthly estimate for the Harry Company.

By September, the new fan had arrived and was shipped to Red Mill by train and moved to the shaft. The motor was trucked in a week later. Clarence Schwerin visited the site on his trips to the valley in April and November. On December 4th, Hoffman contracted with S.J. Harry for $10 per day for use of machinery and costs plus 15% for building the fan walls. On Friday, December 7th, the left air course of second slope broke through to the new shaft, and on the 19th, Hoffman, Huth, and Clark spent the morning inside #6 at the shaft site.

The contractor brought in black laborers for the job just like Lackawanna's contractor did in 1902. Russell Dodson said that there were stories circulating about three or four murders among the workers and that the bodies were thrown out in the swampy ground near the shaft. According to Dodson, three black women and two men arrived in Vintondale and offered him $25 to take them to the shaft. It might be easy to shurg off these stories, but there was an entry in Hoffman's 1923 diary that he attended a black wedding, and the **Nanty Glo Journal** verified one muder. The November 1, 1923 issue reported that Alex Garbit, about 45, was found the previous Tuesday with his throat split. The alleged murderer was Edward Stoddard, 35. The men were employed by the Vinton Colliery Company and had been aid the Monday before the murder. The men lived at a "negro shack" at Red Mill. The paper reported that the men were "ordinarily peaceful" men and presumed that moonshine had played a part in the quarrel over a card game. Supposedly Garbit hit Stoddard with a poker, and Stoddard retaliated by stabbing him. A witness said he heard scuffling outside, but did not see the murder. Stoddard escaped, and there were no reports in later issues of the paper that he had been captured.

The Anderson shaft was built on 25 Right heading. It was a double compartment shaft, twelve feet by twenty-five feet, and was equipped with an escape ladder. There was a twelve by twelve curtain wall in it which split the shaft into air intake and air exhaust sections. The opening was covered with a steel door. The shaft was 413 feet deep.

A third "beautiful shaft," to use Mr. Dodson's words was built along the Strongstown Road in the dip between the Dilltown and the Brush Valley roads. Two deeds from Hugh Altemus, dated March 11, 1949, granted the Vinton Coal and Coke Company a 1.2 acre tract to construct a shaft, escapeway and substation on the land. Included was a right-of-way for an electric line to the tract. There were several temporary buildings on the property which would revert to Altemus when the construction was finished. The Williamson Construction Company was in charge of the work and had constructed a wooden building on the east side of the road. The construction company had the right to remove this building when they completed the job. There was 1,300 feet between the shaft and the bore hole, and the coal company agreed to clean up the area and burn the brush. Mr. Dodson said that the shaft cost around one-half million dollars and was used to try to get a tax write-off. There was talk of using the shaft as a work entrance for the miners. John Biondo said that he and Clarence Schwerin II had discussed the possibilities, but the use of the shaft fell through because of disagreements over how the miners would get to the shaft and who was to pay for the transportation to get there.

CAMBRIDGE BITUMINOUS COAL COMPANY

The Cambridge Bituminous Coal Company had a very short existence in Vintondale. Its central office was in the hard-coal region, but the deeds varied as its exact location. Shenandoah, Ashland, and Frackville were all listed as main offices. In 1902, its president was David R. James; president in 1905 was J.C. Biddle, with J.C. McGinnis as secretary.

The Cambridge mine, later known as #5, was located across the Blacklick from the railroad "y" and the water tower. To get there from the Jackson Township side, a swinging bridge was built across the Blacklick; paths led to #5 and to the Kaiser farm on top of hill. Operating sporadically, the mine was not listed separately by the State Department of Mines in its annual reports.

Cambridge's coal rights, which covered the 479 acre, 49 perch Adam Rand tract in Blacklick Township, were leased in December, 1902 from the Thomas Griffith's heirs. Cambridge's agreement with the Griffith's granted the heirs a royalty of eight cents per gross ton (2,240 lbs.). The first year's production was to be 20,000 tons; the second year's estimated at 35,000; and not less than 50,000 tons for each year after that. The agreement also stated that Cambridge or its successor could within two years purchase the coal at $100 per acre at six percent interest.

According to the agreement, royalty payments were to be made by the 25th of each month to Webster Griffith, executor of the Thomas Griffith estate. If the minimum yearly royalty was not paid, the Griffiths had the right to repossess the coal and mine equipment with a thirty-day written notice. The lease began January 1, 1903, and the royalties were to be paid as of June 1, 1903.

Cambridge had permission to leave pillars to protect the gangways and headings, but had to pay eight cents per gross ton for the coal in the pillars when the other coal was exhausted.

For the coal company's use, the Griffiths granted eight acres of surface. If Cambridge wished to transfer its lease, it had to secure written permission from the Griffiths.

In December, 1904, the Cambridge Bituminous Coal Company sold 7.838 acres of the Griffith coal to the Commercial Coal Mining Company of Twin Rocks, which operated Commercial's #4 mine in Bracken. In addition to purchasing the coal, Commercial obtained the right to build and maintain a drainage ditch through the coal and over the surface. Commercial agreed to pay the eight cents per gross ton for the coal. $960 was to be paid when the agreement was signed, and the balance was to be paid under the same agreement as the Griffith lease. The $960 was for three acres of coal north of the drainage channel, and the eight cents/ton was for the remaining four acres. Also a one hundred foot coal barrier between the two mines was to remain. Representing Cambridge was Barker and Barker, Attorneys-at-Law, Ebensburg. The deed was filed in the Schuylkill County Courthouse.

On March 31, 1905, the Vinton Colliery Company entered an annual lease agreement with Cambridge. For $100, Cambridge leased to VCC a parcel of 13.5 acres in Jackson Township for the maintenance of railroad tracks and tipples. At the time of the lease, Warren Delano was president, and Clarence Claghorn was secretary to the VCC. The Lackawanna Coal and

Coke Company sold .75 acres of surface in Jackson Township to Commercial Coal Company; mineral rights were reserved by LCC.

Cambridge was still in operation in July, 1905 when it entered an agreement with Commercial Coal to furnish compressed air to its #4. The air was to be furnished from Cambridge's power plant (exact location unknown). The compressed air was to be eight pounds of pressure per square inch and paid for with a three cents per gross ton royalty on all coal mined. The transaction was to begin August 15, 1905. By the terms of the agreement, Commercial was to construct a dam on the stream above the powerhouse. Also they were to lay a pipe to the Cambridge air compressor. Expenses for the above were to be taken out of the royalty.

If Cambridge, for various reasons, could not furnish the compressed air for more than five days, Commercial was permitted to deduct $12 per day from the minimum royalty. If Commercial was unable to work for any reason, the royalty would cease. This agreement was not transferable if Cambridge sold the mine.

In the December 8, 1905 issue of the **Johnstown Weekly Tribune**, the Vinton Colliery Company was reported to be taking over both the Lackawanna properties and the Cambridge mine. The paper said that VCC had already been operating #5 which had been idle for some time. Prediction was that up to 100 miners would be hired. Charles Hower, VCC superintendent, had charge of #'s 1, 2, 3, and 5. Mr. Hower met with Cambridge treasurer, J.C. McGinnis of Frackville, in late November and closed the deal. Cambridge's superintendent, F.W. Hamburger of Ebensburg, announced plans to remain in the area and develop coal rights that he owned.

The State Department of Mines did not list any production figures for #5 until 1905; its annual production is listed in the appendix. In its report, #5 is listed as a non-gaseous slope mine. Its fourteen foot Brazil fan was operated by steam and forced 25,000 cubic feet per minute into the mine through two air splits. In 1905, the mine used 500 kegs of black powder and 750 pounds of dynamite. For haulage, seven horses and mules were used.

Room and pillar was the method of extracting the coal in #5; drainage and ventilation were listed as fair in the 1908 report. A twelve-foot Stine fan, revolving 100 times/minute supplied 35,000 cubic feet of air at the entrance. 37,000 cubic feet exited the mine. In February, 1909, #5 was officially abandoned. The mine was drained through #6, and the coal was mined as part of #6.

A formal agreement between VCC and Cambridge was not recorded at the Cambria County Courthouse. On May 2, 1906, an unsigned copy of a letter to J.C. McGinnis from the president of VCC stated that Cambridge had been paying eight percent extra compensation on the coal shipped from "Cambridge Mine" (now "Vinton No. 5"). As of April 1, Cambridge would be charged five cents per gross ton in addition to the eight percent commission.

On October 31, 1911, the Cambridge Bituminous Coal Company assigned the 1902 Griffith lease to VCC for $1.00 and a full discharge of indebtedness. President of Cambridge in 1911 was Algernon T. Burr, treasurer of VCC. Was Cambridge bought out by VCC, or was it a subsidiary? There is no charter of organization recorded in the Schuylkill County Courthouse in Pottsville.

The Griffith lease came to light again in 1926, 1931, and 1933. By 1926, 272.65 acres of the tract had been mined out. In 1931, George Griffith, son of Webster Griffith and future Cambria County judge, claimed that VCC was not paying royalties and sent a letter to superintendent, Milton Brandon, noting that the coal mined on their tract from 1926 to 1931 was 219,270 tons. Griffith's contention was that there were 109 acres remaining to be mined; he questioned Brandon's lower estimate and asked for calculations on the remaining acreage. In 1933, an agreement was made between the Griffith heirs and VCC to reduce the minimum royalty due the Griffith's from 50,000 tons to 25,000. Mr. Schwerin, VCC president, sent the first payment of $200 due on the 1932 royalty. In 1933, much of the surface of the Adam Rand Tract and the Moore Syndicate, 5/12 of which was in Thomas Griffith's name, was sold to the Pennsylvania Game Commission and is presently State Game Land.

After the Vinton Colliery declared bankruptcy in 1940, the Griffiths repossessed the unmined coal. A small portion of the southern end of the tract was leased to Mike Cocho and his Commercial Coal Mining Company of Twin Rocks (not the original Commercial Coal). The Adam Rand Tract also crossed the Blacklick Creek into Jackson Township. The Griffiths received 130 shares of Vinton Coal and Coke Company stock in place of the royalty. When Vinton Coal and Coke was sold in 1957, the Griffiths received some money for the stock. Judge Griffith continued to hold the gas and oil rights for the Adam Rand Tract.

ALGERNON TAYLOR BURR

A name strongly linked with the formation of Vintondale was Algernon Taylor Burr, secretary/treasurer of the Vinton Colliery Company, Cambridge Bituminous Coal Company, Mill Creek Coal Company, Delano Coal Company, Vinton Land Company, and Vinton Supply Company. Burr, the son of Captain Henry and Sarah (Taylor) Burr, was born in Westport, Connecticut on October 29, 1851. Captain Burr, a Civil War veteran, was the postmaster of Westport.

In addition to handling the financial affairs of all the coal companies, A.T. Burr was also in charge of financial matters for the Delanos; he paid bills, purchased stock, booked curises, and even doled out FDR's monthly allowance. (Franklin was usually broke by the end of each month.) Mr. Burr resided in

New York City during the winter, but commuted from Westport in the summer. Due to a childhood accident involving farm machinery, A.T. Burr had a crippled hand. A meticulous person in every way, Mr. Burr demanded the same from fellow employees. Olive McConnell, shipping clerk for Vinton Colliery Company, said she sent a bill of lading to the main office, only to have it returned by Mr. Burr because the figures were not in absolutely straight columns.

Algernon Taylor Burr was married to Clarissa Downes, (December 8, 1857-June 6, 1905). A.T. Burr died at age 77 on November 29, 1928. The couple had four children:

Louis, November 30, 1884-1963.
Perry, September 26, 1885-March 27, 1957.
Julian Penfield, October 11, 1890-November 26, 1913.
Catherine, July 14, 1898-?.

Algernon Taylor Burr - Treasurer of Vinton Colliery Company.
Courtesy of Julian P. Burr.

Louis Burr played his own role in the early history of Vintondale, both historically and romantically. After graduating in 1906 from Yale University's Sheffield Scientific School, Lou came to Vintondale to learn the coal business firsthand. At this time, #6 and its coke ovens were under construction. An ammonia plant, utilizing the gasses from the coke ovens, was planned. Lou's tenure began in the engineering department; then, according to his Yale class history published in 1912, he spent a year in charge of the coal washery and later the coke and ammonia plants. Often Lou was seen sitting on the top floor window of the washery with his feet dangling over the edge. Also, he is credited with designing the bridges which the miners used to cross the Blacklick to #6 and the three company houses on upper Third street known as Burgans, Whinnies, and Thomases.

A Vintondale resident for five years, Burr boarded at the Vintondale Inn. As a company man, he served on the borough council for three years, including one year as president and one year as borough treasurer.

Lou, listed as shy and quiet in his college yearbook, met his future wife in Vintondale. The author's great-aunt, Eula (Lou) Hampson, was one of the few residents in town who knew how to play the piano. Her family owned a beautifully carved rosewood piano. One day, some of the engineering crew working on the streets decided to take a break and asked Eula to play the piano for them. All except Lou went into the Hampson house to listen to the music; he sat outside on a log. The next time Eula saw Lou Burr, he was sitting on the edge of the porrch of the company office (Biondo's and Mesoras' duplex). She was accompanied by her sister Mable and Mildred Rogers. Eula turned to them and said, "Girls, that's the man I want." Mildred laughed and said, "Cry for the moon, young lady. Don't you know that's Mr. Burr's son?" But Aunt Lou was a very strong-willed, patient, young lady. She did not get to meet Lou Burr personally until some time later at a sled riding party. Burr had built a big bobsled which the riders pulled to the top of Chickaree Hill; the momentum of the ride carried them all the way to the train station. On one such ride, Lou sat behind Eula and then escorted her home that evening. After that, they "courted" regularly. A well-meaning companyman in the office sent a letter to A.T. Burr telling him about the romance, saying that it seemed serious and that he did not know if Mr. Burr would approve of Eula, a miner's daughter. Mr. Burr wrote a letter to his son saying that he trusted his son's judgement in such matters. Lou was so angry at that official that he verbally and physically confronted the man in the office, and the office employees had to pull him off.

Louis burr left Vintondale shortly after July 10, 1911 and took a position at the Algoma Steel Company, Sault Sainte Marie, Canada. Eula saw him off at the station. When she got home, her father, Harry Hampson, said, "Well, that's the end of that." Eula replied that it was not because Lou was returning at Christmas, and they planned to get married then.

This family story is supported with some primary source evidence. The 1911 hotel register for the Vintondale Inn lists Louis Burr of Sault Sainte Marie as a guest in Room 7 on Sunday, December 24, 1911. Eula Hampson and Louis Burr were married shortly afterward in the Episcopal parsonage in Johnstown. The pastor did not have any marriage certificates and told the couple that he would send them one, which he never did.

A.T. Burr and son, Perry, came on the night train to attend the ceremony and returned to New York after it was over. Following the ceremony, the bride returned to her home, and the groom returned to the Vintondale Inn. The next morning, the newlyweds caught the train for Canada.

The young people next moved to the Burr family farm in Virginia and permanently settled in Roebling, New Jersey, where Louis worked for Roebling Steel. The Burrs had three children, Helen, Julian and Louis. Eula Hampson was able to get along with "Grandpa" Burr because she was not afraid of him. With his grandchildren, Burr was very generous. Once on a business trip to Vintondale, he even paid a visit to Eula's sister, Mable Huth, to see her new daughter, Agnes, about the same age as his granddaughter, Helen. Eula Hampson Burr remained in Roebling following Louis' death in 1963. Aunt Lou died in 1984 at the age of 98.

SCHWERINS

Clarence Maurice Schwerin I, president of the Vinton Colliery Company, was born August 26, 1881 in New Bern, South Carolina. He attended Horace Mann School, won a Pulitzer scholarship, and graduated from Columbia School of Mines at the age of 19 in 1901. While at school, he obtained his spending money by purchasing slide rules and protractors cheaply from students in the spring and reselling them for a profit in the fall. According to his 25th class anniversary booklet, Scherwin gained experience working in copper, lead smelting, and foundary work until 1904. He worked in the by-produce coke industry from 1904-1907, both at the ovens and demonstrating the product around the country. Next, he was foundary advisor and metallurgist for Central Foundary in 1907 and 1908. Warren Delano hired him in 1908; this was the beginning of his rapid climp to the presidency of the various Delano coal companies. Because of his experience working with coke, Schwerin was first hired as a salesman. His secondary task was to see if something could be done about the high sulphur content of Vintondale's coke. From 1908 to 1911, he was manager of the coke department of the Vinton Colliery Company; from 1911 to 1914, he was general manager; in 1914, he was appointed president. In addition, in 1918, Scherwin became president of the Mill Creek Coal Company, the Delano family's anthracite mining operation.

Clarence Schwerin I, President of Vinton Colliery Company and Delano Coal Company.
Courtesy of Clarence Schwerin III.

In his college 25th anniversary book (1926), Schwerin listed his positions as: President and Director of Vinton-Schwerin Fuel Corporation, Delano Coal Company, Vinton Colliery Company, Graceton Coal and Coke Company, and New Boston Land Company, all companies with Delano roots. He was Chairman of the Board and Director of University Pipe and Radiator, Iron Products Corporation; a member of the executive committee and advisory board of the Central Union Trust of New York; director of the Mechanics Trust Company of New Jersey. In addition, he was a member of the executive committees of Central Foundary Company, Central Iron and Coal Company, Central Radiator Company, and the Essex Foundary Company. In conjunction with the Vinton Colliery Company, he was president and director of the Blacklick Water Company, the Jackson Water Company, Vinton Supply Corporation, Vinton Land Company and the Vintondale Amusement Company.

Eventually Mr. Schwerin bought sole ownership of the Delano Coal Company which was the sales agent for Vintondale coal. After the death of Warren Delano in 1920, there was little direct involvement by the Delanos in their various companies. The Schwerin family also organized the Schwerin Air Conditioning Corporation of which Clarence I was a director and the treasurer.

Roslyn Harbor, Long Island was the home of Mr. Schwerin, and he served as its only mayor from 1931 until his death in 1944. In his obituary in the **New York Times**, he was given credit for being such an efficient administrator that no taxes were needed or collected in 1936 and 1937.

Clarence Schwerin I died on October 28, 1944 after a five month illness. He was survived by his widow, Mary Clothilde Oliver, whom he married September 19, 1905 and three sons: Joseph, Clarence, Jr., and Frederick. Joseph had worked at Vintondale for a time and was president of Schwerin Air Conditioning. In 1940, Joe bought a 600 acre estate in Virginia and lived there until his death in 1980. Frederick, who was born in 1909 and graduated from Yale in 1932, had no connections with the Vintondale mines. The 1981 Standard and Poor's Register listed him as a partner in Schweickart and Company, a brokerage firm in Great Neck, Long Island.

Clarence, Jr. succeeded to the presidency of the Vinton Coal and Coke Company following his father's death and managed the company until his death in 1956. Schwerin died prematurely of lung cancer at the age of 48.

Clarence Schwerin III was serving in the Navy in 1956 at the time of his father's death and was able to get an early release from his term of duty. He decided to find a buyer for the mine, admitting that he had scant knowledge of producing and selling coal. The Schwerin's fifty year association with Vintondale mines came to an end with the scale of Vinton Coal and Coke Company to Tony Collins in 1957. However, Clarence Schwerin III still maintains an interest in the Blacklick Valley. He continues to hold the gas rights to several of the properties in Cambria and Indiana Counties.

OTTO HOFFMAN

Otto Hoffman, the infamous superintendent of the Vinton Colliery Company, was born on January 15, 1874 in Sheboygan Falls, Wisconsin, one of eleven children. His father, a Lutheran minister, and his mother were both natives of Germany. At age 15, Otto Hoffman signed on as a fireman on a Great Lakes ore carrier. Within five years, he was the assistant engineer on the steamship **Maryland**, and by 1897, he was its chief engineer. T.W. Hamilton, general superintendent of the Vinton Colliery Company, rode the ore boats for pleasure; on one of these trips, he met Hoffman. Hamilton was able to persuade Hoffman to spend a winter in Vintondale while his boat was tied up. In spite of what people thought, Hoffman was not the captain of an ore boat.

In 1911, Otto Hoffman settled in Vintondale with the job title of master mechanic. Accompanying Hoffman were fellow ore boat workers, Sam Donaldson, Sam Feldman and Jim Dempsey. Although he had little background in the field of mining, Hoffman's anti-union stance and his ability to carry out company orders gained him the post of superintendent in 1914 after Hamilton departed for parts unknown in the West.

The equivalent of a feudal lord, Hoffman, also known as "Pappy" or "King Otto", kept tight reins on his domain. Like the medieval lord in his castle, Pappy could survey his fiefdom from his manor house at the top of Second Street. Wishing to know what was going on at all times, he attended all the dances and the movies. When possible, he traveled to all away baseball games. Rising daily at 5 AM, he usually went to "see the mantrip off." The first stop on his daily itinerary was #1 mine, easily reached by walking through the orchard and exiting at the gate near #1. After checking on #1, he strolled to #6 where he frequently donned coveralls and accompanied the miners on the mantrip. In fact, most of the time, he wore work clothes rather than a business suit. Almost every day at 11 AM, he met with the firebosses and #6 foreman, John Clark, to discuss mine conditions and production. Hoffman knew each miner and his working place. Pappy then "covered" the flat and spent the remainder of the day and often the evening in his office. Sunday evenings were spent visiting the miners in their homes. For example, Hoffman wrote in the diary on January 28th, 1923, "Visited flats at A.M. at 6:30 started to patrol streets to see what was going on. Many drunks & several poker games." Hoffman frequently settled domestic disputes or played the role of "godfather" in forcing an undesirable suitor out of town. He even made an employee take a formal oath to stop drinking. This was done before the Catholic priest.

First man unknown, Harry Clark #6 foreman, Mike Grosek, Sr., Otto Hoffman, superintendent of Vinton Colliery Company - 1920.
Courtesy of John Huth Collection.

Many foreign miners tipped their hats to Hoffman just as they would have done to their landlords in Europe. Jim Wray told one of them that in the United States, you only tipped you hat to the ladies. Word of this got back to Pappy, and he told Jim that these were "foreign gentlemen" and to let them alone. Hoffman's diary entries show a different attitude on his part toward the immigrants; he frequently called them "Payday Hunks." The hard life of a Vintondale miner could easily be compared to that of a medieval serf.

Hoffman expected the miners to purchase their groceries, clothing, and household goods at the company store. If a miner strayed, he was reminded by the foreman. If it happened too often, the miner was dismissed. During times of strikes and other labor disturbances, company police were posted outside of privately-owned stores to intimidate the customers, both miners and out-of-towners, from entering. If the shopper was brave enough to enter the store, the guards further harassed him by frequently inspecting the parcels when he came out. Lloyd Williams said that even though his father operated a grocery store, foreman Walter Hunter asked him if he was buying from the company store. Lizzie Morey Rabel said that in order to avoid the guards, her mother sent her through the alley to the Cupp Store to buy coffee which was on sale. As a result, her father was dismissed, but later Hoffman relented and hired him back. According to Jennie Simoncini, her husband, Gino, was fired for the same reason and had to find work at Wehrum.

Newcomers to town were strongly encouraged to live in company housing if available. Rose Larish Hubner and her family arrived in Vintondale in 1919 and rented a house in Rexis from the railroad. They soon moved to Chickaree Hill because Hoffman told her father if he was going to work for the company, he was going to live in a company house.

Not only did the VCC employees have to purchase at the company store, they also were expected to buy their house coal from the company. One miner was discharged for working at a small mine on Chickaree on days when Vintondale mines were shut down for lack of orders; he also purchased his coal from that mine. Joe Bennett sold anthracite at his feed mill. Joe Dodson said that his brother-in-law Sam Duncan was given a warning about purchasing Bennett's coal. When Joe's father purchased a load of anthracite, he was fired.

Hoffman's background was ore boating, and he ran the town like a ship. While he was superintendent, the company houses were kept in repair and painted in Sherwin-Williams white. Ollie McConnell said when her family arrived in 1918, the town had the nickname of the "Snow White Mining Town." Many of the painters were actually college friends of Hoffman's only son, Kenneth. These boys from Lehigh and Harvard Law School just happened to be good baseball players. This arrangement was mutually beneficial. Vintondale got good ball players; maintenance taken care of; and the boys were able to make good money at easy summer jobs. One summer, Ken was given one of the hardest, dirtiest outside jobs on the flat. Pete Lazich, a coke worker, felt sorry for Ken and tried to help him. Mr. Hoffman informed Mr. Lazich that he would lose his job if he helped Ken again, that he wanted Ken to learn the value of work and of money.

Another way of keeping the town shipshape was to make sure the town voted the right way: Delano's way, the company way, Pappy's way, the Republican way. On election day, he or the foreman stood at the polls reminding everyone how to vote. In his 1923 diary, one entry said, "The election went our way." Was there any doubt? Some people claimed that the ballots were marked. Others said that a list of those to vote for was included in the pay envelopes. John Morey, Jr. voted his own choice one election day, and when he reported for work the next day, there was a knot in his lamp cord. When Morey inquired about the knot, he was told to see the superintendent; Pappy told him that he had voted the wrong way and subsequently lost a day's work. A few brave souls continued to vote Democratic, especially those with no ties to the company. In the 1920's, the Vintondale vote could be easily predicted at 250 Republicans votes to 35 Democratic ones. It would have been interesting to have seen Pappy's reaction to the voting switch when the union came in.

Because Delanos, Schwerins and Hoffman were strongly anti-union, the town was closely guarded by company police. Union organizers were kept out, and miners sympathetic toward unionism were evicted if they lived in company houses. Those living in privately-owned homes frequently were harassed or even beaten up. Timmons, Butala or McCardle were always ready to keep law and order in Vintondale-company style. A police officer met every train and inquired of the business of the debarking passenger. If the answer did not suit, that person was told to get back on board and head on down the line. Due to fear of a strike in 1923, a policeman was assigned to see each mantrip off. Hoffman also did not want any men dallying on the job. On August 30th, 1923, he wrote in the diary, "I went to #1 & found the car shop & part of the tipple crews loafing. I don't think they will do any more for a while. on to office - Flats . . ." On November 5th of the same year, he wrote, "From 5:30 to 9:30 we had a heavy rain & everyone was drenched, watching the surface and coke crews. I made up my mind they were all faithful." Hoffman wanted a full day's work out of the men with no outside interference. Imagine the scene of Pappy chasing an angry Johnstown Ford dealer off company property telling him that in Vintondale "he will sell his Ford before or after working hours."

In line with the Roman idea of "bread and circuses" to keep the masses quiet, the company tried to regulate the social life of the community. A first class movie theatre was built in 1923 to replace the nickelodeon. There were weekly dances at the social hall, which Hoffman religiously attended. The company booked traveling carnivals and circuses for the ball field during the late spring and early summer. National holidays were celebrated with a community picnic, complete with parades, races and concerts. Clubs such as Boy Scouts and Woodsmen of the World were encuraged. Basketball games were held at the social hall in the winter. And, of course, Vintondale had the traditional baseball team, the better the ballplayer, the better his working place in the mine, or perhaps even an outside job.

There was a humane, sympathetic side to Otto Hoffman of which few residents were aware. Those who worked closely with him said that he was a very nice person once you got to know him. He personally took the miner's welfare to heart. Mike Vereb broke his back in a rockfall in 1917. Hoffman paid Vereb, who was bedfast for over thirty years, a visit every Sunday, bringing him books, ice cream and other such gifts. Some think that Pappy was responsible for preventing the Department of Revenue raids at Verebs during Prohibition. Martha Thomas Peterson was in bed for a year with rheumatic fever and got very excited when Mr. Hoffman arrived carrying a candy box. Much to her disappointment though, the box contained eggs. Hoffman had chickens on his "estate", and son Ken had an egg route in town.

Mary Monyak Drabbant and Rose Larish Hubner both did housework for Mrs. Hoffman. They agreed that both Hoffmans were hardworking and energetic. By the time Rose Hubner reported to work at 7 AM, Mrs. Hoffman had already washed the clothes by hand and hung them out on the line. Mrs. Drabbant said that Mrs. Hoffman would often drive to the Monyak house and ask Mary to come to the house to help her.

Official pesonnel were rarely invitied to the Hoffman house, but before World War I, companyman Nick Gronland occasionally slept there when the weather was hot. His room at the Vintondale Inn was too warm in the summer. However, when Mr. Schwerin was in town, there usually was a bridge game at Hoffman's, but Mrs. Drabbant said they did little entertaining. She remembered only one large party: a birthday celebration for Mr. Hoffman. A dinner party with many New York guests was held at the Hoffman house. Mrs. Drabbant was impressed by the clothes worn by the New Yorkers. After the dinner, there was a square dance at the social hall; Mr. Hoffman insisted that Mary become part of the group and made her square dance, her first and last time.

Occassionally, Mr. Hoffman would stop to visit Mrs. Drabbant on his way to #6. Since the electricity on Dinkey Street was cut off in the daytime, she had to get up before dawn to wash clothes. The ladies on Dinkey Street asked her to talk to Pappy. Teasing her with "Mary, your clothes are already done. What are you hollering about?", Mr. Hoffman ordered that wires be strung for daytime electricity. Also, on one occasion, when he had suggested that there was a good movie at the theatre, Mrs. Drabbant and her young son arrived to find standing room only. She was soon offered a seat; Pappy gave up his.

A first-class theatre and a reliable bank were two of Hoffman's major achievements in Vintondale. The former came about because of his love for the movies; the latter was initiated by his discovery that much of Vintondale's money was being invested elsewhere. Here was a man who made lasting positive and negative imprints on the town.

As a civic leader, Mr. Hoffman contributed to town government by serving as president of borough council for twelve years. His wife, the former Maude Landers, whom he married in 1901, served on the school board for a similar period of time. Yet Charles Mower, supervising principal, said there was no interference with school operations. Only on one occasion did Mr. Hoffman request that the school close for part of a day when a senator of the right party came to town. The Hoffman's also acted as sponsors at the dedication of the bell at the Hungarian Lutheran Church (Hungarian Reformed) in May, 1923. Hoffman was on the county fair board for several years and participated in the fund drives for Memorial Hospital. His social clubs included the Masons, Tall Cedars, Ebensburg Golf Club and the Sunnehanna Country Club. Through these organizations, Hoffman was able to maintain contact with key politicians of the county. Hoffman, a Presbyterian, occasionally attended services in Ebensburg or Johnstown.

On Thursday, March 27, 1930, Otto Hoffman died in Memorial Hospital at 2:20 AM as a result of injuries received in an accident at the #6 tipple. The previous day, Mr. Hoffman was walking on a narrow path under the tipple. According to Joe Dodson, Pappy was on his way to the mine office to tell Allie Cresswell that he wanted threemen spotting cars instead of two. Hoffman bumped his head on a beam and dropped his pipe. When he leaned over the fence to pick up his pipe, he was struck across the back by the crossbar, which carried the mine cars up the conveyor to be dumped. The spotter at the top of the tipple did not see Mr. Hoffman and released the chains and crossbar. Russell Dodson said that Mr.Hoffman insisted on being carried to the mine office on a chair rather than on a stretcher. When informed of the accident, Mrs. Hoffman, a tall, heavyset woman, walked nobly by herself down the hill to the flat.

Hoffman's refusal to lay flat may have contributed to his death. He was taken to Memorial Hospital and died there eighteen hours later of a fractured pelivs, two fractured vertebrae, internal bleeding, and a ruptured bladder.

The funeral was held on Sunday, March 30th at 2

PM at the Hoffman residence. Reverend Doctor Robert MacLeod Campbell of the First Presbyterian Church of Johnstown conducted the service. Harve Tibbott, baritone, sang three hymns. Pallbearers were: Dr. MacFarlane; Herb Daly, master mechanic; Sam Feldman, washery foreman; Mike Mihalik, bank cashier; James Dempsey, surface foreman; John Huth, company engineer. Following the service, the funeral party motored to Johnstown to catch the 5:13 train for Wisconsin. Accompanying Mrs. Hoffman and son Kenneth were Dr. MacFarlane; Mrs. Hoffman's sister-in-law, Mrs. Landers of Lakewood, Ohio; and Captain and Mrs. E.D. Tulian of Lakewood, life-long friend and shipmate of Mr. Hoffman. Burial was the following Tuesday at the family plot in Merrill, Wisconsin with the Masonic burial ritual carried out by the Merrill Lodge.

After the funeral, Clarence Schwerin announced that Milton Brandon, superintendent of Graceton Coal and Coke Company, would assume the position of Vintondale superintendent on the following Monday. Jack Huth was appointed assistant superintendent. Mrs. Hoffman moved to Merrill in early June, and Mr. Brandon moved into the house the following week. Schwerin and son Clarence were in town that week to see Mrs. Hoffman off and get Mr. Brandon settled in.

How do you assess such a person? The following are two opinions made by persons who knew Hoffman personally; the third was made following World War II by a distant relative of the Delano family who was writing about FDR and the Delano connection.

Martha Abrams Young, dauther of #6 forman Abe Abrams, described Hoffman as "*a real czar, physically large, . . . Hoffman found ways of getting rid of good families. He seemed to perfer "trash." A few families did survive.*" (Author's note: This opinion surfaced several times in interviews with other residents.)

Nick Gronland, companyman before World War I and coal salesman afterward, wrote that Hoffman was "*a bit gruff, but fair with miners except in any action dealing with unionism. determined to keep it out and did.*"

Daniel Delano came to Vintondale in 1946 to find grist for his mill that it was the Delano influence that helped make FDR a household name. In his book, **Franklin Roosevelt and the Delano Influence**, Daniel Delano wrote:

> Otto Hoffman was popular with his men; he had always been a close associate of the miners and thoroughly understood their problems. He was always willing to take every risk, even after he became superintendent, and his men relate that he never asked them to undertake a dangerous mission that he was not willing to undertake himself. Every safety aid and device had been provided and the welfare and well-being of the miners had been the rule at Vintondale for almost two decades.

(Daniel Delano obtained his information from various persons in Vintondale, including photographs and research projects from the high school newspaper. However, Jim Balog, newspaper editor at the time, said that Mr. Delano never paid him or returned the photos and other materials he gave him. For over forty years, Mr. Balog thought the book had never been finished.)

Otto Hoffman had his supporters and his detractors and rightfully so. In a small mining town, a man with that much power and prestige earned friends and foes no matter what action he took. As Ollie McConnell said in 1981, "*Mr. Hoffman was good for the company.*"

**Ken Hoffman and Family - 1939.
Courtesy of Huth Family Collection.**

WHO'S MINDING THE OFFICE?

The Vinton Colliery Company employed a large staff to handle the enormous amount of paperwork involved in operating a coal mine. Like the superintendent's position between 1901 and 1915, official personnel and companymen arrived and departed frequently.

Answering the Telephone

Following the construction of a telephone line from Ebensburg in 1894, the switchboard was located in #1 powerhouse. In order to monitor all incoming and outgoing telephone calls, the switchboard was moved to the new company office after 1906.

There were few private telephones in town. Nevy

Brothers' store and Mr. Clement's house on Second Street in 1912 were two of the few private telephones. As last as 1940, outside telephone calls still had to be made through the office, which was inconveniently closed on weekends. To make an emergency call on a Sunday, someone had to locate an operator to open up the office and the switchboard.

Like today, Vinton Colliery had a choice of telephone companies. In 1912, there was the Johnstown Telephone Company and the Central District Power and Telephone Company competing for VCC business. By 1917, the company dealt with Bell Telephone Company.

Telephone operator was one of the entry level positions in the VCC office. Lovell Mitchell Esaias quite school in the eighth grade and got a job in the telephone office. Her first promotion was to the hiring window where she pounded the brass checks for the miners. Another promotion sent her to the tonnage divsion. At the time of her marriage in 1923, she had head payroll clerk. During the 1924 strike, she and her husband, company policeman Dick Esaias, lost their jobs and were evicted.

Cora Bracken Roberts manned the switchboard during the 1922 strike, which was open twenty-four hours a day at the time. She later worked on time cards. Marie Cresswell Mihalik was also an operator before she assumed the position of secretary. Verna Findley started working in the office when it was still on Plank Road. Her first job was making squibbs for shooting coal.

Office secretaries were difficult to retain. Mabel Davis Updike worked as a secretary from 1912 to 1916; she said that there was a lot of friction in the office, especially after Hoffman took charge. She likened him to a tsar. In 1912, there were many personnel changes that had a major impact on later history of the town.

Chief Clerk

In 1912, VCC had difficulty finding a chief clerk. Mr. Rauch arrived on June 26th and lasted until July 27th. Mr. TerBush had been hired on July 23rd, but left without notice, leaving only a note saying that he did not want to go through the motions of applying for a bond. The next candidate met the VCC's qualifications; Lloyd Arbogast was hired on August 5. A native of Mifflinburg, Arbogast remained chief clerk until 1929.

Engineering Department

The engineering department was responsible for surveying the coal tracts, making up-to-date blueprints of the works, and any other odd jobs for which they were trained. Other than Clarence Claghorn, names of engineering department employees before 1906 are unknown. When Delano constructed the #6 mine in 1906, one of the employees with an engineering degree was Louis Burr, son of company treasurer Algernon Taylor Burr. Lou worked at the washery and ammonia plant. Burr left the employment of VCC in 1911 to take a job in Canada.

Vinton Colliery Engineering Crew. Left to right: John Huth, John McGinnis, unknown - 1910. Courtesy of Huth Family Collection.

Arriving about the same time as Burr were John McGinnis and John Huth from the hard coal town of Frackville. McGinnis came with the Cambridge Bituminous Coal Company and brought Jack Huth, then eighteen, with him. Huth, the author's maternal grandfather, started as a chain boy on the surveying crew, passed fireboss and forman's examinations, rose to chief engineer and in 1930 to assistant superintendent. He remained with VCC until he was let go in the early 1950's.

John McGinnis, Engineering Crew - 1906. Courtesy of John Huth Collection.

on the engineering crew before 1930 were: Harry Morgan, who was let go in 1912 for not coming out at 7 AM; Swanson, who began in 1914; James Claghorn, Clarence's son, in 1914; Maltson in 1914; Hannan in 1917; Thomas, let go in 1917. These names were found in the superintendent's mine diaries.

Dan Galbreath worked in the office from around 1914 to 1916; he was responsible for the construction of the large resevoir on Shuman's Run. Galbreath moved on to Wehrum, advanced to the position of superintendent in 1917 and transferred to Kentucky in 1918.

Nick Gronland worked in Vintondale from 1912-1917, but stated in a letter that he had no official job title. The miners called him "Mr. Boss." While in Vintondale, he boarded at the Vintondale Inn. He enlisted in the army in World War I and saw action in France. Upon his return, he decided that he did not want to remain in coal mining and returned to his home in Wisconsin. In the early 1920's, he joined the sales force of the Delano Coal Company and mainly sold their Morea hard coal from the Mill Creek Coal Company.

After 1916, the office staff remained stable until the 1924 strike. Numerous companymen lost their jobs as a result of the strike. (See the eviction list in the appendix.) The next major changes in the office came in 1929 and 1930 with the departure of Arbogast for health reasons and the resignation of John Burgan. Replacing Burgan was Olive McConnell, one of the first, if not the first, female shipping clerk in the industry. In 1930, Otto Hoffman's death had a major impact on the office and the community. Milton Brandon had an extremely difficult task ahead of him. He as faced with the hardships of the Depression and had to try to operate the mines in his own way, not Pappy's.

Source: Warren Delano Papers, Franklin Delano Roosevelt Library, Hyde Park, NY.

George Blewitt, Jr. - Head of Storehouse for Vinton Colliery Company.
            Courtesy of Huth Family Collection.

Vinton Colliery Company office staff c. 1906. L. to R. A.V. Caldwell, R.B. Adams, ?, George Blewitt, Verna Findley Morton, Miss Foster. Taken near the company office on Plank Road.
            Courtesy of Huth Family Collection.

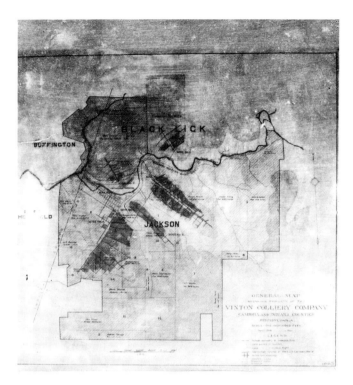

1906 General Map showing the works of the six mines of the Vinton Colliery Company. #1 and #4 are in the lower left corner, on the south side of the Blacklick Creek. #2 is the mine where the word Jackson is found. #3 is to the right of #2. #5 is on the north side of the creek, directly under the words Blacklick on the map. The new workings of #6 are to the left. #5 and #6 eventually combined.
Courtesy of Denise Weber Collection.

#1 Trestle and Tipple - 1920.
Courtesy of Lovell Mitchell Esaias.

Siding at #1, Chickaree Hill at right.
Courtesy of Lovell Mitchell Esaias.

Entrance to Vinton Colliery Company's #1 mine, 1906.
Courtesy of Denise Weber Collection, Pennsylvania State Archives, Harrisburg.

#1 Trestle, 1906.
Courtesy of Denise Weber Collection, Pennsylvania State Archives, Harrisburg.

#1 Mine, Mule Barn in the Foreground, 1906.
Courtesy of Denise Weber Collection, Pennsylvania State Archives, Harrisburg.

#1 Trestle, 1906. Note the absence of a road leading up Chickaree Hill. Houses there were built in 1917.
Courtesy of Denise Weber Collection, Pennsylvania State Archives, Harrisburg.

Vintondale in 1906. #1 rock dump on left, Second Street in background, Barker Street to the right. Union hall behind the house on the corner of Barker and Second Street.
Courtesy of Denise Weber Collection, Pennsylvania Archives, Harrisburg.

Vinton Colliery Company Office, Plank Road. Standing Anson Caldwell, R. B. Adams, George Blewitt. Seated Miss Foster and Verna Findley.

Courtesy of Pauline Bostick Smith and Betty Bostic Biss.

Remains of eight experimental coke ovens built at #3 mine, c. 1899. #3 rock dump in background. Taken May 13, 1984.

Courtesy of Diane Dusza.

#6 Coke ovens, 1906
Courtesy of Denise Weber Collection.

First Street, 1906. Note second stack added to #1 powerhouse. #1 tipple to the right.
Courtesy of Denise Weber Collection, Pennsylvania State Archives, Harrisburg.

Entrances to #6 Mine, 1906.
Courtesy of Denise Weber Collection.

Entrance to #6 mine. Opening at the right was for a furnace which provided air for the mine.
Courtesy of Denise Weber Collection.

Entrance to #6 mine, 1906.
Courtesy of Denise Weber Collection.

One of Vinton Colliery Company's Dinkeys, 1906.
Courtesy of Denise Weber Collection.

Vinton Colliery Company's #6 mine, 1906. Pouring the Footers for the Coke Ovens.
Courtesy of Denise Weber Collection, Pennsylvania State Archives, Harrisburg.

Footers for #6 Washery, 1906. Note tree stumps in the background. Also, Plank Road is in the background. The miners' bridge in upper left was designed by Louis Burr.
Courtesy of Denise Weber Collection.

#6 Washery, 1906-07.
Courtesy of Denise Weber Collection.

Vinton Colliery's #6 Mine, 1906. Foundations for 152 Coke Ovens.
Courtesy of Denise Weber Collection.

Vinton Colliery Company, #6 mine, 1906. Ebensburg and Blacklick Railroad in the foreground. Coke ovens under construction in the center. Plank Road and Maple Street in the background.
Courtesy of Denise Weber Collection.

Vinton Colliery Company, 1906. Building foundations for the Coke Ovens. #2 Rockdump in far background.
Courtesy Denise Weber Collection.

#6 Coke Ovens, 1906.

Courtesy of Denise Weber Collection.

#6 Coke ovens, 1906. Note the trees at the western end of the flat.

Courtesy of Denise Weber Collection.

#6 Coke ovens, 1906.

Courtesy of Denise Weber Collection.

#6 Coke ovens, 1906. Steam in left background is from #3 powerhouse. Houses in the background were built for #3. Goat Hill to extreme right.

Courtesy of Denise Weber Collection.

Vinton Colliery Company's #6, 1906-07. Coke ovens, washery under construction on the right. Center supports will hold tracks for lorry which will load the ovens through the top opening. Electric wiring and water pipes need to be added.

Courtesy of Denise Weber Collection.

Vinton Colliery Company. Completed Coke ovens, 1906-07.

#6 Machine Shop, 1906-07. Foundations in the foreground are for the ammonia plant.
Courtesy of Denise Weber Collection.

#6 Ammonia plant, 1906-07.
Courtesy of Denise Weber Collection.

#6 Powerhouse under construction, 1906.
Courtesy of Denise Weber Collection.

#6 Powerhouse, 1906.
Courtesy of Denise Weber Collection, State Archives, Harrisburg.

#6 Washery under construction, 1906-07. Note the machine on top of the coke ovens.
Courtesy of Denise Weber Collection.

#6 Washery, 1906-07.
Courtesy of Denise Weber Collection.

Vinton Colliery Company #6 mine and coke ovens, c. 1915. Note coal loaded at tipple on left, coke is being drawn and loaded in center. Every other oven is lit. Lorry which loads ovens is in the center; it is run by electricity. Water pipes to dowse coke are on top of the ovens. Washery is on right. Coke and/or coke braize (ash) are on the ground. Horse and wagon are hauling braize to rock dump.
    Courtesy of John Huth Collection.

#6 Washery, c. 1915.
Courtesy of John Huth Collection.

Washery and Powerhouse, Vintondale 1907.

Vinton Colliery Company's #6 mine, second slope. Taken by John Huth, c. 1907.
    Courtesy of Huth Family Collection.

Bracken accident at #6 Runaround. Taken in the afternoon by Jack Huth, c. 1907.
Courtesy of Huth Family Collection.

#6 Tipple - Vinton Colliery Company, c. 1920.
Courtesy of Huth Family Collection.

Brick air vent at Vinton Colliery #6.
Courtesy of John Huth Collection.

#6 Shaft along C&I.

Vinton Colliery Company #6 mine shaft, 1917. Located along C&I tracks about 1 1/2 miles from Rexis.
Courtesy of John Huth Collection.

#6 shaft, 1917, C&I Railroad in background.
Courtesy of John Huth Collection.

Construction of #6 shaft along C&I Railroad, 1917.
Courtesy of John Huth Collection.

Vinton Colliery's #6 mine, Anderson Shaft 1923. Superintendent Otto Hoffman on left.
Courtesy of John Huth Collection.

Yard crew, #6 mine, Vinton Colliery Company, c. 1915.
    Courtesy of John Huth Collection.

Vinton Colliery Company, #6 Powerhouse, c. 1910.
    Courtesy of Huth Family Collection.

Washery Coal Bin #6, constructed 1923.
    Courtesy of Huth Family Collection.

Vinton Colliery Company's #6 powerhouse after tornado, 1919 or 1920.
    Courtesy of Huth Family Collection.

#6 Washery damaged by tornado, 1920.
Courtesy of Robert Cresswell.

#6 Powerhouse damaged by tornado, 1919 or 1920.
Courtesy of Huth Family Collection.

#6 Powerhouse. Right to left: Bailey Stephens, David Stutzman, Wilbur Stutzman, _____ Kempfer.
Courtesy of Dorothy (Stutzman) Stephens.

Vinton Colliery Company #6 mine, 1923-24. Coke Breeze is at the upper end of the coke ovens. Powder magazine is on the lower end of the ovens.

Courtesy of Diane Dusza.

Chapter V
Union Activity: 1894-1930

VINTON COLLIERY VS. THE UNITED MINE WORKERS OF AMERICA: 1894-1930

Although Vintondale was considered a closed company town, the Vinton Colliery Company was prevented from totally eliminating unionism. Granted they utilized Coal and Irons, the National Guard, strikebreakers, and espionage, VCC was unable to prevent the union from acquiring a property in Vintondale. Ironically, the union hall stood directly across from the company store on the corner of Barker and Second.

Several times before the National Labor Relations Act of 1935 (Wagner Act) drove the Vintondale miners into the waiting arms of John L. Lewis, Vintondale miners attempted to assert their grievances through strikes. The usual results were evictions, harrassment, and eventual submission to the company.

The first known newsworthy strike took place in January, 1896. However, neither salary, dead work, or unionism was the cause; it was over a $1.00 payroll deduction for an "imported" doctor. Dr. McCormick had been attending to the miners' ills, but the company brought in Dr. Wilson from Philadelphia and guaranteed him $100 per month. The miners, carpenters, and Italian laborers rebelled and walked out. There was no known success or failure of the strike. The **Indiana County Gazette** reported that twelve cars of coal a day were shipped from Vintondale. Unfortunately, the newspapers provided no additional information. Was the mining done by strikebreakers? A May, 1897 paper only hinted at a strike settlement. Was this the same strike? The answers are not to be found.

The Union Hall

The United Mine Workers of America obtained the property situated on Block A, Lot 16. Purchasing that lot in 1900 from the Blacklick Land and Improvement Company was Judge A.V. Barker, who in turn sold it to P.J. Little of Ebensburg on August 7, 1903. Little immediately sold it for $1,800 to the Executive Board of Sub-District 1 of District 2, UMWA. Serving on the union board in 1903 were Mike McToggart, George Sinclair, Patrick McCarty, John Hogan, Hugh McGinty, Frank Kelly, William Dukes, James Mitchell and Emmanuel Swansbore. On the property was a four room house and a large barn-like structure which served as the union hall.

Vinton Colliery was vigilant in its efforts to prevent the miners from organizing. In a letter to Warren Delano dated November 19, 1905, Superintendent Charles Hower wrote:

While I do not deceive myself with any ideas of false security, I felt nevertheless, that since we have ejected the Commercial men (Bracken mine) that were living in our houses that the Union has met its death in Vintondale. Last night was the regular metting (sic) night, but the hall was dark, and no one showed up for the meeting. I believe our men are very well satisfied with the treatment they are receiving and trust that we can keep in this friendly mind.

The pre-1935 Vintondale UMWA local was #2383, part of District 2. In 1909, District president was Patrick Gilday, an employee in the Morrisdale mines. In the annual secretary's report from December 1, 1907 to November 30, 1908, Local 2383 paid $495.75 in union taxes and assessments.

1909 Strike

The next known Vintondale strike of any length occured in 1909. There had been a nationwide downturn in the economy in 1907. Recovery was slow; and, on the local scene, Warren Delano almost lost his shirt because of the enormous amount of money he invested in constructing #6. (See Delano biography.)

On March 1, 1909, the company posted notices of wage reductions of twelve percent accompanied by a rent reduction of twenty percent. This wage reduction, attributed to poor trade conditions, was met with complaining, resignations, moving out, and eventually a strike. In a March 19th letter published in the April 1st edition of **UMWA Journal**, organizers Thomas King, Domenick Gelotte, and John Lias reported that a local had been formed and that the Vintondale men wanted a checkweighman and the wage cut restored. Vinton Colliery's wage cut was one month ahead of a cut from sixty-six cents per ton to fifty-five cents proposed by the Pennsylvania Bituminous Coal Operators Association of which Vinton Colliery was not a member. The executive board of the UMWA rejected that cut, and in return some mines threatened to close.

Between March 5th and April 2nd, the Pennsylvania Constabluary were in Vintondale to keep order. The company then hired between fifteen and twenty deputies to patrol the town. Forty miners were evicted by Superintendent Charles Hower. The UMWA provided some assistance in the form of rent and groceries for the striking miners and their families, some of whom were almost destitute.

On Monday, April 15th, nineteen more families were "peacefully" evicted by Deputy Sheriff Ed Knee. Sheriff Webster Griffith deputized John Woodchak, C.E. Jenkins, Ed Kruger, and Isaac Kelly to assist Knee. The men completed their task and departed on

Tuesday evening. Some of the strikers moved to Rexis and rented the houses owned by the Vinton Lumber Company. L.H. Davis, superintendent of the lumber company's railroad, rented seven houses to the UMWA inspite of appeals from Vinton Colliery not to rent the houses. (See appendix for receipts paid by the UMWA in 1909.)

In addition to paying the rent, the union purchased groceries from the independent Vintondale merchants. Total weekly bills amounted to as much as $295.00 during the months of May and June. Merchants benefiting from the strike were S.R. Williams, David Nevy, J. Krumbine, Mike Farkas, John Wozniak, and Horvath and Company.

There was no serious trouble in town during April other than threats of dynamiting Hower's new mansion and the administration of tongue lashings to the deputies by the strikers' wives. In spite of the efforts of the striking miners to carry it out, the Vinton Colliery Company returned to work, bringing in strikebreakers. Many were Hungarians, as attested to by Heber Blankenhorn in his book **Strike for Union**. The Finnish population of Vintondale almost disappeared at this time. Since many Finns were attracted to industrial democracy and socialism, it appears that they left town and mirgrated to Nanty Glo, a union town and hotbed of socialism in the county. (Check the list of birth certificates in the appendix for 1906 to 1912.)

The strike continued into May of 1909 with rumors flying that the top brass of VCC were fired. D.R. Pratt, general manager of the Vinton Supply Company was one of the first to resign; he left town on June 1. Others who eventually departed were Charles Hower, Superintendent; Ira Thomas, Assistant Superintendent and Postmaster; A.V. Caldwell, Chief Clerk; and Robert Custer, Chief Engineer. The only companyman rumored to be leaving, but remained behind was George Blewitt, Supply House. Joseph Fitzgerald, married to Cynthia Buchannon whoe relatives helped build Wehrum, left Vintondale for Washington State to lay narrow gauge track for the mines there. His wife and six children remained behind for a year until he was able to send for them.

On Saturday, May 8th, Vintondale's reputation of being a wild and wooly town lived up to expectations with a shooting. Joe Barish and Martin Pinter were arrested along with nine others and taken to the Ebensburg jail. The two were retained and charged with felonious shooting; the others were released. Barish and Pinter, non-union miners boarding at House 33, went to pick up the father of a third party at House 42. A wedding reception was in progress in House 42. When met at the door by striking union miners, Barish allegedly pulled out a revolver and fired several shots. One unnamed victim was hit under the arm and was taken to the almshouse where he was under the care of Dr. F.C. Jones. In this particular case, the union also paid the company doctor for teating the victim.

Domenick Gelotte, Nanty Glo UMWA organizer, attempted to educate the public about the strike situation in Vintondale through letters to the newspapers. He blamed the violence on the strikebreakers.

At the end of May, a house occupied by Hungarian strikebreakers was dynamited at 1:30 AM. The boarders living there were drinking beer at the time of the explosion. One man was injured. The house, which lacked a cellar, was badly damaged; the floor was torn up, walls damaged and windows broken.

The strike dragged on through July, but petered out by the end of the summer. Even the **UMWA Journal** stopped listing Vintondale as a place which miners should avoid.

By December 1909, the strike was over, and the **Weekly Tribune** declared that VCC was busier than ever and paid out its largest payroll in two years. Over half the coke ovens were being fired, the largest number to date. Quality of the coke also improved; perhaps this was due to the efforts of Clarence Schwerin who was hired in 1908 to achieve that task. S. Kedzie Smith assumed the position of superintendent, and Vintondale started on its road to recovery.

Organizing Efforts in 1912

Vinton Colliery won a victory over the union in 1909, but it had to maintain a constant vigilance to prevent a reoccurance. In 1912, the union again attempted to organize the men. This time, the attempt can be viewed through company eyes through the use of the 1912 mine diary. On October 2nd, several organizers arrived in town and were carefully watched by company police and the superintendent. They stayed several days and attempted to contact union sympathizers at night, but were thwarted by the police. On October 4th, the office relayed town conditions to Hamilton and Schwerin in the New York City office. The superintendent also summoned Mr. Tanney, a Pittsburgh detective. Tanney arrived in Vintondale on the 4:14 train, accompanied by "a couple of strong arm men." The company also turned out all the lights at 11 PM. On Saturday the 5th, one of Tanney's undercover men arrived; he was assigned to #1 the following Monday. That Saturday was also payday. The lights were turned off both Saturday and Sunday. The situation appeared to be serious because Hamilton arrived from New York on Sunday, followed by Schwerin the next day.

On Monday, #3 was unable to work, and #1 and #6 had light tonnage, not due to a strike, but to "Blue Monday." All three mines were operating on Tuesday, but production was still down due to the payday weekend. The organizers left town in the morning, but returned in the evening.

Wednesday the 9th, the organizers posted anotice of a mass meeting. The foremen held their own meeting in the office. The next day, Schwerin departed for New

York. The union meeting was held on the 10th. Everyone attending was identified by the company which had installed a light near the union hall entrance. Nineteen Vintondale miners were positively identified and discharged the next day. A second union meeting, which was poorly attended, was held on the 11th. The new company chief clerk, Lloyd Arbogast, was sent to Johnstown to deliver eviction notices to company attorney Davies who was to prepare the legal papers. Several coal and iron deputies were also sworn in. This show of force cowed the miners.

On Saturday October 12, all mines were running, and the miners were being closely supervised. The formal eviction notices for four houses were served by Sheriff Stutzman on October 16th. The miners' goods were attached for costs.

George Blewitt filed a complaint of surety of peace against the organizers with Squire Walters in Ebensburg on Friday the 18th. On Saturday, which was payday, the organizers left for the day. Several more miners were discharged, and one of Tanney's detectives was laid off. Blewitt and the superintendent traveled to Ebensburg for the surety of peace hearing, but the miners were unwilling to testify against the organizers, and the case "fell flat." But, the case was not dropped, the superintendent returned the next day to see about the case. With all the discharges, the mines were short on employees, so George Blewitt was sent to Greensburg to find some miners.

On October 26, the organizers again tried unsuccessfully to hold a meeting. The company let Detective Malory go on the 28th. That was the last diary entry for 1912 which concerned unionization. Once again, by using intimidation, spys, evictions and discharges, the Vinton Colliery was able to prevent the organizing of its miners.

Efforts From 1914-1921

In 1914, things were relatively calm in Vintondale except on October 2nd when two organizers appeared in town to invite miners to a mass meeting in Twin Rocks. The success or failure of the meeting was not recorded.

There were several strikes in 1916 at mines close to Vintondale which had an impact on Vintondale's miners. Big Bend and Commercial mines in Twin Rocks went on strike in mid-March over pick-mining vs. machine-mined. Pay for pick-mined was seventy-two cents per ton; pay for machine-mined was fifty-four cents. The miners claimed that they could produce more coal more easily by pick. This was a major issue which the UMWA had to face; the advent of the various mining machines meant the loss of mining jobs.

On April 1st, miners on the Indiana branch of the Buffalo, Rochester and Pittsburgh Coal Company went on strike to protest the company dominance in their mines. Vintondale's miners went on strike some time between April and October. The only written evidence are the eviction notices. Twenty-five families were evicted, the majority of whom were Italian miners. Very few returned after the strike. Mrs. David Pesci lived in the last house at the end of Plank Road; her mother's neighbor was evicted, and her furniture thrown out on the road. Mrs. Pesci's mother helped the neighbor and brought a mattress into the house. For that deed of kindness, she also was evicted. John Bossolo, father of Josephine Pisaneschi, ran a boarding house in half of the old company office on Plank Road. He was evicted, but was recalled about three months later. The strike was another failure. The majority of the Italian population of Vintondale moved out; those remaining in 1916 were mainly involved with the Nevy Brothers' store.

The union continued its attempts to organize, no matter how dangerous for health and welfare of the organizers or the miners. On January 16, 1917, Tony Mintrella and Philip Destuffor paid a visit to town. Organizers Foster and Gilette came on February 26. Officers Timmons and Butala kept them company from 10:50 to 4:07 and prevented them from talking with theminers. Anothr organizer, S.V. Seveneck paid a visit on Sunday, August 12th, arriving on the 7:50 train. He visited Andy Nevitsky and stayed overnight at the Village Inn and left the next day on the 12:38 train. Are there any doubters left that the company knew **everything** that went on it town? The next week Hoffman spent the day at the flat talking to dissatisfied miners, and also "looked up some union miners."

Companies during World War I had a certain amount of protection against strikes through various government wartime regulations. However, inspite of several pay raises, wages could not keep up with the price increases. The miners were as disgruntled as ever, and a national strike was called in 1919. Newspaper checks gave no indication as to whether or not Vintondale attempted to go on strike. John Brophy's correspondence for 1919 also reveals nothing. Only a letter dated January 20, 1919 rom UMWA organizer Domenick Gelotte to Brophy hinted that the union had failed to organize the men in Twin Rocks and should concentrate on Vintondale, Revloc,. and Park Hill. If the union did try to organize the Vintondale men in 1919, it was a total failure.

But Vintondale's big claim to fame as a non-union town was yet to come.

STRIKE OF 1922

In 1922, Vintondale made the first of her three appearances in the **New York Times** and other national newspapers. Events which unfolded in Vintondale during the nationwide UMWA strike warranted the coverage. John L. Lewis called a strike of all coal miners on April 1, 1922. Union mines complied; many non-union mines also came out, and the miners were rapidly organized. However, at Vintondale, conditions were different. The town took on the look of an armed camp. Each road into Vintondale was guarded by armed company deputies. At each guardpost was a shanty with a telephone hookup to the company switchboard. Six powerful spotlights were positioned in strategic sites around town. One was set up on the Dinkey Track so that it could shine down on "wild and wooly" Sixth Street; another enabled the guards to see as far as Bracken.

Company guards were hired, and five of them were mounted. They wore grey uniforms like the Coal and Iron Police. These guards kept the townspeople in line and kept out unwanted union organizers and anyone else they deemed suspicious. All traffic into and out of Vintondale was stopped. Only those with good reason to enter or leave were given permission. The late Charles McGuire, a salesman from Cresson, was stopped; eventually the guards got to know him and allowed him into town without much hassle. A Schwerin family story had Clarence II, a teenager, assisting with the patrols and even carrying a machine gun.

Vinton Colliery's Company Police. Jack Butala in the center - 1922.
Courtesy of Huth Family Collection.

Vintondale's reputation as a non-union town was well-known in UMWA circles. In 1921, a young volunteer union organizer and Harvard graduate, Powers Hapgood, came to Cambria and Somerset Counties to study non-union conditions firsthand by working in the mines. His experiences were written up in a pamphlet called **The Non-Union Mines-the Diary of a Coal Digger**. It was published by the Bureau for Industrial Research; co-director of the Bureau was Heber Blankenhorn, 1884-1956, AB from Wooster and AM from Columbia. The pamphlet was circulated by various socialist groups. A copy could not be located, but fortunately the section on Vintondale was published in the March 22, 1922 issue of **Survey**.

Mr. Hapgood arrived in Vintondale on April 17, 1921 on the morning train from Revloc. Within five minutes he was stopped by a coal and iron policeman who was dressed in riding breeches and grey shirt, the same color as the state police, but he was paid by the company.

"Would you mind telling me what your business is in this town?"
"Just looking for work."
"What are you, a miner?" "Where are you from?"

The dialogue between the two men was short and to the point. The policeman directed him to the main office. While waiting for the #6 foreman, Hapgood walked up to #1, there he talked to a young man from Patton who was also looking for work. Hapgood was told that all positions in #1 were filled. At 10:30, he saw the #6 forman and was hired if he found a place to board. Powers was able to find a "fine place" for $50 per month. On his return to the mine, Hapgood answered the questions of the foreman and decided to ask a few on his own. He learned that the pay was 72 cents per ton. When he asked about pay for dead work, the foreman replied, "*I guess you don't want a job bad enough to get one here. You might as well go along.*" In spite of Hapgood's protests that he only wanted to find out what company policy was before he started working, he did not get the job. The foreman responded:

Don't make no difference. You might have waited at least until you'd seen your place and got started and found out about it. You don't need a job. You might as well go along.

Powers Hapgood left Vintondale disappointed, but wiser. However, during his short stay he discovered that Vinton Colliery was paying twenty-three cents per ton less than union mines. Even non-union Revloc was paying union scale.

While waiting at the station for the westbound train, he struck up a conversation with two English-speaking Hungarians who had also been told to take the next train out of town. They wanted to interest Vintondale Hungarians in a colonization scheme and received an interview with Mr. Hoffman. However, the men were told that the company owned the town. The Hungarians believed that since the company owned the town, it could keep anyone out of town, just like a man could keep a person out of his house. So they left also.

Mr. Hapgood then caught the west-bound train and stopped in Wehrum where he found the miners working one to two days per week. The Wehrum miners were not interested in the union, neither pro or con. In the article he made the following judgment about non-union miners: "*They seemed, like most other non-union men I've met, too inactive mentally to consider the question.*" His comment was unfair. In many cases the miners were too intimidated to consider going union. Many had families to support and were willing to or forced to work in non-union mines to keep them fed. History proved Hapgood wrong about the Wehrum miners because Wehrum went out in 1922 whereas Vintondale continued to work.

Even before the strike call went out, the Pennsylvania State Police sent a questionnaire on March 18, 1922 to all mines in the soft coal region requesting information on the type of protection the companies might need. Vintondale's response to the questionnaire was found in the State Police files at the State Archives in Harrisburg. Mr. Schwerin sent a reply on March 25, 1922 to State Police Superintendent Lynn Adams. In the letter, Mr. Schwerin indicated that the Vintondale and Claghorn operations, "*of course, expect to operate*." In answer to several questions concerning the ability of miners to obtain explosives, Schwerin replied that no outsiders sold explosives to the miners, and the explosives magazine was not guarded at the moment. However, the Vinton Colliery Company planned to add watchmen who would be responsible for guarding the building. The eleventh question on the questionnaire was: "Can you supply me with the name and description of all known radicals living in the vicinity of your operations?" Schwerin replied that the company knew of no radicals in the Vintondale and Claghorn areas. He added the following endorsement at the end of the letter:

We have always been warm admirers of the State Police and appreciate the good work of the Organization for many years. We do not expect to have any occasion to need your assistance at this time, but, of course, we will call upon you in case anything develops.

When the strike was called on April 1, 1922, Vintondale continued to work steadily. Closed-town conditions caught the attention of news reporters by mid-April. On April 19, three correspondents, one from the **New York Herald**, one from the **Federated Press**, and one from Chicago tried to enter Vintondale.

The reporters were stopped by company guards decked out in the uniform of the state constabulary; the uniform looked authentic except for the helmets. The guards wore 38's at their sides. The reporters were informed that they were on a private road; the coal and irons, using a lot of profanity, declared that they had orders not to let anyone in. However, the reporters were able to get photographs of five of the seven policemen and, at the first chance, telephoned the **Johnstown Democrat** to relay their experiences.

The next day, Bill Welsh, UMWA organizer from Nanty Glo, walked into town with four Vintondale miners. However, Welsh was escorted out of town by the guards, and the four miners were evicted. Their goods were thrown onto the streets. Trucks hauled the furniture to Cresson, and the men were charged $35.

The UMWA took two of the evicted miners, the Pedeck brothers, to Ebensburg to tell their story to Sheriff Logan Keller, who made a visit to Vintondale. A check of the formal applications for eviction notices show that the Vinton Colliery Company did not apply for any at that time.

According to the April 24th **Indiana Evening Gazette**, the sheriff ordered the guards to keep the roads open and not to interfer with the rights of free speech for miners who own their own homes. But according to Heber Blankenhorn, who came to Vintondale a few hours later and was permitted to ride escorted through town, Sheriff Keller soon lapsed into apathy.

Otto Hoffman gave the company's side of the eviction story in the **Nanty Glo Journal**. He denied that the four miners were allegedly evicted. He said ten organizers came into town and only three Vintondale miners accompanied them to the depot, stood around for several hours, and then went home. Vintondale miners refused to go out, and Mr. Hoffman sent an officer to the homes to request the men to go to the mine office to talk to him. He told them they were discharged. The men asked him if they could get a truck for their goods. Mr. Hoffman said he paid the cost of hauling the goods to Spangler and Johnstown.

According to several Vintondale residents, Mike Yelenosky and Mr. Hozik were beaten up by the company guards because of their union tendencies. These men worked at Lackawanna's #3 mine. The Vinton Colliery guards would not allow them to walk to work via the railroad tracks because it was private property. On May 10th, Yelenosky filed charges for assault and battery, false arrest and surety of peace against Otto Hoffman, Ken Hoffman, and Richard Esaias. These charges were filed in Nanty Glo with Robert Harnish. The men waived preliminary hearing and were held for court on $1,000 bail. Ken Hoffman did not appear for the hearing. Then on May 20, according to the **Nanty Glo Journal**, for unknown reasons, Yelenosky went to Ebensburg to withdraw the charges.

While Yelenosky and Hozik were battling the company guards, the UMWA was making arrangements with the fledgling American Civil Liberties Union to make some visits and gather information on violation of miners' rights. Roger Baldwin, director of the ACLU, wrote to District 2 President John Brophy on May 2nd and promised to send in a special newswriter and photographer. Four days later, Baldwin wired Brophy that Crane Gaysz (?) was on his way to prepare a campaign. Organizer Shields of District 2 was sent out into the field to make new inquiries on violations of civil rights. So the scene to "invade" Vintondale was set into motion. The Pennsylvania State Police were also aware that an ACLU visit to Cambria County was planned; it sent memos to various law-enforcement officers in the area.

Vintondale's big headliner came on May 27th. This

time, some distinguished observers attempted to drive into Vintondale to see if the miners' human rights were being violated. Arthur Garfield Hays, a New York lawyer who was representing Samuel Untermeyer of the American Civil Liberties Union, got a taste of Vintondale justice, 1920's style. He was accompanied by Attorney Clarence Loeb of Philadelphia, Attorney Julius Rosenberg of New York City, and Attorney J.J. Kintner of Lock Haven, district attorney of Clinton County and chief counsel of the UMWA. Several newspaper reporters accompanied the three men. The **Johnstown Tribune** on Saturday, May 27th printed a message that came from New York stating that Garfield Hays planned to defy the orders of Cambria County officials and would make a public speech in a strike town. Hays was to be accompanied by bondsmen in case he was arrested. Pennsylvania State Police also sent out flyers to various law enforcing groups encouraging them to keep their eyes on Hays.

Hays' sojourn in Vintondale was well publicized. Besides making the local papers, New York's dialies picked up the story, and by June 14, 1922, **The New Republic** featured an article by Hays. The following description was taken from the article:

You drive from Nanty Glo, Pennsylvania along a highway beautifully set among the hills of Cambria County with quiet woods and mountains on all sides, when suddenly you come to a spot in Pennsylvania, but out of America. It is Vintondale.

According to all the available accounts, this was what occurred. The automobiles carrying Hays' entourage approached Vintondale. A guard ordered them to stop, but they drove past him into town and stopped in front of the company store. Five mounted coal and iron police rode up. One said, *"We know you, you - -. You're organizers from Nanty Glo. Get the hell out of here."* The visitors got out of their car and strolled up the sidewalk; the guards followed them brandishing clubs. The reporters asked what laws they were violating and were told by Chief Clerk Arbogast that they were trespassing on private property and could not use the sidewalks. One of the newsmen was picked up and hustled to the car. Mr. Hays, a cripple, told the guards that they were committing assault, so they started to hustle him to the car. After another warning from Hays, he was released amid profanity from the guards. The visitors then left town while the coal and iron police congratulated themselves on their supposed victory.

Hays returned to Nanty Glo and went to constable Herman Gowan to swear out "John Doe" warrants. Bill Welsh and David Gowan of the UMWA accompanied Hays on the return trip to Vintondale. Their job was to identify the guards.

In Vintondale, they found the same blocked road and a train sitting in the middle of the crossing at the upper end of Plank Road. The driver of one of the cars, named Cooney, was a union engineer and knew the train's engineer. *"Back up, Walter. We're in a hurry."* The train backed up, and the cars moved on to the company store.

The guards greeted them with, *"What the hell are you doing here now."* The constable's response was: *"You're under arrest."* As each of the guards rode up, he was served his warrant. Those arrested were: Richard Esaias; Harry McCardle; John Butala; James Dempsey; and Lloyd Arbogast. David Gowan, union representative, wanted to take the men before a justice of the peace in Nanty Glo. The company officials said that they wanted to confer with the district attorney's office. Gowan called the D.A.'s office and after a delay, talked to a person who claimed to be D.P. Weimer, the District Attorney. He told Gowan to accept bail in Vintondale. Those arrested had their hearing before John Daly. Later that evening, Attorney Loeb contacted Mr. Weimer who denied ever having that conversation. All telephone calls coming in and going out of Vintondale were through the switchboard in the office. It would have been very easy to rig the phone call. During the strike, the company kept the switchboard operating twenty-four hours a day.

Arbogast, and the town justices, George Blewitt and John Daly, met with the constable in the company office. When Hays went up the outside stairs to inquire about a date for the trial, Arbogast ordered Hays' arrest for trespassing. Hays was escorted to the town jail, a 3' x 6' cell littered with paper, refuse and cigarette butts. In the cell was an iron bed covered with a filthy blanket and a toilet. Hays was held in jail about a half an hour while Butala filled out the information for the warrant.

According to Hays, Mr. Blewitt appeared collarless and unshaven and personally paid the $300 bail for the guards. He signed a warrant for Hays' arrest.

Hays: "But I've already been arrested."
Blewitt: "We want to get through with this."
Hays: "I'm ready for trial."
Blewitt: "All right, I find you guilty and fine you $5."
Hays: "I insist on a trial."
Blewitt: "It's all over, but you don't have to pay the $5 if you'll get out of town."
Hays: "I won't get out of town and I won't pay the $5. I want a trial."
Blewitt: "Well, you don't have to. It's all through."
Hays: "But I won't take a verdict of guilty."
Blewitt: "Then you're not guilty."

Court adjourned.

As Hays was leaving the company office, the police were chasing bystanders away. The police chief then followed Hays' car out of town. Hays went to Lock Haven with Kintner to prepare injunction papers against the Vinton Colliery Company. He then hurried back to New York to see Untermeyer who promised to take personal charge of any legal action against the Vinton Colliery Company.

According to the June 1st **Nanty Glo Journal**, a hearing was held that day in the Nanty Glo office of R. Harnish. The defendants: Harry McCardle, James Dempsey, John Butala, Richard Esaias, George Blewitt, Samuel Feldman and Lloyd Arbogast, were accused of assault and battery and were ordered to appear at the June term of court on the first Monday in June. Bail was set at $2,500. Percy Allen Rose of Johnstown represented the defendants. However, Samuel Untermeyer did not appear at the hearing. A large crowd of miners gathered in front of the squire's home and greeted the defendants wih "kidding." Union members presented the guards with copies of the **Penn Central News**, a union paper printed in Cresson. Most of the crowd was anxious to see the infamous Jack Butala.

In response to all the publicity about Mr. Hays' arrest, Clarence Schwerin gave a statement to the press on June 1st. He denied that the justice of the peace was an employee of the coal company and also denied the justice held court on company property. In the press release, he said:

My best advice is that Hays deliberately trespassed. If he feels that he has a case, however, and he and Untermeyer want to take it to the Supreme Court we'll go there with them. Whatever means we adopt to keep undesireables out of Vintondale are taken to protect the non-union men in our mines from the Black Hand letters, intimidation, dynamite and threatened invasion of the United Mine Workers and we believe the laws of Pennsylvania will uphold us in the use of our own police for that purpose.

Mr. Schwerin also denied that the company "owned" the post office and closed it to prevent delivery of union material. He also denied that the company owned the churches and paid the teachers' salaries.

On June 11, Samuel Untermeyer wrote to Governor Sprowl requesting that he evoke the commissions of the Vintondale coal and irons who were accused of assaulting Hays. On June 12th, Arbogast and the four coal and iron police were indicted on charges of assault and battery. Mr. Hays also filed suit in Supreme Court against Vinton Colliery on charges of malicious seizure and arrest and brutal assault.

The UMWA and ACLU filed a petition on June 12th at the Court of Common Pleas asking that the Vinton Colliery Company and its agents be refrained from interferring with public assemblies. The plaintiffs in the petition were: William Welsh, David Cowan, Arthur Shields, Julian Rosenberg, H.S. Cooney, James Marks, John Brophy, Nathan Burch and J.D. Bennett.

The defendants were: Vinton Colliery Company, Otto Hoffman, Lloyd Arbogast, John Butala, Harry McCardle, Richard Esaias, James Dempsey, Edward Carlson, Dewey Rairigh, Jessie McGuire, Harvey Mott, John Cresswell, Ed McDonald, Oscar Dishong, Otto Yank, _____ Yank, Charles Wilson, John Wagner, Lewis Esaias, John Karrish, Anthony Quish, Earl Dewet, Herbert Daly, John Daly, William Evans, Samuel Feldman.

The petition recited the purpose of the UMWA and the ACLU. It also alleged that the defendants had entered into an unlawful conspiracy to prevent free speech and lawful assembly of defendants. It was filed before Judge McCann who notified the attorneys of the plaintiffs that a $2,000 surety bond must be filed. J.J. Kintner, UMWA chief counsel, assured that it would be filed the next day.

Obviously afraid of losing its control over the town, the company undertook a questionable legal action through the Borough council, which it controlled. On June 16th, Council had a special meeting to present an ordinance which would prohibit public meetings, assemblies, gatherings or parades on public highways in Vintondale. Following the first reading, William Abrams made a motion to accept the ordinance, and Sam Feldman seconded the motion. The secretary was instructed to publish and secure 100 copies of the ordinance and enter it into the Ordinance Book.

On June 17th, Judge McCann of Ebensburg granted the UMWA a preliminary injunction to protect their property rights on Second and Main. The injunction restrained the company from interferring with the union's right to free access to their property.

That same day, Justice John Kephart of the State Superior Court, a native of Ebensburg and acquaintance of Otto Hoffman, granted a **supersedeas**, or a stay of all proceedings of the union's injunction until the final hearing in October. Attorney Percy Rose, representing the Vinton Colliery, charged that the injuction was defective. The court held that Rose's allegations were supported by the facts and issued a temporary stay.

However, the stay order was not issued soon enough to prevent a union meeting that evening at 6:00 PM on union property. A motorcade of union officials and Garfield Hays drove through town to the UMWA property. The exact number of residents who listened to the speeches is in doubt. The **Indiana Evening Gazette** said that 300 gathered to hear the speeches. **Strike for Union** said few miners paused to hear the speakers because company guards rode up and down the street trying to discourage any onlookers.

At the rally, Mr. Hays said:

I am particularly intersted in seeing Superintendent Otto Hoffman present, for no man needs education in the principles of the construction or the fundamental principles of life more than Mr. Hoffman.

It is not known if Mr. Hoffman was an observer at the meeting, but rest assured he soon knew the outcome of the meeting.

Union speakers were John Brophy, Mark Cowan,

Bill Welsh and Mayholz. A representative of the New York Public Commission of Coal, McAlister Coleman, also spoke to the miners. He said that the coal operators were successful in public denying-denying unknown facts of coal to all inquiries and denying that the known facts were so. Brophy's entire speech was published in the Indiana paper. When the meeting finally broke up, the party was escorted out of town in a blaze of searchlights.

On June 20th, Borough Council, during a regular meeting amended the new Ordinance 32, striking out some of the original wording and substituting with "regulating street parades, etc." The changes carried, and Ordinance 32 was duly copied into the Ordinance Book on page 113. Unfortunately, this book was lost quite a while ago. With Ordinance 32, Council could use the word "regulating" to continue to ban any type of meeting at the union hall.

That same week, union attorney, J.J. Kinter, sent two letters to John Brophy which discussed the Kephart decision. Kinter wrote on June 19, 1922 that he was satisfied that *"Justice Kephart exceeded his authority in granting a supersedas, and I would like to argue that branch of the case."* On June 26th following a Saturday court hearing in Philadelphia, Mr. Kinter wrote the following to Brophy:

I am also sending you, under separate cover, our brief in the case for Saturday, but it would make no difference if we had had all the authorities in the world, our friend Mr. Kephart had made up his mind to pull the trick and the Supreme Court of Pennsylvania, although they protested, finally suceeded to his wishes. I never saw a man as excited as Kephart was and he said himself that it took an hour before the matter was disposed of by the Court, and I know from that remark that he had much difficulty in bringing them to his way of thinking. I had no chance to argue the case, nor to show the Court the importance of the position. They not only were wrong in their conclusion, but the Supreme Court had no jurisdiction. But it is the old statement, "'What is the Constitution between friends?"' But we have got to fight just the same, and show by publicity what we are up against.

During the course of the strike period, the Vinton Colliery Company was able to obtain the services of the National Guard to keep order in the town. Troops camped out at the ball field and marched up and down Main Street. Part of their daily excercise included a ride to Wehrum, which was on strike. The guards were still in town in August when the **Nanty Glo Journal** reported that the company took part at a military tournament at Camp Little.

The Hays' assault case was continued from the June term of court until the fall term because Attorney Rose was serving as counsel on a murder case. A brief was filed by the UMWA in the Court of Common Pleas of Cambria County for the September term. In it, the ACLU claimed that they had sent Rosenberg and Loeb to Vintondale to converse with the inhabitants and distribute literature which explained the guaranteed civil rights of Vintondale's citizens. In town were posted notices from the sheriff's office which forbid unlawful riotous assemblies. The ACLU considered these notices as unconstitutional. There were additional Vintondalers who joined with the union as plaintiffs. Joe Bennett, feed mill owner, and Nick Burch, barber, claimed that the company activity during the strike led to a depreciation of their businesses and real estate. Many of their customers were intimidated by the company guards and took their business elsewhere.

The Cambria County Court must have ruled in favor of Vinton Colliery in the free speech case because the union presented a petition to the Pennsylvania Supreme Court appealing the Cambria County ruling. The petition was presented in the names of John Brophy, James Mark, William Welch, and David Cowan. Their arguments were the same except that the petition remanded the Supreme Court that when Kephart made his ruling in June, the union lawyers had no chance to be heard or to ask for an amendment. Further results are unknown.

The company guards were eventually found guilty that fall after the strike was over. The injunction was officially dissolved on January 8, 1923. A check of the rcords at the courthouse failed to produce any of the trial records.

As for the $30,000 lawsuit against Vinton Colliery, Hays lost the case in New York City. The jury decided he had provoked the assault. In a footnote in his autobiography Hays said, *"In a sense I had, even though I had merely insisted on my rights."* When Mr. Hays finally got to meet the "czar" of Vintondale, he discovered that Clarence Schwerin was an old college acquaintance who played on the same lacrosse team. Schwerin's response to the czar comment was, *"You're a damned radical, Arthur."* According to Clarence Schwerin III, Hays and his grandfather later became good friends, and Hays dedicated one of his books to Mr. Schwerin.

The efforts of the UMWA to "open up" Vintondale were to no avail. Two weeks after the rally, the circulation managers of the **Johnstown Democrat** came to town on business. They were so harrassed by the guards that they left.

The ACLU did not abandon the free speech effort in Vintondale. Roger Baldwin wired Brophy on October 10th asking if Brophy planned any meetings in Vintondale since the Supreme Court granted the right to hold meetings. However, Brophy responded that due to conditions in Vintondale, there was no advantage for holding a rally there. Because of District 2's efforts in keeping the strike going in Somerset County, Brophy thought that a Vintondale meeting could be postponed until "some opportune time."

Perhaps Baldwin was more persuasive with Brophy because on November 1st, Brophy wired Baldwin that any day during the following week would be satisfactory for holding a free speech meeting in Vintondale. Lucille Milner, ACLU Field Secretary, responded to Brophy on November 6th that the ACLU was unable to secure Bishop McConnell of Pittsburgh for the scheduled Vintondale rally, but promised a "speaker of prominence."

Those speakers were probably Reverend Richard Hogue of the Church League for Industrial Democracy and Assemblyman Patrick McDermott who arrived on November 19th and asked Burgess Evans for a meeting permit. His response was shouted at them, "*It can't be done.*" The men also departed under pressure from the guards.

According to Heber Blankenhorn in the **Strike for Union**, as of August, 1923, no open public meetings had been held in Vintondale. Mr. Hays claimed that when he returned to town, he was given no trouble about a public meeting. However, he did not state the date of his return.

Overall, the 1922 strike was a failure. Most of the non-union miners who went out on strike were let down by the international union. John L. Lewis agreed to make separate contracts with the owners of the old union mines and the newly organized mines, thus weakening the union hold over District 2. Miners found that the mines raised the wages of the "scabs." Many families were so destitute that they had to go back to work.

The 1922 strike made Vintondale a household word in labor circles, but was just one of the many successful battles the Vinton Colliery Company waged against the union. In fact, the strike gave an advantage to the Vinton Colliery which sold coal at a fixed price to the Consolidated Edison. Since the company could supply all the coal that Con-Ed needed, the utility was willing to buy Vintondale coal for years to come. Nick Gronland, salesman for Delano Coal wrote, "*We welcomed strikes at other mines*." The price rose from $2.00 to $2.50 before a strike to $12.50 per ton during a strike. Clarence Schwerin III said that in the 1950's, he received a letter from the vice-president of pruchasing of Con-Ed stating that his company had long ago given full recognition for favors granted in the 1920's.

LABOR ACTIVITY IN 1923

After gaining national attention and making money by not going on strike in 1922, Vinton Colliery officials became even more vigilant the following year. Every rumor was checked out; every train was met by one of the police officers. Even Mr. Hoffman personally met the March 14th train because an organizer was reported to be headed to Vintondale-his trip was in vain.

April 1 was celebrated by the UMWA as the anniversary of winning the eight hour day. Now called Johnny Mitchell Day, it was also the day often used by the union to call a strike. Company officials fully expected a strike on April 1 and began making preparations two weeks ahead of time. On Sunday, March 18th, Hoffman drove to Ebensburg and Revloc for "April 1st information." The next day, he discharged a Gallitzin miner accused of being sent in by the union. That Tuesday, he sent Officer McCardle to Johnstown to "get a look at organizers." Hoffman also made undisclosed arrangements in Claghorn for April 1.

In 1923, April 1 was Easter Sunday, so there was no threat of strike that day. However, the next morning a mounted policeman was stationed at each mine to watch the mantrips enter. They returned in the afternoon when the trip came out. Another officer was sent to meet the early train. All foremen were out in the headings, checking on the miners as they worked. The police surveillance continued daily through morning and evening. Arbogast left for a week's vacation on August 23, the same day Hoffman and the police talked to some organizers. Following the discussion, the organizers drove back toward Twin Rocks. In September, Hoffman refused to hire three hard coal miners who had been on strike.

So again the company thwarted the efforts of the union with constant surveillance. However, they were no so lucky in 1924.

1924 STRIKE

Although the Vinton Colliery Company prevented miners from striking during the 1922 UMWA national strike, Vintondale's miners chose to walk out in 1924. Many residents labeled this episode as the Ku Klux Klan strike. Direct connection between the strike and the KKK is hard to prove, and so the strike will be treated separately from the KKK.

On March 17, 1924, approximately 600 miners and outside employees of the Vinton Colliery Company walked out when the company posted a wage reduction of up to $2.50 per day. This amounted to a 20-33% pay cut. Led mainly by companymen on the flat, strikers wanted to maintain the 1917 wage scale. In 1924, the Vinton Colliery Company was supposedly paying union scale, but the miners were only working two to three days a week. The company's response to the strike was to proceed with evictions. The first normal eviction notices were applied for at the Cambria County Courthouse on March 19, the last on July 2. (See appendix for the list of evictions for 1924.)

UMWA organizer Bill Welsh of Nanty Glo appeared in Vintondale and began to organize the men. This strike was at first very peaceful compared to the unionizing efforts of 1922. The **Indiana Evening Gazette** reported that Vintondale had "provided many exciting incidents during the last coal strike."

On March 18, John Brophy, president of District 2, immediately telegraphed Governor Gifford Pinchot informing him of the company evictions, which probably took place without the formal notice. The telegram said that striking miners were given twenty-four hours notice to move out of the company houses. Brophy asked Pinchot to "take steps to prevent such action." Pinchot reportedly did not plan to act until he heard the company's side. Efforts of the newspapers to get confirmation of the evictions were fruitless.

On March 19, the **Gazette** reported that the Vinton Colliery Company advised Pinchot that he had been misinformed about the evictions. Only twelve miners, alleged instigators of the strike call, were evicted. Brophy, attending a UMWA convention in Altoona, continued to assert that some strikers were given twenty-four notices. A copy of an eviction notice was sent to Governor Pinchot, who then sent a telegram to the coal company challenging their facts. He informed the company that he had an eviction notice, signed by Hoffman, dated March 17 with the miner's name and house number on it. The tenant was to leave by March 18. In his telegram, Pinchot stated, *"The evidence of wilful misrpresentation would seem to be complete. I would be glad to have your explanation, if any."* The company's reply was not made public even though Hoffman indicated that the governor could go public with their statement. No public statement was made by March 29.

In another telegram to Pinchot, Brophy reported that twenty miners were given eviction notices. Pinchot did send a special investigator to check on the reports that fifty miners were evicted. Vinton Colliery admitted to evicting twelve. The total eviction notices applied for by VCC were forty-eight. The "Vintondale News" section of the **Nanty Glo Journal** on March 27 reported that the Swartz's, Wilson's, Leatherman's, Kanich's, and Joe Rafas were leaving town. Even the Millers, who managed the Vintondale Inn, were planning to move to California. The following commentary was included in the report: *"By the looks Vintondale will soon be deserted if people keep moving out."*

Grant Gongloff said that his wife sassed the "pussyfoots", and they were evicted. They became permanent residents of Nanty Glo. John Galer, who worked in the powerhouse, was evicted; his family shared a house in Rexis with the Yahnke family. Mr. Galer later bought the Blacklick Inn in Wehrum, only to loose everything when that town closed.

The miners stayed out over a month, and the strike remained peaceful at first. Perhaps this was because the company had fifteen to twenty policeman patrolling the town. Many of the men employed as company police in the 1922 strike joined the strike in 1924 and received eviction notices. Richard Esaias, a World War I veteran and policeman in 1922, received an eviction notice. He had his wife, the former Lovell Mitchell, lived with her mother in the house next to the Baptist Church. The evening before they were to leave, Esaias suffered a stroke, but the family was still evicted the next morning. Mr. Esaias survived the stroke, and the couple settled in Johnstown.

In late April, twenty-five men returned to work to mine coal to run the boilers and pumps. At that time, the **Indiana Gazette** reported that Cambria County officials were investigating a dynamite charge which went off on one of the hills. The company blamed the union, and the union denied the action, saying it would injure their cause. No mention was made in the paper as to the involvement of the KKK in this incident.

Trouble did break out on Friday, April 25 when some striking miners and their wives tried to prevent the strikebreakers from going to work. Those arrested for rioting and assault and battery were Mr. and Mrs. John Muzik, Mr. and Mrs. Andy Leguish, Mr. and Mrs. George Eleske, Mrs. George Kurtenecz and Mrs. Ann Dancha. All the men and Mrs. Leguish were charged with rioting. All the other women were charged with assault and battery. Arrest warrants were sworn out before Justice Charles Rowland of Ebensburg. The defendants were released on bail pending court hearings in June. The company claimed the working miners were attacked with stones and pokers.

Mr. Muzik was unable to appear at the hearing because of the beating he received at the hands of the company police. Two officers, J.M. Farr and Charles Brosch were arrested on charges of assault and battery with intent to kill. The two policemen were arrested by Nanty Glo Constable Allen Russell; Justice of the Peace S. Campbell refused bail and committed them to the Cambria County jail. The charge of intent to kill was added by Campbell after he read the physician's statement on Mr. Muzik's condition.

Mrs. Ann Dancha recalled her role in this episode and gave the author a colorful account of the events. She lived in a privately-owned house on Plank Road and had to walk to Yuhasz's store (corner of Maple and Plank Road) for a loaf of bread. To intimidate Mrs. Dancha, Jack Butala rode up behind her and pulled the reins so that the horse reared up on his hind legs. She ran behind a telephone pole, and he chased her around the pole six or seven times. She threatened to hit him with a poker if he did that again. As she was coming out of the store, she saw one of the other policemen hit Mr. Muzik across the head with his stick and split his head open. Mrs. Dancha said that Mr. Muzik did not say a word to anyone, that he was just going to the store for tobacco.

The next time Mrs. Dancha ventured to the store, she carried a long poker. Jack Butala was hiding behind one of the double houses watching the activity on the street. As she walked up the street, Butala followed her, harassing her as before. This time she was prepared and hit the horse on the rump with the poker. The horse took off toward Maple Street with Butala hanging on with all his strength. Mrs. Dancha said if she had hit Butala on the leg, she would have broken it. She ran into the store and hid from him, returning home by way of the alley.

The same day, a neighbor came over and told her,

"Annie, get ready." She wanted Mrs. Dancha to get some eggs and peppers in her apron and help them stop the strikebreakers from going to work. Because her husband was away working in West Virginia, and she had two young children at home, Mrs. Dancha refused to go with the women. When the miners refused to go back to their homes, the women pelted them with the eggs and peppers and threw their lunch buckets in the creek.

Jack Butala continued to harrass Mrs. Dancha by throwing stones at the house and breaking her windows. He also rode the horse through her yard and tried to push in the door with the horse. Mrs. Dancha said that a union representative visited her and told her they would take care of the damages.

When her assault case came before court, Mrs. Dancha was afraid that she was going to jail, but the union representative told her to tell exactly what she saw. Mrs. Dancha said they only asked her a few questions, and she won her case. The policemen who were also charged did not receive jail sentences but had to pay fines.

These Plank Road and Maple Street strikers were harrassed by the company police because they lived in privately-owned houses. The company could not evict them as they had the other strike leaders, so they had to resort to harrassment.

There were no references in the local newspapers as to the official end of the strike. Two things are certain: the strike and the union organizing effort failed. Many families left Vintondale permanently; a few were recalled and remained residents for the remainders of their lives. The UMWA and the miners locally and nationwide were in for some very lean times in the next decade. It is ironic that Warren Delano's nephew, Franklin Delano Roosevelt, through the Wagner Act, brought unionism to Vintondale.

No. 3 Vinton Colliery Comapny tipple. Bert Cramer and Pete Cramer.
Courtesy of Hazel Cramer Dill.

Vintondale Ammonia Plant, c. 1907.
Courtesy of Pauline Bostick Smith.

Vintondale Colliery's #1 Mule Team. Clarence "Socks" Stephens second from right. He often had to take the mules to Clawson's Blacksmith Shop in Belsano for re-shoeing.
Courtesy of Dorothy Stutzman Stephens.

After Five Days Return to
VINTON COLLIERY COMPANY
CLUB HOUSE
VINTONDALE, - PENNA.

Chapter VI
Vintondale Starts to Grow

VINTONDALE BEFORE WORLD WAR I

In 1892, the creation of a new town whipped up an air of excitement and speculation throughout Cambria and Indiana Counties. Unsubstantiated rumors, such as an English syndicate developing the properties, floated throughout the area. Even the name of the new town was in doubt. Newspapers referred to planned community as Vinton, Vinton City, Baker, Barker, Barker City, Vintonville, and not inappropriately, Bark City. Vintondale emerged as the victorious title because Vinton's mail was becoming confused with that of nearby Vinco. The official name of Vintondale retained its Barker heritage by borrowing the middle name of Judge Augustine Vinton Barker, the person most responsible for the town's existence.

Vital to the survival of a mining town was cheap transportation. Judge Barker, with his connections, was able to convince the Pennsylvania Railroad to extend their Cambria and Clearfield branch to Vintondale. The new subsidiary was called the Ebensburg and Blacklick Railroad. As work on the rail line progressed, ground was broken for the new town on July 19, 1893. Thirty men were employed in clearing the land, not an easy task considering the rocky terrain and thick woods. There were some areas in the vicinity which had been cleared previously during the iron furnace period or by the Pringles because Cora Bracken Dill, daughter of Twinan Bracken and mother of Watson Dill, picked blackberries in Vintondale when she was a child.

As early as August 3rd, 1893, Vintondale recorded its first casualty. Allen Graham cut his knee with an ax while clearing ground and was confined to bed. The same month there was an unexplained work stoppage in both the Vintondale and the railroad jobs.

Area residents arrived to "take in the sights." Buffington Township residents George Kerr and Earle W. Graham were two of the first visitors to the new site. Other satisfying their curiosities came from Belsano and Greenville.

C.R. Kleghorn (Claghorn), chief engineer of the Vinton Colliery Company ordered maps of the town made in November, 1893. In the spring of 1894, the Blacklick Land and Improvement Company offered lots for sale. Judge Barker and his son Fred purchased numerous lots on Main Street and Plank Road. His brothers Constantine, Florentine, and Valentine, operating as Barker Brothers, purchased lots, mainly on Plank Road and Maple Street. The Barkers were prohibitionists, which accounts for the anti-alcohol provisions in the deed when they sold a property. (See biography of Judge Barker.) Various members of the Thomas Griffith family also purchased a few lots.

The streets were laid out and graded, and curb stones were hauled in and installed. Thirty-five houses were built quickly; six in each block. Local residents profited from the construction boom. J.D. Martin of Kimmel sold shingles to Claghorn, who was building a large shingle-sided house at the southwest corner of Lovell and Second Streets. His house was completed in the summer of 1894. George Kerr and Earle Graham, who had farms between Vintondale and Strongstown, were back in Vintondale in August 1894, moving railroad ties. John Davis of Grismore found employment as a teamster; Joseph Conrad worked on a pipeline. J.B. Graham and crew successfully bid a contract to build fifteen additional houses. Decker Brothers of Penn Run won the painting contract for the houses.

Many of Vintondale's first permanent settlers were Strongstown and Pine Township residents. One of the first lot purchasers not involved in the land deals which created the Blacklick Land and Improvement Company was Blair Shaffer, who erected a much-needed planing mill at the west end of Barker Street, Block AD, Lots 2 and 4. His own house was built in 1895 on Lot 2, Block AD. (See Lumbering in Vintondale.)

Lodgings at the Depot

Shaffer's brother-in-law, Dr. M.B. Shultz, Strongstown physician, saw Vintondale as a good investment site. In 1894, he had a hotel constructed on Lot 2, Block AB across from the railroad station. An ice house was added in 1902. Shultz, who eventually moved his practice to Johnstown, leased his hotel in 1902 to Joseph Shoemaker, Buffington Township stockbreeder. In 1903, Dr. Shultz sold the hotel to Shoemaker for $3,800. Then T. Stanton Davis bought it. By 1907, Fred Bittorf was the proprietor of the Village Inn and retained ownership until the hotel was sold to the Nemishes in the 1940's.

Village Inn, c. 1910.
Courtesy of John Huth Collection.

Village Hotel, Fred Bittorf, Proprietor, c. 1910. Bill Clarkson is sitting on the steps.
Courtesy of Gloria Risko.

Thomas Morris, an uncle of Burgess Walter Morris, arrived in Vintondale in 1894. He purchased Lot 12, Block AB and built and operated the Morris Hotel, later known as the Red Hotel. He moved his family to Ebensburg in 1906. Taking over proprietorship was John Myers. Later, Mrs. Elsie Hunter ran the hotel.

Joe Bennett, another Strongstowner, built his house across from the railroad station. False rumors were flying that he was going to erect a glass and brick works, but he did operate a feed mill behind his house. Charlie Bennett, a relative of Joe's, ran a livery stable behind the building known as Mary Nancarvis' for a while and then moved to Indiana. Harry E. Miller, engine hostler for the PRR and Justice of the Peace from 1907-1918, also conducted a boarding house and restaurant in Joe Bennett's property, which was across the street from the Village Inn.

Local men were not the only ones to take advantage of the economic growth. In 1896, Maggie Caufield, daughter of James Caufield of Buffington Township, opened a dressmaking shop in Vintondale. Three years later, she purchased Mrs. West's millinery and dressmaking store in the Farabaugh Building; then in December, 1899, she married William Huey. Local newspapers offered no indication if Mrs. Huey continued to operate her shop. Her father sold his farm to the Vinton Colliery Company in October, 1899 and moved to Armagh.

Jobs Aplenty

Jobs as laborers were plentiful at this time. L.J. Armond of Penn Run gave a glowing report on the new town to the **Indiana County Gazette** in June, 1896. A plasterer by trade, Armond had just completed plastering thirty-two houses and signed a contract to finish an additional eighteen. According to Armond, two mines were opened, and a third was in the planning stages. A powerhouse was completed to supply the mines and the town; a hotel and a number of private residences were completed. In Armond's opinion, "*the company houses are cozy and comfortable but not large and expensvie.*"

However, not everyone had such glowing reports about Vintondale. Tom Hoffman, Vintondale weighmaster, quit in September, 1899 to take a job in Johnstown with Cambria Iron Works. Life in the young town was difficult; the streets were nothing but a sea of mud. Board sidewalks were built along Barker Street to keep the pedestrians out of the mud. Ticks, which clung to the legs, abounded and had to be burned off. Snakes, especially rattlesnakes, were plentiful. Ann Petrilla Dancha's father worked at the Barker Brothers' sawmill and helped build some of the houses on Plank Road. The family lived in a shanty one winter until their house was finished. There was no glass in the windows, so Mr. Petrilla covered the windows with carpets to keep out the snow and wind. Mrs. Petrilla fashioned a mattress from gunny sacks, cut the long grass along the railroad tracks, dried it and stuffed it into the mattress. This was placed on a framework of 2x4s along the beams of the shanty to keep the children from being bitten by a snake.

Mrs. Dancha also claimed that there was an old Indian who lived in a hole at the top of #6 hill who made belts and rings from snakeskin. When #6 mine was being developed, she said his son took him away.

Barker Brothers ran a supply store in town, but sold it to undisclosed buyers in April, 1897. The construction boom continued its busy pace into 1899. John Dick contracted with John and Lowry Grow to build him a livery stable; the property, stock, buggies and sleighs were sold at public auction in March, 1902.

Homer City contractors F.C. Laney and James Robertson successfully bid to lath the new houses at #3 mine, which opened in August, 1899. Working with them were Bert Myers and John Marshall. In 1900, they plastered seven houses for the Vinton Lumber Company. Fifteen more houses were completed by J.B. Graham and his crew in April, 1902. Decker Brothers of Penn Run were back painting at the same time.

The town continued to grow as changes were made in the management of the coal company. More area residents sought new employment. Some worked in Vintondale during the week and headed home for the weekend. For instance, Ollie Ruggles worked in the Vintondale supply store, and his visit home to his parents in Berringer was newsworthy enough to make the local column in the **Indiana Progress**. Mable Davis Updike worked in the office from 1912 to 1916 and boarded the "Mountain Goat" each Friday for home in Ebensburg and returned on the Monday morning train.

Other area families arrived and assumed a role in the town's civic affairs. Paul Graham moved to Vintondale in 1904 and was employed in Ruffner's store. In 1909, he purchased Sam Brett's store and operated a meat market for many years. He served on the Vintondale School Board and was treasurer of Vintondale Borough for eleven years. In 1904, he married Ivy Rairigh of Indiana; their four children, Chester, Viola, Geraldine, and Dorothy were born in Vintondale. Allen S. Graham, who was no relation to Paul Graham, also operated a meat market in the same building. He had originally gone into business with Charlie Douty, but bought him out. Allen Graham bought cattle by the carload in Pittsburgh and had them shipped to Vintondale where he had a slaughterhouse on the family farm. Allen Graham then sold out to Walter Morris, who had been a butcher in Wehrum. Morris went to Arizona when Wehrum closed in 1904 and returned to Vintondale in 1909. During the New Deal era, Morris served as burgess and Works Progress Administration co-ordinator in Vintondale while daughter Katherine served as postmistress and son Barry taught at the high school and coached boys' basketball.

Happy events in Vintondale occassionally reached the society section of the local papers. In July 1900, VCC's electrician Robert Clyde married Noona Brodney of Mahaffey. The couple were married July 4th in Ebensburg by Rev. Lancaster. Attending were D.L. Berkey, John Stoop, Fanny Hullihen, Nina Gaster, and Bertha Lynch, all of Vintondale. Also present was Hattie Hullihen of Mahaffey. A wedding dinner was held at the Metropolitan Hotel. In another announcement, the wife of W.H. Adams, blacksmith, gave birth to a ten pound boy in July, 1902.

A lengthy article in the June 28, 1907 issue of the **Johnstown Weekly Tribune** listed the following merchants:

Vintondale Inn - Albert Fleitzer, one of leading hostelries in Cambria County
William D. Kephart - photographer
George F. Geyer - agent to PRR and BR&P
Village Hotel - Fred Bittorf, near depot
Hotel Morris - John Myers, proprietor, near depot
Frank McFadden - bowling alley and cigars
J.M. Jones - postmaster
Jacob Brett - general store, Barker Street
H.E. Miller - justice of peace
H. Barron - dry goods, clothing, Barker Street
S.R. Williams - general store, Barker Street
H.L. Sher - jeweler, Barker Street
Thomas E. Morris - groceries, ice cream, Barker Street
Dowdy (Douty) Brothers - meat market
Samuel Brett - groceries, Barker Street
E.L. Davis - barber
Vintondale Supply Company - meat market, general store
F.I. Farabaugh - wholesale liquor
F.J. Ruffner - general store, Barker Street

FIRST FAMILIES

Goughnours

One of the earlier settlers in the vicinity of Vintondale was John A. Goughnour, born April 19, 1859, son of Jonas and Hannah (Benshoff) Goughnour. On February 24, 1879, he married Christean Bracken, who was born in the family home, lived there her entire life, and died there in 1939. The farm which belonged to Christean Bracken's family was known in Vintondale as Goughnour's farm. (Mrs. Goughnour's brother, Jake, who lived along the Stone Pike near the Summerhill Road, served in the Civil War and had his chin shot off.)

John Goughnour was a farmer and also an employee of the Vinton Colliery Company for forty-seven years. A carpenter, Mr. Goughnour helped build much of the town. The family farm, located about a mile above Maple Street, was accessible by a dirt road, but there was another road over the hill which connected with the Chickaree road. On the farm were three houses; Mr. and Mrs. Goughnour lived in the main house. Below the farmhouse was a small one where son Charles lived. Another son, Dave, built a third house down the road closer to town. The couple's third son, William, worked in Wehrum and lived on the former Mack farm above Wehrum. Edith Boyer, only daughter of John and Christean Goughnour, lived in Vintondale for a time and then moved back to the farm. In her later years, she lived on Maple Street.

Because of the hilly terrain, farming was difficult. The Goughnours did raise some grain and hay, and there was an excellent apple orchard on the farm. Mrs. Goughnour had a large garden and sold her produce to the company store. Several cows were kept for the family use. In the basement of the house was an excellent spring. Milk produced on the farm was kept cold in the springhouse. Attached to the roof of the spring house was a large dinner bell. Grandson Bill Goughnour said he could hear the bell announcing mealtime any time he was out picking blackberries. The bell was stolen, but recovered and is now in the yard of Bill Goughnour's daughter in Brush Valley.

Goughnours kept a team of horses and used them to deliver props to the mine. Because of the hilly terrain, logging was also difficult; the men had to put chains on the wheels to act as sprags. One winter, the snow was so deep that the fences were covered, so they drove the team over the fences.

There was a mine on the farm, which Bill Goughnour said belonged to the company. A railroad track crossed the farm and may have connected the outcroppings of #2 and #3. This mine possibly could have been the mysterious Panama Drift which was mentioned, but never pinpointed in the mine diaries. The drift would have entered the outcrop of #3 and would have probably have been along the old road which led from the farm to Chickaree. Perhaps the street in front of the Goat Hill houses was that same road. Traces of the roadbed of the railroad remain, but the strip mining in the 1940's removed almost all evidence of where the tracks led.

The farm was abandoned following the death of Mrs. Goughnour in 1939. It was too hard to get out in winter or summer. Storms often washed out the road, and it was frequently impassible in the winter. In the 1950's and 1960's, the author occasionally hiked to the farm with her grandmother, Mable Huth, who had been a childhood friend of Edith Goughnour Boyer and stayed overnight at the farm. By that time, the farm was overgrown, and the farmhouse was in disrepair.

Shaffers and Dalys

Some permanent residents were drawn to the area from the hard coal region of eastern Pennsylvania or the soft-coal fields of Tioga County around Blossburg and Arnot. John Daly, a foreman in #6, migrated with his family from Arnot around 1900. The John Daly home in Vintondale was across the street from the Vintondale Inn. His son, Herbert, became a company electrician and married Blair Shaffer's daughter, Ella. The younger Dalys first settled in Rexis, living in the former family storeroom. After Blair Shaffer's planning mill was dismantled, the Herb Dalys built a house on that site. After Blair Shaffer' Sr.'s death in 1902, his widow and several of the children remained in Vintondale, living in the family homestead on Lot 2, Block AD. Clarence Shaffer stayed on as an electrician; Blair Jr. was a school teacher in Vintondale in 1900, and Homer "Jack" worked for the company and was a borough official until he retired. Mary Shaffer lived at home until her death. Merritt purchased the lot next door, and a small store was built there. The building at various times served as Shaffer's Store, Nevy Brothers, and Averi's Store.

Blair Shaffer home. Shaffer's Planing Mill on the right. Behind fence: Lester, Tillie (Reed), Zorah, Merritt Miles, Emma B. (Schultz), John, Samuel. In front of fence: Ella, William, Blair W., Homer "Jack", Bert, Logan.

Courtesy of Lynette Daly Tarr.

Blewitts

When Clarence Claghorn arrived in Vintondale as chief engineer and superintendent, one of the first persons he hired to work for Vinton Colliery Company was George Blewitt, a former employee at Claghorn's Connell Breaker at Bernice, Sullivan County. George Blewitt, Jr. purchased two of the first lots in Vintondale, Block I, Lots 9 and 11. A bungalow was built on one of the lots; the other was left vacant. After a fire, the house was enlarged into a two story structure. George Blewitt, Jr. was mine foreman at #1 when it first opened and later served as the storehouse clerk. He and his wife, Jenny, had two sons, Bill and Don. Bill also worked in the storeroom and lost his job as a result of the 1924 strike; he and his wife, office employee Mable Reese, settled in Johnstown. Don, who was born in Vintondale, was known as the "snakeman" because he roamed the hills above town looking for snakes. George "Grandpa" Blewitt, Sr. lived across Griffith Street in a company house built by Blair Shaffer.

"Auntie and "Uncle" George Blewitt, Sr. born in England - came to Vintondale from Bernice, PA.
Courtesy of Huth Family Collection

Hampsons and Huths

George Blewitt and Harry Hampson, both natives of England, lived side by side in a duplex in Bernice. When the Hampson's daughter Mable was a baby, the family moved to the "White House", which was built to accomodate company officials who made periodic trips from Philadelphia. Clarence Claghorn's father, J. Raymond Claghorn, was the owner of the mine and made occasional visits to Bernice. He was usually accompanied by a female companion who was not his wife. When the officials were present, dinner parties were given where imported French champagne was served. Any unfinished champagne was reclosed and used later by anyone who wished to have some, even though it had lost much of its fizz. An office worker, Jimmy Gillighan, had never tasted champagne, so Mrs. Hampson, housekeeper at the "White House", told him to come in after the dinner party for some leftover champagne. When he got up to leave, his legs went out from under him, and he fell flat on the floor. Jimmy did not think that anything so good could make you drunk.

There was a wine cellar in the house which was kept locked; however, the laborer who cleaned out the furnace ducts discovered how to get in and gave himself away by his inebrious singing.

Outside was an ice house and chicken yard. The Hampson's cow was able to open the gates with her horns and raid the neighbor's gardens, so a board was

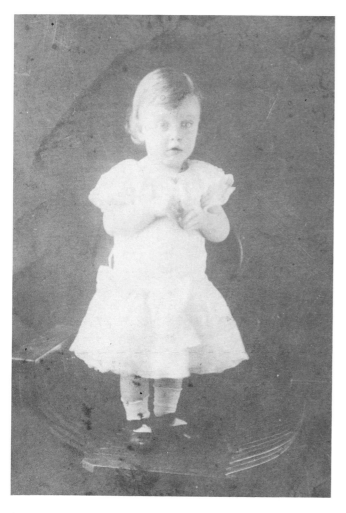

Donald Blewitt, son of Mr. and Mrs. George Blewitt, Jr. born in Vintondale in 1906.
Courtesy of Huth Family Collection.

devised to fit over her horns which stopped the raids. The cow was fed chopped pumpkins and turnips with the mash. Perhaps that accounted for her rich milk which was frequently made into ice cream. Mrs. Hampson's speciality was lemon ice cream topped with lemon syrup.

Harry Hampson 1856-1950 - Weighmaster at #3 and #1

Courtesy of Huth Family Collection.

J.R. Claghorn lost the mine some time after the Panic of 1897. When the mine changed hands around 1900, all former employees were terminated. The White House was torn down and replaced with another building. After both Hampsons lost their jobs, "Gramp", almost in desperation, wrote to George Blewitt to see if there were any jobs in Vintondale. He received an invitation to become the weighmaster at the new #3 mine. One of his jobs was to load the tenders of the steam engines which stopped at #3 for a refill. There was a trestle built across the track which led to the #1 mine; on the trestle was a hopper which was kept filled at all times. Hampson had to answer the alert whistle any time day or night if an engine required coal.

Harry Hampson, wife Eliza, and daughters Eula and Mable, (the author's maternal grandmother), arrived in Vintondale and moved into the company house on upper Main Street. Here they were welcomed with bedbugs. When the houses were completed at #3, the Hampsons moved into the one occupied today by Bruno Cassol. However, Mrs. Hampson had little time to get to know the town or watch it grow. She died of tuberculosis in 1903. When #3 closed, Hampson moved on to #1 as weighmaster until his retirement in 1914. The family moved back to Main Street and then to Third Street. Occassionally, he was called out to assist in checking cars for dirty coal. Gramp remained in Vintondale following retirement and died at the ripe old age of 94 in 1950.

Eula Hampson married Louis Burr, son of company treasurer A.T. Burr in 1911, and eventually settled in Roebling, New Jersey. Mable Hampson married John Huth in 1912, and raised a family of seven daughters: Agnes, Aileen, Margaret, Claire (Tood), Christine, Jean and Carol. The family has one lasting memory of the Bernice days. J.R. Claghorn had given Mrs. Hampson a copy of a lithograph found in his father's art collection in Philadelphia. This print still hangs in the living room of the Huth, now Michelbacher, home.

John Huth - lithograph on wall was a gift from J. Raymond Claghorn to Eliza Hampson in the 1890's.

Courtesy of Huth Family Collection.

"White House" in Bernice. Eliza Hampson was housekeeper for the Claghorns c. 1890.

Courtesy of Huth Family Collection.

Roberts

John Roberts and his family arrived from Scranton around 1902. He worked in the #1 powerhouse; after it was closed, he moved to #6 working in the shop until his retirement. Most of his family also remained in Vintondale. Son, Roy, worked at the flat as a plumber and also operated a billiards parlor for a while; daughter Ruth was the postmistress from 1922 to 1933; son, Bill, operated a popular confectionery; a third son, Jim, also worked in the mines. The family home across the Main Street from the bank is still occupied by Roy's son Bill, the third generation of Robertses in the house.

Wrays

Matilda and William Wray purchased a lot on Fourth Street in 1899 and built a double house there. They also owned a lot and house below the train station which they purchased in 1896 and sold to the Vinton Lumber Company. Mr. Wray, who came from Homer City, was a blacksmith. His sons stayed in Vintondale. Frank and his wife, Zetha Brown, raised their family in a house at the corner of Fourth and Main. Oscar "Pete" worked in the shop. Jim Wray did odd jobs around town, frequently driving truck for Mike Farkas and "driving" for Harry Reitler.

Edward Findley Family

Edward Findley, originally a native of Buffington Township, arrived in Vintondale from Blairsville in either 1892 or 1893. Because times were hard and he was out of work, Findley applied for a job in Vintondale. When asked if he could pull stumps, he hastily replied yes even though he knew nothing about the job. So, Ed Findley bought a horse and a stump puller and went to work. He purchased the last house in Vintondale from the Blacklick Land and Improvement Company, Lot 4 in Block AE. After he purchased the house, he opened a bakery behind the house. Family members delivered the bread to their customers by horse and wagon.

Findley and his wife Sara had five children: Maime, married to John Empfield; Pearl, married to Edward Bostick; Verna, married to Aubrey Morton; Will, an employee of the Vinton Lumber Company; and Murdick, married to Isabelle Dunlap.

Verna Findley started working in the company office making squibbs, black powder measured onto a square of paper and twisted at both ends and used to shoot coal. Progressing quickly in the office ranks, Verna soon was doing the payroll, counting out the money for the pay envelopes and paying out on pay day. If any of the miners failed to pick up his pay, Verna had to carry it home with her overnight. So, she bought a small revolver and carried it in her purse to protect herself. Because she had to take dictation and minutes at company meetings, Verna developed her own style of shorthand.

Edward Findley's and later William Clarkson's house c. 1905.
Courtesy of Pauline Bostick Smith.

Edward Findley Family c. 1915. Seated: Verna Findley Morton, Sara Findley, Murdick Findley. Standing: Will Findley, Pearl Findley Bostick, Edward Findley, Maime Findley Empfield.
Courtesy Pauline Bostick Smith

During the 1924 strike, a group of union men tried to march into Vintondale. They were met at the iron bridge by mounted company police who beat them with night sticks and billy clubs. The next day, one of the bosses asked Verna what she thought of the foray the day before. She replied that if it had been her brother, she would have shot them. Shortly afterward, she lost her job in the office. Because the Findleys lived in a privately owned house, the company could not evict them, but Verna was forced to find work in Wehrum offices. She walked back and forth daily to Wehrum. After the Wehrum shutdown, Verna and her mother sold the house to William and Mary Clarkson and moved to Indiana where Verna found a job in the business office of the Indiana Normal School. After moving from Vintondale, she and Aubrey Morton, after courting for over fifteen years, finally married. Both had been taking care of their parents, and Verna enjoyed being the independent working woman. She always had her hair and laundry done, but also was very generous with her nieces and nephews, who always looked forward to Aunt Verna's Christmas boxes. Verna Findley Morton retired from Indiana University in the 1960's and remained in contact with her Vintondale friends.

Mrs. Charlotte Morris holding Billy Morris. Standing: Margaret Morris. Seated: Mrs. Barry, mother of Mrs. Morris, Claire Huth, Mable Huth holding Christine c. 1921.
Courtesy of Huth Family Collection.

Verna Findley and Murdick Findley c. 1918.
Courtesy of Pauline Bostick Smith.

As Vintondale was entering its third decade, the town underwent a third growth spurt. World War I was tearing Europe apart, and demand for war material, even coal, jumped. To meet the demand for coal, Warren Delano, in 1916, created the Vinton Land Company to build new houses in town and expand the coal holdings of the Vinton Colliery. (See chapter on Vinton Colliery Company.) Life in Vintondale between 1917 and 1930 can be seen first-hand through the eyes of Frances Wojtowicz Pluchinsky whose "Memories of Vintondale" is found in this chapter.

Mable Hampson Huth, Emma Cousil, Mildred Rodgers and unknown c. 1910.
Courtesy of Huth Family Collection.

TOWN IMPROVEMENTS

Following the demise of the Blacklick Land and Improvement Company and the short ownership of the town by Lackawanna, the Vinton Colliery Company partially owned and definitely controlled the town. Repair crews kept Vinton Colliery Company's houses in order on a regular basis. These included the ten houses on Goat Hill and the five houses close to #3 mine. On Chickaree Hill were another twenty-five company houses, built in 1917. Houses on Fourth, Fifth, and Sixth Streets and half of Dinkey Street were also company houses, as were most of those built above Griffith Street. The majority of the company houses were simple five room single dwelling houses, equipped with running water and electricity. There were also some twelve-room duplexes, several of which were divided into four apartments. Some of the company houses, singled out for foremen and higher officials, had indoor plumbing.

Early residents who purchased lots and built their own homes often kept to the same plan as the company houses; a little architectural flair was shown in the houses built by Blair Shaffer and in the Farabaugh Building. Many of the main street buildings were multi-purpose, a storeroom on the first floor and an apartment on the second. Private homes, especially those on Plank Road and Maple Street, were kept under close scrutiny by the company police.

Bobby Morris, Charlotte Morris behind wheel, Mable Huth in background, Walter Morris, Margaret Huth c. 1921.
Courtesy of Huth Family Collection.

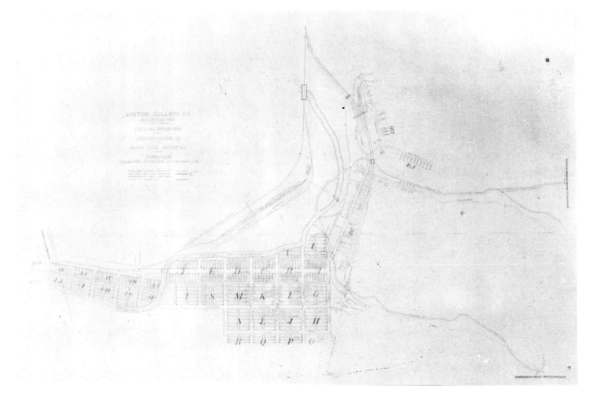

Vintondale Town Plan, 1906.
Courtesy of Denise Weber Collection.

Keeping the town orderly led to cleanest yard contests to encourage tenants to do regular maintenance on their yards and fences. Trash was hauled twice a year to the rock dump by "Huffy", the company deliveryman and "go-fer." The mine superintendent diaries give insight not only into the day-to-day events at the flats, but also the repairs and changes made in Vintondale in 1912, 1913, 1914, and 1917. In the 1923 diary, town repairs were not recorded. A summary of improvements follows:

1912

Water was connected at Gaal's store on Saturday, March 23. VCC purchased the Farabaugh building, the former liquor wholesale store on Barker and Second; included in the April 23 sale were its extra lots. Cement sidewalks on upper Main Street were the major improvement job in the spring and early summer. Businesses which received new walks were: Vintondale Inn, fireman's hall, Edward's, Callet's, and Gaal's. New screens were installed on the office windows on May 24. Cows were rounded up on the flat on the 31st. Many cows roamed freely through town and were brought home in the evenings for milking.

House painting proceeded on June 11. Re-roofing the doctor's house was completed on July 2. Installation of a furnace in the superintendent's house began on July 18 with a trial run on the 24th. Steam pipes from #1, which previously heated the house, were torn down on the 26th. Streets were cleaned on November 11, and installing daytime electric service to the house below the Vintondale Inn commenced. Water was put in the superintendent's residence. Meters were installed at the Village Hotel and the PRR depot; a separation was made on light and power in the company store on December 5. New incandescent lamps were installed on December 6, and work was done on Shoemaker's furnace. (The Vintondale Inn).

1913

The company opened a reading room in the Farabaugh Building on January 6. In addition to making books and newspapers available, there were also board games, such as chess and checkers, to play. A meter was placed on Krumbine's house on March 1. The painters hired by the company demanded a new agreement; instead of the agreed-upon thirty-five cents an hour, they wanted thirty-five cents for painting and thirty cents per roll for wallpapering. The total hours they worked were 244. In May, the company built a fence at shipping clerk John Burgan's house on upper Third Street and painted the Vintondale Inn. June improvements included blowing the sand out of the valve at Williams' store; ordering Sam Adams to put wooden eave troughs on the hotel and superintendent's porch; and painting the inside of Jack Huth's house.

In July, the porch of the company store manager Clement's house (formerly the superintendent's house on Second Street, later Jack Huth's) was repaired. A concrete basement floor was poured at Huth's house (later Dan Thomas') on upper Third Street. Nick Altemus completed the spouting for the hotel, doctor's office and house, and superintendent's house on July 22.

A water closet was installed at Burgan's house in August; Galbreath's house was wallpapered. Water was shut off in Mike Farkas' house #267 on September 4 due to non-payment of the bill. Altoona Water Closet Cleaners arrived on October 15 to clean the toilets. Mr. Smith, the closet cleaner completed his job on November 1, was paid off and left town. George Rodgers was ordered on November 5 to vacate a company house because he was not a company employee, and housing was scarce for miners. Mrs. Kisch purchased house #100 from Barker Brothers the same day. The next day, Mr. Callet, wholesale liquor dealer, appealed Rodgers' case at the office. In December, the left storefront in the Farabaugh Building was remodeled for Bill Roberts' Confectionery Store.

1914

Barker sold one of his properties, house #99, on January 30; at this time all houses in town were occupied except two apartments on Sixth Street. On February 7, Officer Timmons and Hoffman checked the borough lines and marked places for future installation of No Tresspassing signs. New boards were put on the office fence; Mike Meisner was kept busy repairing fences in later March. The borough ditch on #1 hill was also cleaned out.

On April 4, water was hooked up in the old schoolhouse across the alley from the Baptist Church; it was remodeled into houses #162 and #163. The doctor's house was reroofed on the 6th. Superintendent Hamilton visited Bennett, Findley and Meisner on the 11th and rented garden patches to them. All houses in Vintondale were rented on April 18. In May, Hamilton made an inspection visit of houses, yards and streets; the borough ditch was cleaned out again. On May 21, house #81 won first prize for keeping the nicest yard.

In June, Bennett was ordered to dismantle a fence which he constructed on company property. Streets were again inspected on the 30th by Hoffman and Arbogast.

On July 9, Hoffman visited the houses on Goat Hill where new posts were being installed for the foundation. (These houses were built on posts with no stone or block foundations. In winter, these homes were impossible to keep warm.) The task of felling the dead trees in town was finished on July 14. The honeydippers arrived on the 15th to "clean some of our closets." (See "Memories of Vintondale.") Wastepaper cans were put on the streets; burnt cinders were spread on the sidewalks from the hotel to the cement bridge. Red dog from the rock dumps and cinders from the powerhouse were inexpensive and plentiful paving materials, as were the ashes from the coal stoves which were dumped onto the alleys by residents. On July 28, the small schoolhouse was sold to John Risko. A decision was made to use materials from Steve Monborne and from the #3 barn to build a house. Demolition of the barn started on July 29, and excavation of the foundation began on Plank Road the next day. The masons began laying the foundation on the 31st.

In early August, a town bandstand was finished in the park above Clement's on upper Second Street. This was the future site of the high school. A fire plug was installed in front of Callet's wholesale liquor, formerly Jacob Brett's mercantile. To help ease the housing shortage, Hoffman inspected the vacant Vinton Lumber Company houses in Rexis on the 12th, but found them unfit to live in. The company teams began hauling the #3 barn lumber to the site of the new house, and the carpenters began framing it on the 21st. Risko paid cash for the schoolhouse on the 20th. The schoolhouse behind the iron furnace was rented to a miner; the seats were moved out, and the company was going to ask the school board to remove its furniture.

In September, fire hydrants were flushed, and water lines were installed in Mike Farkas' houses on Plank Road. Farkas promised that he would rent his five houses to tenants suggested by Hoffman. Four inch water pipe was put on the Plank Road main. The new house was completed on the 11th, and a two-family dwelling was planned with material ordered through the storehouse. a contract was let to Walter Schroth of Carrolltown to paint the houses on Second Street; several days later, an additional fourteen were added to his contract. Good weather in October enabled him to complete his work in November. (Mr. Schroth's son, Walter, was a supply clerk for VCC and married Charley Bennett's daughter Margaret. The family eventually settled in Indiana.)

Mr. Jones came to town on September 17 to bid on installation of spouting for company buildings. Because the Vinton Lumber company was remiss in its water bill payments, the company shut off the water to Rexis on the 19th. Mr. Davis of VLC settled the accounts right away, and the water was turned back on.

In October, partitions were installed in the school behind the furnace to make it into a dwelling. Bill Roberts' old store between Nevy's and Freeman's was converted to a house. Henry Meisner was assigned the task of digging up the sewer in the superintendent's yard. On October 26, water taps for Bennett's two new houses were installed, and a new wash stand was put in the "Greek priest's house" on the 27th.

In November, heaters were changed in the office, water installed in house #180 on Plank Road, and electric lights installed by tenants in houses #14, #99, and #100. Piles of garbage collected from company tenants were hauled away by the company teamsters, presumably to be dumped on the rock dump. The new waste can installed at the corner by Louie Rose's store was smashed by Callet's runaway team.

1917

This was a banner year for construction in Vintondale. The Vinton Land Company was formed to handle all of VCC's real estate, including the new houses which were built on Chickaree, lower Main Street and Sixth Street. Others were planned for Claghorn. Contractors and salesmen were coming and going on a regular basis.

Hess Brothers of Johnstown obtained a contract to build the houses and a coal bin at #1. Trouble with their carpenters began in mid-January because of illness. The foreman and two laborers left town on the 12:38 PM on January 19. Six others quit on January 30 because boarding at the Vintondale Inn for $1.00 per day was too high for them. Hess Brothers representative, Mr. Lamb, arrived to discuss construction plans in Claghorn. In April, Arbogast was consulting a Mr. Bird about taking over the Hess contract. Timmons served notice to Hess Brothers that VCC will finish the houses if "he don't get some action within the next 3 days." Apparently this threat was successful because Hess Brothers were plastering the frame houses in May. At this time, Delano suggested that some of the lots on Barker Street be used for houses. Hoffman and Huth followed up on that suggestion. Judge Reed was consulted about the Hess contact on May 31 and July 6.

In June, Mr. Burr of Ebensburg was hired to paint the engineer's room, the telephone room and Mr. Burgan's office. Porches were started on the ten four-room houses on Sixth Street (the western side). In July, Meisner began excavating for water closets on the houses between the office and the social hall. Walter Boswell, representative for Sinisep Closets, called on July 13 and returned on August 7, at which time, he received an order for a carload of twelve. Four were to remain in Vintondale, and the rest were for Claghorn. Boswell called again in September to give Huth instruction on how to install the toilets. In December, the VCC crew began building toilets between Cameron's and Williams' houses.

Hess Brothers representatives, Blair Hess and Mr. Johnston, made several visits to Vintondale in July and August. On August 28, Hess Brothers unloaded the first car of lumber for ten additional houses. The next day, Mr. Larvey of Red Plastic Roofing reported to the office that Hess Brothers credit was no good. On the 30th, Eihler Lumber Company informed Hoffman that they were going to place a lien on the lumber that was sold to Hess Brothers and delivered to Claghorn. The inspector of the Indemnity Bond Company paid a business call concerning the Hess Brothers contract, but Hoffman did not reveal the results in his diary entries.

Houses on Goat Hill received some attention in the fall of 1917. After accompanying John Meisner on an inspection of these houses and those at #3, Hoffman ordered repairs. Because the Goat Hill women were complaining about a red fox stealing chickens every day, Timmons began setting traps. Coal sheds were built there in December.

Webster Griffith contacted Hoffman about purchasing and repairing four of the Griffith houses. It was decided to approach Schwerin on his next visit. The ten new houses were completed in December, and Mr. Johnson of Hess Brothers conuslted Hoffman during the evening of the 6th concerning painting of the houses. The first tenant moved into the lower Barker Street houses on December 19.

The Vinton Colliery Company moved closer to becoming a total company town through new construction and purchase of privately built houses, and why not? World War I was raging; the demand for coal was high; wages were rising; and available housing was scarce.

**Vintondale, 1907.
Courtesy of Huth Family Collection.**

First Street (Dinkey Street) 1906. #1 Powerhouse in the foreground, excavations for #6 in the background.
Courtesy of Denise Weber Collection, Pennsylvania State Archives, Harrisburg.

**Lower Maple Street, 1906.
Courtesy of Denise Weber Collection, Pennsylvania State Archives, Harrisburg.**

Lovell Street. Doctor's house on left, Russian Church and Community Park on right. Taken after 1909 and before 1926.

Courtesy of Huth Family Collection.

Plank Road, c. 1918.
Courtesy of Huth Family Collection.

Superintendent's house and yard, c. 1920.
Courtesy of Huth Family Collection.

Vintondale 1923-24. Chickaree Hill and #1 in foreground.
Courtesy of Diane Dusza.

"MEMORIES OF VINTONDALE"
BY
FRANCES WOJTOWICZ PLUCHINSKY

INTRODUCTION

Some time ago, as I was working on a crossword puzzle, my pencil accidentally dropped on the newspaper. As I picked it up, my eyes caught a news item datelined November 11, 1915, Vintondale, PA-- Sixty years ago-today "The opening of the new Vintondale Inn was one of the biggest events of the season here. More than 100 people coming from a score or more towns within a radius of 50 miles, were served a turkey dinner by Mr. and Mrs. Joseph Shoemaker, who were conducting the new hostelry."

Well, that was the end of the crossword puzzle, for in no time at all, I was entertaining a mental picture of all those people having that turkey dinner, and enjoying the waltzes of the day and perhaps some fox trots, too. It could have been that some of the tunes they danced to were "Let Me Call You Sweetheart", "It's Three O'Clock in the Morning", or other current popular hits of the day. Then just as quickly, my mind flashed back to the present, thinking in reality, that same Vintondale Inn stands no more. A modern V.F.W. stands in its place. A bit of nostalgia enveloped my mind. The hotel is gone and forgotten, the memories left behind are forgotten, for I myself, can't remember when the hotel was razed.

So I cut out the clipping from the paper and put it away. Some time passed before I came across that clipping, and after reading it over and over again, years and years flashed before me. So I began to jot down notes to myself, with the hope that I could relive some of the past activities and remember some people of the town. As I went back into time, I decided to dwell mostly on the 1920's and 1930's.

So I wish to share some of my memories. I have no writing talents and some of my facts may be slightly incorrect. But this is the way I remember them and I'm expressing these facts as I experienced them.

CHAPTER ONE

At the turn of the twentieth century, when the Roosevelts married into the Delano family in New York City, they had no idea that indirectly they were going to have an effect on my life. As it came to pass, the Delanos and the Schwerins formed a merger, either by marriage or by business transaction, and came to Pennsylvania to buy coal lands. They bought up Graceton, Claghorn, and Vintondale. It was in this town of Vintondale where I was born, grew up, attended school, made friends, and married.

One of my most memorable and vivid pictures of Vintondale was, as a very young child, our country was at war. It was World War I. I keep remembering how the townspeople went to the train station to see the young men who were going off to camp, board the train. They carried flags and banners, and patriotically sent them off, most of them going away from home for the first time in their lives. I remember especially well the end of the war. I can still hear the mine whistle blowing and blasting, sirens blaring, the church bells ringing. At first we didn't know what was happening.

In a matter of a day or two, the town leaders organized a parade. All the school children were dressed in their Sunday best, the style of the day was middy blouses and skirts for the girls, and military suits for the boys. Someone gave each of us a flag, and we paraded, grade after grade down Main Street. We waved our flags so proudly. We were so happy because the adults were happy. Other people joined in the parade. Many of the young wore a white band with a red cross on the left arm, a symbol of the Red Cross organization. Some of the girls dressed as nurses, and were so happy to be parading. It was a holiday. The whole town took part and it was a happy time. I didn't understand too much about it all, but to the older people, it meant relief and peace. Families could be reunited, there would be more food on the tables; it meant that soon they'd be able to buy white flour. That was November 11, 1918, Armistice Day. In the next few weeks, the young men would be returning. The train would bring them back home instead of taking them away.

I can remember the "flu" epidemic and later, a diptheria epidemic. Unfortunately, I fell a victim of diptheria. After it was all over, the health officers came to our house on Dinkey Street to fumigate it. I can remember so well, it was a beautiful spring day--a Saturday--so while the house was being de-germed with a sulphur burning chemical, we all had to stay outdoors, all day long. After that, our family was allowed to remove the quarantine sign from our front door. Then we were free to go to the store, church and school. During the sickness, only my father could leave the house, go to work and come straight home. The store would send the delivery man to the front gate to deliver the groceries and take the order for the next day.

Dinkey Street was our street, so named after the small locomotive that rode the tracks in front of the houses. The track was narrow gauge in construction and very important to the growth of the mines. The dinkey and its cars carried supplies back and forth to and from No. 1 mine to No. 6 mine, and again from No. 6 mine to No. 1 mine. Both mines flourished, so the dinkey made many trips past our house every day. When we children would hear the engine coming, we would run out to the front yard to watch it, and wave to the man in the window. Sometimes we even got brave and put rusty nails on the rails, then after the engine passed, we'd pick up the nails and marvel at how the engine flattened every one of our nails, and today, I can't even remember when that dinkey stopped running.

CHAPTER TWO

I will try to give the reader a picture of our town in the 1920's; so I'll start with our homes first. Built at a small cost by the Vinton Colliery Co., they were made very plain, with large rooms. Most houses had four rooms, the ones built later had five and six rooms. Of course, we all had a front porch and a back porch. Along the back porch wall, and alongside the back wall of the house, were hung washboilers, wooden tubs, galvanized tubs, buckets, and even dishpans. There was nowhere else to store these items, yet they had to be nearby as they were used every day--the tubs for the laundry, for the miners' baths, and of course, for the family Saturday night baths. Some houses had shanties attached to or nearby where the miners could bathe away the coal dust in the summer, and the women did their laundry. Everyone had an outside toilet some distance from the house so that made for a long walk on cold wintry days or the dark scary nights. Everyone had a coal shanty nearby, since coal was used in winter and summer. The coal bucket had to be filled frequently, and kept handy on the back porch all year round. So the back porch was dirty almost all the time from spilt coal. It was the boys' chore to chop wood that was hauled from the woods, pile it into a box on the back porch, handy for the mother every morning to start the kitchen stove. The boys also had to keep the coal buckets filled.

Everyone had their own garden, which was usually fenced in to keep chickens, dogs, and other animals away. In fact, everybody's own yard had a fence--some were fancy picket fences, some were wood slats, and many were of plain chicken wire nailed from post to post.

And everyone had a front gate, Oh, the front gate! How we used to swing on it, until the screeching annoyed our parents.

There was no grass, so every Saturday one of the girls or the mother swept the yard in the summer; yes, it was some sort of ritual. The front porch, the back porch and the toilet got scrubbed, after the dust settled from sweeping the yard of the stones, glass and small sticks. One wonders how the glass and the stones came about. Well, the small girls would play "house" in the back yard with neighbor girls or friends. By using stones as borders, they knew which house was their house, while other stones divided rooms in each others' houses. The glass was the broken pieces of china, picked up from the back alley, which served as our "china" as we entertained our friends at our "house". The biggest pieces were plates, and they usually had the largest floral design. Little did we know that we had our most enjoyable moments entertaining our friends with broken Dresden, Haviland china, carnival glass and hand painted porcelains. What happy times they were.

Usually, every street had several cow barns in the back yard. We were one of the lucky families who had a cow; her name was Jenny. I can remember helping my mother take Jenny to pasture on a meadow next to the creek below Drabbant's house each morning. Then again in the evening, she and I would go after her, put her into the barn, feed her with store bought chop, then I'd watch my Mom milk Jenny. She said Jenny gave rich milk. I didn't know what that meant, but it must have been pretty good, because we had several milk customers. I delivered the milk each evening. It was this extra money my Mother made that bought yard goods for our dresses and brothers' shirts. I can remember when we were getting ready to move to Ohio, we had to sell Jenny. I saw my Mother cry.

Describing the interior of our houses, I'd have to say they were very plain. Kitchen floors were mostly bare wood with perhaps a handloomed carpet scattered by the sink and stove and entrance. In the sink, the spigot was raised high (no one had hot water spigots) enough for a bucket or tall coffee pot. Around the sink was a cretonne curtain, hung by a thin rope or usually store string. This curtain made a hiding place for the dishpan, scrub bucket and scrup brush. A small wooden cabinet held our dishes, while the bottom drawers kept towels, our stockings and our undergarments. These were handy because it was in this kitchen where we got our Saturday night baths, usually after supper. It was in this kitchen every Saturday night that our moms would put up our hair in long strips of rags while our hair was wet, so that Sunday morning we had "banana curls" sporting a large ribbon. In the winter time, the big round tub was usually placed beside the kitchen stove, with the oven door open to furnish heat. And the first one to have a bath got all the clean water. As each next child stepped into the tub, more water from the steaming kettle was added. And so on. Each kitchen had a tea kettle and a blue granite coffee pot on the stove constantly steaming. Every family had their own coffee grinder--no coffee ever came ground. I can remember Arbuckles brand coffee was the most popularly used. In each pound purchased, premiums were given, or packaged right in the coffee; cups, glasses, and the children used to collect the picture cards inside with historical information. Today these are a collector's item of high value.

In the summertime, fly paper usually rested on the kitchen table, and, of course, removed at mealtime. (We had no screen doors or screens till later.) With all the chicken shanties, cowbarns, etc., flies were plentiful. This flypaper was about 9 inches by 12 inches in size, and was topped with a gooey, gluey substance, giving off an odor to attract the flies. At mealtime flies were all over, so Mom would ask me to "chase flies". So with a large towel in each hand, I'd swirl and swirl everywhere and out the door. It was a useless and unending job for the flies always came back.

The curtains on the windows were usually handmade, some were crocheted or embroidered, while some were sewed from flour and sugar sacks. The luckier women, who could afford the lace curtains, ordered them from the Sears, Roebuck or Charles Williams catalog for 59 cents a pair. And every doorway had curtains, these were called "portieres". They were usually long panels, very narrow; some were made of fancy ropes and beads; some from a soft flowery cretonne with a soft sheen. Even the door at the foot of the stairs had "portieres".

In the living room a long couch was the first that you would see. The father usually rested here after supper, after a hard day at the mines. A large heavy library table, usually solid oak or mahogany, stood in the corner or along a wall. A library lamp topped this table. Every parlor had a heating stove, and it stood here winter and summer. It sent heat upstairs through a large hole in the ceiling directly above it. The arm chairs were solid--very heavy, made of wood except the padding. This was genuine leather or heavy brocaded flowered velvet. An axminster rug or a linoleum covered the floor. A 9 x 12 linoelum could be purchased at the company store for $1.98 and the colors usually lasted only a couple of years.

The two bedrooms were very plainly furnished, each with metal beds and a dresser. Clothes were hung behind doors, as no house had a clothes closet. The parents had one bedroom and the children had the other. It was the same arrangement whether families were large or small. In the winter, the bedrooms were so cold; so, every family had feather ticks for covers or heavy laps made from old woolen coats. So many families kept boarders who shared the bedroom with the children.

The mother worked hard, cleaning this modest house, taking care of the family, garden and so forth. But the laundry was the biggest chore. Monday was the traditional washday. The women removed their copper or galvanized or tin boilers from their nails outside, filled them with water and placed them on the coal stove to heat. This was their hot water system. The first boiler fill of hot water was used in the tub of white clothes, to which P. & G. soap or Star soap was cut into chips. The clothes were rubbed by hand on a washboard, then run by a hand wringer into the cold rinse water. All clothes were hung outside, except, of course, in bad weather. On rainy days, both porches had clothes hanging every which way, and in the winter it was not unusual to see stiffly frozen sheets waving in the wind, along with frozen long underwear, and the rest of the family laundry. So many mornings the mothers had to thaw out frozen spigots first, before preparing breakfast.

CHAPTER THREE

Now that I gave you a small idea of family life, I'd like to picture the little town itself, the streets, the stores, and some of the people. Beginning at the corner of Maple Street and Plank Road, the Wozniak family owned and operated a meat and grocery store (later the business was sold to Adam Antol.) Directly across the street was the Yuhas store, which along with Adam Antol's was quite convenient for the people at that end of town, Maple Street, upper Plank Road and Goat Hill which had two rows of houses. A short distance down the same side of the street was Joe Shander's store and further down in the middle of Plank Road was Vargo's store. Further down was Alex Sagadi's place of business. These four store owners were Hungarians, and they catered to the Hungarian people who settled in the upper end of Vintondale. There were Berishes, Hornats, Verebs, Yuhases, Geros, Delaneys, Gulyashes, Shadors, Monars, Antols, Vargos, Marksues, Serotskys, to name some of them.

I can remember Alex Sagadi's place well. We lived on Dinkey Street; Sagadis was directly across the rock dump from us. We were having a christening early in January 1919. My sister was baptized, and, as is custom among the ethnic people, friends and relatives were invited for a big dinner and a party. It seemed my mother ran out of cookies for the children, so she gave me money to go to Alex Sagadi's for more. When I brought them home and I bit into my cookie, a worm wiggled. I had a fig newton. Those days, perishable foods were hard to keep.

At the West end of Plank Road, Andy Jacobs had a garage in the late 20's and early 30's. And between the creek and the railroad section of tracks stood the town's social hall. This is where so many of the social functions were held: church bazaars, school dances, basketball games, summer dances and mini medicine shows.

Now going down Main Street we find the Nickelodeon, where the silent movies played and where "Skooky" McPherson played the player piano. Everyone in town knew "Skooky". He was an important person. It was he who chose the fast, fast tunes for the wild west scenes of galloping horses, and Indian fights. And he knew what piano rolls to put on the piano when there was a movie about an intense love story--beautiful, soft, low, enchanting music. And we got all this for an admission of five cents.

Next door was the company store. Oh, the company store! It was a large cut-stone building with bold gold letters "Vinton Supply Company, Ltd." above the entrances. As a child it worried me that I didn't know what the Ltd. meant, but all those gold letters fascinated me. The store was the pride of the town and its center of activity. It housed the meat market, and the grocery store, but it could be called a department store, too. Here is where one could purchase everything from augers, thread, meat, candy, hats, underwear, shoes, in fact, almost anything and everything. I'll cover the company store in another chapter. But upstairs in this same building was the company office where all of the mine business was taken care of. Mr. Otto Hoffman who was superintendent of the mines had an office here, along with Mr. Burgan, Mr. Arbogast, and of course, the office girls. In the later 20's and 30's, these are the girls who worked there: Ruth Hopfer, Marie Cresswell, Dorothy Krentzberger and Alice Fetters. Olive McConnell was the telephone operator, and Mary Kerekesh did the night shift at the switchboard.

On the outside of this store structure were two sets of stairs, one up from the alley, (the back stairs) to the landing porch, and down the side of the building to Main Street. It was this set of stairs that the men used to get their pay on pay day, as they were leaving the mines. The men would go up the back alley entrance to the landing where the windows were opened to receive each miner's statement. As each man presented this statement to an office girl, she, in turn reached into the file for his pay, and he proceeded down the front way onto Main Street. During this period of time it wasn't unusual to see a cripple sitting at the foot of these steps begging for a coin to two. This was also where some organization was selling tickets or some gadget. This was an "every pay day" routine.

Directly across from the Nickelodeon, was Krumbines'. Mr. Krumbine was the undertaker, but he also sold hardware, wallpaper, paint and such items. In one section of the building, on a downstairs floor at street level, Isabelle Krumbine was postmistress. I can remember her waiting on me. In the back area of this same house, Mr. Krumbine kept a hearse and some caskets, and also used this for storage. Some years later the post office moved down the street a few doors directly across from Roberts' store. Ruth Roberts was postmistress for many years. In a little room alongside the post office area, with its own entrance, was the town dentist's office. Dr. Cottom was what the town needed. He commuted by train from Altoona once a week.

There was a tall three-story building across the street that still stands today; but in the 20's and 30's housed Roberts' Ice Cream Parlour and a barber shop. I can't remember an earlier barber but I do remember Mr. Beechey leaving and Jim Samatore renting that space, barbering for years. Later on, another barber, Nick Burch, left Vintondale and sold Jim the small red brick building he owned. Here, Jim cut the people's hair until his death. But I can well remember that first shop. During the Depression, 1928-1932, when business was bad, no one had work nor money, Jim rented the back room of his shop for a very small fee, to a group of young fellows. They used this room for playing cards, meetings, and just old fashioned gatherings. They had nothing else to do. They called this their "Idle Hour Club". I'm sure any of the fellows living today could tell you of the great times they had there.

Now, next door, in this same building was Roberts' Ice Cream Parlour, a delightful little shop, where you could come in for a 5 cents ice cream cone, or penny candy, cigars, cigarettes, or even a few small drugstore items. Or, you could just sit at one of those round ice cream tables surrounded by four wrought iron ice cream parlour chairs, and order a banana split or a milk shake or an ice cream sundae, if you could afford it. Here you could enjoy the delicacy, and chat away with friends. Julia Brozina worked there for years. Bill Roberts who had been badly crippled by a childhood disease and was pitifully bent and disfigured, was at the store most of the time to help wait on customers. Bill was a most jolly person and he had a smile for everyone. His laugh was so hearty it gave you a pleasure to hear him laugh. He was happy with the world. Everybody knew Bill Roberts.

Going down Main Street, next is the Farkas Hotel. The tavern part remained closed for years during the Prohibition, but opened up again for business after the repeal of the 19th Amendment. This was during Franklin Delano Roosevelt's term of office. After Mike Farkas died, his widow operated the business for years. Next door was the Nick Burch Barber Shop I mentioned before. Then, Beerman Store is next. They sold clothing, dry goods, shoes, a little hardware and carpets. I can remember Mr. Beerman being all business. He always seemed so serious, but his wife was most pleasant and attractive. Next door was Paul Graham's meat and grocery store, the next door was Butch Morris' meat market, although I think he did sell a few fresh fruit and grocery items. As a child I can remember seeing huge blocks of ice being placed in sawdust in the upper area of the walk-in-cooler, here at Butch Morris'. They explained to me it was the same ice that was frozen in the creek in the winter. The ice plant at the mines also furnished these meat stores with ice.

Now in one of these spaces our town actually had a bowling alley after one or both businesses left the town. Then I can remember Mrs. John Burch having a restaurant here for a few years followed later by Rafases Tavern.

Next door was Louie Rose's, a large wooden frame building which also is not standing today. Louie rose was a small man in stature, but he did have a business head. His store was a department store. He clothed the whole familiy and furnished the needs of every household, with such items as dinner buckets for miners, chicken wire, furniture, dishware, curtains, blankets and carpets. I always respected Louie as a child, and everytime I saw him I'd greet him with a pleasant "hello". Louie was a most generous person, though. He helped so many people out by giving them credit, and their payments were extended out into years. Louie always owned an expensive automobile which he called a "Puick" and he used to enjoy going to the movies. When the Nanty Glo Theatre got "talkies" he'd often take someone along with him. And often he would be accompanied by the office girls from the mine office. Now Abe Solomon worked for Louie for years, and the Solomon family was well liked and respected. Also working for Louie was Anna Kanich after she graduated from high school.

Directly across the street was the Sam Williams

building, with the store on the street floor, and the family on one side of the upper floors, and a church room on the other side. Mr. Williams delivered groceries in the late 20's with his horse and Kramer wagon; so the name Kramer stayed as a nickname some of his classmates gave Lloyd, one of Sam's sons.

Across from Williams on the same side of the street--corner of Main and Third Streets, stood another tall building where they bottled soft drinks, and if I'm not mistaken, beer also. I can't think of the name of the owners, although the name of Labelle comes to mind. In its place, now stands a red brick building which Joe Hassen built.

Across the street, nestled among tall maple and poplar trees, stood Dr. MacFarlane's office. He was the company doctor for many, many years. He not only healed the sick, but he delivered hundreds and hundreds of Vintondale's babies right at their homes. He was paid by the company, then in turn the company deducted one dollar each month from the miners' paychecks to pay his salary. So in turn, any member of any miner's family could go to the doctor's office any time of the day, or call him anytime of the night. Dr. MacFarlane left town in the early 30's, Dr. Jerome Cohen took charge of the office, followed by Dr. Philip Ashman.

Next door to the doctor's office, was a store owned and operated by Gaal's followed by Nevy Brothers' ownership. Nevy Bros. sold meats and groceries, Italian speciality foods, but they did carry other merchandise for the family and home. This building was another of the several three-story buildings in our town. The David Nevy family lived in the upper floors, above the store area. While David and his brother, Ralph, spent most of their time in Cumberland, Maryland, managing their macaroni factory, Louie and Charley managed this local store. Then later, Joe Pioli and Frank Pioli operated the store for years.

On the same side of the street further down was Sam Freeman's, another clothing and department store. My Mother and Dad used to buy almost all of our clothes from Sam in the 1920's. My Dad has great respect for Sam, a tall, gentle, gray-haired man. I can't remember if he died in Vintondale or whether he left town. Guerrino Averi bought the building and store and operated a grocery store here for years.

Next door our new bank building stood. Mike Mehalik was the first bank manager. I can remember how proud the town was to have this new building and a bank of our very own. I can also remember that Mr. and Mrs. James Roberts and son, Jimmie, were the first tenants to rent the upper floor.

Alongside the bank stood our prominent Vintondale Inn. Three stories high, in its day, it was a beautiful building with ivy growing on its outside walls, and tall trees providing shade in the front of the structure. The front porch stretched along the complete front of the building, and in the summertime, wicker chairs and setees were arranged here and there for the comfort of the tenants. It was a pleasure to sit here and watch the townspeople pass by. This hotel accomodated travelers who stopped off at the train station, mostly salesmen or suppliers who came once a week or once in two weeks. These people sold to the merchants and also to the mines. Some stayed only overnight, but some were steady guests here. So many time, peddlers came to town via the train, and carrying their wares on their backs, they made house to house visits selling to the housewife. If these men, mostly Jewish, made friends, they would be invited to stay at a home. But mostly they, too, were guests at the Inn. Years later, as the Depression set in, business was bad, guests were fewer and fewer, the school teachers moved into the Inn and rented rooms from the Coal Company.

Upon entering the Inn, the large foyer or hall was quite attractive, with its leather covered furniture and desks. Off to the right was a reception room, with dropped windows and large ferns cascading in their wicker ferneries. The furniture was a bit more elegant, upholstered with a tapestry type of fabric. The sofas and chairs were placed in such a manner that the room made reading the newspaper for the visitor a comfort. And when a visiting dignitary, such as a Schwerin or some other mine official stayed at the Inn, this room was held for business meetings.

On the other side of the foyer, was the large dining room. I've been told that as many as a hundred people could be served at supper at one setting. The large kitchen had an enormous sized stove, work tables and sink. I've often wondered whatever happened to all this equipment and furniture after the hotel closed. And, do you know, I can't remember when this Vintondale Inn was torn down.

I must call to attention, too, that the heat for this Inn was provided by the steam plant at the mines. It was projected via overhead pipes from the mines to the Inn, to the Company Store and offices, and on to the Social Hall.

Beside the Inn stood our new theatre. It was built after the Nickelodeon was torn down. Our own Vintondale Theatre! What joy and pleasure that theatre brought to its town people. In the late 20's and early 30's, for a dime, you could see Tom Mix in westerns like, "Riders of the Purple Sage", with his wonder horse, Tony. Or you could see Hoot Gibson and Buck Jones who were also our cowboy heroes. We'd see Charlie Chaplin, Marie Dressler, Wallace Berry, Mary Pickford, Our Gang comedies, and even comedians such as Chester Conklin, and Steppin Fetchet. We'd see a segment of the world news, showing the trade mark of Pathe--the news people--the cock crowing. We'd see comedy shorts and also a chapter of a continuing serial, every Saturday after Saturday. The child stars that were popular were Mickey Rooney, Jackie Cooper with Shirley Temple. When the "talkies" came into town, Skook McPherson didn't need to play the piano. So he acted as assistant manager, taking care of getting the films in the mail, returning the features that had just recently been shown. He installed posters on the outside marquee and the glass cases inside the entrance hall of the theatre. He did the custodial duties and even performed as an usher. Many times he even had to "shoosh" the children in the front rows when they became too noisy. Paul Smidga was the cameraman who projected these movies and my Dad helped him a few years. Now this theatre had a stage which was used for the Junior and Senior class plays, for the Senior commencement exercises, and later on for talent shows. It was later used, too, for bank night drawings, with the large cage on stage, some child would draw a number and money was presented to the winner. It was a sad day when this theatre closed its doors. And I can't remember when that happened, either.

A few yards below the theatre was Mr. Hubner's shoemaker shop. For a half dollar you could get your shoes soled and heeled. In those days, you grew out of your shoes before you got new ones. They were repaired over and over.

Down the street further on was the Baptist Church. It was in this Baptist Church where baccalaureate services and even commencement exercises were held before the new theatre was built.

Beside the church was a tall building that had once been a school building before the Delano school was built on top of Fourth Street. The place stood empty for years. In the 30's it was divided into apartments. Two doors down was the Balog Bagu building, where Mr. Bagu sold meats and groceries. This was quite convenient to the people in the lower end of town. Further down the street was the Hungarian Reformed Church, which in the 20's and 30's had a sizable parish. I can remember Reverend Hunyady serving the people and I can remember when he brought his bride to town.

Passing across the bridge, the next visit is the P.R.R. Station. Mr. Clarkson was the agent. His job was to sell passengers train tickets, take care of the dispatches using Morse Code, take care of incoming and outgoing mail and small parcels. Adjoining the station was the express office. Mr. Lockard was here to take care of the large parcels of express, mail orders of bulk, and he even delivered these items. In the 30's Mr. Clarkson was replaced by D.P. Dutra at the station. Later express business discontinued as the train stopped coming to town. I can so well remember the trains, the train whistles annoucing the 7:30 a.m. and 10:30 a.m., 4:00 p.m., and 7:30 p.m. trains. The 7:30 a.m. and the 4:00 p.m. were eastbound toward Cresson, while the 10:30 a.m. and the 7:30 p.m. were westbound towards Wehrum.

I can remember especially well the morning train; several times I boarded the 7:30 a.m. train to Nanty Glo! One time I went on the four o'clock train and came back on the evening train, just to go for an Easter hat in Nanty Glo. The fare was 25 cents roundtrip. The hat I bought was a pretty straw thing, with a cluster of bright red cherries on the side, trimmed with a red ribbon. I was so proud that I did that all by myself--ride a train and get to choose my own hat. What an experience!

Our train was a link with the rest of the world for all of us. It brought us visitors from other towns, it brought us our newsapers, our mail, our orders we received from Sears, Roebuck or Charles Williams catalogues. It also brought the priest in from Twin Rocks every Sunday at 10:30 to say the eleven o'clock mass. I can still picture one of the Rager men meeting Father Howard, then Father Hebner, and later Father Charles Gallagher at the station on Sundays. This train brought business people to board at the Vintondale Inn, or at Bittorf's Hotel across the street from the station. It was a link with other towns such as Wehrum on down the line a couple of miles. When railroad travel began to dwindle, with the oncoming of an automobile, I can remember a one car passenger coach would arrive at 7:30 in the evening, and people named it the Toonerville Trolley after a popular comic strip in the newspapers. They also called it the "Doodle Bug". The train schedules were usually accurate. You could almost always check your clock with the train whistle and they'd coincide.

Now, Fred Bittorf's Hotel was across the street from the station. It, too, was a three-story building, with a front porch the whole width of the structure. Of the two entrance ways, one went to the tavern area, and the other into the hotel reception room. Fred's business prospered until the Depression.

Down the street on the same block was another hotel called the Red Hotel for it was painted red. I don't remember who owned it, but I do remember of them having a small store in the front part, where they sold grocery items and candy. It was very close to the baseball park, so customers, mostly children, would stop on their way to and from baseball games to purchase their candy or ice cream cones.

Down the street, at the end of the next block, was Rosaches' store, another small store, and quite convenient for the people at that end of town. Down a few doors, and the last house at the west end of town, is where Mrs. Clarkson operated a small restaurant in the late 20's and 30's. She had converted a large living room into a dining area for customers. She called her place the "Dew Drop Inn".

I must also include our churches in describing the town. Our Catholic church stands on top of Fourth Street, and the Greek Orthodox is on top of Third Street. But in the 30's, there was a church called "Christian Endeavor" on Jack and Mary Shaffer's property. So recounting the churchs mentioned before, we had seven places of worship.

Our school buildings, located on top of Fourth

Street, served elementary and high school until the new high school building was built in 1927.

I have given the reader a general lay out of our town, and small details of the operating businesses. They flourished and progressed until the Depression came along. Little by little, the smaller ones disappeared, then later the company store closed its doors one Good Friday, never to open again. Everyone was shocked at this. All the Easter hams, kohlbassi, and special foods spoiled: That's how sudden the news of the closing came. The buildings stayed empty for years, although the office space was still being occupied. The one night this beautiful building caught fire and burned to the ground. It was quite a loss, a lot of important mine files and papers were lost.

CHAPTER FOUR

Whether one grows up in a large city or a small town, every person has some memories, good or bad, some faint, some very vivid of the happenings of one's childhood. Living on Dinkey Street, I can remember of a horse-drawn wagon going up and down the street. It was the junkman yelling "Rags, bones, paper, silver", all in broken English, for he was a Jew. All the children would run with their junk they saved since his last visit. They would get a penny, maybe two pennies, and would be so happy with those coins. Some of the things we saved were balls of string--string used to wrap groceries in the company store. We'd roll this string into balls. We'd pick up lead wrappers from chewing gum and cigarette packages off the street and roll this into a lead ball, too. It didn't matter whether the ball was small or large, you would still only get a penny from the junkman. So we wised up and made smaller balls and had more. I can remember the scissors grinder coming to town a couple of times a year, someone who needed scissors sharpened, or a knife or a saw. The men had axes and hatchets also to be sharpened, so a visit was quite profitable. Pots needed soldered and wash boilers needed repaired.

I can still remember, as a small child, the pretty display windows in the company store and how fascinated I'd be, especially at Christmastime when all the toys were on display in the middle window. And how my Dad would send me to this store every Saturday with a nickel to buy his two "Philadelphia Handmade Cigars". It was his only luxury. And he gave me the paper ring wrapped around each cigar, and I'd wear it on my finger so proudly. It was my only jewelry.

I can still picture the Hurdy Gurdy that came to town a couple times each summer. We'd hear the music way down the street, and we'd wait anxiously for it to come to our street. We were not allowed to go off our street, except by special permission, and we obeyed our parents. What seemed ages, finally the wagon came. It was boldly decorated, with pictures of girls with flowing hair and colorful gowns. Some wagons were painted with pictures of wild animals, and reminded us of a merry-go-round at a carnival. Some of the wagons had a calliope type, shrill sounding music coming from its pipes. But whatever wagon came around, it usually had popcorn for sale, or peanuts and sometimes ice cream. Sometimes he'd have a parrot that really fascinated all of us. The man used a stick, held it to the parrot and the bird jumped onto this stick, he'd pick his beak into a box of cards and for a penny you had your fortune told. To have a fortune told was a happening! And we believed everything that little card said!

I can remember medicine shows, and the exciting carnivals at the baseball field. The whole town was in anticipation. I can recall my Dad winning a kewpie doll at one of the stands and my Mom placed it in a prominent display on the library table in our living room. A kewpie doll (I didn't know that as I do now) was an invention of Rose O'Neill, and she became a popular item and a big hit in the early 20's because of her winning smile, fat cheeks and roly-poly belly. So the doll was not a plaything; it was a display item. She was usually dressed with a full skirt, trimmed with tinsel or ribbon. The doll itself was made of chalky plaster of Paris, very delicate, so it was easily broken. We children used the broken pieces for chalk.

And, growing up, we played the children's games of Hide and Seek, Blind Man's Bluff, Follow the Leader, Run Sheep Run, Tisket a Tasket. As we grew older, we girls jumped rope while the boys played marbles. Sometimes the girls played a marble game called "Piggy in the Mud Hole", where we made a hole in the ground with the heels of our shoes, and from a distance threw marbles toward the hole. If you got a marble into the hole, you got another shot, each time shooting with your middle finger. The last one who got the marble into the hole, got to keep all the marbles. We'd keep our marbles in cans, and as we would add to our collection, what few we had, each evening counting our marbles. After this inventory, we'd know if we were winners or losers.

In the winter the boys played "shinny" with a smashed tin can and a stick. It was their version of hockey. But the summer games were the best of all. We hated to hear our parents call us home at the end of the evening and spoil all our fun. I know the adults were glad when the games were broken up. I'm sure they were tired of hearing "applies, peaches, pumpkin pie, who's not ready holler I". The "it" post was usually the corner lamp post. It was so much fun, we were so happy, no worries in the world. We never stopped to think about all the foul air that covered then whole town.

Gas fumes from all the coke ovens that burned brightly all night and day long, added to the fumes coming from the rock dumps, polluted the air. We lived in this day after day, month after month, year after year. We never heard the word "pollution", yet lived right in the midst of it. We thought it belonged, as a part of life, and we lived with it, endured it, never knowing of its harm to the body.

Part of growing up in Vintondale was the thrill of going to the company store. This store carried popular items such as Arbuckle's Coffee, Gold Dust washing powder, Star soap, Pond G. soap, Old Dutch Cleanser, Camel Cigarettes, Bon Ami, and so many brands that have disappeared from our way of life like CruBro and Premier.

In 1924, when I was in sixth grade, the Coal and Iron police were the law of the town. They tried to stop miners from unionizing. This is so vivid in my mind, the bad word of "scab" given to the miner who went to work and ignored the ones who wanted union. Also vivid is the memory of how so many people moved out of town by force, sometimes their furniture was stacked out on the street as they were forced out of their homes. This was the time the Ku Klux Klan took liberties to burn crosses, mostly on top of No. 6 Hill. But one night they burned a cross on top of No. 2 Hill, very close to where we lived in a Jacob's owned house, back of Plank Road where we had recently moved. I can still hear the explosive boom, which preceded the burning of the cross. The ground shook bringing fear to the whole town. Then the cross began to burn right above the tipple behind the social hall. All of us children cried, we had experienced a fear we would remember all our lives.

Growing up, I can remember the park on the upper part of Second Street which the coal company had built for the townspeople. It had a bandshell, a circular shaped structure with a roof. On special occasions, bands played for the enjoyment of adults. But the children had another form of entertainment-athletic. There were swings scattered around the park, also a maypole, see-saw, sliding boards and more. It was a lively place, green grass and flowers; but as the years went by, it became neglected and badly run down. So this lot was available for the new high school buildings. This was especially beneficial to me for in 1925 we moved to Second Street, and this meant just a short walk to school for me.

When this school was dedicated and opened in 1927, I was in my Sophomore year. Our class was the very first Sophomore class in the history of Vintondale, and when our class graduated in May, 1930, we were the first four-year high school class. Before this our high school was only called three year high. To finish their education, former graduates went by train to Ebensburg to complete their education.

CHAPTER FIVE

When something happens in a small town everyone becomes aware of it. When a wedding takes place, there's excitement in the air, because a wedding in Vintondale is a rarity. As a rule, a young couple would "run away" to Cumberland, Maryland to be married, where proof of age or parental consent were not needed. So when the young couple, newly married, came back home, all the kids had a shivaree. A shivaree is what is called "serenading" of the newlyweds. The kids would gather tin cans, buckets or even small tubs, pound them with a stick, the louder the better, continuing this racket until the young couple had to come out of the house. This was a sign to acknowledge the serenading, thanking the children, then treating them sometimes by throwing candy kisses up in the air to those outstretched hands. Sometimes chewing gum or pennies were thrown and there was so much grabbing. The penny grabbers would run to Roberts' store to spend their pennies on candy.

I can remember one instance exceptionally well. Harry McCardle was our town policeman, and he married the third grade teacher, Miss Delevette. Upon their return from their honeymoon, on the 7:30 train one Sunday evening, they went to the house where they were going housekeeping on Third Street. This house was formerly the Blewitt house. And we were ready for them! All around the large lawn was a picket fence and we were busy all afternoon hunting up tin cans from the back alley, and placed one tin can on each picket. When a few minutes passed after the newlyweds' arrival, we started our serenade. What a racket we made! And what fun we had making this racket. It was well worth it, because we got a treat from Roberts' Ice Cream Parlor. Everyone os us took turns at the candy counter showing what we wanted. Then, of course, the adults had their way of serenading the couple. They took the newlyweds for a ride in the old fire engine up and down Main Street and Plank Road. While all this was going on, never in my wildest dreams did I imagine that in 1935 when Joe and I were married we'd go through all this too. Sirens were blowing, everybody on their front porches watching! The firemen hung a chain swing on the back of the firetruck, placed us on the swing, then drove us up and down. I was so embarrassed, they just "took us" for a ride. Later, the men were given cigars much to their pleasure. After the children serenaded us in their favorite manner, Joe sent them all to Roberts' store for an ice cream cone and told Julia Brozina to keep the "tab", then he paid later.

But, back to the children's fun, we had another kind of fun, called people watching. From 1920 to 1933 our country had Prohibition. So, at our town being no different from any other town, some people made moonshine whiskey and sold it. These homes were called "Speakeasies". So it was nothing unusual to see a drunk staggering on any street in town. And the kids would follow him, too, as if they were playing "follow the leader". He'd turn around and with his forefinger pointing, he'd try to "shoo" the kids away, but as soon as his back was turned, they were marching behind him, mimicking his staggers and of course, laughing too. Sometimes a drunk would get violent, beat up his wife, or maybe another man. He'd be placed in the town jail. The jail was at the foot of Second Street in back of the company store. The drunk was usually let out the next morning after he sobered up, no fine or no charge against him. But the experience was embarrassing for him because the whole town knew who was in jail that night.

People watching was a game with us. Peddlers came in town with their good strapped to their backs, and farmers came to town in the wagons huckstering produce to townspeople who didn't have gardens. And we would follow these wagons hoping the farmer would toss us a bruised apple or pear. That farmer was our "best friend" if he did. The housewives looked forward to the farmers' visits because they liked to bargain and make deals with them. And, then too, we would follow the ice man when he made ice deliveries twice a week to the people who were fortunate enough to own an ice box. And we hoped too, this ice man would throw us a chunk of ice.

And then we'd watch a certain group of men at work, who were called "honey dippers". About every second or third summer, the coal company contracted to have the outside toilets cleaned, and this was a job usually done by 4 or 5 Negroes. This was the only time we ever saw Negroes, so it's no wonder all of us kids thought that toilet cleaning was the only thing Negroes did. But we'd follow them up and down the alleys as they worked from one to the next, cleaning then spreading lime around the area just worked on. We didn't mind the smell, for we had all kinds of strange odors from rock dumps and coke ovens. We lived in pollution, never even having heard that word.

Now, a short description of our outhouses, shows the interior with a seat with two holes. Families with small children had a lower a seat with a smaller hole. Every toilet had an outdated catalogue to be used as toilet paper. These catalogues were our department stores, and when the new issue came in the mail, these old ones were put to a different use. You could almost always tell who was the last one "in" by what page in the catalogue was open. If the page showed toys such as trains, etc., the small boys were in, if curtains or sheets or utensils showed, the mother had the last use, or if the pages showed girdles, ladies underwear, ladies hoisery, it was the older boys or the father of the house who enjoyed viewing those pages. And that's the way it was.

And sometimes we'd go up the hill for fun. One summer, maybe when I was 10 or so, and the elderberry blossoms were out in full bloom on the hill above Dr. MacFarlane's house, a gang of us girls would gather armfuls of these blossoms and have a "pretend" wedding. We chose a bride and groom, then we strung blossoms on a heavy cord for the bride's tiara or crown and she carried an armful of these flowers during the wedding procession. The rest of us girls--perhaps eight in all, would be bridesmaids and we also carried big armfuls of these yellow-white blossoms, and proceeded with the procession . . . "Tum-tum-ta-tum, Tum-tum-ta-tum", as in a real wedding. How we cherished those happy moments. Afterwards, we danced and had a pretend banquet. One day in the midst of our joyful celebration we were interrupted with a "Ha, ha, ha, ha"" from a nearby tree top. Without us knowing, there were two Main Street boys watching us. They climbed down from the tree, and one of them had a box camera, and told us their camera "Can see right through you." Frightened, all of us ran different directions and headed homeward as fast as our legs could carry us. For years, I had a dislike for those boys, even though later I found out their camera had no film nor could it "see right through us."

As we became teenagers and went to high school, we had our weiner roasts at Mule Field, and sometimes at Devil's Elbow on the road to Wehrum. We didn't mind the long walks, because a weiner roast was a happy occasion. Even adults had their weiner roasts at these locations. No one ever heard of a weiner roast in one's back yard!

And sometimes, a gang of us girls would go to Burches on Sunday evenings for they were the only ones nearby who had a player piano. And we'd sing those tunes to the top of our voices following the words written out on the piano roll. We heard later that neighbors really enjoyed our singing.

And the adults had their games too. Once in a while there was a birthday party, or a bridal shower, but quite often the ladies had a Larkins party. Larkins was a firm that sold household items such as vanilla, spices, toiletries, talcum powder, along with other useful household items. The hostess would get a premium, usually a set of dishes or a lovely piece of silver, and guests could add up their coupons also toward china and so on. The Larkins pattern on china was very pretty, dainty pink roses. Today's collectors of these dishes find them priceless.

Sometimes the ladies held stork showers. The guest of honor wore her best Hoover apron, which was the popular maternity wear of the early 30's.

And there were times when the men were invited to those parties, so they too participated in adult games such as "Spin the Bottle", "Postoffice", "Pitching Cherries", "Poor Pussy", and also the popular "Musical Chairs".

In the mid 30's a group of men formed their own little band called the "Little German Band" and had their own parties. The "oompah-oompah" music was quite delightful and entertaining.

Meanwhile, the ladies had their quilt parties, mostly during the winter months. There was always a quilt in a frame at someone's home, and they all helped each other at the same time they enjoyed the togetherness.

Those were the years--the 20's, when we were learning new things and building our characters. At home and at school we learned to respect others; we were taught manners and etiquette. The boys tipped their hats to the ladies, following examples of their fathers. We were taught to say "yes, sir" and "yes, ma'am". We learned our basics from stories in the McGuffey's Readers at school. And, from newspapers and also the Pathe News at the theatre we learned about people living in other parts of the world. In 1927 we were all so happy at the wonderful news of Charles Lindbergh crossing the Atlantic Ocean. On May 20, he landed in Paris after taking off from New York. Everyone in town talked about this achievement, and such excitement was repeated over the world. We remember well, also, the summer of 1928, when Herbert Hoover was running for president, his motto, was "a chicken in every pot, and two cars for every garage." But in Vintondale, there were few garages.

In the early 30's our town had a good baseball team--a winner. If a young man came to town to the mine office looking for a job, and he could play ball, he was given a good section in the mines and he earned more money. That's how highly respected baseball players were. So every Sunday afternoon, practically the whole town went to the ballgame. We cheered and rooted for the home team! Sam Butala was our best pitcher and we really cheered him. Some of the other players were Joe Lonetti, John Risko, Enoch Hallas, Andy Chekan, Carl Michelbacher, and Charley Butala. Our rival team was Wehrum, but teams from as far away as Blairsville came to play here. After the game was over, you'd see a parade of people walking uptown to their homes--to relive the game again.

And our town made itself a name from the accomplishments of our basketball teams. Mr. Mower, upon his arrival as principal of the schools, built and developed strong and fighting boys' and girls' basketball teams. An old newspaper clipping dated February 12, 1929, headlines read "Vintondale High Wins From Bolivar Tossers--Girls Take 14th Straight Victory of The Present Season". The score was 34-0. The lineup was: Michenko, S. Sileck, Kaltriter, A. Huth, V. Daly, Morris. Substitutes were: A. Balog for Michenko, Bepplar for Kaltriter, McPherson for Huth, Bracken for V. Daly, Swanson for Morris. The boys won that night 21-15. The lineup was: Jendrick, Frazier, M. Nevy, E. Nevy, Morey. Referee was Lee Cresswell.

In 1929 the girls varsity first team players were Florence Kaltriter, Eleanor Swanson, Mary Balog, Virginia Daly, Viola Bracken, Harriet McPherson, Anna Sileck, (Agnes Huth), and Anne Minchenko. The 1929 boys team players were: Merle Nevy, Roy Kuchenbrod, Jack Frazier, Robert Morris, Lloyd Williams, Ricco Bonetti, Ed Nevy, Bill Jendrick, Julius Morey, George Hozik and Geno Davalli. As the years went by, the Vintondale High School basketball teams kept winning games again and again up to a point where the girls' accomplishments got a nice write-up in the Pittsburgh newspaper sports section. Basketball put Vintondale on the map and the whole town was proud of the teams. The folks really patronized their home games with so much enthusiasm, and many followed the teams to their championship tournaments at St. Francis College.

CHAPTER SIX

The coal company built its townspeople a large social hall, located at the west end of Plank Road alongside No. 1 tipple and railroad siding. The creek ran alongside this building. In the 20's and 30's, indoor shows were held here such as meets, side shows, civic and church affairs. It was some big social affair when the Catholic church held their winter bazaar. There was dancing, good food and refreshments and all sorts of raffles and wheel of fortune games. It was a nice gathering place for old friends and their familes. Then every fall, the Hungarian Reformed church and the Hungarian people held a grape dance. It was really a harvest dance, following a custom brought over from the old country. Crates and crates of grapes arrived at the train station express office, and workers spent all afternoon tying ropes or wire across the large dance floor of this social hall. They then tied bunches of grapes over these ropes. It was a beautiful sight to see these blue, purple and red grapes dangling above the heads of the dancers. The young man chose his partner, she chose the bunch of grapes, he paid for them and only then would they dance. Hungarian gypsy orchestras would furnish the music, and they featured violins. John Morey from our town played at many of these grape dances. He was a violinist. The violins made beautiful music and the dancers would glide merrily around the hall. All Hungarian dances are lively and rythmic, but the Hungarian Czardas is a grand sight to watch. I can still picture Mr. and Mrs. Steve Neggie, both of them somewhat heavy, yet they'd dance so gracefully to the Czardas, it was a pleasure to sit and watch them. The younger people learned how to do this dance by watching the older couples. This was a dance the whole town waited for. Most came to dance, but many just to watch. Everyone asked everyone else all day long, "Are you going to the dance tonight?"

And every May this social hall was gay again with the whole town attending the graduation dance. Right after commencement exercises were over at the theatre, practically everyone went to this dance. Those who didn't have admission tickets or who couldn't afford a ticket, got to watch the dining from the outside when several large doors were open. I can remember my own graduation in 1930. For the commencement exercises I wore a lovely all white crepe dress that my Mom bought at Glosser Bros. store in Johnstown for the priceof $1.98. But I had a more expensive dress for the dance, an orchid colored flowery chiffon with two tiers of ruffles afterward, this was my Sunday dress. My commencement dance was a grand affair, and everyone had a memorable time. We danced to the tunes of "Stardust", "Old Spinning Wheel", Red Sails in the Sunset", and othe popular songs of the day.

I must mention that after our commencement, our class went by automobiles to Washington, D.C. for a glorious week. For four years we worked and saved for this with raffles, bake sales and even a dinner at the Vintondale Inn. On this trip we were privileged to see Herbert Hoover on the balcony of the White House reviewing the Memorial Day parade.

All year round, some organization or group held dances, sometimes there was a dance a week. In the 30's small dance bands became well known, many starting at small town halls like our own. They imitated the well known big bands that were touring the country, they copied their sounds and rythmn styles. Some had vocalists who imitated the singing of Rudy Valee. We were all singing and dancing in 1930 to the tunes of: "Dancing with Tears in My Eyes", "Something to Remember You By", "Embraceable You", "What Is This Thing Called Love?", "Beyond the Blue Horizon", "Three Little Words", "Stein Song", and "Sunny Side of the Street" By now such popular tunes that were outdated from the previous year were "Sunny Side Up", "Pagan Love Song", "Am I Blue", "Tip Toe Through the Tulips", and others. And the bands across the country were popularizing more hit tunes such as "When the Moon Comes Over the Mountain", "My Silent Love", "Good Night Sweetheart", "Peanut Vendor", "Dancing in the Dark", and "I Love a Parade".

During these years young men and ladies from other towns came to our dances, became acquainted and there were courtships. Those in love danced with no one but each other. The girls were dressed in their best clothes, and their "hair do's" were almost perfect. This was the era of the "bobbed" hair look. Later came the Croquinoles of the beauty shops. Then the girls learned to wave set their own hair, first with spit curls using lots of soapy goop, then later with the miraculous invention called "wave set". It made a change in hair styling. The girls did each other's hair, a process they called "fingerwaving". So it was with these fingerwave hair styles that the young girls made "eyes" with and flirted with the young men at the dances.

And so, as all things pass by, so did the passing of the old social hall. One hall burned, another replaced it, but it was torn down and dismantled. The buildings are gone, but the beautiful memories linger on. They cannot be taken away.

CHAPTER SEVEN

Like all other small coal towns, Vintondale had its share of Europeans, who left their families and came to a new world. They brought over so many of their customs and traditions. Some of these things I can still visualize. I can remember at Eastertime the special foods prepared, placed in a basket and taken to church Easter Sunday to be blessed by the priest. There was the braided Easter bread, dyed eggs, horseradish, ham, kholbassi, a special cheese, and beets. Every food had a meaning and was a symbol. I didn't know what they meant but our parents did. Then every Easter Monday was "Ducking Day", when the young men went out to sprinkle the girls. It was a tradition among the ethnics that Easter meant a new spring, a new enthusiasm, a new vibrance, so this was a first step to courtship. It got to the point that for young people, Easter Monday was more important than Easter Sunday. As the boys went out to find the girls, the girls tried to hide, but usually they were urged by their parents to come out of hiding. It was really a honor for a young girl to be sprinkled. The young men who could afford to buy perfume, mixed it with water and carried this in bottles. Others would make a mixture of talcum powder and water and that was their perfume for sprinkling. And so it passed, every Easter Monday morning the young men in their finery, made their rounds sprinkling and exchanging Easter eggs.

Another custom was to marry within your own nationality. Slavs married Slavs, Polish married Polish, Hungarians dated only Hungarians, and so on. Once in a while some couple would break tradition and marry out of their nationality, then had many a finger shaken at them. A marriage out of your own culture was a "no, no".

The ethnic people brought over their food specialities. It was not unusual to see a mother make the sign of the cross with the knife before cutting the first slice of a freshly baked bread. This gesture meant a thankful and grateful sign to our Lord for giving them this loaf of bread. The special foods for special holidays are such a nice memory. Each mother had her own speciality, and my Mother's was raised doughnuts. She always made more than we needed, and she shared them too.

The Hungarians had their grape festival and Christmas customs. They had their bacon roasts and hog roasts, and they invited neighbors to help celebrate. And the Croatians and Serbians from Sixth Street had lamb roasts and made parties from these affairs.

Then at Christmastime, the Russian Orthodox people would celebrate on January 7th, with their traditional midnight mass, ringing their church bells, chiming and ringing and echoing again. Then on the afternoon of the seventh, some men of the parish would walk all over town visiting every Orthodox family wishing them glad tidings of the holiday. One man was dressed as St. Nicholas, others wore white robes and they carried a manger set. They sang Christmas Carols at these homes, then drinks were passed to the carolers, as messages of good cheer were exchanged. Before the day was over most of the men were quite full of "good cheer".

Another tradition was the Catholic blessing of the homes after Epiphany. The priest went to all the homes, blessed them and marked over their door 19+C+M+B+30 (or whatever the year was). These were the initials of the Three Kings, Caspar, Melchior and Belthazar.

And then there was Green Sunday, which was actually Pentecost. The father of the family went out to the woods late Saturday evening and cut down green leafy branches from young saplings and nailed them to the posts of the front porch. This was a symbol of Christ's appearance to the Apostles, a sign of new life, a new spring, so the branches signified a fresh start and a new beginning.

Another custom was the traditional fall butchering, popular mostly with the Hungarians. I can still hear the pigs squealing as they were being stabbed. The men made huge fires outdoors, to singe off the hair. Then all helped each other in cutting up and sectioning the pork. Most of them had smoke houses, or shared with each other. They made their own pork sausage, seasoned their own hams and bacon, and made their customary "hurka". This was a rice sausage using the blood and the organs of the pig. That "hurka" was really delicious. Many times I watched my Mother stuff casings with sausage meat and this rice combination using a sausage stuffer. Some of the people cleaned out the intestines of the pig and used them for casings, but mostly they purchased the casings from the company store butcher shop.

CHAPTER EIGHT

Every town has special people, these people have either done something out of the ordinary for the citizens of this town, or have left a special imprint on them for their achievements. Some were just there doing their duty. One of these was Otto Hoffman, the superintendent of the mines. He was held in great respect. The miners liked him and respected his discipline; he was their "Boss". The women liked him for many reasons, he was fair in helping them find a house. People moved around so much those days, and the women always wanted a better house. The Hoffmans lived in a large house on the upper end of Second Street. To us it was a mansion, with large lawn, fruit trees, garden and gardener, tennis court, stable, and plumbing indoors. They had a son, Kenneth, who was tall as were his mother and dad. Mrs. Hoffman was a lady, she carried herself proudly, walked with an "air". Hoffman's gardener was Mr. Jendrick. It was here that a bee sting caused his death.

Across the street lived Dr. MacFarlane. The Doctor spent most of his time in his office on Main Street. Office hours were all day long and every evening. He delivered hundreds of Vintondale babies at home. He served the people for many, many years. He was paid by the company, which in turn took one dollar a month from the miners' pays for that service. This was the only fee the doctor received. I can remember the spinster lady who was their housekeeper. Her name was Anna Bell, and she was a friendly and good person.

There was Jack Butala, the town policeman, a strict disciplinarian. When he coughed, or cleared his throat, children knew he was nearby and so they scattered homeward. When night time came and the quarter to nine curfew whistle blew at the mines, kids were off the street. If not, Jack Butala made sure they were scampering home.

The movie was always over at 8:45 and the kids went straight home, there was no loitering. Our town was a good town, never any crime. Maybe once in a while a drunk was put in jail overnight to be sobered up, and once in a while at a dance, some young men would get hot tempered and have a fight. Jack Butala was there to separate the scrappers, scold them and shame them into leaving. Sometimes, too, there'd be a fight on Sixth Street, over a card game or an argument at a speakeasy. Fistfights produced bloody noses and Jack was called to quiet the men. And of course, there were the domestic fights when our policeman was the peacemaker.

And I have to mention Lawrence Hoffman. Everyone called him "Huffy". He was an all-around handy man, delivery man, ice man, errand boy for the company. He had a room in the company barn across from the company store, on the lower end of Second Street. "Huffy" took care of the company horses and wagon. He delivered coal to homes, furniture, in fact everything. People moved on, so it was "Huffy's" job to move them free. Some families moved twice or more a year. The company paid his wages, and for this he worked 14 and 16 hours a day. When freight arrived at the railroad station, he delivered it. When mine supplies arrived, or plumbing repairs, he delivered them. "Huffy" was a jolly well-liked person, but I remember him mostly from his fantastic vocabulary of cuss words.

We also remember Dan Thomas, the town justice of the peace for years. Thomas' had a large family. Then there was Perry Wilson, the store manager after Mr. Garrity left. Then was Jim Moyer, the company store butcher before Joe Pluchinsky came to town in 1929. And John Barclay was assistant to the manager and handyman at the company store. John trimmed the windows, stocked shelves, did repairs and sometimes delivered orders. Then there was John Lonetti, whom everyone knew as "Blues". He was the policeman after Jack Butala left town, and he was most helpful with United Mine Workers Local. Blues was everybody's friend. Then there was Patsy Codispodi who organized and trained a small band in our town. Following him were band directors, Jim Rutherford and Bill Jendrick. It was under Bill Jendrick's leadership that the United Miners Local decided to sponsor this band. They bought them new uniforms, attractive red cadet styles, which made the band quite striking as it marched in parades.

When Mr. Hoffman died, M.F. Brandon replaced him as mine superintendent. Then we remember when Mr. Mower came to town, who quickly gained respect of the people. Through his basketball team, the whole area heard of Vintondale and its high school. Mr. Mower was a strict disciplinarian, and the students obeyed. Even today Mr. Mower is an honorary citizen and is held in high esteem with the people of our town.

And I must mention Grandpap McConnell, who was the night watchman for so many years. He carried a lantern, and it was his duty to check out everything. He looked after the No. 6 mines, walked to the station, up Main Street, checked the theatre, the jail house-council room, the company barn, the company store, walked to No. 1 mine, looked over the tipple and supply room offices then. Bad weather or good weather, Grandpap was on his job.

Then there was Mike Hihalik, our banker, who opened up our brand new bank. Mike smoked a big cigar and wore an immaculate white shirt, and was there to help with money problems.

Then there were men like Harry Ling, the insurance man, Jim Samatore, the barber, Mr. Hubner, the shoemaker, whom we cannot forget. But we must remember our teachers whom we also respected. Miss Burkett, Miss Jones, Miss Condie, and Miss Jenkins. There was a Mr. McChesney who had such a strange way of teaching at our high school. One thing I did learn from him was the Battle of Hastings was fought in 1066.

CHAPTER NINE

In construction, the company store was a large stone building, huge for Vintondale. It housed the store and butcher shop (no one called it a meat market) on the street floor, and the mine offices and company office on the second floor. A telephone central was also on this floor. The meat department was on the front corner of this building, and it was the butcher's duty, not only to cut, wrap, sell meat, but also to handle all produce, cheeses, eggs, fish and even yeast and butter, lard and such items. He was responsible for a monthly inventory and an accurate record of all sales slips, invoices and bills. The bookkeeping was done by the store manager. In the mid 20's, meat was inexpensive compared to today's standards, but again wages were small, very small. Chuck roast was 14 cents a pound, round steak was 27 cents a pound, hamburg was 15 cents a pound, frankfurters were 18 cents, but bacon was a luxury at 45 cents a pound. For one quarter, you could get a good sized bag of mixed fruit. Eggs were 18 cents a dozen. Every Friday evening the butcher made preparations for the coming busy Saturday, by cutting some meat ahead, and placing and arranging the fruits and vegetables on their racks. Sometimes this work continued on till midnight, at no extra pay!

Now the store sold just about everything: hardware, tolls, clothing, work clothes and dress clothes, drugs, cigars, cigarettes, candy, household furniture, and many, many other items besides groceries. It had many departments, sectioned into several rooms. There was a grocery area, the shoe room, the drug room, a notion counter, a men's clothing and underwear area, a back room for storage of feeds, large bags of flour, sugar, potatoes and crates and crates of groceries. I have first hand knowledge of this store for I worked there from 1928 to 1930 as a part time worker on Saturdays and summers, then after graduation May 1930 to July 1935 as a full time clerk. I worked side by side with John Barclay, Kathryn Makepeace, Howard Gould, Pearl Gallandy and Joe the butcher. A short time later, Agnes Huth joined us.

We worked hard. Nothing was packaged, we did it all. If a customer wanted a peck or half peck of potatoes, that meant we had to go down to the cellar, fill up the bag, carry it up to the scale for the proper weight, then tie this bag with twine that hung constantly over our heads. What a struggle if was for us to wrap six cans of milk or six loaves of bread. Paper bags were scarce and flimsy. The wrapping paper was also flimsy and it was a miracle if it lasted the trip home. How many potatoe spills there were out on the street, how many loaves of bread fell to the ground. We each had our counters to work from, and there wasn't a day pass by that our small fingers didn't get all cut up with the wrapping twine, and we had to run to the medicine cabinet for bandages. Behind these shelves were a number of pull out bins. These contained dry lima beans, kidney beans, navy beans, Roman beans, and rice, and each bin held the contents of a 100 pound sack. Many a time I'd open a bin, as a customer wanted a pound of beans or rice, and a mouse or a rat would jump out. Very handy were items that were frequent sales such as cigarettes. We handled Lucky Strikes, Camels, Chesterfields, Old God, Philip Morris, Herbert Tareyton cigarettes. They were 15 cents per pack. During the Depression, we added "76" and "Wings" that sold for 10 cents a package. Also popular were tobacco sales, Five Brothers and Cutty Pipe, came in 15 and 10 cent packages. Our cigars were in a glass enclosed showcase, and of course, we handled the best brands. All types of pipes were on display cards, corn cobs were 25 cents each. And, of course, we sold ever so much Copenhagen snuff at 10 cents a box.

In the early 30's, there was a federal law passed that miners and other industrial workers must wear safety toe shoes. We had such a busy week selling these shoes to the miners, hundreds of pairs. And, of course, all the miners thought that was such a silly law!

Above the grocery shelves, were more shelves, with neat displays of coal buckets, miners' dinner pails, cooking pots, tea kettles, and of course, the chamber pots that were used at that time. I didn't care too much about climbing a ladder for any of these items if a customer needed them.

At the end of the store against the whole length of the wall were the notions, dry goods, ladies wear, infants wear, and of course towels, curtains, blankets and bedspreads. You could buy unbleached muslin by the bolt, 10 yards for 59 cents. Cotton lisle hoisery was 15 cents a pair. Children's hose were the same price, because all children wore long stockings. Silk hose were too expensive at 98 cents so they didn't sell too well. You could buy a ladies bra for 25 cents, but it was plain, not form fitting. Some bras had rayon flower designs, these were the better ones for 29 cents. A beautiful item was a crepe de chine chemise, sheer and lacy, a one piece of underclothing for summer wear under a sheer blouse, only of course, for the dainty lady. And . . . we sold bloomers . . . all kinds, white sateen, black sateen, pink sateen, nainsooks, came in all sizes, and these were summer wear. In the winter flannel ones were the best seller. The store sold boxes and boxes of long underwear, these were called unionsuits. Everyone wore these. These showcases displayed infant sweaters and lovely dresses, the lace curtains and fancy brocaded bedspreads, and lace table cloths. Linen dresser scarves, madeira lace doilies, and Philippine embroidery needles, thread, lace, elastic, and everything else that the home sewer needed.

Directly opposite the household showcases was the men's department, where the latest styles of shirts, ties, socks and sweaters were on display, also in glass showcases. There was a cabinet with a glass sliding door that held nothing but men's hats, for the dressed up gentleman. The store boasted its latest styles and fads. Several weeks before Easter, the management arranged to have a factory representative, with a knowledge of tailoring, come to this store with his samples. After choosing the fabric, the gentleman was measured up for a suit, and an order was placed. These suits all came in time for Easter. This was quite a profitable sales arrangement for the store, for the paying arrangements were quite ideal, $5.00 per month until paid off.

Down in the basement we kept tools and such related items. I had to learn early and quickly the difference between a 6d nail and 10d nail. I had to know the different kinds of shovels and picks, the different screws, nuts and bolts, cotter pins, etc. I had to know what an auger was, or a bit or an awl, the different hatchets and axes.

The upstairs contained a storage area for screen doors, screening, galvanized wire for gardens, and chicken yards; for galvanized buckets and pails; but, we also had a few furniture items.

On one special occasion, I can remember showing Mr. Schwerin, the New York City V.I.P., a light oak swinging cradle. And he was fascinated so much by it, he cranked and cranked the spring, just to watch it go back and forth, the 1930's system of automation.

Our company store window displays were quite an attraction. The fall window was full of colorful leaves, hunting suits, guns, hunting caps, shells, all arranged around stuffed pheasants, a big black bear, squirrel, and a deerhead or two. But the Christmas window display was always my favorite. As a child I'd be so excited that I stood out there in the cold, with my lips next to the glass, and how many times my breath froze my lips to that window. I wanted to get as close as I could to see the dolls that closed their eyes when laid down, dolls that said, "Ma, Ma", the stuffed Teddy bears. I wanted to see the wind-up trains on tracks, the trucks, trolleys and little red wagons and sleds. There were cast iron banks of all sorts, dancing Negroes on a wind-up box, balls, dishes of good china and also a tin for the little girl all placed on a miniature table with chairs. There were checkers and dominoes. There was always a little broom. It was such a popular sales item, and the little iron and ironing board. All this dream world was enough to excite any little child. And all this was around a decorated Christmas tree.

The other window had to show the popular Parker pen and pencil sets in gift boxes; there were lead pencils on a velvet ribbon a young lady wore around her neck. Also shown were fancy lingerie, nighties and kimonos.

Now this store sold a lot of candy from a special display case, and sometimes a special candy order would come that would offer a free gift with every pound. Along with a barrel full of marshmallow peanuts came elephant banks. With a barrel full of gum drop "Nigger babies" came gum balls to each purchaser of a pound. Our busiest candy sales were from children going and coming from school, noon times and afternoon. They came with their pennies, and the poor clerk would have to wait till they made up their mind on what they wanted. When jelly beans arrived every spring in barrels, we'd count them out ahead of time as spare time allowed, 15 for a penny, and the tiny bag was ready. There were boxes and boxes of penny candy, long licorice sticks, ice cream candy cones, caramels, maple sugar in a dish with spoon. There were coconut stripes, crystal candy, little glass trains filled with candy pills, along with glass pistols filled with the same. All the favorite chocolates were there in the candy case also. My favorite was Schraft's maple nut cream.

But the most enticing display of candy was the Christmas candy arrangement. Counters were built in the center aisle of the store, and bucket after bucket of candy were out in the open. Of course all this was enclosed by chicken wire, for there would be too many samplers. Some candy came in fancy tin boxes and fancy cans that could be used as gifts. My mouth waters even now as I think of all those goodies. Chocolate drops were 15 cents a pound, sugar cream candies and peanut brittle were 19 cents a pound. The good chocolates were 25 cents a pound. I can still see the fancy cleartoys, in all colors, the white, pink, yellow and chocolate coconut bonbons, striped coconut bars, peanut clusters, peanut butter bars, all colors of gum drops, the old fashioned rock candy. For a dollar you got a very generous supply of Christmas candy. Anyone who spent more than that was considered extravagant.

Sometimes the manager made a good deal with a drug company, and he'd get a big order of Dr. Caldwell's cough syrup. So a supply of advertising came with the deal. A window display followed, showing life sized cardboard figure of the bearded Dr. Caldwell, surrounded by three sizes of cough syrup. Of course there was a sale, and the result was quite satisfactory.

Our store had a very good assortment of drugs. We carried all kinds of liniments, pills, laxatives, alcohol, syrup of ipecac, Lydia Pinkham's liver tablets, all sorts of herb powders. We had colognes, shaving soaps and Bay Rum. There were bandages, epson salts, breast pumps, boric acid and sulphur powders. And too, we sold lots and lots of bed bug powders and roach repellents, and there were moth flakes. Baking aids such as vanilla extract, other extracts were on these shelves, side by side with blood builders, castor oil, tonics and the most popular aspirin tablets.

I have given you an idea of the outlay of the store, but must also mention the business part of it. Perry Wilson took care of the books, but it was Katherine Makepeace who took care of all office transactions. It was here that the customer came for a coupon before any purchase. Before Katherine could issue a coupon, she rang a crank phone connected to the upstairs mine office. At the other end, one of the girls would check each miner's work card in their files. If the miner was not overspent, the coupon was written out. There were white coupons with a five dollar value, yellow was worth three dollars, pink was two dollars. During the Depression, a new one dollar blue coupon was introducted. And many times even a dollar's worth of scrip was hard to come by.

So as a purchase was made, the total amount was deducted from the coupon value. The clerk itemized the items purchased, on a saleslip. This saleslip and coupon were sent by overhead pulley car to the office. This overhead carrier rode on wires over our heads and after the slip and coupon were safely clipped onto it, a rope on a pulley was pulled to ship it on its way to the office. There were a few charge accounts at the store, but most of the business was transacted by using coupons. The butcher shop, too, had the same system. So with the slips and carriers whizzing by, this was quite an active and very busy place.

In the late afternoon, when the office girls upstairs quit for the day at five o'clock, you'd see each of them carrying one of the file boxes downstairs to the store office so Katherine would have the records to refer to. The store remained opened till eight o'clock on weekdays and nine o'clock on Saturdays.

EPILOGUE

And so the years passed by, it is December, 1939, all of Europe is in a turmoil, we read the newspapers and listened to radio news, never dreamed that war would affect our lives. We kept busy as usual. Bill Jendrick drilled and drilled the Union Band, practiced them twice a week at the social hall, preparing them for the great United Mine Workers Concention in Columbus, Ohio. The date was January, 1940. Their brand new uniforms looked most striking as they marched before the town. They marched very well. They had no idea that their band would be chosen to greet John L. Lewis, U.M.W.A. president, at the railroad station in Columbus, upon his arrival. They escorted him in parade to the convention hall where the Union members were celebrating the 50th anniversary of the U.MW.A.

And so as the year ends, as we sit beside our radio listening to Guy Lombardo play his traditional "Auld Lang Syne" at the Roosevelt Hotel in New York City, we can hear shouts outside all around us, people downtown celebrating, shouting, drinking, dancing the New Year in! The year is 1940.

VINTONDALE AND VINTON COLLIERY GO TO WAR

As 1917 dawned, it seemed more and more ominous that the United States would be drawn into the Great War. In spite of Wilson's campaign pledge of "He kept us out of war", most Americans were well aware of the economic impact of the European war. On the local scene, miners were working six days a week trying to keep up with orders. Prosperity peeked around the corner in Vintondale. The economic situation encouraged Warren Delano to gamble again. Between 1916 and 1917, he purchased additional coal lands for #6, bought the abandoned town of Claghorn from Lackawanna, and built houses on Chickaree Hill, Sixth Street, and Upper and Lower Main Street.

The life in Vintondale during the war was revealed in the 1917 mine superintendent diary. Not only did the superintendent have to keep up production demands, but he had to be up-to-date with new and sometimes conflicting federal regulations which came from the various war offices. Replacing miners who volunteered or were drafted became a major headache. Then he also had to deal with disgruntled miners who felt that the pay raises did not keep ahead of consumer price increases. At the same time, the super had to maintain the patriotic mood of the townspeople.

A summary of wartime-related events of 1917 are as follows:

January 8: Posted notices of 10% bonus for employees from January 1. Hoffman talked to Schwerin three times that day about the bonus.

February 15: First Aid meet held. (This event could have been for the mine rescue or for civil defense.)

February 22: A second First Aid meet.

March 8: Another First Aid meet.

March 29: First Aid meet.

April 5: First Aid meet in borough building.

April 6: (War was declared, but no mention was made of it in the diary.)

April 16: Hoffman spent all day in the office putting a 10% bonus on all employees.

April 26: First Aid meet.

April 27: School children had a flag parade. Raised a flagpole at the head of Barker Street.

May 25: Flag raising day. Almost entire town there. Warren Delano, Samuel Lemon Reed, Jones and Rev. Puhinsky of the Russian Church gave speeches. Four-year old Agnes Huth portrayed the Goddess of Liberty. (Her only memory of the event was that she had to wear a gold sash across her chest. Agnes was told not to finger it, so she did.)

June 5: Mr. Alten enlisted in the Navy and died in Philadelphia, buried him from here today. Corporal here to take out several boys who enlisted. Later corporal took out nineteen boys and three had previously gone.

June 21: Red X meeting in Hoffman's office to make arrangements for a rally June 29 and 30. Nick Gronland tendered resignation effective July 15 to enlist.

June 29: Red Cross meeting at the flagpole.

June 30: Pay Day, Red Cross making collections.

July 12: Joe Ellis received notice to report to National Guard on the 16th.

July 14: Joe Ellis, National Guard, leaves to join the army.

July 16: Nick Gronland departed to enlist.

July 23: Galbreath, superintendent of Wehrum, called to go over labor situations.

July 25: Galbreath back, had trouble with his motormen; they demanded Vintondale scale.

August 6: Sixteen of "our boys" at Ebensburg for medical exam - drafted.

August 14: #6 lost thirty men.

September 3: Abe Abrams and Dr. MacFarlane to Ebensburg for the First Aid meet.

September 7: Abrams and other called meeting of Home Guards for 7 PM in Social Hall.

September 18: Seventeen of our boys left for Army Camp. School children and citizens were at depot to see our boys off.

September 19: Friends of our boys that went to training did not turn out for work.

September 20: First Aid meet and Compensation meet in Social Hall.

September 24: Sent Mike Farkas to mine between Cardiff and Nanty Glo for three Hungarian miners.

September 30: Crowd went to Johnstown to hear Roosevelt.

October 10: Mr. Madigan brought in three laborers.

October 13: Teams brought miner from below Wehrum to our #318, starts in #6.

October 22: Foremen's meeting on Liberty Bonds.

October 25: Liberty Bonds meeting at 7 PM in the office.

October 26: $23,500 in bonds subscribed the night before.

November 1: Berkey, Abrams, Holt, Dr. MacFarlane used Hoffman's car to drive to Johnstown for a YMCA War meeting. Put Dr. Garfield's war rate into effect.

November 2: Many of the monthly men dissatisfied with the 10% increase.

November 8: YMCA War Work Council met in the borough council rooms. The cash collection taken among foremen and company officials amounted to $187.

November 22: Rally Round Table meet and First Aid meet held in the council rooms.

December 21: Hoffman spent most of the day in the council rooms claiming industrial exemptions for his employees. His evening was spent in his office working with the Exemption Registrant Board.

December 22: Hoffman spent most of that Saturday night in the council rooms claiming industrial exemptions.

December 27: Rally Round Table and First Aid meet.

Vintondale did its patriotic part in the war effort. Her people bought bonds, practiced wheatless and meatless days, and grew Victory gardens. The company complied with its growing list of federal regulations which bombarded it, but also sought us many exemptions as possible so they could keep up the production required. The miners received what seemed to be generous pay raises, but were disgruntled because they hardly kept up the price increases for basic commodities.

Some of her finest marched off to war, many seeing combat service on the western front in France. Known Vintondale residents who saw service during World War I were: Charlie Lynch, Allie Cresswell, Paul McCloskey, Primo Cassol, Joe Garvis, Vince Wilson, Joe Giazzon, Hugh Pisaneschi, Metro Karol, Murdick Findley, Homer "Jack" Shaffer. Russ Dodson enlisted from Cresson and spent part of the war in Germany as a prisoner of war. He and his fellow prisoners were put to work and tried to sabotage the machines as much as possible. Roy Hunter enlisted from Windber; he was killed in France when a bullet ricocheted off a tree. With him at the time was Jack Shaffer. Charles Mower, high school principal, 1927-1960, was also a veteran of World War I. Pete Toth, who arrived in America in 1907, was drafted into the army, but only reached either Fort Monmouth or Fort Dix in New Jersey before the Armistice.

Nick Gronland returned from France and was met in Philadelphia by his Vintondale "sweetheart", Ruth Roberts, and her sister. He stopped in Vintondale on his return to Wisconsin and decided that the coal mining life was not for him. Later, he joined the salesforce of the Delano Coal Company and sold mainly their Morea hard coal mined at the Mill Creek Coal Company mines in Schuylkill County.

Through the efforts of Vintondale's miners, production at the Vinton Colliery mines reached the highest level to date. With the new prosperity, Delano could see his dream of making marketable coke finally fulfilled. He had a coke expert as company president, stability in the office, and plenty of housing in town for his miners. Things could not have been better in 1918 when the Armistice was announced.

Roy Melvin Hunter. Born January 9, 1896. Was killed in action October 11, 1918 in World War I. The Hunter-Weaver V.F.W. post in Windber, PA is named for him. The Weaver sons were killed in World War II.

Courtesy of Merle & Winnie Hunter.

Ruth Roberts, Ada Kempfer, Lovell Mitchell, Jennie Kempfer, Anna Bell, Marie Cresswell, Ruth Kempfer, Ima Mitchell, Lualle Shoemaker, Sarah Williams, Minnie Gaal, Gladys Whinnie, Margaret Crawford, Amanda Krumbine, Sarah Noel, Isabel Krumbine, Hazel Evans, were out of this, c. 1917.

Courtesy of Lovell Mitchell.

VINTONDALE'S ETHNIC HERITAGE

The rapid spread of industrialization in the United States following the Civil War led to major changes in the workforce and the ethnic makeup of the population of the United States. New technologies in the steel industry eliminated the need for the skilled ironworker and demanded a cheap, unskilled labor force. The coal industry, supplying a vital ingredient for steel manufacturing, boomed between 1890 and 1920; complementing all this industrialization was a rapid expansion of the railroads into the new mining and mill towns.

To meet the demand for cheap, unskilled labor, many companies sent agents to Southern and Eastern Europe to recruit workers. There they found many eager to migrate to the United States. The idea of migrating to find work was not a new concept for the Eastern European peasant. Each spring, many of the men in the villages left their wives and children behind and traveled to Germany, Poland, and other countries to find work on the large estates and in the factories. They returned home in the fall, using their hard-earned money to repair the farm or buy new furnishings to outdo the neighbors. For those living under the autocratic rule of Emperor Franz Josef of Austria-Hungary, Wilhelm II of the German Empire, or Tsar Nicholas II of Russia, America appeared to be the promised land, the land of milk and honey, the land where the streets were paved of gold.

Universal Conscription

But in spite of the hardships encountered in reaching America, the immigrants continued to pour into Ellis Island and other ports of entry. Many were single young men trying to avoid the mandatory universal conscription into the army. All three of the Eastern European empires had large standing armies, the bulk of which was made up of young men of peasant stock. In the Austro-Hungarian Empire, all men under 21 had to register for military service. The recruits, of which there were three classifications, were called up by village. These classifications were:

Class One: The young man passed the physical and was inducted into the army for three years on the first call.

Class Two: The prospective recruit was rejected on the first call and re-examined at age 22.

Class Three: He was rejected two times and subject to a third examination at age 23.

If the young man was rejected three times, he was permanently excused from military service. Anyone deferred was required to pay a tax called a military **porciju** of three **zlaty** per year for twelve years. In 1900, a **zlaty** was worth forty-five cents.

To leave Austria-Hungary legally, the potential emigre applied for a General Pass. Upon receiving the pass, he presented it to the **shandars** (border police). Those trying to avoid the guards and cross the border illegally, were often caught and returned for prosecution. If a deferred man wanted a pass, he had to pay the twelve year tax in a lump sum. Physically handicapped persons were not permitted to leave the empire. Many immigrants reached the United States by using an assumed name and borrowed pass. Austria-Hungary's passports did not require a photograph, so thousands of passes were mailed home and re-used by a family member or friend. By emigrating and then returning to the old country, a man was still liable for the draft. For instance, Dora Lazich said her father returned to the Sebrian section of the Austro-Hungarian Empire and was immediately drafted and had to serve his three years.

Industrial Feudalism

Many families emigrated out of economic necessity. Serfdom had not been outlawed in Eastern Europe until the end of the 18th century; Russia was the last in 1861, two years before our Emancipation Proclamation. Most Eastern European peasants had struggled for centuries under the medieval manorial system of farming. Until the 1800's, the peasant not only had to pay a portion of his crop to the lord, but he also had to work several days per week (robot) for the master. In Eastern Europe, the peasant was paid annually when the harvest came in. However, in America, the pay, small as it was, arrived every two weeks or once a month.

In American company towns, feudalism and manorialism had its counterpart in industrial feudalism; the company hierarchy filled the positions found on the medieval feudal pyramid. Substituting for the lord of the manor was the company superintendent, who lived in a suitable manor house. The super's knights-at-srms wore the guise of the coal and iron police. In a mining town, the freemen of the feudal pyramid were the independent merchants, if there were any. At the bottom of the pyramid were the superintendent's serfs, the miners, whom he benevolently protected from the unions. The annual payment of a portion of the crop and the **robot** of several days per week, were substituted with electricity, the doctor, and rent. For those who were independent-minded and pro-union, their stay in a company town was often short and/or unpleasant.

Birds of Passage

Many of the emigrees were married men who had intended to work in America for several years, save money, and then return to the old country and live well. Many returned as planned and remained in the old country. Some, often called birds of passage, made numerous trips across the Atlantic because their wives often refused to move. One of the author's fraternal great-grandfathers, Louis Horvath, made four different trips to Pennsylvania. He mined coal for two years on each trip and then returned home. His wife Rose (Saboyna) Horvath did not want to make the journey. Daughters Helen and Margaret and son John emigrated to the United States; a son Louis remained in Hungary. John Horvath, became a UMWA organizer; he died in Cardiff (Nettleton) in 1930 as a result of an attempt to break up a domestic squabble. Punched in the jaw, an infection from a bad tooth spread throughout his body and killed him. He left a wife and children in the Old Country. Helen Balog Oravec said that her father, Mike Balog, deliberately emigrated to avoid the draft and that her mother wished to return, but illness changes their plans. Mike Beres wanted to go back to Hungary but his wife Ethel refused, realizing how much better off she was here than in Hungary. Considering the upheaval of World War I, World War II, and the Communist take-over, many were thankful that they had remained in America.

Joseph Gresak's Memories

Joseph Gresak, 1878-1950, in his autobiograhy, explained in detail what life was like in the old country and what it was like to be a "greenie", or greenhorn, in America. What makes this autobiography unique is that Joseph Gresak lived in Vintondale for a short time in 1902. He saw the new town through the eyes of the European newcomer.

Like many of Vintondale's families, the Gresaks were of Rusian (Ruthenian) descent and lived in the Carpathian Mountains of what today is Czechoslovakia and western Russia. Gresak arrived in 1901 on a borrowed pass. From his hometown, he traveled to the harbor city of Bremen, Germany where he purchased a steerage ticket to the United States, underwent a physical, and received the necessary vaccinations at the shipping company headquarters. The cost of his tickets from Bremen to Horatio, Jefferson County, Pennsylvania was 100 **zlaty** or $45.00, leaving $18.00 to sustain himself until he found a job. In Horatio, he boarded with an uncle, but suffered from culture shock. Joe was called a "greenor" by his own people. He had to learn English to survive. Those who were able to speak German or Hungarian had an easier time adjusting to the new life.

For Gresak, the first major adjustment came in housing. Most of the immigrants were from agricultural communities and resided in a family cottage. Living with strangers in a boarding house took some adjustment. Adapting to American clothing was the next hurdle. Most men bought themselves an "American" suit right away, and the regional peasant dress disappeared, only to be brought out for special occasions. Mr. Gresak followed that pattern by traveling to nearby Punxsutawney to purchase a suit for $6.00, a hat for $1.00, a pair of shoes for $1.50, two shirts for $1.00, and a necktie for twenty-five cents.

The next adjustment was the language. The greenhorn had to learn enough English to survive in the mines. Gresak discovered that there remained one strong link to the old country: the food. Immigrants were reluctant to change their eating habits.

Gresak's Sojourn in Vintondale

Mike Gresak arrived in Vintondale in 1902 from the Jefferson County mining town of Frostburg. He described Vintondale as larger and more beautiful than Frostburg. At the time, Vintondale sported a post office, company store, three taverns and three mines. Hired at #3, he worked ten hours a day for $2.00 per day. He also said that the #3 was more modernized than Frostburg. (#3 was Claghorn's longwall mining experimental mine.) Work was plentiful at the time since Wehrum was just opening, and the miners moved often searching for better working conditions and salary.

In Vintondale, Gresak lived with another uncle and paid him $15.00 per month room and board. Mike was forced to deal with the company store, using coupon books whose value ranged from one cent to one dollar. Naturally, his purchases were limited to his earnings. However, Mike found it difficult to socialize with the young ladies of Vintondale. Single girls were a scarce commodity, and at dances, he was turned down by the girls. Most of the Vintondale's young people of immigrant origin spoke English, including those who had only been in Vintondale a few years. Harassed as a greenie by his own cousin, Mike moved to Dunlo where he had some friends in September, 1902.

Elaborating on the Adjustments

Switching to western clothes was the only practical alternative for a miner or steelworker. The high boots and baggy pants worn in Eastern Europe were hardly a suitable outfit in which to dig coal or pour steel. One tradition which the men refused to abandon after arriving in American was the long drooping mustaches.

The women maintained more of the traditional clothing than the men. The babushkas, braided hair, the large aprons, the gold looped pierced earrings, and the cotton stockings, were part of the standard everyday dress. Ear piercing was usually done using a needle and black silk thread or a strand of hair. Having pierced ears carried a stigma as being immigrant; this custom was certainly not the trendy thing to do in the American culture until the 1960's.

Boarding Houses

In a mining town, many immigrant miners kept boarders to help make ends meet. The wife, or boarding houses Missus, worked as hard or harder than her husband. Up before dawn to pack buckets and fix breakfast, she also had to wash the family's laundry and the boarders' work clothes by hand. A copper boiler full of hot water for use in bathing and washing clothes was on the coal stove daily; the wife had to keep it burning all day, winter and summer. When the men came home from their shifts, she had to have hot bath water ready. If lucky, the missus had a bathhouse built next to te outhouse in the backyard. Otherwise, the kitchen was used for the miners' daily bath. In addition, either she or the children usually had to haul water from one of several springs in town because there always seemed to be a shortage of water in Vintondale. The coke ovens had top priority over the people. Occasionally, the missus could obtain hot water from the company steam pipe valve near the company store.

The immigrant wife baked her own bread; storebought bread was shunned by the foreigners. She kept a vegetable garden. She canned vegetables. She butchered and plucked the chickens for "chickee" soup. She made her own noodles (nood'LEES). She made the sausage and blood pudding. She often made her own butter and cheese. She brewed the beer and distilled the moonshine. She sewed clothes for the family, frequently using the multi-purpose flour sack. She milked the cow, if the family could afford one. She fed the chickens and the pigs. She was staunchly religious, attending church weekly and observing all the fast days. She was pregnant almost every year; large families were the norm; infant mortality rates were high. In many cases, she was subject to beatings from her husband, who often drank to excess. Needless to say, these women worked hard and made great sacrifices for their families.

If her boarders were on shift work, the missus had to try to keep the young children quiet so that the men could sleep during the day. Boarders often paid reciprocal roles as babysitters. Frequently there were squabbles among the boarders which resulted in serious fights. (See chapter on crime.) Many fights were caused by alcohol use, but there were squabbles over the amount of groceries purchased by the boarder. Some paid a flat monthly fee, but others bought their own groceries and had the missus cook for them. Many boarders followed their mister from one mining town to another. Others frequently moved from one family to another in town. In many households, the boarders remained with their families until their deaths.

Vintondale's Colorful Bachelors

Vintondale had its array of colorful old bachelors. Many older residents remember Handlebar Pete or Fat Charlie Arbuckle, a Farkas border who died after falling off the rear second floor porch of the Farkas Hotel. What about Pete Pierondi, Bianucci's boarder? The author remembers many a time seeing Pete sitting at the bar in the Farkas Hotel carrying on an animated conversation with the glass of wine. Pete supposedly was engaged to get married, but got cold feet. Upon being jilted, his fiancee broke the news to her mother crying, "Mamma, Mamma, fat boy jump fence."

How many girls in town were afraid to walk near Faish's boarder, Andy Sabo? One never knew when he might pinch you or swat you on the rear. He was especially dangerous when he was standing in the checkout line at Nevy Brothers' store.

The author's fraternal grandmother, Helen Dusza Farkas Firko kept many boarders when she helped operate the Farkas Hotel. When she moved across the street to the duplex above the old post office, John "Paney Bacci" Paney moved into the basement where the post office had been. Paney Bacci was a special part of the family; he had boarded with Mrs. Firko for years, following her from Onnalinda near Beaverdale to Nanty Glo to Vintondale. Paney was responsible for saving the life of Steve Dusza. While playing cowboys and Indians in Onnalinda, Steve's playmates acutally strung him up. Paney happened by in time to get him down. The Dusza children became his adopted grandchildren. Every month when his Social Security check came in, Paney struggled up Third Street carrying a half-gallon of ice cream or a watermelon. Each Duzzie received $5.00 annually for his/her birthday.

One day, Paney picked a bunch of snowball flowers for Agnes Dusza; she rooted the stems and planted them in the backyard; today there is a living memorial to Paney Bacci.

Remember Jim "I Can Get It for You Wholesale" Buckeye? He was always wheeling and dealing and not caring if he made any money as long as he was doing business. During the Depression, he formed a lime company in Buffington Township near John Wagner's farm. Working with him were "Big Alex" Oro and Steve Faish. They built a lime kiln to burn down the lime and were to split the profits. While they were working, Buckeye was in charge of feeding the men. In the end, Buckeye's partner could not pay for the lime. After #1 closed, he batched (set up housekeeping) in one of the remaining mine shacks. The author's father frequently sent supper to him. Buckeye died at the county home.

What about Whiskey John? He built himself a shack along the Dinkey Track above Sixth Street. Kids in the 1950's steared clear of that section of the Track. Delivering meals to him was a mission of mercy that most of Dusza children tried to duck. However, John was appreciative of the food.

John King, "Keedeye Bacci" lived with the Moreys for years. Vintondale's last immigrant bachelor, Joe Silagyi, played violin with John Morey Jr.'s gypsy orchestra. He made his home with John Morey Sr. and later his daughter, Liz Morey Rabel; Joe died in 1984. German John was another bachelor who arrived in the 1920's and lived for a while with the Moreys. He later lived in one of the houses on Goat Hill and raised carrier pigeons as a hobby.

English-speaking immigrants who could translate in both German and Hungarian were invaluable to the coal companies. They could explain company policies to other workers and recruit miners when there were shortages. In Vintondale, John Morey, Sr. and Mike Farkas filled those roles many times. Before he left town in 1909, merchant Jacob Brett was often called on to be a translator for the company.

Ethnic Foods

Ingredients, preparation and religious significance of some foods were a carryover from the old country. The blessing of the traditional Easter foods of eggs, paska, ham, kolbassi and Easter cheese has not died out, but rather is undergoing a revival which has attracted many families who have no ties with the Eastern European culture. America has awakened to the intricate art of **pysanky**, Ukrainian Easter eggs. The lowly pierogie has now risen to be the new low calorie, high carbohydrate safe diet food, sans the butter of course. The pierogie has assisted in the survival of Sts. Peter and Paul Russian Orthodox Greek Church in Vintondale. Holupki and pierogie help keep the people coming back to the Vintondale Reunion, enabling it to turn a profit every year.

The use of the lowly noodle took on new meaning in

immigrant households. Nourishing and filling, it was served almost every day with soup, the main dish. The homemade noodles were cooked separately and added to the broth only when ready to serve. Drop noodles or spaetzels complemented the Hungarian goulash and chicken paprikash, or the re-discovered delicacy, haluski. Who would have dreamed that sauteed cabbage and dumplings would be one of the first items sold out at ethnic fairs. So the food gave Mike Gresak and many like him a taste of the old country.

The Italian pastas and sauces have also become an intricate part of American cuisine. Vintondale's Nevy family made an important contribution to popularizing Italian cooking with its Alpine Eagle macaroni and spaghetti.

Native Language

At home, the native tongue was usually spoken. At the Russian and Hungarian churches, summer school classes included instruction in hard Russian and Hungarian respectively. Intermingled in the classes was religious instructions. Immigrant children were eager to learn English so that they would be accepted by the other children. In some cases, they were ashamed of their background and did not pass their native language on to their children. Only now do many realize that it was a mistake to deny their ethnic heritage. Of course, in many cases the language adjustment was never fully made by the first generation immigrant. Many had difficulty expressing themselves in the English language even after fifty or more years in the United States.

VINTONDALE'S ETHNIC MINORITIES

The Finns

The earliest ethnic minority to arrive in Vintondale in any appreciable numbers were the emigres of the Grand Duchy of Finland, part of Russia's tsarist Empire. Some of the Finns who emigrated were from the west coast of Finland where they spoke Swedish. In all, about 250,000 Finns migrated to the United States between 1860 and 1930. About 2,500 Finns settled in Pennsylvania. An estimate of those living in Vintondale is unknown, but the majority lived on Sixth Street, which until 1917 only had houses on the east side of the street.

Lutheranism was the state church of Finland, and the Vintondale Finns had a meeting house behind the Village Inn. It is not known if this building served as a bona-fied Lutheran church. According to the clerk's 1907 minute book, the Baptist Church gave the Finns permission to use their church for services.

Since drinking was a major problem among miners, many Finns joined temperance societies. Their societies and meeting halls were centers of Finnish culture. Dances, concerts and plays were held there. Vintondale had its own chapter of the temperance society, **Urhojen Lutlo** (League of the Brave).

Many Finns became involved in the Socialist movement which was growing at the turn of the century. The socialists stressed government ownership of the major industries. Many supported socialism as a means of bettering the conditions in the mines. Pennsylvania eventually had ten chapters of the Finnish Socialist Federation. Nearby Nanty Glo was a hotbed of socialist activity well into the late 1920's.

Based on evidence from the birth certificates from 1906 to 1921, the majority of the Finns appear to have moved from Vintondale in 1909. There were only two birth ceritfcates registered in Vintondale after 1909 that list the parents' birthplace as Finland. The most probable reason for the departure was labor problems. Many Finns were pro-union, and there was a long strike in 1909. Families in company houses were evicted. Many of these ex-Vintondale employees settled in the unionized mines of Nanty Glo. A section of western Nanty Glo is called Finntown, and a Findlanders' Cemetery is still maintained there today.

The Hungarians

Probably the largest ethnic group in Vintondale, many Hungarians arrived as strikebreakers in 1909. The Hungarians, or Magyars, differed from the fellow Eastern European immigrants in that they were not Slavish, and in fact aided the Hapsburg monarchs in ruling the Slavs. The Magyars invaded the Danube Basin from Central Asia in the tenth century and settled there, slowing gaining control of the various Slavish groups in the region. The Hungarian language is most closely related to Finnish. In the Hungarian section of the Austro-Hungarian Empire, the Magyar language was taught in schools, rather than the Slavish languages. Most of the government positions were filled by Germans or Magyars. There was little love lost between the Hungarians and the various Salvish groups.

Religiously, the Hungarians were divided into three groups. Some followed the Latin Rite of the Roman Catholic Church; some were Uniates or Byzantine Catholics; and others were Hungarian Reformed (Calvinist). Vintondale and Nanty Glo offered churches of all three denominations to tend to the religious needs of the Hungarian population. Many Hungarians attended Immaculate Conception Church and donated the cross which stands below the church. Some traveled to St. Nicholas Byzantine Catholic Church in Nanty Glo. The largest number of residents belonged to the Vintondale Hungarian Reformed Church, organized in 1916. This large congregation spread the Hungarian culture through summer schools, dinners, and dances.

A large number of Hungarians owned their own homes on Plank Road and Maple Street. Others lived on Chickaree Hill and lower Main Street. Most Hungarian families kept boarders.

In addition to the usual card games and drinking

bouts, the Hungarians had a special cultural tradition, the bacon roast. A chunk of slab bacon was roasted over hot coals. As the bacon browned, the drippings were caught on a slab of homemade bread. When edges of the bacon were crisp enough, they were sliced onto the bread along with onions, cucumbers, and green peppers; the chunk of bacon was then returned to the fire for the next sandwich. All of this was washed down with beer. By the end of the evening, the men were in the mood to harmonize in Hungarian.

The Italians

Before the strike of 1916, Vintondale had a fairly large population of Italians who lived mainly on Goat Hill and lower parts of Main Street. However, most were evicted during the strike and did not return. One exception was John Bossolo, the father of Josephine Pisaneschi, who worked at #3 and ran a boarding house in the old company office on Plank Road. He was evicted, but was called back three months later. Many of the Italian families who remained between 1916 and 1922 were affiliated with Nevy Brothers store.

Arriving from Europe in the 1920's was Gino Simoncini; his first stop was Butler where a relative had a tailor shop. His brothers, Joe and Bruno, were miners in Vintondale, so Gino settled here. He married Jennie Grassi, an employee at the Nevy Brothers' store. For several years, Gino was forced to work in Wehrum because he lost his job for buying groceries at the Cupp Store.

Hugh Pisaneschi was born near Florence, Italy and came to the United States when he was four. His father arrived first and then sent for the family; they settled in Weedville in Elk County; later the Pisaneschis moved to Wilkes-Barre. Hugh came to Vintondale on his own when he was twenty-five and worked on the flat. He married Josephine Bossolo; the couple first rented a house on Plank Road from Jack Butala's father. In the 1924 strike they received an eviction notice and moved to Twin Rocks and then Nanty Glo. Later the Pisaneschis were called back. When the company sold its houses, Hugh purchased the one originally occupied by Grandpa Blewitt on Third and Griffith.

Pietro Biondo came to the United States in 1913 and returned to Italy in 1918. He came back to Somerset County in 1922 and arrived in Vintondale in 1924. His wife, son, John, and daughters, Katherine and Toni, landed in New York City on Christmas Day, 1924. Son, Joe, "Beppy", was the only one of his four children born in the United States. John began working in the mines in 1929 at the age of fifteen. The family moved into the old company office around 1930. Today, John Biondo and wife, the former Elizabeth Simon, live on one side; and Katherine and husband, Joe Mesoras, reside on the other.

Other Italian miners who lived in Vintondale were Joe Giazzon, Bruno Bianucci, and Premo Cassol. Former Wehrum residents who settled in Vintondale were Guerino Averi, Mr. Avalli and Patsy Codispodi. Mr. Averi moved from Wehrum in 1926 and opened a store in the Shaffer building and then moved uptown in 1930 when he bought the Freeman store building from Mrs. Beerman. Mr. Codispodi, who directed the Wehrum band, became Vintondale's band director in the 1930's and later settled in Johnstown.

The Carpatho-Russians, Red Rus and/or Ruthenians

A large number of interrelated families from present day Czechoslovakia and western Russian settled in the Blacklick Valley. This area was part of the Austro-Hungarian Empire until 1919 when the boundaries of Europe were redrawn following World War I. Most of these immigrants were from the provinces of Maramaros and Weleklucki. They lived in company houses on Chickaree Hill and below Fourth Street. In many cases, it was difficult to identify the correct family. Out of seven Balog families, three of them had Michaels as heads of household. The wives' maiden names were used to differentiate the families. The names of the children invariably were George, John, Helen, Charles, and Mary. Helen Balog Oravec was frequently called by Mr. Mower to the high school to identify the correct Balog when he received a request for a transcript.

The majority of these new residents belonged to Sts. Peter and Paul Russian Orthodox Greek Catholic Church and faithfully followed the Julian Calendar. This occasionally had an adverse affect on the working schedule of the mines as the men refused to work on feast days. The church split in the early 1920's also had a profound effect on the unity of the Carpatho-Rusin Vintondale. (See chapter on the Russian Church.)

Serbians and Croatians

A smaller number of immigrants arrived in Vintondale from the southern, less developed regions of the Austro-Hungarian Empire. These were the provinces of Serbia and Croatia. The settlers spoke Slavish dialects which were different from their northern cousins in Slovakia and Poland. In addition, there were language differences between the Serbians and Croatians. There were strong religious differences also. The Croatians were loyal to the Roman Catholic Church, and the Serbians were Orthodox, having a Serbian Orthodox Church in Woodvale in Johnstown. Most of the Horwatts, as they were called, lived on Sixth Street, which earned a widely-known reputation for being wild and wooly. The double houses there were divided into four apartments; each one was jammed with families and boarders. Fights, especially those using knives, were common on payday weekends. Sixth Street probably had the largest number of speakeasies in town. (See chapters on union organization and crime.) Lamb roasts were a speciality on weekends, as were card and drinking parties.

During the labor troubles of the 1920's, the company kept a spotlight shining nightly on Sixth Street. Some of the families living on the street were: Eli Abramovich, Pete Page, Mr. Ploaich, Nick Oblackovich, Pete Lazich, and Mr. Valent.

The company encouraged each ethnic group to settle on certain streets. Company houses on Second and Third Streets, especially above Griffith Street were off limits to the foreign miners. There were privately owned lots on lower Third Street which did fall into the hands of immigrants. However, the ethnic clusting in Vintondale did not break down until the company houses were sold in 1943.

Steven and Mary Buchey - old homestead at 241 Plank Road, Vintondale.
Courtesy of Charles Sago.

Mike Farkas' General Merchandise Store, c. 1910.
Courtesy of George Lantzy.

Steve Buchey, about 12 years old - 1929 his First Holy Communion.
Courtesy of Charles Sago.

Steve Hubner in front of his shoemaker shop, c. 1923.
Courtesy of Goldie Hubner Britsky.

Britsky Family, taken at the Red Hotel, c. 1920. Back row: Charles Britsky, Pearl Britsky, Bertha Ezo Cook, _____ Paul Britsky. 2nd row on left: Bill Ezo, Betty Ezo. Front row: 3rd from right, Bill Ezo. Courtesy of Goldie Hubner Britsky.

Standing left to right: ___, Marge Pruskey, John Morey and Joe Silagyi with violins, Mr. Nagy, John Gallo, Mike Farkas, Helen Farkas, Mrs. Anna Szerkeresk, Emma Toth, Marge Toth. Middle row, seated on chairs: Mrs. Weges, Mr. Pruskey. Seated on ground: Guisa Lokos, Mr. Prunar, Mrs. Sabo, Mrs. Babincsak, Mrs. Onderko, Mrs. Bukovich, "Big Alex" Oro, Mrs. Kristian, Frank Sabo. Taken in 1940 in Munday's Corner by Hicks Photography, Maple Street, Vintondale.

Courtesy of James Morey.

TOWN TIDBITS

The Twenties were an exciting time to be growing up in Vintondale. New inventions, such as the radio, opened the window of the world to Vintondalers. The airplane was coming into its own with air mail and passenger service. Why even an airplane passed over Vintondale on a special trip to take photographs of the town! Train service was still available, so those without Model A or anything more luxurious, could still reach the outside work-maybe even beyond Nanty Glo. But Vintondale in the 1920's was also a rough, tumble, and sometimes violent decade with several murders and two serious labor situations.

The more pleasant memories of the town can be brought to light through the "Vintondale News" columns in the **Nanty Glo Journal**, which began publication in 1921. There seemed to be an occasional blackout of local news, but for the most part the town gossip played its role in the daily drama and will be repeated here for interest's sake.

1921

In May 1921, the company was reported to be in the process of fixing porches and making other general repairs. William McCracken, instructor of the town band, had practices on Monday and Friday; two concerts were given at the carnival grounds. P.J. Houchak was the director of the Vintondale choir. Irma Mitchell and Hazel Evans graduated from the Altoona Hospital Nursing School. Attending the ceremony were Mrs. M.B. Mitchell and Mr. and Mrs. William Evans. Messrs. Hoffman, Clement, Lockard, Burgan, MacFarlane, and Arbogast attended a shad dinner in Ebensburg.

The Decoration Day program was set up with a Servicemen's Parade at 9:00 followed by morning speaking, ball games, band concerts, races and a dance from 9-1. Winners of the races were:

50 Yds. - 16 and under: Charles Gongloff - $1.00
100 Yds. - 16 and under: Charles Duncan - $2.00
50 Yds. - girls under 10: Nora Gaylor - $1.00
100 Yds. - girls under 16: Beatrice Baldwin - $2.00
100 Yard Dash - over 16, $1.00 entrance fee, winner takes all - Charles Duncan
100 Yard Dash - girls over 16, $1.00 fee, winner takes all - Catherine Morris
Sack Race: Bill Evans and Clarence McPherson - $2.00

Students returning home in June for the summer vacation were Kenneth Hoffman of Mercersbury Academy, Barney Morris of St. Francis, and Esther Swartz, Helen Beechey, and Thelma Clemens from Indiana Normal School. Francis "Doc" Thomas improved his speed as company messenger boy by purchasing a new bicycle. Mrs. Dan Thomas (Mary Ann) returned home following a six-week stay at West Penn Hospital. The upcoming Fourth of July schedule included a parade at 8:30 AM, addresses at 9:30, ball games at 10:00 and 2:00, and concert and races at 6:00. Refreshments sold benefitted the Boy Scouts. Mary Riggle, former assistant stenographer of VCC, returned to her home at Kelly's Station; she and Allie Cresswell were married later in the month. Mr. and Mrs. John Burch, proprietors of a Main Street restaurant, purchased a player piano.

1924

In February, Bill Blewitt left Vintondale for a job in Youngstown. The stork visited the Allie Cresswells in Claghorn, and Mrs. John Cresswell went to see her new grandchild, Robert. Abe Solomon returned from Mt. Clement, Michigan after taking treatments for rheumetism. Clarence "Skooky" McPherson planned to seek a job in Florida after quitting at Roy Roberts' Empire Bowling and Billiard Parlor. Paul Kempfer, Vintondale's pool wizard, traveled to Indianapolis for a tournament and reported by wire that he had been successful.

1929

David Nevy and Dr. MacFarlane sailed on the **Majestic** from New York on July 10th for a ten-week tour of Europe. In addition to visiting the British Isles, France, Germany, and Switzerland, they planned to pay a visit to the "Nevy Villa" near Parma in Italy. While MacFarlane was on vacation, Dr. Smith of Blairsville covered his practice. MacFarlane made his return trip on the **S.S. Franconia** which arrived in New York on August 29th. Davey Nevy returned October 25 aboard the **Berengaria**; he was met in New York by his brother Gene.

Charles Mower and Gene Nevy constructed a new tennis court on the "Hoffman estate". This clay court was located between the superintendent's house and Second Street.

1930

Jack Butala left in June for Europe for a six week trip aboard the **S.S. Bremen**. Following their purchase of Verna Findley's house at the western end of Main Street, Bill and Mary Clarkson opened a "quick lunch stand" in their house. The new restaurant was named the Dew Drop Inn.

Lackawanna Boy Scouts, Vintondale, PA, c. 1905.
Courtesy of Huth Family Collection.

Nick Gronland, Lovell Mitchell, Ruth Roberts, Isabelle Krumbine and Phil Mains at Wehrum Dam, c. 1916.
Courtesy of Lovell Mitchell.

Clarence Stephens (left) and George Straw in mine prop wagon in front of Joe Bennett's and Village Inn, c. 1910.
Courtesy of Hazel Cramer Dill.

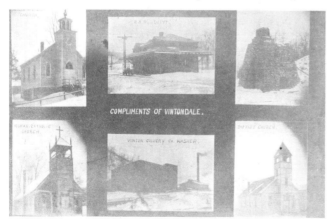

Vintondale, c. 1910.
Courtesy of Daniel and Helena Ploaich.

Vintondale. Freight Station-Pennsylvania Railroad, 1919. Morris or Red Hotel in center of photograph, Charlie Bennett's building is to the right of the hotel.
Courtesy of Huth Family Collection.

Chapter VII
Public Services

WATER COMPANIES IN VINTONDALE

"Water, water everywhere, and all the boards did shrink. Water, water everywhere, nor any drop to drink."

These lines from Samuel Taylor Coleridge's poem, "Rhyme of the Ancient Mariner" adeptly describe the water situation in Vintondale since its founding in 1894. Supplying sufficient clean water for the residents has been a major problem since the water lines on the town plan were drawn. All lots surveyed between 1892 and 1894, were to have access to water mains. To provide this water, a subsidiary of the Blacklick Land and Improvement Company entitled the Vintondale Water Company was created. Two small resevoirs were built, one on Bracken Run and one on Shuman Run.

The Vintondale Water Company was dissolved in 1900 when the Lackawanna Coal and Coke Company was organized. Lackawanna's legal representative, Judge Barker, advised that the Vintondale Water Company be dissolved and four new companies be created in its place. Charters for the Blacklick Water Company, the Jackson Water Company, the Buffington Water Company and the East Wheatfield Water Company were filed in the county courthouses in December, 1900. Each company was to exist for 99 years. Investors were Fred Barker, Ebensburg; Clarence Claghorn and H.B. Douglass, both of Vintondale; John A. Scott, Indiana; and Warren Delano, East Orange, New Jersey. Each owned two shares worth $100 each, making the capital stock of each company $1,000. J.P. Higginson, Treasurer of Lackawanna Iron & Steel Company, was the treasurer of the new companies; each company deposited $100 in cash with him. Directors for the new companies were Barker, Claghorn, Douglass, Moses Taylor of New York, and Henry Wehrum of Elmhurst, Lackawanna County. Stockholders of the Vintondale Water Company held a meeting on December 13, 1900, at which they authorized two officers to sell and convey the rights and franchise of the company to Lackawanna Iron and Steel. Deeds to execute the appropriate properties to the Blacklick Water Company for $1,750, to the Jackson Water Company for $12,750, and to the Buffington Water Company for $500 were authorized by Lackawanna on February 27, 1901.

When Delano was negotiating the repurchase of #1 and #2 in Vintondale in 1905, much of the discussion centered around the division of the water companies and the use of the water in the North Branch, which was still clean. When finalized, the Blacklick and the Jackson Water Companies became the subsidiaries of the Vinton Colliery Company. Price paid by Delano for the Blacklick Water Company was $750 and $12,750 for the Jackson Water Company.

Vintondale's two resevoirs which served the town were owned by the Jackson Water Company. The monthly fee of $1.00 was automatically deducted from the miner's pay. For that $1.00, the miner got cold running water, occasionally. Inside plumbing was a luxury, meant for only a few houses above Griffith Street which were occupied by companymen. A few private homes, such as Sam Williams' and Jacob Brett's, had indoor bathrooms.

In spite of two resevoirs, with a small additional dam above Sixth Street, which may have been part of Griffith's sawmill, a water shortage was an annual occurance. The main reasons for the shortage were the coke ovens and the powerhouse. Both needed clean water to operate properly. Originally, VCC had a pumping on the North Branch below the iron furnace which supplied water to #6. After the Colver mine was opened, the stream joined the ranks of other polluted waters in the vicinity. VCC tried to build a filtration system at the powerhouse to clean the sulphur water. John Galer, powerhouse engineer, earned a patent for his invention of a water filtration system.

In 1914, through the efforts of Dan Galbreath of the engineering department, a larger resevoir was constructed on Shuman Run at the site of the present dam. Clearing the underbrush began in May; and throughout the summer, every available cokeman or washeryman was sent up the hill to work on the dam. Galbreath and/or coke boss Dempsey supervised the men. In July, the ditch from the spring to the dam on Bracken's Run was cleaned out. Several bad water leaks were also found in Rexis which required immediate attention. The pipes and valves for the new dam were ready on July 24th, and Goat Hill was without water on the 30th. In the meantime, company teams began hauling sand to the resevoir site, and half of the concreting was completed on August 30th. At this time, only a one inch steam of water was flowing into the Bracken resevoir. A new four inch pipe was installed on Plank Road. On September 22nd, Hoffman took the inspector of the State Water Commission on a tour of both resevoirs and the pumping station in the North Branch.

The new resevoir was completed just in time in October because the water in the North Branch was so low that the work schedule at the washery was affected. Water from the new resevoir was turned into the town pipes on the 14th. Low water continued to plague the company; on October 23rd, the pumps on the North Branch had to run all night to fill the tank and the boilers at the powerhouse. The first good news about water that year came in November when water was reportedly going over the spillway of the new dam, the first time since it was built.

In the 1920's, the borough paid the Jackson Water Company $20 per month in water rents. Fire hydrants were maintained by the water company, but one wonders how adequate the supply was in case of fire. In 1930, the Hungarian church and the Delano building were destroyed by fire; lack of water due to a drought definitely prevented saving the school.

But even a new dam was not enough. Almost every summer, Vintondale residents had to carry water from nearby springs to supply their needs, while continuing to pay the $1 per month charge. A spring on Main Street between Second and Third Streets had to be sealed around 1900 due to typhoid germs found in the water. Residents could haul water from springs in what is now Dancha's Field behind Maple Street, behind the grade school, and above Sixth Street. Steve Oblackovich recalled cutting himself badly on a broken jug in 1922 when he scurried out of the way of the National Guard who came to water their horses. The guard was in town to "keep the peace" during the UMWA strike. Another reliable source of water was Three Springs, located halfway between Wehrum and Vintondale. Townspeople still use these springs today for their drinking water.

1930 was a year of an especially bad drought. The lines at the springs built up again, as did tempers. Even operation of the washery was affected by the lack of water. One year, residents even had to haul water in the winter. Thus, Vintondalers have had to wage a never-ending battle for potable water; most of the time, the townspeople were the losers.

VINTONDALE BOROUGH COUNCIL

Until 1908, the village of Vintondale was governed as part of Jackson Township. How much control the Vinton Colliery Company exerted over the township supervisors is unknown. The twelve years between its founding and its incorporation as a borough remain a mystery. Sources providing facts on early government are also scarce. Occasionally a local paper carried a news story; the mine diaries provide a few facts, and the extant borough minutes date to 1921. By gathering together these fragments, town government in the second and third decades of existence can be pieced together.

From its earliest existence, the borough council can be called the tool of the company. Representatives were Replican companymen. Louis Burr, son of company treasurer A.T. Burr, worked in Vintondale from 1906 to 1911. During that time, he also served on town council. The borough council president frequently was the superintendent. After succeeding to the position of superintendent in 1915, Otto Hoffman served as council president until his death in 1930. During those years, the polls were watched by council members, and those businessmen of Democratic persuasion were often urged to see the light. In a typical election in 1915, the vote was 83 Republican to 8 Democratic. This lopsidedness did not alter until the mid-1930's when Warren Delano's nephew, Franklin, signed the Wagner Act, which legalized unionization.

In February 1913, an effort was made by a few townspeople to open up council to more public representation by enforcing the Act of Assembly of May 18, 1907, which forbid company employees to sit on council. The unidentified writer of the 1913 mine diary decided that Mr. Krumbine, the undertaker, was behind this move. The matter did not die an instant death.

In August 1915, a mini-revolution struck Vintondale. Disgruntled by a continued lack of representation on council, a number of townspeople signed petitions to have council members show by what right they were on council. The group convinced District Attorney Greer to present the writs to court. Papers were filed on Burgess William Berkey and Councilmen Dan Thomas, William Abrams, John Morton, Lloyd Arbogast, Otto Hoffman and Sam Feldman; these members, obviously all companymen, were elected November 4, 1913. Exempt from the petitions were Tax Collector, Edward Wright, and assessor, George Blewitt.

The company furnished the heat, water, lights, and building for council meetings. All council members were VCC employees; and in violation of state law, the company provided laborers for the borough.

As a result of the disagreement, the old council resigned, and a new one was appointed. In the meantime, the company discontinued furnishing the lights, building, coal, etc. for the council and the jail. In addition, Mr. Burkey resigned as burgess, and there was no successor, so the town was without lights, burgess or jail. Final outcome of the dispute is again unknown.

The earliest existing borough council minute book started in September, 1921. At that time, Otto Hoffman was the council president; secretary was John Burgan, treasurer was Paul Graham, non-company. Tax collector was Dan Thomas. Part of the business enacted that fall was to set the pay for town policemen, Jack Butala and Harry McCardle, at $35.00 per month. The company, through council, was able to control the number of vendors who came in town. Each cart or truck was required to pay $5.00 a day for a vendor's license. Each additional person had to purchase another license. The borough no longer sold a half-day license. The monthly average acquired during the winter was $25; in July and August, the total was as high as $120. In the late 1920's, Norman Hyatt, P. Shinhefrin, D. Menosfick, B. Rubin were all granted peddlar's licenses on a fairly regular basis. The borough also collected $4 per day for the organ grinder's license. Another duty of council was to grant gasoline tank permits. (See the section on automobiles.)

One building permit was granted in September, 1922. Nick Burch applied for a permit to build a barber shop adjacent to the Farkas Hotel. Council delayed granting the permit until it received a blueprint of the proposed building, which was required by ordinance. Burch constructed a new brick barbershop and received permission to erect a sign 12"x24". The old wooden barbershop was moved two feet away from the new building.

In 1922, Henry Misner was appointed street commissioner, and Grant Gongloff took a seat on council. The borough retained Charles Evans as solicitor for $100 per year. In 1922 the UMWA strike brought notoriety to Vintondale, and the company, through council, attempted to prohibit public meetings, assemblies, gatherings, or parades on the public highways. This was done by Ordinance 32. (See chapter on 1922 strike.)

In August 1923, council passed Ordinance 33 which vacated lower Fourth Street. No reason was given in the minutes, but this was the site chosen by Schwerin and Hoffman for the new movie theatre.

New members sworn in on January 7, 1924 were Arbogast, Feldman, Swanson, and Hoffman, who was unanimously elected president. Due to the KKK Strike of 1924, there were some changes on council. John Morton resigned in March; Herb Daly was appointed to his seat and then resigned in June. Tom Whinnie and Walter Hunter replaced Hilding Swanson and Grant Gongloff, who "moved out of the community." These men actually did not move out on their own free will. They were fired for their strike activity and may have been evicted. (See list of evictions in the appendix.) In September, Russell Dodson took the vacant Morton seat, and Allie Cresswell succeeded to William Abrams' empty seat.

To stay afloat during the 1920's until the tax money came in, council borrowed money on short-term loans from First National Bank of Ebensburg and then Vintondale State Bank after it opened in 1924. For instance, in May 1923, council borrowed $1,700 for four months; it paid $735 on the note in September and renewed the $1,000 note for four more months. In March 1924, the note was transferred to Vintondale State Bank. Money was borrowed on a regular basis between 1926 and 1930.

In 1925, Dan Thomas resigned his commission as borough tax collector, and Roy Roberts was appointed to the post. Council also passed a motion to purchase new chairs and seats for the playground equipment.

John Clark became the new council member in 1928; he replaced Sam Feldman, who became a justice of the peace that year. Harry Hampson resigned as assessor in December. In January 1929, council passed motions to petition the court to appoint Dorothy Kreitzberger and Cora Roberts to fill vacancies created when Dr. H.V. Cottom and A.H. Tipton moved from town. Marie Cresswell's name was petitioned for assessor. Council authorized the purchase of six blankets for the jail cells. In May, O.P. Wilson took a seat on council replacing Lloyd Arbogast, who had suffered a breakdown. John Burgan, who was moving to Ferndale, tendered his resignation as borough secretary; Ruth Hopfer was elected to fill his position. Council voted to give Burgan a vote of thanks for nineteen years of service.

Council members in 1930 were Otto Hoffman, Walter Hunter, Allie Cresswell, Russell Dodson and Perry Wilson, again all companymen. Jim Dempsey was appointed street commissioner, and Louis Nevy was reappointed to the Board of Health.

Due to Hoffman's death in March, council did some scrambling to regroup itself. Allie Cresswell became chairman; Harry Hampson was elected to Hoffman's seat and then to the council presidency. Hampson resigned in May; Cresswell became chairman; Frank Colbert was elected; John Clark resigned in August, due to moving from borough; and Milton F. Brandon, new superintendent, was appointed to council and the presidency because Cresswell wished to remain a member only.

In 1930, council did approve a motion to have the garbage collected and carried away to the rockdump. A desk from the company office in Claghorn was moved to council rooms with the borough paying for transportation costs.

Council underwent some upheavals in the 1920's, but the worst was yet to come, the Great Depression. Borough bank accounts would at times amount to less than $10. Politics underwent a complete flip-flop too in 1932 and 1936. Council was no longer to be a tool of the company. Now company and union had to cooperate to keep the town going.

MEDICAL CARE IN VINTONDALE

One of the most demanding occupations in a mining town was that of the doctor. He made daily house calls, had office hours several times a day, had to be on the alert for mine accidents, and was called out at all hours of the night to deliver babies. Rarely did he get a full night's sleep. As long as the company controlled the town, the doctor was chosen by the company. His monthly salary came through the company who deducted $.75 to $1.00 per month from each employee. Out-of-town patients often paid the doctor in kind: produce, eggs, vegetables, and meat.

In Vintondale, the doctor's office was a two-story building at Third and Main adjacent to the Nevy Bros. store. It had a small waiting room, an examining room, and an operating room downstairs. Upstairs were six to eight beds which could be used as an infirmary if necessary. About the only time the beds were occupied was after the doctor removed children's tonsils. They stayed overnight as a precaution.

In the hard times of the 1920's and especially the 1930's, being a company doctor actually was a lucrative occupation. The doctor's fee was still deducted from the paycheck even if the mine only worked a few days that month. That meant that in Vintondale, the doctor had a guaranteed income of several hundred dollars a month, perhaps as high as $500 per month.

In addition, the doctor had the option to rent one of the largest houses in Vintondale at the nominal rent of $22 per month. The Clarence Claghorn house on the corner of Second and Lovell became known as the "doctor's house" until the company sold all their houses in the mid-1940's. Since labor was so cheap, most of the doctors had housekeepers or full-time cleaning ladies. Dr. Phillip Ashman said that he hired a cleaning girl for $5.00 per week. The house was so large that even his St. Bernard dog had his own room.

Vintondale doctors performed few operations because of lack of facilities. Emergencies, such as mine accidents, were sent to the hospitals. If a patient was sent to the hospital, the case was referred to a Johnstown physician because the company doctor did not have the time to drive to Johnstown to check on his patient.

Transportation to the hospital was also a problem. Roads were treacherous even in the summer because state roads were not paved until the late 1920's. Many times the patient suffered additional injuries by being bounced around in the vehicle.

Many miners frequently came to the doctor's office, sick or not, because they believed that they were paying for the doctor, so they should take advantage of his services. To satisfy this type of patient, Dr. Ashman said that he distributed a lot of pink aspirin tablets during his tenure in Vintondale.

At the same time, some people, especially immigrant women, were reluctant to go to the doctor. Many women chose to have a midwife deliver their babies. Before 1920, Vintondale had three known midwives: Ann Petrilla, Agnes Such Toth and Mary Shisler. Mrs. Petrilla's daughter, Annie Dancha, said that her mother had midwife papers from Slovakia and was not paid for her services. Mrs. Shisler, who had a boarding house on Plank Road, also on occasion practiced denistry. Mrs. Dancha, as a child, observed Mrs. Shisler pull a boarder's tooth with pliers. She told the man she was going to kill the pain with tobacco and then yanked out the tooth before he knew what was going on.

The local newspapers are main sources of medical news in the early years of the area mining towns. In 1898, Dr. Merritt Shultz of Strongstown, amputated the left leg of Henry Shultz, who was suffering from "senile gangrene". Assisting Dr. Shultz was Dr. D.S. Rice of Hastings. Mr. Shultz survived the operation and rallied quickly. Dr. Shultz also treated forty-two cases of typhoid fever between June, 1897 and October, 1898. Only two of these cases resulted in fatalities. In 1901, the **Indiana Progress** reported that there had been several cases of typhoid fever in the Vintondale/Strongstown area. Catherine Mahan Wray's father, Bert, a carpenter who helped build many of the houses and the train station, contracted typhoid shortly after moving to Vintondale. Verna and Pearl Findey had critical cases of typhoid. Dr. Gile nursed them back to health, often spending the entire night at the Findley house. Any time one of the girls would spike a fever, he would immerse them in a tub of cold water. After they recovered, they worked at Dr. Gile's house; one took care of the kitchen, and the other helped with the children. One source of the typhoid germ was a spring on the Main Street lot across from where the Farkas Hotel was later built. This spring was sealed off to prevent further outbreaks. In January and February 1903, there was an outbreak of smallpox in the valley. Because there were several cases in Wehrum, the schools were closed, and the town was put under quarantine. In Vintondale, a known smallpox case was Joseph Bennett. While working at a Vintondale hotel, Alfred Detwiler of Nolo contracted the disease. The D.M. Henry lumber camp near Strongstown was also quaranted. By March, the disease had run its course, and there were no reported deaths from smallpox.

Town Doctors

The first known resident doctor was Dr. McCormick in 1895. The company replaced him in 1897 with Dr. Wilson; this caused a labor stoppage. (See section on union activity.) The **Indiana County Gazette** briefly noted that a Dr. Wilson of Vintondale had paid a visit to Indiana. Dr. Merritt Shultz practiced in Strongstown and purchased a lot on lower Main Street, but it is not known if he had a regular practice in town.

Another resident doctor, B.C. Giles practiced in Vintondale at least between 1898 and 1900. Sadie Cameron, employed as a child's nurse by the Giles, resigned in 1900 and took a job at Mrs. Lundy's boarding house near the Vinton Lumber Company mill in Rexis. Dr. Gile moved to Wehrum to assume the position of company doctor and may have taken care of both towns. In Wehrum, he lived in the large house at the top of Broadway. In July, 1902, an Indiana paper reported that Dr. Giles operated on Ben Duncan of Strongstown removing portions of two ribs. Dr. Giles probably left Wehrum during the shutdown because on May 24, 1906, his wife, Ysable Gile, now Ridley Park, Delaware sold Lot 2, block D to the Vinton Colliery Company for $150 on May 24, 1906. This property on Fourth and Barker had been purchased in 1900. Another house which the Giles owned in Vintondale was Lot 2, Block L, present home of Stephen Dusza. The Giles sold this house, which may have been used as the Catholic Church rectory, to Warren Delano in 1906. Mrs. Gile, who was of Spanish descent, was a sister to the famous actress, Faye Templeton, who paid occasional visits to her sister in Wehrum.

In 1906, there was a Dr. I. Pollum who had addresses of both Vintondale and Wehrum. The next known

Vintondale company doctor was J. Alvin Comerer, who practiced in Vintondale from 1906 to 1912; he may have come to Vintondale earlier than that. During his tenure in Vintondale, he married Laura Lynch, and three of their children were born here. Upon leaving in 1912, Dr. Comerer moved to Bethlehem where he specialized in opthomology.

Replacing Dr. Comerer was not a simple task as the 1912 mine superintendent's diary attested. In June, Dr. Boyd of New York came to Vintondale to "look at the situation". That summer two cases of diptheria developed, but it is not known who treated the patients. In August, Dr. Johnson arrived, followed by Dr. Humphrey in September. Humphrey apparently agreed to accept the job, but did not show up as expected in December. In the meantime, Dr. Nix of Wehrum took charge of the Vintondale employees.

As of January 1, 1913, the town had no doctor. Nanty Glo physician, Dr. James Patteson MacFarlane paid a visit to town on January 15th and took over the position of company doctor for the Vinton Colliery Company on February 15th. Dr. MacFarlane was born near Glasgow, Scotland on August 28, 1878. Upon emigrating to the United States, his father settled in Renovo and eventually became superintendent of the Blossburg mines. A graduate of Mansfield State College and Jefferson Medical School, Dr. MacFarlane worked for the Weaver coal interests in Idamar and Nanty Glo until 1912. He married Alice Jose of La Jose in 1908. (The town was named after her family who had a hotel and lumber interests there. Their son James remembers being at the hotel in 1936 the night before the ill-fated logging raft went down the West Branch of the Susquehanna; he said all the lumbermen were drunk.)

Upon his arrival, Dr. MacFarlane faced a measles epidemic with 100 cases reported on March 2, along with a case of diptheria. George Blewitt was sent to Ebensburg to retrieve some anti-toxin. The measles epidemic lasted at least until May because the diary reported that John Burgan was off work with the measles. Hazel Cramer Dill said that her sister had a case of the croupous membrane (diptheria); the treatment was to use a gold tube to open the breathing passage and to administer a dose of anti-toxin.

As company doctor, MacFarlane was provided with what VCC called a hospital, complete with a new drug cabinet that had been purchased in March from the Johnstown Planing Mill. MacFarlane moved into the doctor's office on the corner of Third and Barker Streets on May 12, 1913.

**Dr. James P. MacFarlane.
Courtesy of Jim MacFarlane.**

Substitutes

Occasionally, the doctor was able to leave town for business or vacation. One of the doctors from Wehrum, Expedit, or Nanty Glo handled the emergencies. Once in a while Dr. MacFarlane had an assistant to cover for him. In most cases these doctors had just graduated from medical school and worked as assistants for a year or two to gain for practical experience. Then they left to open their own practices or returned to medical school for additional specialization. In 1914, Dr. St. Clair served as an assistant, but left that October to relocate to Alexandria. In June of 1914, Dr. MacFarlane broke his arm when he was thrown off Officer Timmons' horse. The saddle was ruined, and the horse was "skinned up quite a bit". Dr. MacFarlane's injury required several stays in Memorial Hospital and a trip to a Chicago facility. Dr. MacFarlane was able to leave the practice in St. Clair's hands. Another assistant was Dr. H.C. Thomas in 1916. Dr. Paul Blake was

**James MacFarlane.
Courtesy of Huth Family Collection.**

in Vintondale in February, 1917, followed in May by another assistant, Dr. Kim Curtis. Dr. H.F. Gockley assisted in Vintondale in 1920. In July 1923, a goiter and skin disease specialist educated in Scotland, Dr. H.C. Brown, came to Vintondale to be the assistant. An unnamed doctor turned out to be a drug addict.

Dr. MacFarlane was also gone for extended periods of time in 1925 when he traveled to Mt. Clemens, Michigan for treatment of rheumatism. The **Nanty Glo Journal** reported that Dr. M. Rosenzweig of Pittsburgh was his replacement. The author contacted a member of the Rosenzweig family in Pittsburgh to see if this Doctor Rosenzweig was a relative. Most likely, Vintondale's Dr. M. Rosenzweig was Maurice Rosenzweig, (1898-1977), who graduated from high school in 1916, attended Pitt's pre-med school, graduated from Jefferson Medical School and interned at West Penn Hospital. According to his brother, Aaron Rosenzweig, Maurice spent a short time around 1925 working as a company doctor in some mining town. Later, he specialized in pathology and then hospital administration in Boston, Milwaukee and Eleria, Ohio. After his retirement, Dr. Maurice Rosenzweig returned to Pittsburgh and assisted at the Veterans Administration Hospital. All this circumstantial evidence leads the author to conclude that Dr. Maurice Rosenweig was the doctor substituting for Dr. MacFarlane in 1925. In 1928, Dr. Mac. was off duty again when he sailed home to Scotland for a visit. When he fractured a bone in his foot while making a house call in 1933, Dr. Prideaux of Twin Rocks covered for him.

Dr. Mac As Diagnostician

Using a mortar and pestle, Dr. MacFarlane ground his own medicines, and his diagnoses were usually very accurate. Mabel Davis Updike, who worked in the office from 1912 to 1916, said Dr. MacFarlane diagnosed her nausea and headaches as eye strain. Not believing him, she went to an eye doctor anyway and received the same diagnosis. In another case, Mary Monyak Drabbant developed an infection in her leg from a bug bite. After examining her leg at her home, he took her in his car to his office. He told her that the treatment would burn badly, but the infection cleared up. Some of the miners grumbled that MacFarlane's medicines were no good. However, many of them did not follow the directions. John Leminos, one of Mrs. Drabbant's boarders, had a packet of powder to take three times a day. She observed him dump the entire packet into his soup. When she quetioned him as to why he did that, he said that if he had to take it three times a day, what was wrong with taking it all at once? Luckily, the medicine did not harm him.

The flu epidemic hit Vintondale as hard as most other communities. Lovell Mitchell Esais recalled that she had it "real bad" and was confined to her bed at home, next to the Baptist Church. She stated that she did not know if MacFarlane had any other doctors assisting him. From her bed, she could watch the patients being carried in and out of the school building (the Old Building) on the hill. Joe Dodson said he helped carry bodies to the school, which served as a hospital and morgue. Wooden caskets were built in the company machine shop. Ann Sileck Pytash credited MacFarlane for saving her mother's life; Mrs. Sileck caught the flu after giving birth to a son, Joe. The infant survived because a neighbor became his wet-nurse.

During the flu emergency, many useless cures were touted. The liquor industry printed propaganda that whiskey was a valuable part of the treatment for the flu. The US Army Medical Corps countered that claim and declared whiskey detrimental to recovery. In fact, they urged people to avoid public places such as saloons. Because World War I was still being waged, there was a shortage of caskets in the states. The War Industries Board eased up on regulations to hasten the construction of caskets of the simplest type.

According to many of the women in town, MacFarlane was an excellent obstetrician and pediatrician. Dr. MacFarlane delivered Catherine Mahan on the family farm near Cardiff; he also delivered her future husband, Oscar "Pete" Wray. Their first daughter, Lavone, was one of the last babies delivered in Vintondale by Dr. MacFarlane. Sara Williams George threatened to go to Indiana to have Dr. MacFarlane delivery her last child. However, Bill George was born at the Mendenhall Maternity Hospital in Johnstown.

Agnes, Aileen, Margaret Huth - quarantined for chicken pox, c. 1920.

Not all Vintondale residents were satisfied with Dr. MacFarlane, especially those with union leanings, in particular, those evicted in the 1924 strike. Lovell and Dick Esaias, office employee and company policemen respectively, were given an eviction notice. The night before they were to be evicted, the couple was in the process of packing when Dick sat down on the edge of the bed, said "Oh" and fell back on the bed. He had suffered a massive stroke. By the time Dr. MacFarlane arrived, Esaias was unconscious. The doctor asked what they had for supper and if it made her sick also. He told Lovell that if Dick lived until morning, he would take him to the hospital. Otto Yankhe helped Lovell care for Esaias that night. In spite of her husband's stroke, the eviction took place as scheduled. Mr. Esaias, who had been gassed in France during World War I and spent time recovering at Walter Reed Hospital in Washington, D.C., survived the ordeal. As a result of the stroke, Esaias was paralyzed on one side. The couple settled in Johnstown, where he died in 1929.

Changing of the Guard

When the union was accepted in 1935, the miners wanted no reminders of the old closed company town days. Since the doctor's wages came from their pay, the miners believed they were entitled to choose their own doctor. Their replacement for Dr. MacFarlane was Dr. Jerome Cohen. In spite of a certain amount of hard feelings between the two doctors, Dr. MacFarlane did not depart Vintondale broken-hearted as many thought. Jim MacFarlane said that his father had planned his retirement for several years, and even owned the house at 830 School Street in Indiana for two or three years before they moved into it. Dr. MacFarlane opened a practice in a room added to the house; the dining room served as an examination room. However, his retirement was short-lived as he died of a stroke on September 7, 1937. His widow Alice, rarely seen without a pearl choker necklace, remained at that address and graciously welcomed old Vintondale friends. Mrs. Otto Hoffman was a close friend of Mrs. MacFarlane, and they often visited back and forth after both left Vintondale. Mrs. MacFarlane died in February, 1976. Son James, a 1935 graduate of Vintondale High School and a law school graduate, set up a practice in Butler.

Shortly before Dr. MacFarlane's departure, his office burned to the ground on the night of March 12, 1936. Only a medicine chest and Dr. MacFarlane's diploma were rescued. The fire was blamed on the electric wiring in the basement. A temporary doctor's office was set up above Bill Roberts' Confectionery in the Farabaugh building.

Dr. MacFarlane's successor, Jerome Cohen, had interned at Memorial Hospital. Dr. Cohen and his wife, the former Freda Zeff of Johnstown whom he married in 1935, moved into an apartment on Main Street until the doctor's house was vacated in the spring of 1936. Although Dr. Cohen and his Vintondale patients had a good working relationship, he decided to specialize in opthamology. In 1937, Dr. Cohen and his family moved to New York City while he took a two month course at the Ear and Eye Hospital. After studying in Europe, he opened a practice in Johnstown, and many Vintondale families, including the author's, became patients. Dr. Cohen died in the fall of 1981.

Dr. Ashman

Replacing Dr. Cohen as the union doctor was Dr. Phillip Ashman, a former classmate and fellow intern at Memorial. In 1937, Dr. Ashman had a practice in a small Pennsylvania town. Because of the Depression, the practice was not very good, so he took the Vintondale job. Dr. Ashman made housecalls every day, starting at the top of Chickaree, working his way out Plank Road and then down Main Street. He kept a little black book in which he wrote down his stops. After the visit, he checked off the name. Almost everyday, he stopped at Joe Lutsko's house on Chickaree. Because of the large family, at least one of the children needed medical attention that day. Some patients came to him at least once a month to make sure they were getting their money's worth.

Ashman held office hours twice a day and once on Saturdays. In winter, he made housecalls because few people could get their cars up Second or Third Streets. Although early tensions with Superintendent Brandon eased, Ashman had little contact with the company. Since Ruth Hopfer handled the check-off system in the office, she was the only company personnel with whom Dr. Ashman dealt. For miners, all fees were covered by the wage check-off system, except deliveries and veneral disease treatment. A delivery was about $20, and the VD treatment was extra because of the series of shots. For those not on the check-off system, payments were made in cash, in kind, and sometimes not at all.

Because of the difficulty of getting to Johnstown and the inconvenient phone system, Dr. Ashman also turned his hospital patients over to another doctor. Few families had a telephone, and if you wanted to make a call out of Vintondale on weekends, you had to get someone to open up the office.

Dr. Ashman's busiest times were the Saturday nights after paydays. He had the task of mending and stitching up the brawlers who were busy fighting their way up and down Main Street.

One non-medical memory remained with him. His son was born in Vintondale, and as a tot, Jerry quickly learned how to find Mabel Huth's house where Mom and the girls would give him cookies and other goodies. When he came home for supper, he would not eat. So Mrs. Ashman began pinning a "Do Not Feed" sign on his clothes. One day, he wandered as far as #6, insisting that he was going inside to work that day.

In 1940, Dr. Ashman joined the service. After the war, he opened a practice in Johnstown. In the early 1980's, he was semi-retired and worked at the Hiram Andrews Rehabilitation Center. Dr. Ashman was a very welcomed guest at the 1984 Vintondale Homecoming. He died the following year.

David Dusza and Jerry Ashman, June 1940.
Courtesy of Agnes Dusza.

Dr. Prideaux

Dr. William A. Prideaux was a very familiar figure to the miners of the middle Blacklick. He was the resident physician for Commercial Coal, Big Bend Coal, Jackson Colliery, Bethel Colliery, and Blacklick Colliery. Born in Smith Mill, Clearfield County on February 7, 1873, Prideaux was a graduate of Juniata College and the Medico-Chirurgical College of Philadelphia in 1899.

His first practice lasted seven years in Cherry Tree, and in 1907, he moved to Expedit (Twin Rocks) with his wife, the former Anna Grumbling. Prideaux remained Twin Rocks' only physician until his death in May, 1953. During his career, Dr. Prideaux traveled 350,000 miles, wore out ten horses and twenty-six cars, and delivered 4,000 babies. Each day, he made rounds at Commercial's patch village of Bracken. As a young doctor, he would hop the fences which separated the yards and shout into the kitchen, "Everyone OK?", and then proceeded to the next house. Dr. Prideaux even saved the life of a Wehrum youngster when he performed an emergency appendectomy on the family's kitchen table.

In honor of his long service to the Valley, a testimonial dinner was held in Twin Rocks in 1951. A year after his death, a fund-raiser was held to raise $5,000 to furnish a room in Dr. Prideaux's memory in the new Memorial Hospital edition. The money was raised within two months thanks to an anonymous donation of $2,000; the room was a fitting tribute to a man who dedicated his life to the well-being of the people of the Blacklick Valley.

EDUCATION IN VINTONDALE

Providing a quality education has always been a priority in Vintondale. Hundreds of well-educated students have graduated from the halls of Vintondale High School and sought careers as doctors, lawyers, teachers, scientists, managers, skilled craftsmen, etc.

Before 1908, Vintondale's schools were administered by the Jackson Township School District. Buildings in town serving as school houses were a room behind the company store, attached to the jail; a large two-story wooden building across the alley from the Baptist Church; and half of the large white building which stood behind the iron furnace. When the Buffington Township school at Lackawanna #3 burned in 1910, the iron furnace building was converted into two rooms: one for Buffington students, and one for Vintondale students. That school was remodeled into three apartments in 1914 with its desks moved to the abandoned #1 powerhouse. Charlie Shestak was one of the miners who lived in the schoolhouse along with the Shooks.

In 1908, Vintondale became a borough, and the Jackson Township School District released the wooden school building and rights to three lots in Block E to the new Vintondale Borough School District. Cost of the transfer was $1.00.

Until 1916, Vintondale schools provided an education only to the eighth grade. Those who wished to continue with high school had to catch the earliest train in the morning to Ebensburg, attend the Ebensburg high school during the day, and return on the evening train. Eula Hampson Burr, Ada Kempfer, and Jennie Kempfer attended high school in Ebensburg around 1901 or 1902. The Kempfer girls were from two separate families. Some of Vintondale's students of the pre-high school years attended college. In 1909, Joe Fleitzer, son of the Vintondale Inn proprietor and Sam Sher, son of the jeweler, were students at the Indiana Normal School, Indiana, PA.

Students who attended the school beside the jail often wore their coats and boots during class because the room was so cold. It was heated by the steam pipes from the company store. Martha Abrams Young, whose father served on the school board, attended that school. She said one day the students heard a moaning in the jail and discovered that evening when they went home that the noise came from an inmate that had committed suicide by hanging. (See death certificates.)

Because of the increase in the number of pupils, the school district decided to construct a new school. School board secretary, Robert Clyde, transferred the four lots in Block E, including the wooden school house to the company. In exchange, VCC granted the district six lots in Block N. This undeveloped area was located above the Dinkey Track between Fourth and Fifth Streets. A new six-room brick school was completed in 1913 on that site. With a commanding view of the town, the new school had two rooms on each floor with a boys' restroom in the basement and a girls' restroom on the first. All classrooms had a second emergency exit. Students later identified this schoolhouse as the "Old Building".

Vintondale Elementary School, built 1913. Known as the "Old Building". Currently in the process of being dismantled.
Courtesy of Huth Family Collection.

At times, the student population taxed the capacity of the building, and extra classrooms were opened in the Russian church basement, at the back of the company store, and in the old school by the Baptist Church. Annually, there was an increase in student population, especially girls, because in the past many had been kept at home to help their mothers. Many of the older boys were already working in the mines.

The 1916 **Department of Public Instruction Annual Report** listed Vintondale as having five schools, (classrooms perhaps?), with one male teacher and four female teachers. In those days, there was a definite discrimination in salary between the male and female teachers. The monthly salary of the male teacher was $92.50 while the women drew $61.25. Pupils enrolled were 146 boys and 111 girls; the attendance rate was 89%. In 1917, an additional male teacher was added; enrollment was 167 boys and 149 girls with the attendance rate dropping to 78%.

The total receipts of the district in 1917 was $8,880.57; expenditures were $7,968.40. The largest expense, $3,677.26, was for salaries. $553.97 was spent on textbooks and $254.68 for supplies. Enforcing the compulsory attendance laws cost the district forty dollars. Mary Monyak Drabbant said that her father frequently kept her at home to work and paid the truancy fines.

Until 1919, the length of the school year was 8 months; then, it was increased to 8.85 months. Four more teachers were added in 1919 bringing the total to ten. Accompanying the increase in teachers was a large salary hike. The lone male teacher received $160 per month; the females earned $92.77. Student population rose also, 214 males and 208 females; attendance reached 93%. District receipts in 1919 amounted to $33,099.87 with expenditures at $31,337.65.

The elementary curriculum was standard, judging from Agnes Huth Dusza's report cards. In third grade, students were offered arithmetic, language, geography and penmanship. The handwriting method taught at this time was the Palmer Free Arm Movement Handwriting; this became the standard method used in all Cambria County schools during the tenure of County Superintendent M.S. Bentz.

First High School

Vintondale boasted the first high school in the Blacklick Valley. It originated with the ninth grade class in 1916. The class met in one room of the "Old Building" and was under the direction of supervising principal, Professor Montgomery. As the high school enrollment rose, it occupied two rooms in the adjoining brick school known as the Delano Building. (This building burned and was replaced in 1930.) Professor Peters was the next supervising principal.

The Vintondale High School was a three year high school, which meant there were no sophomores, only freshmen, juniors, and seniors. Since Vintondale had the only high school, students from Twin Rocks (Expedit), Nanty Glo, Wehrum, and Buffington Township attended. Maurice Shadden, who later became an attorney, attended the Vintondale schools for two years. Ysabel Bostick Frazier, daughter of Edward and Pearl Bostick, Cherry Tree, R.D., lived with her grandparents, Edward and Sara Findley for two years so she could attend high school in Vintondale. There was no secondary school in her home district in Montgomery Township. In the first graduating class of eight, seven were from Vintondale and one was from Nanty Glo. (In 1953, the class ring of Mildred Kerr Myers, Class of 1920, was found in front of Rabel's ice cream parlor in Nanty Glo. The ring was lost in 1938. Betty Rabel gave the ring to Mr. Mower who traced the owner through school records.)

The curriculum offered to the freshmen included English, Latin, one-half year of Civics and one-half year of History, General Science and Algebra. In the senior year, the standard courses were: English, Latin IV (Cicero), Problems of Democracy, Physics, Solid Geometry and Health. Later a business course was added.

Between 1920 and 1926, the high school faculty consisted of J.R. Jones, supervising principal; Miss Laura Burket, English and Latin; Mr. S.G. McChesney, History; and Miss Maud Jenkins, Science.

Writing the weekly school news for the **Nanty Glo Journal** developed into an English class assignment. Each week, two new students became the reporters. In 1925, the **Journal** listed the school averages for the top students in the Vintondale High School. They were:

Class of 1925 - Seniors
Five Semesters
Ruth Woodward - 87.4%
Thelma Thomas - 87.3%
Helen Thomas - 86.7%
Class of 1926 - Juniors
Three Semesters
Mary Sileck - 90.6%
Mary Burgan - 89.2%
Mildred Evans - 86.7%
Class of 1927 - Sophomores
One Semester
Alfer Robanke - 93.1%
Bruce Lybarger - 90.0%
Lavon Lockard - 90.0%
Walter Simons - 90.0%
Anna Kanish - 89.6%

The **Journal** also identified the names of eight previous graduates who had averages of 90% and above:

1919: Benton Williams, Kenneth Hoffman, and Charles Clement.

1920: Elizabeth Beechey.

1924: Nina Carson, James Beechey, Joseph Sileck, and Katherine Morris.

As can be seen, the grading system at the Vintondale schools was demanding; no one had a 100% average, not even the superintendent's son.

In the 1920's, the Vintondale School District was in excellent financial condition. In 1925, the board was able to pay off $12,000 worth of bonds four years early. Board president at the time was Mr. Lockard, and Mrs. Hoffman was vice-president.

From 1920 to 1926, J.R. Jones was the supervising principal of the Vintondale schools. Born January 23, 1893 in Millerstown, Jones was a graduate of Shippensburg Normal School in 1910; he also attended several summer sessions at the University of Pittsburgh. This teaching experience included three terms at Mercersburg, Millerstown in 1913, two years in Center City, and from 1916-18 with the U.S. Bureau of Education in the Philippines. During World War I, Jones served ten months in France as a member of Company E, 11th Regiment Marines. Mr. Jones, a Republican, accepted the Vintondale position in 1920; he married Ethel Patterson of Johnstown in 1921. The couple, who lived in the Thomas house on Third Street, had a son Wayne, born February 9, 1923.

During Jones' first year, the students raised funds to purchase a piano for the school by holding socials, girls' basketball games, and public entertainments.

The commencement in 1921 was held June 2nd in the Baptist Church; valedictorian was Ralph Smith of Twin Rocks, and salutorian was Winifred McCracken. Other known graduates were Hugh Harrison and Gazel LaBalle. A class picnic was held on Rexis Hill, followed by a banquet a few days later. A class picture was taken by Deck Lane Studio of Ebensburg.

Charles Mower

In 1926, Jones resigned to accept a similar position in Summerhill. The school board lured Charles Mower, the "popular and energetic" vice-principal of the Reed High School in Blacklick Township to accept the post of supervising principal. Mower was a graduate of Shippingsburg Normal School. Before starting his teaching career, he sailed with the Merchant Marine as a common sailor, making port in Germany, Sweden and England. A college friend, Mr. Nettinger, taught eighth grade at Blacklick and convinced Mower to teach in Blacklick Township. His salary at the time was $180 per month, considered a good wage. There were two teachers at the Reed School; each taught a class of about twenty students. Mower was in his second year of teaching when the county superintendent sent letters to area teachers notifying them of the Vintondale vacancy. The candidates' names were submitted to the Vintondale Board by the county superintendent. Serving of the board in 1926 were: John Burgan, John Roberts, Herbert Daly, John Lockard, and Mrs. Hoffman. When Mower contacted the board, the job was given to him. At that time, the position of supervising principal also included some classroom teaching. Mr. Mower's preference was Math, but he was also certified in Latin, General Science, and English. During the summers, Mr. Mower attended school to pick up needed credits for a four-year degree; eventually he received a Master of Arts from Arizona. Mr. Mower took up residence at the Vintondale Inn and boarded there until 1942 when he entered the service in World War II.

Mr. Mower stated that he had no pressure from the coal company in running the school. Only once, Mr. Hoffman "requested" that Mower close the school for a day because a senator of the "right party" was coming. That day off school may have been in October, 1926 because the **Nanty Glo Journal** listed a group of Republican state candidates who were going to meet at Hoffman's house and tour towns in Cambria County. One of the candidates was John Fisher of Indiana, future governor of Pennsylvania. Fisher's law firm eventually became the legal representatives of the Vinton Colliery Company. Fisher was associated with John Scott, lawyer for Lackawanna Coal and Coke Company.

New Four Year High School

With the school district financially in the black and the high school desperately needing better facilities, the board voted to expand to a four year high school and erect a new high school building. In Mower's first year as principal, there was no graduating class as the school expanded to four years. All but one of the seniors of 1927 returned to complete the additional year. In Mr. Mower's opinion, this class of 1928 was the most outstanding in school history. The three top students in the class: Bruce Lybarger, Mike Hozik and Lavon Lockard attended Indiana State Teachers' College and graduated in the top three there. The order of their rankings is not known. A story circulated in town that Mike Hozik was called into the office of the Dean, who wanted to know what kind of school we had in Vintondale that could produce top students like them.

The class of 1928 was also the first to graduate from the new high school. The site chosen for the new school was the park at the corner of Lovell and Second Streets, Lots 10 and 12, Block J on the town plan. Construction started in August, 1927 and was ready for occupancy on January 23, 1928. To save labor costs, Lloyd Williams and other students worked on the school. Gig Hubner and other mine employees were sent to assist in the construction because things were slow at the mines.

The new school had six classrooms, including a laboratory with $500 worth of equipment and a double room with a small stage for classes or assemblies. The library had seven tables with a seating capacity of forty-two; they arrived in early February. The collection consisted of 600 volumes, including the purchase of 100 new books at a cost of $150. Library rules in March 1928 banned students from talking in the library; those found guilty were banned for two weeks and had to make up the time after school.

On March 24, 1928, the Department of Education accredited Vintondale as a four year high school. The class of 1928 had nine graduates, and the total enrollment for 1927-28 school year was 65.

In order to give townspeople an opportunity to tour the new school, an open house was held inconjunction with an art exhibit loaned by the Nelson Company. The senior class sold refreshments to raise money for the class trip. Each evening of the exhibit, students provided entertainment, which varied nightly. The exhibit brought the school a net profit of $110 which was used to purchase pictures for the new building.

When school opened on August 27th for the 1928-29 school year, the total enrollment for grades 1-12 was 500. The high school had 75 students. The teachers were:

Mr. Mower - Principal
Latin and English - Laura Burket
Science - Maude Jenkins
History and Health - Isabel Condie
Eighth Grade - Ruby Meyer
Seventh Grade - Alice Dormon
Sixth Grade - Jennie Whiteknight
Fifth Grade - Margaret Breth
Fourth and Fifth Grades - Olive Owens
Fourth Grade - Virginia Cunningham
Third Grade - Thelma Hurlbert
Third Grade - Helen Crouse
Second Grade - Mary Flick
First and Second Grades - Ruth Garner
First Grade - Anna Johnson

Changes for the 1928 fall semester included a new flagpole in front of the grade schools, and a new course in chemistry taught by Miss Jenkins. Fair Day was Wednesday, September 5.

The tenth commencement of the Vintondale High School was held in May, 1929. There were eleven graduates. Valedictorian of the class was Sarah Beechey; salutorian was Ricco Bonette. The Nanty Glo school orchestra provided the entertainment.

When school started for the fall semester in 1929, the enrollment was 490, a loss of ten. Numbers at the high school rose to 83 pupils. A new sewing machine was procured for the sewing class, which made a curtain for the play. However, Rose Williams' speed on the machine caused the high school reporter to remark that "a cooling system must be installed to make it efficient."

The class of 1930 was the first true four year class to graduate from Vintondale High School. This group, which had excellent basketball players, was in ninth grade when the decision was made to add a year. It was also the last year that any Wehrum students attended high school in Vintondale. The Wehrum closing notice was posted in 1929. Several families remained so that their children could finish their schooling. Among the 1930 graduates from Wehrum were Bernice Craig and Helen Kuchenbrod; Jack Frazier had attended for three years, but moved from Wehrum before graduation.

Jack Burgan, Vintondale's own novelist, attended Vintondale High School for three years and graduated from Dale High School because his father left the employment of the VCC in 1929. As a junior, Jack made the School News headlines as a "hunter de luxe" by mistakenly shooting a dog for a rabbit.

The class of 1930 also made a trip to Washington, D.C. The money was earned by serving sauerkraut suppers at the Vintondale Inn and other kinds of fundraisers. At graduation, the students could not afford caps and gowns, so the girls wore inexpensive white dresses. 1930 was also the first year in which the Vintondale School System held an eighth grade commencement. On May 22nd, salutatory address was given by Julia Zoltansky, and the closing speech was made by Jim Dempsey.

Faculty for the Vintondale schools in 1930 were: Charles Mower, Principal; High School, Theodore Melcher, Maude Jenkins, Queen Keating; Grades, Jennie Whiteknight, Mary Beechey, Margaret Breth, Jean Owens, Virginia Cunningham, Emma Francis Fry, Helen Crouse, Anna Johnson, Catherine Williams, and Ruth Owens. In the fall of 1930, the seventh and eighth grades began deparmentalization with the high school.

On October 19, 1930 fire erupted in the four-room brick school house built in 1917. Because of a severe water shortage caused by a drought, the school burned to the ground. The Nanty Glo Fire Department was called to bring extra hose but even laying one-half mile of hose to the Blacklick was not enough.

The loss of the building and fixtures placed a severe hardship on the district. The school and contents were valued at $15,000 and were covered by insurance. In fact, Lloyd Williams tried to save some of the desks, and as he threw them out the window, someone threw them back into the flames. With the insurance money, all new desks, etc. could be purchased. Temporary classrooms were set up in the "Old Building", the high school, and the basement of the recently rebuilt Hungarian Church. Rebuilding plans were initiated immediately; the end result was the Delano Building, a four-room brick school complete with modern fire doors and fire-proof stairwells.

Student Activities

Students at Vintondale had a variety of activties in which to participate. In addition to the sports teams, there were class plays, picnics, dances, contests and a school newspaper.

A patriotic entertainment in honor of George Washington's birthday was given by students at the Baptist Church in February, 1916. The program consisted of chorus selections, solos, recitations, dialogs and speeches. On behalf of Warren Delano, Otto Hoffman presented the school a 12'x18' flag. County Superintendent M.S. Bentz and School Board President Abe Abrams gave speeches to the packed house.

The school paper, **Vintondalian**, started in May, 1923 through the efforts of juniors, Joe Sileck and Jim Beechey. The first edition, a twenty page commencement number, consisted of school items, jokes, literary articles and advertisements. The paper was printed on a small hand press, one page at a time. Assisting Sileck and Beechey were: Katherine Morris, Alex Molnar, Chester Graham, Arthur Buterbaugh, Susanna Lewis and Viola Graham. Copies were sold at school and at Bill Roberts' store. The paper did not appear to have been published again the following school year.

To raise money for their senior class trip, the class play was staged annually in February or March. The girls' chorus provided entertainment between acts. The highlight of the evening was the presentation of the Lincoln Essay Award, given to the student who wrote the best essay on Abraham Lincoln. Mr. Geraty, company store manager, displayed the award in the store window. The award was granted from at least 1925-1929. The only known winner was Catherine Williams in 1928; Mike Hozik won second place. Judges for the 1929 contest were Miss Crouse, Miss Whiteknight, and Miss Condie.

The social hall was the setting for the 1925 class play. **The Path Across the Hill** was presented on February 27th with the usual choral entertainment and Lincoln Essay Award. In 1926, the senior class play **Pygmalion & Galatea** was presented in March at the movie theater; the between-acts format was the same as usual.

In a local oratorical contest on the Constitution in 1927, Lavon Lockard placed first and won the right to represent Vintondale at the county contest. Second place went to Bruce Lybarger. A student council, complete with a constitution and by-laws, and an honor society were added in 1928.

During the 1927-28 school year, an effort was made to organize a Vintondale High School Alumni Association. Between 1919 and 1927, eighty students had graduated from Vintondale. A banquet and dance were held in July, 1928 at the Mundays Corner Cafe. Speakers were Mr. Mower and Thomas Swope, a Nanty Glo attorney. Traveling farthest for the reunion was Lucille Piper of Denver. Musical entertainment was provided by Mario Biancotti, Wehrum's and later Vintondale's wizard on the accordian. Cosmopolitan Orchestra played for the dance.

Not all school activities were company or Republican Party oriented. At an assembly in November 1928, Lloyd "Speedo" Williams gave a campaign speech for the Socialist candidate.

The highlight of the school news column on November 20, 1930 was that Julius "Bab" Morey attended school two days in a row without being late.

Charles Mower remained supervising principal, except for a period of service during World War II, and returned to Vintondale as supervising principal, retiring from the Nanty Glo-Vintondale Joint School District in 1966. After retiring, Mr. Mower remained in Vintondale, maintained an apartment in the "doctor's" house and toured the country by travel trailer. Nicknamed the "military man" by the male students in the 1920's, Charles Mower ran a well-discipline school and was respected by the students.

VINTONDALE VOLUNTEER FIRE COMPANY

Over the years, Vintondale has been the scene of some spectacular fires: house fires, forest fires, and fires at the flat. Credit for preventing more serious calamities goes to the Vintondale Volunteer Fire Company. The first organized company was founded on October 17, 1911. Its fire apparatus consisted of a two-wheeled hand drawn cart on which were attached two tanks. This "fire truck" was used until 1923 when a Dodge truck equipped with chemical tanks was purchased. The contract for the new truck was awarded to Hour Apparatus Company for a bid of $2,775. According to borough council minutes for June 19, 1923, the truck was paid off.

Interest in the fire company slacked off in the 1920's, and it was reorganized in June, 1928. John Huth, a charter member of the old company, was elected president; Mike Mihalik was treasurer; and

Butala; serving as assistant fire chief was Roy Roberts. At that meeting, donations, which had been solicited from local businesses, were acknowledged.

On October 22, 1930, borough council, which had a lot of control over the fire company, authorized Mr. Brandon to make a $50 donation to the Nanty Glo Fire Company who assisted Vintondale firemen when the school burned. Vinton Colliery Company also made a $25 donation. A total of $100 was donated to the Nanty Glo Company. Funds were badly needed by the Nanty Glo company because its monetary assets were tied up in the failed banks in Nanty Glo.

Vintondale's fire company was able to survive and thrive because its volunteers lived and worked in Vintondale. They were ready for action within a few minutes. The major problem faced throughout the 1920's was lack of sufficient water to fight fires. This was especially true during the drought of 1930 when the Hungarian Church and the school burned. (See chapters on the Hungarian Church and Education.)

UNITED STATES POST OFFICE

Vintondale's first recognition as a bona-fied post office came on June 18, 1894 when Clarence Claghorn was appointed the first postmaster. At that time, the town was called Barker City, then Vinton; the post office was named Vintonville. This was changed quickly to Vintondale because its mail became confused with that of Vinco. On its first official business day, fifty-two letters were cancelled.

The rurual route between Vintondale and Nipton (Red Mill) was handled by H.G. Duncan and/or his brother John. Mail on that route was delivered three times a week. An unnamed rural carrier was robbed on June 7, 1895 by two footpads. The carrier fired several shots, but the robbers pursued and captured him. One held the carrier at bay while the other rifled through the mail. Authorities kept this robbery quiet hoping to make arrests.

Rexis was the site of a Fourth Class Post Office from 1902 to 1909. It was located in the building occupied by Krumbine's furniture and undertaking business. Notices for a civil service test were posted in March, 1909. The job, which began on June 19th, paid $135 per year. The test was open to male citizens over twenty-one and female citizens over eighteen. No applications were received, and a petition for a rural route from Wehrum was circulated and forwarded to Washington, D.C. Roads in the area also needed immediate repair if there was to be rural service. The Rexis post office ceased to exist in 1909.

In 1907, Vintondale's postmaster was J.M. Jones; he was succeeded by Ira Thomas, assistant superintendent. The 1909 strike had a direct impact of the position of postmaster in Vintondale. Most of the top company personnel resigned or were fired as a result of the strike. As of July, 1909 Mr. Thomas had resigned and headed to Atlantic City for a vacation. New VCC superintendent, S. Kedzie Smith, made it known that he would not be appointing a successor. Petitions of appointment for postmaster were circulated by George Blewitt, Sam Williams, and Jacob Brett. In addition, the Vintondale post office was reduced to a fourth class post office because of falling receipts. This demotion may possibly be contributed to the strike. S.R. Williams, a Republican, received the appointment and served as postmaster until 1913 when the Wilson Administration came in. John Krumbine, a Democrat, followed as postmaster for the next eight years. In 1921, when the Republicans under Harding regained the White House, Ruth Roberts became postmistress and held the job until 1933 when the Roosevelt administration came in and turned politics in Vintondale upside-down.

Before 1930, the Vintondale post office had several homes. The first was in a small building nestled between what is today Vasilko's and Averi's. After the post office was moved to Williams' store on Block B., the former post office was remodeled to house Bill Roberts' Confectionary. When there was available space at the Farabaugh Building, Bill's store moved there, and the little building was remodeled and rented out as a house. It was torn down in the 1950's.

Mr. Williams remodeled his store so that there was an entrance for the store and one for the post office. Krumbine then moved the post office to his large building in Block A. The next home for the post office was a former store and dentist's office directly across from the Farabaugh Building. Harry Ling, Prudential insurance agent, and Abe Solomon, clerk at Louis Rose's store, lived in the duplex above the post office This remained the home of the post office until the 1950's when it moved to the former bank building.

Vintondale High School Class of 1930. Row 1: Harriet McPherson, Margaret Thomas, Bernice Craig, Anne Sileck, Agnes Huth, Frances Wojtowicz, Ella Jendrick. Row 2: Miss Isabelle Condie, Mr. Charles Mower, Bernice Jones, Maude Jenkins, teachers and principal, Phoebe Michelbacher, Bill Jendrick, Emma Nevy, Gino D'Avalli, Helen Kuckenbod, Jack Frazier, Rose Williams, Mr. Theodore Melcher, Mary Balog.
Courtesy of the late Agnes Dusza.

Chapter VIII
Businesses

VINTONDALE SUPPLY COMPANY

A company store was a fact of life in a mining town, and Vintondale was no exception. Michael Gresak, who lived in Vintondale briefly in 1902, remarked in his memoirs that there was a well-stocked company store in town. The store probably was located in Block A on Barker Street between the present post office and Shamko's house. This wooden building was later used as a nickelodeon until 1923 when the Vintondale State Theatre was opened at Fourth and Main.

When Delano re-purchased #1 and part of the town from Lackawanna Coal and Coke in 1905, part of his plan included construction of an impressive two story cut-stone company store. The Vinton Colliery Company sold lot #15 and part of lot 13 to the Vintondale Supply Company, Ltd. for $1.00 on March 1, 1907. The structure was completed that year. The full basement was used for storage of supplies for the store; first floor was occupied by the company store; and Vinton Colliery moved their office from Plank Road to the second floor. On the east side of the building was a large two-way wooden staircase that served as the office entrance and the pay window. Miners lined up at the rear of the building, climbed the stairs to the pay window, and descended to Main Street.

A full description of the company store with layout, merchandise and employees is found in Frances Pluchinsky's, "Memories of Vintondale".

Construction of the Company Store, 1906. Old Company Store (later Nickelodeon) on the right.
Courtesy of Huth Family Collection.

In 1916-17, the Vinton Colliery was meeting the new demands for coal by buying new coal rights, building additional houses and forming the Vinton Land Company to handle these new properties. On March 15, 1917, the investors of the Vintondale Supply Company held stockholders' meetings to renew the company charter for ten additional years. Its capital stock was increased from $30,000 to $60,000. Of this $60,000 half was to be paid in book accounts and merchandise, $20,000 in cash and $10,000 was to be a leasehold taken over and owned by the company.

The capital stock of $60,000 was divided into 1,200 shares at $50 per share. The partners and the amounts they invested are as follows:

Warren Delano, Red Hook, N.Y. - $28,500
Algernon Burr, Westport, Conn. - $16,500
Robert Cook, New Brunswick, N.J. - $5,000
Clarence Schwerin, New Rochelle, N.Y. - $5,000
William Clement, Vintondale, manager - $2,000
Otto Hoffman, Vintondale - $2,000
Lloyd Arbogast, Vintondale - $1,000

The charter listed the place of business as Vintondale, with a branch at Claghorn. In 1920, when Delano purchased the Graceton Coal Company, he added

Vinton Supply Company, 1906-1907.
Courtesy of Leona Clarkson Dusza.

another company store to his holdings, the Graceton Supply Company. Steve Morey, an employee at the Vintondale store, was sent to the Graceton store. He eventually moved to Windber and became an employee of Berwind-White's Eureka Store.

The Vintondale Supply company's charter was to last until March 31, 1927 and could be dissolved by a majority vote of the shareholders. When a member died, the heirs had the right to continue to hold the shares or sell them to the remaining partners, who had a ninety day right to purchase them.

Robert Cook died in 1919, and his widow took over as executor and shareholder. Warren Delano died in 1920, and his son Lyman and brother Frederic were the executors of his estate. Mrs. Jennie Delano was entitled to one third of the personal estate, and the five children each received two-fifteenths. It is unknown if any of the children retained their company store stock.

The letterhead of the Vintondale Supply Company Limited listed the company address as 50 E. 42nd Street, New York. Its stores were located in Vintondale, Claghorn and Graceton; general manager was Walter P. Geraty of Vintondale. A letter at the FDR Library showed that in 1924, Frederic Delano of Washington, D.C. purchased fifty shares of preferred stock at $100 per share and 50 shares at $50 per share. The letter implied that the Vinton Supply Company was a new organization, and that A.T. Burr, treasurer, was selling some of his personal stock to Frederic Delano. Perhaps Mr. Burr, who was well into his seventies at the time, was retiring from the company and divesting some of his assets. To support the contention that the Vinton Supply Company was a new organization, the June 14, 1924 **Indiana Evening Gazette** listed the Vinton Supply Company as a successor to the Vintondale Supply Company.

In the 1920's, the Vintondale Supply Company annual turned a profit. Financial records found in Frederic Delano's files in the FDR Library show that in 1922, the Vintondale Supply Company made a net profit of $11,609.31, and in 1923, it was $18,582.99. The 1923 inventory of Vintondale was worth $42,393.63; at Claghorn $7,429.74; and at Graceton, $17,294.17.

The other available financial records for the store are 1925. By this time, the Claghorn store was closed. The capital stock remained at $60,000 but 600 shares of common stock worth $30,192.30 were also listed. The value of the inventory at Vintondale dropped to $39,666.47; Graceton's remained about the same at $17,090.40. The net profit of $6,596.17 was realized for 1924, year of the KKK strike.

The company store continued to operate during the lean days of the Depression. Many families owed the store bills of varying amounts and received "blacksnakes" instead of any money on payday. A black S scrawled on the payslip meant that the miner owed the company more in store bills, utilities, rent, doctor, tools, etc. than he earned. A few families escaped their oppressing store bills by packing up and leaving town in the middle of the night. They left no forwarding addresses with their neighbors or relatives so that they would not be caught and have to return.

The company store closed permanently and unexpectedly on March 15, 1940 when the Vinton Colliery Company went bankrupt. (See chapter on the bankruptcy and in fate of the building in Volume II.)

Mr. Clement - store manager - sitting in driver's seat in front of company store, c. 1920.
Courtesy of Huth Family Collection.

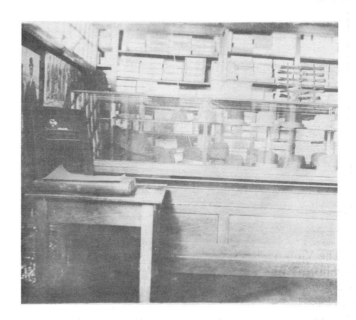

Vintondale Supply Company, c. 1925.
Courtesy of Aileen Huth Michelbacher Ure.

Vintondale Supply Company, c. 1925.
Courtesy of Aileen Huth Michelbacher Ure.

Vintondale Inn, c. 1906.
Courtesy of Tom Shook

THE VINTONDALE INN

The Vintondale Inn, one of the finest of its kind in Cambria County, was constructed around 1896. The hotel was built on Lots 11, 13, 15, Block C which were purchased for $1,150 from the Blacklick Land and Improvement Company on September 22, 1896. Buyers of the lots were Dr. Abner Griffith, Theodore Bechtel, and Warren Delano. The hotel, which is described in Mrs. Pulchinsky's "Memories of Vintondale", was built to provide lodging for the many salesmen and businessmen who came into town. In the pre-automobile days, it was necessary to provide overnight accomodations for these people. The hotel provided a bus which carried guests to and from the train station.

Dr. Abner Griffith died on December 17, 1896. Griffith's widow, Alice, sold her undivided 6/10 interest in the property to Delano for $7,200 on September 16, 1906. Delano also gained control of Bechtel's shares and continued to privately operate the hotel until 1915 when he conveyed the property to the Vintondale Inn Company for $1.00.

The hotel was operated on a lease basis. Early known proprietors were J.D. Wentworth who retired in 1900. His successors were W.F. Griffith and then Samuel Dorsey Griffith. In 1904, S.D. Griffith sold his furniture, fixtures, good will, and stock of liquor to Albert Fleitzer of Greensburg; cost of the transaction was $6,500. Albert Fleitzer leased the premises from Warren Delano from 1904 to 1912. The mine diary entry for January 25, 1912 hinted that Delano was raising the rent and that Fleitzer probably could not afford the increase. However, he did sign a three year lease on March 7, 1912 Fleitzer was to pay Delano $150.00 monthly, payable at the end of the month and keep the hotel in good condition paying all heat, light, water, phone rents and any other charges. The lease also hinged on the fact that the hotel be granted a yearly liquor license. If the license was refused any year, either party could terminate the lease by submitting a thirty day written notice. The barroom of the hotel was located in the basement. The new lease was to run between April 1, 1912 and March 31, 1915, but on June 10, 1912, Fleitzer transferred his title and rights of the lease to Joseph A. Shoemaker who had formerly rented the barroom in the Farkas Hotel and the Village Inn before that. Millers then ran the hotel after the Shoemakers.

In 1911, a room cost $2.00 per day, but it is not known if that price included meals. The hours for meals were as follows:
Breakfast - 6:00-8:00 AM
Dinner - 12:00-1:30 PM
Supper - 6:00-7:30 PM
Sunday
Breakfast - 8:00-9:30AM
Dinner - 12:00-1:30 PM
Supper - 6:00-7:30 PM

Many of the companymen boarded at the Inn when they first arrived in town. Two of those early boarders were John Huth, the author's maternal grandfather, and Louis Burr, son of Algernon T. Burr, Treasurer of the Vinton Colliery Company. Mabel Davis Updike, a company secretary, said she and Leila Packer, a teacher, boarded at the hotel from 1912-1916. Mrs. Updike took the train home to Ebensburg each weekend. Nick Gronland, who worked in the office, also kept a room at the hotel. In the 1920's, Mr. and Mrs. Arbogast had a suite of rooms in the hotel.

In the early years, many townspeople went to the hotel for dinner on Sundays or for supper on Saturday evenings. On several occasions, the author's grandmother's (Mabel Hampson Huth) and great-aunt's (Eula Hampson Burr) names were on the hotel register for dinner or supper. Mrs. Theresa Kerekish, who worked at the hotel, said that Eula Hampson often played the piano in the evenings. The hotel had one of the few pianos in town, and Aunt Lou was one of the few persons in town who knew how to play the piano.

Hotel Registers

The hotel registers for 1911 and 1922 were found in the attic of the John Huth home. By checking these registers, one can discover the frequency of the visits of company dignitaries from New York. In 1911, Warren Delano visited Vintondale on April 5 and had dinner at the Inn. He stayed overnight in Room 7 on July 26 and August 29. Mr. C.M. Schwerin accompanied him on August 29th and stayed in Room 9. Both had breakfast the next morning. Mr. Delano was again registered for Room 7 on October 3. Mr. Schwerin was also back in Vintondale in Room 11 on Monday, November 13, His brother, Joseph Schwerin, was a guest on December 8, 1911.

Other distinguished guests in 1911 included:

A.T. Burr - Room 7 - Tuesday, April 18, 1911.

Webster Griffith - Saturday, December 3, 1911; Friday, September 22, 1911; Saturday, November 12, 1911.

Otto Hoffman - Room 21 - Sunday, October 15, 1911.

Mr. Hoffman's address was listed as Cleveland; he began working for the Vinton Colliery in 1912.

On August 31, 1915, Warren Delano deeded the hotel to the Vintondale Inn Company for $1.00. The Inn underwent some remodeling, and a grand re-opening dinner was hosted by the Shoemakers on November 11, 1915. In 1916, the Vinton Colliery Company sold Lot 9, Block C, the future home of the Vintondale State Bank, to the Vintondale Inn Company. The Company owned the lot as a result of the 1906 Lackawanna Coal and Coke sale to Vinton Colliery.

However, in November 1920, after Delano's death, the Vintondale Inn Company re-sold the hotel to the coal company. The Inn sold for $25,000: $17,000 for the building, $7,000 for the fixtures and $1,000 for the lots. Lot 9 was transferred for $1.00. The deed listed the officers of the Vintondale Inn Company as: Otto Hoffman, President; W.H. Clement, Secretary; and A.T. Burr, Treasurer.

1922

The hotel saw many out-of-town guests in 1922. Mr. Schwerin was registered on January 30, 1922. The next day, the Vinton Colliery Company gave a banquet at the Inn. The reason was unknown, but according to the **Nanty Glo Journal**, the dinner was held at the late hour of 10:00 PM. Mr. Schwerin and Mr. Burr attended from New York City. Thomas Fleeson of Pittsburgh also came. Ebensburg guests were: A. Kinkead, Herman Jones and Leonard Jones. Attending from Claghorn were Evan Dunlap, Allie Cresswell and Mr. Buck. Milton Brandon, Mr. Edwards, Mr. Heath, Mr. Phillips, Mr. Doherty and Mr. Rankin all came from Graceton. Vintondale residents in attendance were: Otto Hoffman, Lloyd Arbogast, John Burgan, Walter Geraty, George Blewitt, Jack Huth, Dan Thomas, William McClune, William Evans, Walter Carlson, H.I. Clark, W.H. McCracken, Howard Williamson, J.H. Breckons, Isaac Baldwin, Robert McGee, James Dempsey, John Daly, Herbert Daly, Dr. MacFarlane, Jack Butala, Harry McCardle, John Wagner, Samuel Feldman, John Galer, A.S. Kempfer, William Abrams, Grant Gongloff, Doyle Williams, George Cook, A.H. Tipton, Henry Misner and Tom Whinnie.

Mr. Schwerin was also at the Inn on Monday, April 10, 1922; with his son on Monday, April 24, 1922; Saturday, May 6, 1922 with Frederic Delano for breakfast; Tuesday, May 23rd; Monday July 24th; and Monday, September 11 with Clarence Jr. and F.W. Schwerin of Great Neck, Long Island. W.S. Arbogast of Mifflin, was a guest at the Inn on May 18th, July 13th, and November 23rd.

In April, 1922, the national coal strike was at its peak, and company was doing its best to prevent Vintondale miners from unionizing. The Vintondale Inn played host to four state policemen for breakfast on April 22nd; these officers may have only been in town to delivery the payroll. Mr. Hoffman had numerous meals at the Inn during the strike alert. His name appears on the register for breakfast and/or dinner on Tuesday, April 4; Wednesday, April 5; Thursday, April 6; Tuesday, April 11; Monday, April 17 with his wife;

Wednesday, April 19; Thursday, April 20; Monday, April 24 with Mr. Schwerin; and Sunday, April 30.

Hotel Employees

Mrs. Theresa Kerekish started working at the Vintondale Inn in the winter of 1907-08, shortly after her arrival in Vintondale from what is now Czechoslovakia. She worked as chambermaid, waitress, cook, desk clerk, and even hotel manager when the Fleitzers went on a trip to Europe. After her marriage, she remained on for a while. Other early employees were Mrs. Ann Dancha and Mrs. Mary Shedlock.

Mary Monyak Drabbant worked as head waitress for several years during the early 1920's. Since there was still several companymen, such as Swanson and Tipton, boarding at the Inn, Mrs. Drabbant's day started at 4:30 AM. She had to be dressed in her uniform, hair fixed, lunch buckets packed, and have the dining room door unlocked by 6:00 AM so that the men could have time for breakfast before the mantrip left. She said the men were glad she was working because the previous waitress often did not have the door opened on time, and they missed breakfast.

As waitress, she was responsible for taking orders, as many as six at a time without writing them down. Also she had to clean off the tables and set up for the next meal. Her workday started at 4:30 AM and ended when supper was over in the evening. Mrs. Drabbant received an hour or so off every afternoon. She worked seven days a week for $7.00 a week.

The Millers were the proprietors at the time, and Mrs. Drabbant was often in charge of the dining room and the desk when they went away. She said it was hard to keep track of the salesmen who stayed overnight and had to catch the first train the next day. Many of them tried to sneak out without paying.

At that time, the dining room served two types of soups and two meat courses at dinner and supper. Each boarder or regular who came in for meals had a regular seat. The only person who gave Mrs. Drabbant trouble about seating arrangements was the dentist who always sat at different places and wiped his hands on the corner of the tablecloth instead of the linen napkin. She checked with Mrs. Miller who told her to handle the problem. So the next time he came in, he sat in the wrong seat. Mrs. Drabbant put his glass of water at his usual seat and refused to wait on him. She pointed to his regular chair, and he finally moved to it, but he still wiped his hands on the tablecloth.

Mrs. Drabbant said she got to know many of the company men, such as Arbogast and Hoffman, quite well because at the time, the waitresses were permitted to converse with the customers. Late they were not.

In the late twenties, the hotel was open only during the school year because the teachers boarded there. Mrs. Kerekish returned as the cook for part of that time. She was followed by Edna Richardson of Portage. Later the teachers had to do their own cooking.

Rose Larish Hubner worked at the Vintondale Inn during the late 1920's for nine months of the year, seven days a week with two hours a day off if her work was done. During the other three months, she worked for Mrs. Hoffman. Her title was "upstairs girl", even though she had to clean all three floors. Other chores included sweeping and hosing down the porch. When hard times caused a cut in employees, Mrs. Hubner also became a waitress in addition to her other duties.

During this time, a person could not go into the Vintondale Inn and order a meal. If any guests, such as Mr. Schwerin, were expected, the office had to let the cook know in advance how many extra meals to prepare. She would then notify the company store butcher what to cut for dinner. The only unscheduled meals served were for the State Police who brought in the payroll; they were given breakfast. Lunch and dinner had a common menu. The meals were served individually; then Mr. Arbogast changed the meals to family style. The only choice on the menu came on "ice cream night" - Sunday night. Two flavors of ice cream were served.

There was one known attempted robbery at the Inn. In June, 1921 at 3:00 AM, someone tried to rob a teacher who had just been paid, The **Nanty Glo Journal** reported that a "BVD Police Force" of thirteen members had been organized to protect the guests.

Charles Mower was one of the teachers who boarded at the Inn from 1926 until he joined the service in 1942. It was closed when he returned in 1947. He rented an apartment in Roy Robert's house, now Betsy Mesoras Shandor Helsel's, where he lived until the fall of 1990 when he moved to a retirement home.

There was also a funeral home in the Inn during the early 1940's. Thomas Reade used the basement for a mortuary and lived in an apartment in the Inn. Later he moved to Johnstown.

After the company store burned in 1943, the hotel became the company office. All the surviving company records were kept on the upper floors, along with all the pigeons, only to be destroyed when the hotel was torn down to make way for the new VFW in 1973.

In the 1950's, the kids called the Inn the union hall. Here Local 621 passed out the Christmas treats, which consisted of an orange, a popcorn ball and a small box of hard candy and creme drops. Another time of the year, either Memorial Day or the Fourth of July, cheerio ice cream bars were given out.

The basement of the Inn also served as a Teen Canteen in the late 1950's. A committee of parents worked hard to get the place in shape. It died out because of lack of interest from the teenagers and the age-old problem of not enough workers to keep it open.

So the Vintondale Inn has become another one of those "Memories".

VINTONDALE'S JEWISH MERCHANTS

Most Vintondalers who have grown up since World War II probably do not realize that in its earliest years, Vintondale had a sizeable number of Jewish merchants. Most of these men had emigrated from the Jewish region of Russia called the Pale. They came to escape the pogroms (organized persecutions) of Tsar Nicholas II's Cossacks. Many settled with relatives in the ghettos of the large cities like Philadelphia and New York, but some became wandering peddlars, going from town to town selling their wares. Others saw the advantage of settling down in a mining town, and establishing a mercantile business or a grocery store. Many built up successful businesses and then expanded to the larger markets in the nearby cities. Most of Vintondale's Jewish merchants had relatives in Altoona or Johnstown who operated department, clothing, or jewelry stores.

The Brett Brothers

Four of the Jewish families who lived in Vintondale between 1900 and 1915 were directly related. Brothers Jacob and Samuel Brett each had mercantile stores on Barker Street. Their sister was married to Daniel Myers, who had purchased, along with Louis Rose, Barrons' store on the northeast corner of Third and Barker. A second sister was married to a Kass who also had a store in Vintondale.

Jacob Brett migrated from Russia in 1892 via Ireland; he originally planned to go to South Africa with a cousin, but changed his mind. His father's family had relatives in Altoona, so Jacob, who spoke no English, decided to take a train to Altoona and look for his relatives. Within two years, he had made enough money as a peddlar to bring his father to the United States. However, his father was upset with Jacob's lapse in following the rules of his religion and wanted him to return to Russia. His father went back, but eventually Jacob earned enough money to bring over his parents, two sisters and brother.

Jacob Brett married Ida Block on September 18, 1900 in Altoona. For years, Ida's family operated Block's Department Store in Moxham in Johnstown. When Brett brought his bride to Vintondale, it was sight-unseen and at night. The conductor held a lantern for them to safely debark the train. They made the half-mile trek uptown to his store building in the pitch black; Vintondale had no street lights in 1900. Mrs. Brett told her daughter Miriam that she would have never married Jacob Brett if she had seen the town first. Jacob promised his wife that if they had children, they would leave Vintondale when the children became of school age.

Jacob Brett's Store offered clothing, boots, shoes, furniture and mine hardware. For each dollar purchased, Brett gave the customer a "check". For each $25 of checks redeemed, a "handsome premium" was awarded.

His store, located at Main and Third, was in a large three story building which extended to the alley. There also was an ice house on the property. The rear of the building was used as a storeroom. The Brett's apartment on the upper floor had indoor plumbing, one of the first in town along with Sam William's house, directly across Third Street. Across the front of the second floor was a large, wide porch; chicken wire was stretched around the bannisters to keep the children from falling off the porch. The Brett children were not permitted on the street to play; so the porch became their play area, and in the children's eyes, their prison. After breakfast, the children, often with their cousin Edna Kass, were put out on the porch to play. Hating it, they frequently acted up to get attention. One trick was to undress; then the customers would come in to tell Mr. Brett. Another play area for the children was the back apartment where platforms of carpeting were kept. On one occasion, the children stripped and were dancing in the buff on the carpets when their father brought up a customer to look at some carpeting.

Vintondale: Main Street in 1907. Jacob Brett's store is on the right. #1 powerhouse is in the background.

Courtesy of Lloyd Williams.

Cousin Rachel Myers, who adored Miriam Brett Kranich, took her to school one day without asking Mrs. Brett. The teacher allowed her to stay, and Mrs. Brett, in a panic, had the entire town looking for her daughter. All Miriam received from the experience was head lice.

Two of Jacob Brett's local employees were Mary Lynch Clarkson, bookkeeper and salesperson, and Frank Wray. Frank's daughter, Betty, also worked many years for the Kranich family in Johnstown.

Jacob Brett also hired several men from Altoona to help in the store. These men refused to board at the hotel because they strictly obeyed the dietary laws, so Mrs. Brett had the added burden of cooking and doing laundry for the boarders. When Mrs. Brett returned from Altoona after a visit, she found the house like a pigsty. The boarders did not wash any dishes, and when they needed clean ones, they went down the stairs to the store to get them.

The Brett Brothers were party to several property transactions while they lived in Vintondale; these are worth pursuing. On October 17, 1901, Jacob Brett purchased Lot 2, Block C from Mary E. Davis of Nanty Glo who had bought it from BL&IC in 1898 for $2,600. This lot and house is the one now owned by the Jansuras. He also purchased Lot 11, Block B from the Barker Brothers, who had purchased it in 1900. Price paid was $1,000. For unknown reasons, Jacob Brett purchased a one acre lot in Jackson Township in 1905 from Charles Strausbaugh for $30. He later sold it to Charles and H.E. Douty for $200. In January 1908, the VCC sold Brett Lot 5, Block B for $750; Mike Farkas purchased it for $4,300 in July that same year. By that time, judging by the price increase in six months, Farkas' Hotel had been built on the lot. In August, 1907, Charles Douty purchased Lot 11, Block B for $5,000, $500 down and $1,000 per year plus interest. A second deed dated June 14, 1909, sold the same building again for $5,000 to Douty and cancelled the article of agreement.

Sam Brett also bought some of the Main Street stores. In 1906, Clarence Claghorn of Pierce County, Washington, sold by Sheriff's Deed Lot 3, Block C to H.L. Sher for $1,000. This is the site of the former Nevy Brothers' store.

Sher sold the store building to Sam Brett for $3,000 August 2, 1907; the next month, Brett sold it to Joseph Gaal for $2,800. Webster Griffith, Sheriff, sold all title and interest in the Gaal building to Sam Brett in March 1, 1909, who then sold the building to Lena Gaal that October for $1,200.

The coal company also sold Sam Lot 7, Block B in January, 1908 for $750 and Lot 9 for the same price in December. Benton Edwards, who had moved to Ebensburg, sold Sam Brett his building on Lot 7, Block C in 1915 for $4,250. The brothers also jointly bought a lot in Rexis.

In 1909, Jacob Brett sold his building to Tobias Callet and Harry Silverstone of Johnstown who opened a wholesale liquor business there. Brett bought a building in Indiana, but then decided to settle in Altoona. Forming a partnership with a man named Snitzer, Jacob Brett opened Brett's Department Store. His daughter Miriam married Sam Kranich, who joined his family's business, Kranich's Jewelry. During the Depression, jewelry sales were way down, and Jacob Brett trained Sam in anticipation that he would open a store in Johnstown. The store they opened became Bretts, known for their quality clothing.

Samuel Brett remained in Vintondale for a few years longer. He was married to Lena Levison, sister of Nanty Glo merchant Leon Levison. By 1917, all the Bretts were living in Altoona and working for the family. Sam Brett died in Altoona in January, 1943.

Louie Rose

Louie Rose and Daniel Myers had a partnership for a short while. Then Rose ran the store himself until about 1930. Located in the former Barron building, Louie sold all types of clothing and shoes. Like Jacob Brett, he also started his career as a peddlar, visiting mining towns all over the area. The author's father, Steve Dusza, remembered Louie Rose peddling in the patch town of Onnalinda near Beaverdale.

Clerking for Louis was Abe Solomon, who lived in one side of the duplex above the postoffice. Rose, who had an apartment above the store, ate all his meals at Solomon's. Minnie Solomon followed the kosher rules. She served the main dish at noon, and supper was usually soup, hard-boiled eggs, pickled herring and the like. Rose Larish Hubnar lived with the Solomons as a "mother's helper" and said that Louie would always wipe his hands on the tablecloth and not the napkin. She reprimanded him, but he did not change his ways.

Townspeople figured that Louie had quite a bit of money set aside because he kept the store going in the Twenties when business in town dropped off. Abe Solomon was kept on even though some days, the total sales may have been no more than a dollar. His only family was a sister and a nephew to whom he sent $100 a month, quite a sum at the time. His taste in cigarettes was a bit expensive; he smoked English Ovals which could not be obtained in Vintondale. Louie asked Mabel Huth to pick up the cigarettes for him any time she went shopping in Johnstown.

After Louie Rose closed the store and moved to Altoona around 1930, the Solomons moved to Johnstown in 1931 and then on to Detroit. This way, Abe Solomon was closer to the mineral springs at Mt. Clements, Michigan where he frequently went for treatments for arthritis. The Solomons kept in touch with several families after their departure.

There also was a clothing store run by Millers; it was located across the street from the Farabaugh Building. Later, the post office moved into the building.

Louie Rose was the last of Vintondale's Jewish merchants. Today, many of the major clothing stores in Altoona and Johnstown can claim Vintondale ties.

SAMUEL R. WILLIAMS

Sam Williams was a local Indiana County young man who saw opportunities in Vintondale and chose to make it his permanent home. He arrived with his family from Pine Township in 1900. There were several reasons for this move. He and his brother Ben shared the family farm near Strongstown. Because the

Sam sold his share to his brother, who also had business interests in Indiana. (Ben sold the coal rights on the farm during World War I and also bought the Studebaker franchise in Indiana which later became the Mercedes-Benz dealership. Until recently, his descendents still operated the business.) The other reason for moving to Vintondale was to be closer to the care of a doctor. During one winter, three of the Williams' children died of croupous membrane (diptheria) within thirty-six hours because they were unable to get a doctor.

Samuel Williams' Family, c. 1913. Row 1: Lloyd, Marshall and Rose. Row 2: Samuel R. Williams, Mary, Margaret Sides Williams. Row 3: Benjamin, Paul, Sarah. Family tree in frame - now hangs in Lloyd's living room. Sides family organ - used in Mr. Williams' Church.
Courtesy of Lloyd Williams.

When the Williams' family arrived, they rented half of the double house owned by the Thomas Wray family (directly below the Catholic Church). Mr. Williams worked as a carpenter and helped construct the houses on Dinkey (First) Street. In 1900, Williams purchased Lot 16, Block B (southeastern corner of Third and Main) from A.J. Miller of Loretto; the selling price was $350. Miller had purchased the property from the Blacklick Land and Improvement Company in 1895. The **Indiana Progress** reported in November, 1900 that the ground was being cleared and excavated for Williams' large building. The carpentry work, using local lumber, was done by Bert Mahan. The beams in the basement were 2' x 12' hemlock. In April, 1902, the **Progress** wrote that A.W. Strong and Son were painting the house and storeroom. The house, which included a street level store, a church on one half of the second, and living quarters in the rest of the house, cost $1,900. The building was equipped with electricity and indoor plumbing.

In 1907, Mike Petrilla sold a house on Plank Road to Sam Williams for $600. Judge Barker had sold the property to Petrilla that same year; Barker's no alcohol sales stipulation applied to this sale. This was the property that Williams rented to the United Mine Workers during the 1909 strike. (See appendix for 1909 Strike Receipts.)

In addition to operating the store, Mr. Williams never lost his love of farming. At the sheriff's sale in 1915, he purchased a ten acre tract near the Shuman Run resevoir for $25. He kept a large garden there for years.

At his store, Mr. Williams kept the usual staples of food. He had a double bookkeeping system. A customer's total purchase was recorded in the big ledger. An itemized list was kept in a small black book. Each customer had a black book. Sam Williams received some harrassment from the company police, but not as much as other merchants. His son Lloyd thought it was because his brother Bennie was a good friend of Ken Hoffman and helped Ken with his math homework. In 1914, however, he acquiesced to the company and agreed to stop selling black powder and oil to the miners.

The Williams' store delivered groceries to Rexis, Bracken and outlying farms. One of the children would take the orders in Bracken one day and then deliver them by horse and Cramer wagon the next. One year, the wagon was used in a Halloween prank. Some boys took the wheels off the wagon and hoisted it onto the roof of the barn behind the store. Then the wheels were replaced. The next morning, Mr. Williams only reaction was, "I wonder how that got up there?"

Sam Williams had the reputation of being honest and trustworthy. Mrs. Miriam Brett Kranich said that if her father needed change at his store across the street, Mr. Williams would hand him the change bag and tell Jacob to help himself to what he needed. As a butcher, Sam would often fill the role of rabbi on religious holidays for the Jewish familes in town.

Mr. Williams operated the store until 1937 when it was taken over by his daughter Sara and her husband, Del George. After twenty-five years of broken service with the railroad, Del decided to quit to run the store. In 1947, George's Economy Store was closed for three days of remodeling. When it reopened, it was a cash and carry market with a twenty-foot aisle and checkout counter on the side door. Pushcarts and baskets were purchased for the use of the customers. In their advertisement in the **Nanty Glo Journal**, coffee was 45 cents per pound; Jello was 8 cents; and center cut pork chops were 69 cents.

In 1952, the Georges sold the store to her brother Lloyd, who had been a fireboss in the mine and had injured his back. Speed (a.k.a. Cramer) ran the store as an Economy Market until 1966 when he closed the store to assume the position of Vintondale's postmaster. He was the second generation of Williamses to be postmaster and borough council member. Daughter, Sandi, is the third generation to be employed by the United States Postal Service.

Samuel R. Williams and Margaret Sides Williams, c. 1935.

Courtesy of Lloyd Williams.

BILLIE'S CONFECTIONERY

A popular hangout in Vintondale was Billie's Confectionery operated by Bill Roberts, who was handicapped and had difficulty walking. By the age of twelve, he was able to walk with crutches and worked as messenger boy between Lackawanna #1 and #3. John Daly and Mr. Clements of the company store decided to set Bill up in his own business. Through a collection, they were able to raise $50, which was enough to buy some equipment for a candy store. The old post office building next to Gaals was empty. At first the store was spartan; the candy boxes were on long wooden boxes; there were no showcases. Bill did have a couple of ice cream tables. The ice cream was kept frozen in metal cans which were packed in ice. The cans were placed on wooden tubs on a covered back porch.

As business improved, Bill was able to add showcases and later a soda fountain. When the left side of the Farabaugh Building had a vacancy in December, 1913, the company remodeled the store, and Bill opened there. In the 1920's Bill's place became a popular loafing place for the young gentlemen of town. Eventually, a group started calling themselves Sons of the Restaurant. Dutch Shaffer was the first president, but resigned because the rigors of going to the post office were too demanding. The major task of the president was to go for the mail daily. Shaffer was replaced by Dewey Esaias.

Family members helped Bill when they could, especially after he was injured in a automobile accident in 1930. Bill was pretty much bedfast at home after that. Julia Brozina Ure managed the store for Bill. Ruth Roberts also helped out between the time she lost the job of postmistress in 1933 and when she was hired as school secretary in 1935. Bill left the store to her. After his death in 1943, she sold the confectionery to the Glowas.

Vintondale Post Office, c. 1902. Later Billie's Confectionery.

Courtesy of Lovell Mitchell Esaias.

NEVY BROTHERS

Before World War I, there was a large influx of Italian immigrants into Vintondale; this created a demand for Italian speciality items such as salamis, olive oil, pastas, etc. which the company store normally did not carry. These needs were supplied by the general merchandise store named Nevy Brothers.

Five brothers: David, Charles, Louis, Henry and Gene, arrived in America from Bergoto, a village in the Italian province of Parma. Settling first at Cokesville near Blairsville, the eldest brother, Henry, worked for a year in the mines, but did not like it. So, the brothers opened a bakery and delivered baked goods and later groceries to Italian customers in the surrounding mining towns like Homer City, Black Lick, Coral, Graceton, Vintondale and Colver.

In 1910, the brothers transferred their store to Vintondale, occupying the former Shaffer store building on lower Main Street. They moved uptown into the building which later was known as Averi's. In 1923, the brothers located permanently in the building formerly occupied by Gaal's store. Sher's Jewelry store was in that building before Gaal's. That building was moved to the back of the property, and the present two-story building erected in its place.

Nevy Brothers General Merchandise, lower Main Street.

Courtesy of George Lantzy.

Several brothers opened a branch store near Colver in a patch called Jewtown. In addition to having trouble with stray goats and other animals wandering into the store, the house was extremely cold. The store was abandoned after one of the wives died of pneumonia.

The entire Nevy family resettled temporarily in Vintondale and put forth a full family effort to operate the store. For a while, they lived in the rooms above Callet and Silverstone's wholesale liquor store (previously Jacob Brett's store). David's wife Mary, known as a great cook, assumed the role of head of the family and made sure family and guest alike were treated properly. Harry Silverstone frequently took his meals with the Nevys. Charley Nevy purchased a house on Third Street in 1923 from Cambria Savings and Trust, Administrators of the estate of Jesse Donahue. Other Nevy family members lived above the store.

The last of the Nevy family to emigrate was Paulina, who arrived with her mother in 1914. They traveled on an Italian liner, and Pauline said she was seasick the entire seven-day voyage. She also went to work in the store. (See chapter on the Black Hand.) When Paulina married Frank Pioli, her brothers said they did not have time to close the store for a wedding. The couple were married in a quiet ceremony at St. Anthony's Church (Italian) in Johnstown. Witnesses were her mother and her sister-in-law. After a honeymoon of a few days in Pittsburgh, the couple returned to Vintondale and lived in an apartment behind the store. Frank, who had been a miner, worked at the store filling the orders and delivering them by horse and wagon and later truck until felled by a stroke in 1946. Frank's brother, Joe, arrived in Vintondale in 1922 and joined the team at Nevy Brothers. He continued working there until the store was sold in the late 1960's.

According to Paulina, her brother Davey always had a business mind; he moved to Cumberland, Maryland and opened the Cumberland Macaroni Factory. At first Davey had difficulty with the macaroni splitting, but soon found a man from Italy who solved the problem. The macaroni had to be manufactured at the correct temperature and humidity to prevent splitting. After that situation was solved, the company became very successful. All the brothers except Louis moved to Cumberland to run the factory. Henry Nevy, the oldest, worked in the factory for a year, but did not like it. He opened a liquor store in Cumberland, but was forced to sell it when he developed lung cancer. Before he died, Henry sent for his wife, who had remained in Italy.

The macaroni factory did very well during the 1920's and 1930's since pasta was a very filling, but inexpensive food item. The Nevys were taking trips to Europe; sending their children to private colleges: Merle to Kiski Academy, Emma to Seton Hill, and Gene to St. Francis; investing in gas and oil wells in West Virginia; and driving new Studebaker cars at a time when the Vintondale mines were working one day a week. During those lean years, Nevy Brothers granted a lot of credit to the miners.

The blue box with the Alpine Eagle lable was a standard staple on the shelves of the Nevy Brothers store. The quality of the macaroni was so good that it was reportedly served at the White House.

Nevy Brothers store remained open during the turbulent years of the 1920's where several other food stores failed. Vintondale's police would not let people into town to shop and intimidated shoppers coming out of the store by inspecting their packages. Much of Nevy Brothers business at this time came from out-of-town customers to whom Nevy's delivered the orders.

FREEMAN'S

Samuel Freeman operated a clothing store on Main Street in the early 1920's. Philip Wayne and Sam Campbell were employed at the store, which was located what is today the Averi building. If you could not find what you wanted, or if they did not have your size, Freeman's could order it for you. Olive McConnell tried on a blue suit with long Eton jacket and box pleats, but was reluctant to purchase it because everyone else would have one like it. Nan was told that if anyone else got a suit like it, she would not have to pay for it. She saw the same suit in Sam Goldman's window. When she informed Phil Wayne that she saw the same suit, he investigated for himself and told Nan that she did not owe him for the suit.

Like many other stores in Vintondale, Freeman's closed in 1925 because of poor economic conditions. The merchandise was offered at a Sheriff's Sale in late December. Purchaser was a Mr. Henderson of Altoona who planned to dispose of the stock there.

CUPP STORES

Chain grocery stores were a developing item in the first two decades of the twentieth century; a local chain which eventually had sixty stores in Western Pennsylvania was the Cupp Grocery Company. The founders were H.H. and George Cupp of Johnstown. The chain, which was founded in 1914, attempted to operate a store in Vintondale in 1927 in the former Freeman store. This grocery, which offered lower prices than the company store, had two known managers: Donald Zimmerman, who was sent to Ebensburg in April, 1927 and Joseph Barr, who was the youngest manager for the Cupp chain. However, the store only remained opened for two additional months and cited poor business as the reason for closing. Considering the harrassment of the company guards to those who tried to shop there, it was no small wonder that the store survived only a few months. John Morey, Sr. and Gino Simoncini both lost their jobs at the mines for a while because their wives had purchased groceries there. Morey was recalled, but Simoncini had to seek work in Wehrum.

The Cupp chain continued to prosper in other towns however. There were two stores in Nanty Glo and one in Twin Rocks. In January 1929, the American Stores purchased the sixty stores of the Cupp Grocery Company. The Acme Market remained in Nanty Glo, with former Wehrum reside at Mario Biancotti as manager until the chain closed operations in the early 1970's.

VINTONDALE STATE BANK

In the midst of the lean times in the coal fields in the 1920's and on the eve of the KKK strike, Vintondale proudly hailed the grand opening of the Vintondale State Bank. The establishment of a bank was in response to a need: to keep Vintondale money in Vintondale. To conduct their banking, families had to travel to Nanty Glo, Ebensburg, Johnstown, or Indiana, a nuisance to those without an automobile. If one wanted to save money, the only alternatives to traveling to a bank, were to stuff the money in a mattress or purchase postal savings stamps. Because Prohibition stopped the manufacture and sale of legal liquor, certain individuals in town made large sums of money from the sale of moonshine and beer. Many families purchased the postal savings stamps with their ill-gotten profits. Postmistress Ruth Roberts Anderson said that some Sixth Street residents brought in as much as $5,000 in cash in a single trip to purchase bonds or stamps. Postal regulations required each bill's serial number to be written down in duplicate and forwarded to Philadelphia. When she ordered money from Philadelphia, the main post office often sent her $1,000 bills. The spark which ignited the bank movement was a **Johnstown Tribune** article which reported that Vintondale residents had $100,000 deposited in the Postal Savings Plan. That's when Otto Hoffman decided that Vintondale money should stay at home.

Efforts to organize a bank began in March, 1923 with Otto Hoffman meeting with Ebensburg attorney Randolph Myers (1889-1989) to discuss the feasibility of a bank. A second meeting with Myers followed on April 20. A successful town meeting was held on April 25th where $22,500 was subscribed right away. In addition to "talking bank" to the miners as they descended the office steps on paydays, Hoffman and others went door to door selling the shares, $55.00 each. The late Josephine Pisaneschi said people bought them because they were afraid to say no to Pappy.

The bank charter application was signed on May 15 by Otto Hoffman, Lloyd Arbogast, Mike Farkas, Sam Freeman, and George Balog. The signees composed a cross-section of the community: company, private business, labor, and ethnic minorities.

Other local bank presidents called on Hoffman during 1923. Alvin Evans of the American Bank of Ebensburg made a visit to quash rumors that his bank was weak and would close. Lester Larimer of Patton also paid an official visit in October, 1923.

Block C, Lot 9, east of the Vintondale Inn was chosen as the site for the bank. Vinton Colliery, owner of the lot, conveyed it to the Vintondale State Bank for $1.00. The construction bid was awarded to Carstensen and McLain of Johnstown. The project architect was Hersh of Altoona, who designed a two story brick structure with the bank on the first floor. The interior had white tiled floors, mahogany trim and "delicately tinted" walls and ceilings. The bank floor plan consisted of cashier's and teller's cages, a director's room and a vault. Built of concrete and steel and measuring eight feet by twelve feet, the vault had a door seven inches thick and an enclosed time lock mechanism. The entrance to the vault was re-enforced by a heavy steel gate. In the basement was a large fireproof vault and additional space for another vault.

On the second floor was a six room apartment with bath. First tenants were Mr. and Mrs. James Roberts and son. Later occupants were Bill and Beatrice (Moore) Jendrick; Beatrice operated a beauty shop there.

At the official opening of March 1, 1924, the Vintondale State Bank listed its capital at $50,000 with a cash surplus of $5,000. The first depositor was John Roberts. The board of directors in 1924 included Hoffman, Arbogast, Freeman, Balog, Farkas, and David Nevy of Vintondale; rounding out the group was I.E. Lewis of Ebensburg. C. Randolph Myers of Ebensburg was the solicitor.

Mike Mihalik

Cashier was Michael Mihalik, who had been the assistant cashier at the Union National Bank of Nanty Glo. Mr. Mihalik was born in Brisbane, Clearfield County on October 10, 1887. At the age of ten, he worked as a doorboy for fifty cents per ten-hour day. When the union came in, his salary increased to sixty-two cents a day. Much of his business expertise was gained working at Greenwich Supply Company, Emans; Big Bend Supply Company, Expedit; and the Eureka Supply Company, Windber. When the A.L. Anderson and Sons of Altoona was awarded the contract to build the Cambria and Indiana Railroad between Belsano and Nanty Glo, Mike signed on as a timekeeper. During World War I, he was drafted into the US Army and served for a year in France. Upon his discharge in 1919 after serving with the Army of Occupation in Germany, Mihalik became a teller at the Nanty Glo bank at a salary of $30 per month. The directors of the Vintondale State Bank elected him cashier, and Mike Mihalik assumed the position of bank cashier on December 1, 1923.

Mr. Mihalik's investment expertise kept the Vintondale bank in the black during the 1920's and 1930's. His wise counseling on investments, stocks and bonds aided several townspeople as well. While many families were scraping to find where the next meal came from in the Thirties, Mike Farkas made money from his investments, especially in steel. Because of Mihalik's business practices, the Vintondale State Bank was one of the few banks between Pittsburgh and Harrisburg to remain open during the banking crisis in 1933. The **Nanty Glo Journal** reported in its March 16, 1933 issue that Vintondale State Bank, First National Bank, and American National Bank of Ebensburg were reopened for legitimate banking business, but were not permittted to pay checks drawn from banks that had not yet received consent to reopen. Two of those which never reopened were the banks in Nanty Glo. The fiscal reliability of the Vintondale bank in 1933 was a great source of pride to Mihalik and bank president, Dr. MacFarlane.

Assets and Liabilities

In its annual statement for December 31, 1927, Vintondale State Bank listed assets and liabilities of $286,853.98. The Board of Directors chosen at the Annual Stockholders' Meeting were Hoffman, Arbogast, Farkas, David Nevy, Louis Nevy, William Roberts, and John Huth. Hoffman remained president; Arbogast was vice-president and Milahik, cashier.

The major depositors in the bank were the Vinton Colliery Company, Hoffman, Arbogast, and Farkas. The bank's major assets were in bonds and cash; it did not deal much with mortgages, probably because Vintondale was partially a company town, and real estate prices were at rock bottom. However, in 1929, the bank did purchase the house of the late Sam Butala on Plank Road for $800 from Jack Butala. The bank took a $100 loss on the transaction when it sold the duplex in 1936. Joseph Kemevich bought the easterly half for $325, and Andy Skvarcek purchased the westerly half for $375.

The Move to Nanty Glo

To comply with federal regulations, the VSB insured its accounts up to $2,500 through the Federal Deposit Insurance Corporation. However, by 1934, it was obvious that the bank had to be moved because there was not enough money in Vintondale to keep it operating. Nanty Glo's Union National Bank and the Miners and Merchants Bank failed, so the town was the prime location for the transfer. For two years, the Nanty Glo Lions Club waged a campaign to get a bank in Nanty Glo. A committee of Herman Donofsky, A.A. Dietrich, and Thomas Bello, with the approval of the directors and stockholders of VSB, made trips to Harrisburg and Pittsburgh to seek state and federal approval for the move. The state secretary of banking held the assets of the defunct Miners' and Merchants' Bank and gave his approval for the exchange of bank buildings and $10,000 in cash. A deed recorded October 8, 1936 granted VSB the Nanty Glo bank building for $10,000. For $1.00, the VSB building was transferred to the receiver. In late October, 1936, Vintondale and Nanty Glo representatives made a final trip to Harrisburg to complete the state approval for the transfer.

A director's meeting of the Vintondale State Bank was held on November 19, 1936 in Vintondale to finalize the transfer arrangements. The state-required stock subscriptions were made and paid in at the meeting. The only 1924 director remaining on the new board was Mike Farkas. Dr. MacFarlane, bank president, retired and moved to Indiana in 1936. Two other Vintondale directors also retired from the board. Otto Hoffman's replacement, Milton Brandon, remained on as Vice President of the newly-named bank, the Nanty Glo State Bank. The three new directors were President Joseph Delisi, Attorney Maurice Shadden, and Garfield Wilkens.

After six years without a bank, Nanty Glo had a solvent one thanks to the hard work of dedicated citizens. The moving date was November 28, 1936, and on

November 30th, the bank opened for business in the former Miners' and Merchants' Bank building. Mike Mihalik remained as cashier. Jesse Edwards, who had been a clerk at the Miners' and Merchants' Bank, became assistant cashier. On the first day, new accounts totaled five figures. The first checking account was assigned to A.A. Dietrich, and the first savings account was opened by Abe and Bessie Connor. Congratulatory floral arrangements were sent by the Nanty Glo Lions Club, Capitol Theatre and Mellon National Bank.

After the move, the banks issued its year end statement:

ASSETS
Cash on Hand	84,453.12
Loans and Discount	24,051.55
Bonds	112,928.30
Banking House	15,000.00
Total	236,432.97

LIABILITIES
Capital	25,000.00
Surplus	25,000.00
Undivided Profits	5,758.39
Deposits	180,674.58
Total	236,432.97

In 1952, the bank's assets and liabilities were worth $2,581,187.53; in 1954, they had risen to $2,672,681.85. By that time, Mr. Mihalik was president; Brandon was Vice-President; Jesse Edwards was cashier; and Tom Sabo was the assistant cashier.

The Nanty Glo State Bank merged with the United States National Bank. Tom Sabo advanced to assistant vice-president and manager of the Nanty Glo branch. In 1982, Nanty Glo and its surrounding neighbors share of the assets of US National grew to $28,000,000.

For those who retained their shares of Vintondale State Bank stock, the payoff was well-worth it. Mrs. Pisaneschi said that she had purchased two shares of stock in 1923, and by 1982, those two shares had multipled to 160 shares of US National Bank. Another resident had six shares in the late 1930's and sold them to Milton Brandon for $65 each. Brandon told him that the stock was worthless and encouraged him to sell. Since times were hard, he sold the stock to Brandon, and the money was used to buy furniture. Today, the seller realized that the stock was much more valuable than Mr. Brandon implied.

The Vintondale State Bank building is the only company building, or at least company organized building, remaining in Vintondale. The company store, hotel, company hall and others have either burned or have been razed. The building has had several owners, Mr. Averi for one and the borough for another. For years, it housed the US Post Office until the new one was built on the site of the old company store. Today some of the borough offices are housed there.

VINTONDALE AMUSEMENT COMPANY

In the first two decades of the Twentieth Century, the town of Vintondale mushroomed out of the rugged forests of the Blacklick Valley and sprouted into a town of 3,000 inhabitants. By 1920, this new town offered: a choice of three hotels, of which the Vintondale Inn was considered first class for its time; five to seven churches at any one time from which to choose; thirty-some businesses, including a large, well-stocked company store; electricity and running water in the houses (sometimes); a company social hall for dances and basketball games; and before Prohibition, Farabaugh's and then Callet's and Silverstone's liquor wholesale business and afterward, plenty of speakeasies in which to quench one's thirst. However, Vintondale was lacking one facility which was near and dear to the heart of Superintendent Hoffman: a first-class movie theatre and auditorium.

Yes, Vintondale had a nickelodeon in the wooden structure beside the company store; Mr. Burgan, shipping clerk, was the proprietor from at least 1917 to 1923. Evidence confirming his ownership came from the 1917 and 1923 mine diaries. On November 5, 1917, Shilling and Duncan showed ten reels of mining movies at the theater. Hoepfully, those ten reels were only for the miners. Burgan closed his "movie house" in 1923.

But, Vintondale did not have a stage large enough for variety shows, pageants, and graduation ceremonies. The social hall was not the proper setting for a solemn occasion such as a graduation; the churches were not large enough to hold a large crowd; and the senior high school auditorium was not built until 1927. So, Vintondale needed a theatre and got it, or was it that Pappy got his theatre?

A contract for architectural plans for the new theatre was negotiated with Mr. Myton of Johnstown on March 9, 1923, five months before the final site was chosen. On April 10, Mr. Myton paid a business call to show Mr. Hoffman the theatre plans, or as Mr. Hoffman referred to it in his diary, the "picture show". Mr. Burgan, proprietor of the Cozy Theatre, closed his operation until later notice. On August 22, Schwerin and Hoffman selected the site for the theatre, the undeveloped plot at the foot of Fourth Street, adjacent to the Vintondale Inn. On the town plan, this site was to be an extension of Fourth Street, so borough council enacted ordinance #33, dated September 26, 1923, to vacate the property. Vinton Colliery then transferred the parcel to the Vintondale Amusement Company for $1.00. The transferral was not recorded until August 20, 1924.

Several construction companies placed bids on the project. Those mentioned in the diary were: Nup and Bee of Starford, Hildebrand, and Carstemin and McLain. The diary gave no indication of which company was awarded the contract, but it probably was

Carstemin and McLain because they also built the bank. Excavation began on Friday, October 5, 1923; on his daily route, Hoffman checked the progress of this project and of the bank which was under construction at the same time.

Mr. Hoffman also spent time interviewing theater sales representatives; on October 17, a theatre seat salesman called. Since this was the era of silent movie heroes and heroines like Valentino the "Shiek", the "It" Girl Clara Bow, and Cowboy Tom Mix, a piano was needed to convey the mood of the film flashing on the screen. Mr. Hoffman personally made shopping trips to Johnstown and Barnesboro to find the right piano. On Saturday, November 24th, he purchased the piano in Barnesboro.

The author recommends Frances Pluchinsky's "Memories of Vintondale", for a first-hand description of the theatre. Originally, movies were shown five times a week, but in 1928, the theatre was only open three times per week. In January, 1943, the movie theatre was sold by the Vintondale Amusement Company to the Vinton Coal and Coke Company for $1.00. At that time, Bill Kissinger was president of the Amusement Company, and Aileen Huth was the secretary. Since these were the years when the coal company was reorganized and divesting itself of its houses and its subsidiaries, the theatre was sold to Ralph Lonetti and renamed the Vintondale State Theatre. In February, 1943, the boys who worked at the theatre formed a "union" called the "Reformers". They promised to pay dues, to attend church regularly and not to swear. President of the group was Ralph Lonetti; Vice-President, Leland Hagens; Secretary, George Dancha; and Treasurer, Joe Minerovich. A successful party, chaperoned by Thomas Reade, Vintondale's undertaker, was held in March, 1943; another was planned in April.

In February 1950, the theatre was owned by James DeMauro who had it completely remodeled. A new RCA sound system was installed as were new seats. The office was moved to the second floor, leaving room for a new snack bar in the lobby. An additional renovation was the construction of a new marquee.

In the 1950's with the advent of television, the drive-in theatre, and poor economic times, patronage at the theatre dropped so low that the owner could not make a profit. The Vintondale State Theatre closed its doors to its movie customers in the late 1950's. The last film that the author remembers viewing there was, **A Night to Remember**, a black and white classic about the sinking of the **Titanic**. The theatre continued to be used on rare occasions in the late 1950's, especially for high school and eighth grade graduations. The last high school graduation held there was in 1957, even though at that time, the Vintondale schools were part of the Nanty Glo-Vintondale Jointure. The final eighth grade graduation took place at the Vintondale Theater in 1959; the author's eighth grade commencement exercise in 1960 was the first combined eighth grade program in the jointure, which was held at the Capitol Theatre in Nanty Glo.

As a child, the author always heard rumors that the theatre was built over a mine shaft and that the floor could collapse at any time. Not until researching this book she did discover that those rumors were true, at least the mine shaft part. The main heading of Vinton Colliery's abandoned #4 mine was under the building and Fourth Street. That accounts for the cracks and occasional sinking of the paving on Fourth Street.

The Vintondale State Theater sat empty until 1973 when it and the old Vintondale Inn were razed to make room for the new Veterans of Foreign Wars Post.

Vintondale State Bank Grand Opening, March 1918.

Courtesy of Huth Family Collection.

Nevy Brothers Store, c. 1925. Front: Frank Pioli, Louise (Pioli) Pisaneschi, Louis Nevy. Rear: Joe Pioli.

Courtesy of Anna Pioli and Richard Pioli.

Chapter IX
Churches

VINTONDALE BAPTIST CHURCH

On April 14, 1895, Reverend E.E. Parker of the Crozier Seminary held services in Vintondale. Afterward nine persons: Mr. and Mrs. George Berkey, Mr. and Mrs. M.A. Davis, M.J.H. McCormick, Mrs. Harvey Berkey, Mr. and Mrs. R.R. Cunningham and Reverend Parker agreed to form a regular Baptist Church. Rev. Parker was asked to serve as a pastor half-time, since he was already ministering to a church in Ebensburg. Before 1895, several of the above members attended the Bethel Baptist Church, located half-way between Belsano and Ebensburg. The first service of the newly organized church was held on June 19, 1895.

To incorporate as the First Baptist Church of Vintondale, the members petitioned the Court of Common Pleas, Judge A.V. Barker presiding, on September 2, 1895. The purpose of the corporation was to worship publicly according to the constitution of the Baptist Church. The petition had been previously advertised in the **Cambria Herald** and the **Mountaineer Herald** for three weeks in August. Judge Barker approved the petition the same day it was presented.

Also on September 2nd, the members asked the churches of Indiana Association to advise them of the steps to be taken to gain recognition as a regular Baptist Church. On September 18th, a council of delegates of the Indiana Association, led by Rev. B.B. Henshey and James McAnulty, was held, and Vintondale was accepted unanimously. Rev. Parker and Mr. Davis went to Cookport for an Association meeting, and the Vintondale Church and Sunday School were received as members of the Association.

The first service as an incorporated member of the Association was held at Barker's Hall on October 13, 1895 with a Christian Endeavor Meeting at 7:00 PM. November 4-15, 1895 saw the first of many revival services in Vintondale. Rev. J.J. Bullen of Canada was the guest revivalist. According to the church minutes, the services "resulted in an awakening of spiritual interest in the community and the conversion of a number and the reclamation of a few from a back-slidden state."

Mrs. G.W. Berkey and Pastor Parker acted as a soliciting committee in January, 1896 to raise money to build a church. However, Rev. Parker resigned as of April 1, 1896. That same month, Mr. and Mrs. J.B. McClure were invitied to hold special meetings. They stayed for ten days, and the special collection taken up for them amounted to $16.00.

That summer, arrangements were made to build the church on Lot 15, Block D, which was donated by the Blacklick Land and Improvement Company. The conveyance fee was $1.00. The deed, recorded in 1897, listed the trustees as George Berkey, Harvey Berkey, William Shuman and Gust Rager.

The costs of building the church were:

Barker Brothers - lumber, bricks	$116.00
I. Brown - dry lumber	33.10
B. Shaffer - for planing	86.64
H. Bantley - hardware	94.00
W.J. Rose and Son - windows and glass	36.00
D.W. Brallier - carpentry work	161.67
J. Davis - shingles	46.00
organ and pulpit	33.00
seats and chairs	33.00
stoves and pipes	18.00
painting, paint and oil - Dufton	45.00
boarding - Berkey	68.00
common labor	16.00
hauling for church	60.00
	$859.91

The building was finished and opened for services on January 23, 1897. The initial service was preached by Rev. F.I. Sigmund of Ebensburg who was then asked in March to preach once a month for a stipend of $5.00. In March, Rev. Bullen also returned for a two-week revival and gained two professed conversions.

The formal dedication services were held on October 3, 1897. Delegations attended from Ebensburg, Pine Flats, Bethel, and Dilltown. Rev. Sigmund preached the sermon, and Rev. Elias Rowland of the Bethel Church recited the prayers of dedication. On that day, $163.00 was raised to help defray the costs of the church. $427.62 had already been raised before and during the construction of the church.

Rev. Sigmund resigned on October 26, 1898 and preached his last sermon on November 21. In March 1899, Brother R.B. Dunmire was asked to preach. C.T. Cornman of East Run preached for three consecutive days and then was asked on May 20th to be pastor at a salary of $800 a year. He accepted in June and a schedule of communion services and business meetings were set up. Rev. Cornman served until June 1, 1902. During his pastorate, a cantata was presented at Christmas in 1901, and a social to raise funds for the church was held on February 2, 1902.

Supply pastors served the spiritual needs of the Vintondale Baptists between 1902 and 1904. Some of these men were the Reverends: Dr. J.D. Feltwell (Altoona), Griffith, Postun, Bowler, Evans, Bradford (Baltimore), Dr. Wilson, and R.C. Morgan. The next fulltime pastor was C.W. Sheriff of New Jersey who served from November, 1904 to July 31, 1905 when he accepted a pastorate in Blairsville. During the same year, the interior of the church was painted. The next pastors were Brother Rees, J.M. Hunter (1906) and Reverend James (1907-08).

On February 13, 1907, the church board passed a motion which allowed the Finlanders to use the church after the Baptist services concluded.

The church was in financial trouble from 1907 to 1909 with cash on hand amounting sometimes to twenty-six cents. In 1907, the church had to spend $45 to get the foundation repaired and basement drained. These fiscal problems co-incided with poor economic times at the mines, which were suffering from the Depression of 1907. During those years, the miners were often paid only in scrip.

The records are not clear as to whether the church had a full-time pastor between 1908 and 1914. C.W. Teasdale preached part of the time in 1910, and an Indiana Association missionary, Rev. J.T. Davis, frequently preached in Vintondale between 1908 and 1914. In the summers of 1912 and 1913, several Bucknell students served as intern pastors. S.J. Cummings served as pastor beginning in March, 1914 and was credited with sixty-six conversions. However, he resigned June 14th because of ill health. In the summer of 1914, the church was served by the Reverends: Davis, Griffith, Evans, Roberts, Sheriff, Nupp, Stewart, Smith, M.M. Smith and Lamber. A supply pastor, H.F. King arrived in November, 1914 and held regular services until January 3, 1915. During that time, he did a complete census of the congregation. In April of 1915, Reverend W.M. Jennings, who had a church in Nanty Glo, was called to be a half-time pastor. A Dr. Sigmund also served the church in 1915 to 1917. The church minutes stopped for six years until revived again in 1921.

Vintondale Baptist Church, c. 1915. Taken before a tornado tore off the steeple.
Courtesy of Leona Clarkson Dusza.

Renovations were begun in 1921. On March 13th, motions were made and carried at a board meeting to excavate the basement, pour a concrete floor and repair the basement walls. In addition, a new chimney was built and the roof replaced. The old belfrey was removed and rebuilt. No costs were listed for the projects. The repairs to the steeple were necessary because in 1920, a tornado blew it off onto the roof of Mitchell's house next door.

Also in the early twenties, the church, for unknown reasons, severed its membership with all Baptist associations. Extant church records stop in 1921 and were not revived until 1936.

CATHOLIC CHURCHES IN THE VALLEY

At the advent of the twentieth century, the Blacklick Valley was considered missionary territory by the Roman Catholic Church. Between 1890 and 1915, the mining towns of Vintondale, Twin Rocks, Nanty Glo, Wehrum, Dilltown and Colver surfaced on the county maps. An influx of Catholic immigrants from eastern and southern Europe helped populate these towns. In 1890, the closest Roman Catholic churches to minister to the needs of these newcomers were St. Patrick's in Cameron's Bottom and Holy Name in Ebensburg.

Until recently, the hierarchy of the Roman Catholic Church in the United States consisted of Irish-born priests. In many cases during the early years of this new immigration, the Irish bishops did not understand the customs practiced by the Church in Eastern Europe, especially those of the Uniate churches. Many misunderstandings arose between the clergy and the parishoners. Like the bishops, almost all of the priests serving the Roman Catholic parishes in the Blacklick Valley between 1895 and 1930 were natives of Ireland.

As early as 1895, Rev. James J. Deasy, pastor of Holy Name from 1890-1900, said mass in a boarding house in Twin Rocks. The first known priest assigned to Vintondale was Rev. Patrick O'Connor of the Pittsburgh Diocese. Fr. O'Connor, ordained in 1895, served in Vintondale from October 6, 1900 to December 6, 1900. He was then transferred to Alpsville and was pastor of St. Peter's Church in McKeesport at his death at age 79 in 1950. The reason for his short stay is not disclosed in the records of the Pittsburgh Diocese, and his name is not included in the Priest's List of the Altoona-Johnstown Diocese.

In June 1901, the Altoona Diocese was officially created; Rt. Rev. Euguene Garvey was consecrated on September 8, 1901 as its first bishop. Due to a shortage of priests in the new diocese, Archbishop Ryan of Philadelphia granted permission in November, 1901, for one of his newly-ordained priests, Fr. Thomas J. Hurton, to assist Fr. Ludden, pastor of Holy Name. Part of Fr. Hurton's pastoral ministries included the Catholics of the Blacklick Valley. Fr. Hurton set about raising funds for a new church in

Vintondale. According to a January 22, 1902 letter from Walter Scranton to Henry Wehrum, Fr. Hurton had requested that Lackawanna Coal and Coke Company grant the Altoona Diocese a property on which to build a church. Mr. Scranton favored granting the request and had a company lawyer, Mr. Higginson, draw up a deed. Property granted included lots 7, 9, and 11 on Block K (Fourth Street); price paid to Lackawanna Coal was $1.00.

A white frame church was constructed at a cost of $3,000. On June 22, 1902, Bishop Gravey laid the cornerstone and afterward confirmed a large class of students. The dedication was attended by several hundred people, some of whom arrived by two special trains. Included in the entourage were the pastors of churches in Gallitzin, Lilly, Wilmore, St. Augustine, Carrolltown, Patton and Ehrenfeld. The sermon was preached by Rev. Matthew Smith, Chaplain of Mt. Aloysius Academy in Cresson. Guests at the dedication were served dinner by the women of the Vintondale parish.

The new church was almost debt-free from the start, and from 1902 to 1908 the priest resided in Vintondale, presumably in the house on the corner of Third and Griffith. It was first owned by Nicholas Altemus, then the Vinton Lumber Company, and Dr. Gile, who sold to Warren Delano in 1906. Today it is owned by Stephen Dusza. Because the Giles had a large house in Wehrum after 1902, they may have donated the use of the house to the priest. It is thought that until the church was completed, mass was said in the house. There was a rectory in 1906 because William Hunter and Mary McCloskey were married there on February 15, 1906 by Fr. W.F. Davies.

The priest also liked to have music at mass; so he paid the author's great-aunt Eula Hampson Burr, a non-Catholic, a generous stipend of one dollar per Sunday to play the organ and sing at mass.

Fr. Hurton, another native of Ireland, not only served Vintondale, but missions in Twin Rocks, Nanty Glo, and Wehrum as well. He often signed the marriage records with "Administrator of Blacklick Missions" and described his work in a long lost pamphlet, "A Sketch of the Blacklick Valley and its Catholic Missions". In July, 1904, he returned to the Philadelphia Diocese and served parishes at Minersville and Frackville in the hard-coal region. At the time of his death at the age of 71 on July 22, 1948, Fr. Hurton was rector of St. Theresa's in Philadelphia.

Fr. Hurton was succeeded by Rev. William Davis (Died 4/13/1946), who was then replaced by Rev. Timothy Goblen (Giblin) in 1907. (Died 1913 in Dunlo). In 1908, because Nanty Glo attained the status of a parish, Vintondale and Twin Rocks became missions under Fr. Michael O'Connor. In 1912, Holy Family Parish was created in the new mining town of Colver. Twin Rocks was then served by St. Mary's in Nanty Glo while Vintondale's Catholics were ministered to by the Colver priests. Fr. Ignatius Herkel was assigned to Colver until August, 1915 and was followed by Fr. Edward Daly. A native of Thorndike, Massachusetts, Fr. Daly was ordained in 1915 by Bishop Garvey. His later assignments were Dunlo and St. John Gaulbert in Johnstown.

The Colver priests traveled to and from Vintondale via the C&I Railroad; often the Colver choir accompanied the priest. Claire Bearer of Colver remembered making the trip to Vintondale on Sundays. In Colver, the priest received a stipend from the Ebensburg Coal Company, but it is not known whether the Vinton Colliery gave its local clergy stipends.

In 1915, the Immaculate Conception parish had two collections as its sources of income: a weekly "Penney Collection" and a monthly collection. The weekly amount ranged from a low of $2.51 to a high of $5.98, and the monthly varied from $36.00 to $75.00. The collection record for April through July, 1915 were found in one of Jack Huth's notebooks. The records, as found in the ledger, were as follows:

Priest's Account
4/24/1915 - Received from Henry Rager
Flower money - $13.05
5/21/1915 - Penney Collection - $3.00
5/9/1915 - Penney Collection - $3.33
5/9/1915 - Monthly Collection - $73.90
5/16/1915 - Penney Collection - $3.86
5/23/1915 - Penney Collection - $4.04
5/30/1915 - No Church
Cash on hand from April - $83.26
Total - $171.39
June 1915
6/6/1915 - Penney Collection - $4.25
6/13/1915 - Penney Collection - $3.02
6/13/1915 - Monthly Collection for May - $73.80
6/20/1915 - Penney Collection - $3.73
6/27/1915 - No Church
7/18/1915 - Monthly Collection for June - $48.00

In 1917, St. Charles Church in Twin Rocks became an independent parish, and Immaculate Conception became its mission. The facilities at St. Charles consisted only of a basement, which was built in 1910 by Fr. O'Connor on land donated by Fred McFadden. Rev. William H. McCook (1879-1948) served as first pastor of the newly independent parish. Ordained in 1913 by Bishop Garvey, Fr. McCook served from October 30, 1917 to May 21, 1919 when he transferred to the Philadelphia Diocese. At his death on August 2, 1948, he was rector of St. Aloysius in Pottstown.

Fr. Joseph Howard succeeded Fr. McCook and built a rectory in Twin Rocks in 1919. Masses were held in Vintondale each Sunday, alternately at 9:00 AM and 11:00 AM. Again, the priest usually arrived by train, had dinner at Walter Morris' house after mass and was driven back to Twin Rocks by Mrs. Morris.

In 1923, Fr. Charles Haley served several months as pastor at St. Charles and then was transferred to St.

Columbia's in Johnstown. His replacement was Rev. James Hebron, former assistant pastor at Renovo. During his pastorate, St. Charles Church, a veneered brick structure, was completed by C.R. Dilling Construction Company at a cost of $15,000. The cornerstone was laid by Bishop McCort on October 24, 1924.

Fr. Hebron was born in Johnstown on May 7, 1887 and was ordained in 1913 by Bishop Garvey. He served Twin Rocks and Vintondale from 1922 until January 10, 1927 when he was sent to the new Visitation Parish in the Roxbury section of Johnstown. He was also assigned as the chaplain to Mercy Hospital. In 1951, Fr. Hebron was killed in an automobile accident enroute to the funeral of another priest, Fr. James Tolas of Tyrone.

Fr. Charles Gallagher, born January 3, 1890 and ordained on June 29, 1916, was the next assignee to Twin Rocks and Vintondale. He took charge in January, 1927 and served during the lean years of the Depression. An avid baseball enthusiast, Fr. Gallagher played third base for Prospect in the City League. Consequently, he set up a baseball team in Twin Rocks called the St. Charles' Athletics. His previous pastoral assignments were: St. John's in Bellfonte and Rockview State Prison, St. Peter's in Somerset, and St. Columba's in Johnstown. Father Gallagher's accomplishments in Vintondale will be written up in Volume II of **Delano's Domain**.

Immaculate Conception Catholic Church, c. 1902. Rectory in the background on the right.
Courtesy of the late Agnes Dusza.

CHURCH OF GOD

The Vintondale Church of God was founded by Samuel R. Williams around 1900. Mr. Williams planned his house/store on the corner of Third Street and Barker Street with the church in mind. The west side of the first floor was constructed as a single room which extended the entire length of the house; width of the room was seventeen feet. Here church services and Sunday school were held. Instead of pews, chairs were used to seat the congregation. Music for the services were played on an organ which had belonged to Margaret Sides Williams' family. The first organist was Mrs. Williams; later, daughter Sarah assumed the post.

The Vintondale church was affiliated with the Church of God, headquartered in Anderson, Indiana. There was never a resident minister, but one from Nanty Glo usually conducted the services.

The Church of God does not believe in having a bishop as a head of the church, keeping a church roll, and casting out members. To become a member, one has to make a public profession of faith in the church; you then become a son of God. One church ordinance is the belief in washing of the feet. This is done monthly at the communion serivce. The communion is a closed communion, meaning church members only.

Sunday school was taught by Mr. Williams, who was self-educated and versed in the **Bible**. A devout man, Mr. Williams considered even the Baptist Church too worldly. He lived his religion daily. The family started the day with a "family altar". They got down on their knees and prayed before breakfast. Mr. Williams read from the **Bible**. His son, Lloyd, said he never heard his father swear or lose his temper, even though he could get angry occasionally. Lloyd said that his father refused to sell tobacco in the store, and if you asked for it, you might get a lecture on the evils of tobacco. Mr. Williams got along with everyone and was respected by all. The Church of God disbanded about a year before Mr. Williams' death on July 4, 1943.

SAINTS PETER AND PAUL RUSSIAN ORTHODOX GREEK CATHOLIC CHURCH: THE LITTLE CHURCH THAT COULD!

Ethnic Background

Among the waves of immigrants who passed through the gates of Ellis Island between 1890 and 1914 were former residents of the eastern slopes of the Carpathian Mountains and the Western Ukraine. The most accurate name for this portion of the Austro-Hungarian Empire was Red Russia. (Not red as in communist, but in contrast to the region called Byelorussia, or White Russia). Almost all of these newcomers to America were farmers by occupation and were closely tied to the church. Originally, these churches were part of the Orthodox Rite, but joined the Roman Catholic Church in order to secure some legal rights within the Austro-Hungarian Empire. The churches which swore allegiance to the pope were known as Uniate or Byzantine Rite churches. The ligurgy remained basically the same, but the loyalty shifted from the patriarch to the pope in Rome. When the immigrants from Red Russia, designated Ruthenia by the Roman Catholic Church, reached the United States, there were no Byzantine Rite churches here. The Catholic hierarchy in the United States did not understand the Eastern Rite and were determined to absorb it. Due a misunderstanding on both sides between the Bishop of Minneapolis and a Byzantine priest, a schism resulted. That Byzantine priest and his parish joined the Orthodox Church; soon they sent missionaries into the cities and mining towns of Western Pennsylvania organizing many of the immigrants into Orthodox churches. The split was also accompanied by a decree in 1890 from the Roman Catholic church which confirmed clerical celibacy; whereas, the Orthodox Church allowed her clergy to marry before ordination. These schisms had little to do with doctrine or the liturgy, but were due to the lack of education on the part of the clergy and parishoners and a lack of control by the bishops.

Most Carpatho-Rusins were uneducated peasants, and the church was the center of their lives. The priests, usually European-born, controlled the congregations and were very slow to Americanize and helped slow down the process of integration of their people into American society. The various schisms and shifts in allegiance had a profound effect on the families in the Blacklick Valley. Some families remained Uniate and formed St. Nicholas Byzantine church in Nanty Glo. Others were Orthodox in Europe and practiced Orthodoxy in the United States. A third group were members of the Orthodox Church both in the old country and here, only to split away completely to become Jehovah's Witnesses.

Sts. Peter and Paul Church

Sts. Peter and Paul Russian Orthodox Greek Catholic Church was first organized in Wehrum some time between 1902 and 1904 through the efforts of St. John the Baptist Church in Conemaugh. From the earliest days, Sts. Peter and Paul Church's loyalty was to the Metropolitan of Moscow. Lack of a church charter and church property in the town of Wehrum hinder pinpointing its founding. Actual location of the church in Wehrum is unknown, but the most likely site was A Street where the Methodist Episcopal and Catholic Churches were built. It may have first occupied the church where the Catholic parish of St. Fidelis was formed in 1912. (Records for St. Fidelis are almost non-existant.)

When the town of Wehrum shut down indefinitely in 1904, all the miners were told to vacate the town. Many moved to Vintondale. There, members of the church purchased Lot 11 Block J in 1907 for $1.00 from the Vinton Colliery Company. The location of the new church was to be on the corner of Third and Lovell Streets. The company deeded the lot to Archbishop Platon, Archbishop of North America. A white frame church was built in 1907, and the cornerstone was laid on November 28, 1907. A photograph was taken at the dedication hangs in the vestibule of the church.

Sts. Peter and Paul Russian Orthodox Church Dedication, c. 1909. Charles Tegsa, cantor, on the priest's left. Mr. Totin third from the left on porch, Mrs. Totin and son Mike at foot of steps.
Courtesy of Sts. Peter and Paul.

In February 1911, the congregation purchased Lot 9, Block J from Joamiky Kraskoff for $300.00. A former resident of Wehrum, Kraskoff had purchased the house in 1907. Both deeds prohibited the buyers from selling or keeping for sale any intoxicating beverage. Also the owners were forbidden to build a mill, slaughter house, stable or factory for manufacture of glue, soap, candles, or starch on the lots.

Two of the early founding families were the Jacobs and Silecks. Mr. and Mrs. Sileck were donors of one of the church icons. Some of the other icons and the chalice came from Russia. Tsar Nicholas II's treasury annually budgeted $77,850 for the Russian Orthodox Missions in America. Some members believe that some of the money used to furnish the church had filtered down from the tsar.

The Vintondale church adherred to the Julian Calendar, so their Christmas celebration was on January 7. Easter, the most important church holiday, was fourteen days later than Easter on the Gregorian Calendar. Following a strict forty day fast, the parishioners celebrated the Resurrection of Christ with singing, processions and blessings of the Easter foods.

Even funerals became colorful spectator events. If the deceased had been a parish notable, the coffin was carried from the home to the church, accompanied by the mourners and the large processional banners. Following the funeral liturgy, all walked to the cemetery, a long hike no matter which cemetery.

Because there are no surviving church minutes and the **Nanty Glo Journal** only began publishing in 1921, there is little known about the church between 1911 and 1921 except from deed books, one mine diary, and the cross in front of the church. There was a change in bishops in 1915; Bishop Platon Rozdestvensky, former bishop, turned over all church property in North America to Bishop Alexander Nemolousky.

In 1917, the resident pastor in Vintondale was Rev. Puhinsky, who played an active role in the community. He, along with Warren Delano and Judge Reed, gave a speech at a patriotic flag-raising ceremony in May, 1917. When a member of the church committee was discharged at the mine, Rev. Puhinsky was willing to approach Mr. Hoffman about restoring the miner to his job.

In 1917, the congregation raised $250.00 for a granite cross which was placed between the church and the parsonage. The inscription and names of the donors are in hard Russian and were translated for the author by Father Joseph Raptosh. The inscription is as follows: "This cross is erected to the glory of God for the remembrance of those of our brothers who have died in the war of 1914-" Below the inscription is the name of Vasili (Charles) Tegsa, church cantor. Donating $50.00 were M. Dovbinich and F.I. Bogdan. Peter Dancha contributed $40.00. $20.00 donors were: I. Popp, I. Dovbinich, M. Marichnich, M. Pakanich, V. Draguski, M. Ankricho, Ila Blanda, I. Lach, P. Gaidur, I. Diak (Daok?), B. Babich, C. Roman, Elean Roman, I. Kanchi, V. Stoyka, A. Babich, I. Babich, M. Lach, G. Kerekes, I. Bogdan, I. Karol, M. Livka, N. Rosocha, U. Garasti (Harasty), and Zuba all donated $10.00. Contributing $5.00 were A. Laskovich, I. Spaschuk, V. Leleck, M. Kerekish, S.H. Kostan, and V. Goborich.

Until the spring of 1921, Sts. Peter and Paul Church was prospering. Russian summer school was offered; children were taught their **Bible** lessons while learning to read and write Russian. The priest and sometimes the cantor served as teachers. People traveled in from neighboring towns to attend services. (Carefully observed by the town police of course.) Church bazaars and dances were well attended. The parish even had two paid cantors, Charlie Tegsa and Charlie Kovach. The cantor also had the responsibility of ringing the church bells.

In April or May, 1921, a new priest was assigned to Vintondale, Reverend Dzwonchick (means bellringer in Russian). Within a short time, he had organized a choir which was going to entertain at an ethnic dance. Three months later, the church was in the throws of its own schism. Suggested reasons for the split were dislike for the new priest and rivalry between the cantors. What ever the cause, the sides were drawn both physically and verbally. Each side hired lawyers to represent them in court.

At the time of the split, the church officers were Mike Glowa (brother of Steve Glowa), President; Charles Volosin, Secretary; Mike Babich, Treasurer. Other directors were Joseph Holod, George Stoika, Charles Kovach, Mike Dzuba and Joseph Tenchak.

According to a document in the Cambria County Equity Docket, the church officers filed for an injunction against Charles Harasty, George Babich and Charles Risko. These three men, along with their families and some friends, staged a sit-down strike in the church, claiming that they did not like the new priest.

Plaintiffs in the injunction petition were the church officers and church members. Randolph Myers acted as their attorney. (Mr. Myers died in 1989 at the age of 100. He informed the author in 1983 that he was not taking any new clients.) The three defendants were represented by attorneys Walter Jones and F.J. Hartman. The petition, two injunction affadavits, and bond were filed on August 20, 1921. A preliminary injunction was awarded to the church officers. Harasty, Babich, and Risko were restrained from interfering with the use of the church, contents or property of Sts. Peter & Paul by the priest and parishoners. The injunction was first continued until August 26, then until further notice. The defendants were also restrained from paying out any funds belonging to the church, from destroying or mutilating church books, records or property.

The defendants filed an answer to the injunction on September 19th. On December 14, 1921, court set a hearing for January 9, 1922; at which time, testimony was lodged. By March, 1922, the church had a new pastor, Rev. John Kozetski. He made a court appearance in the civil suit in early April with John Rubish, Charles Kovach, Joseph Holod, Harry Stroyer, Charles Babich, Mike Glowa, John Hazey and Charles Valosin. In July, a petition was filed with acceptance of service. An answer was filed on August 22. The judge ruled that all church offices held by lay members were delcared vacant. All church records, money, papers and property were turned over to court-appointed officers. The new officers met the approval of the Archbishop and were sworn in by the priest. The new warden or president was John Rubish; Vice-Warden,

Mike Drabb; Secretary, John Popp; Assistant Secretary, Harry Stroyan; Treasurer; Charles Harasty; Assistant Treasurer, Mike Glowa. The make-up of the new court-appointed board was a mixture of plaintiffs and defendants in the original petition.

Undocumented stories of fistfights and fights using fence pickets circulated for years. The once thriving church was severely, but not mortally, wounded. Its treasury was depleted with most of it going for legal services. Many families stopped going to church completely.

The Russellites

Slavish-speaking evangelists from Brownsville came to Vintondale to conduct **Bible** study classes. About twelve families left the church to join together in a **Bible** study group which became known as the Russellites. Charlie Tegsa, former cantor and butcher for Walter Morris, was the leader of the group. The Russellites were named for Charles Taze Russell of Northside, Pittsburgh who attracted followers as early as the 1870's by teaching that Jesus was not God, but the son of Jehovah. Christ was a crown prince waiting to take the reins of power. The group's magazine, **The Watchtower**, was first published in 1879. In 1884, Russell organized and served as first president of the corporate arm of his religion, the Watch Tower Bible and Tract Society. In 1914, Russell taught that Jesus Christ was crowned in heaven and cast the devil and his followers to earth where they used their wiles to initiate World War I. Russell's Witnesses believe that they are citizens only of Jehovah's kingdom, but they do obey all temporal laws except those which contradict their teachings, such as voting, fighting in wars, and saluting the flag. Russell's teachings were spread through door-to-door evangelizing; the Brownsville group was very successful in Vintondale.

In August 1923, the group of Vintondale dissidents, identified as the second party of the Russian Church by the **Nanty Glo Journal**, purchased Lot 4, Block K on Third Street from David Kempfer for $4,000. On the lot were two houses; the second floor of the upper house was used for **Bible** study. Listed as owners of the deed were: John Popp, Metro Rosenka (Rosacha), Mike Babich, Charles Harasty, John Krisfolosky, Mike Kudrince, Metro Karoly, John Kovach, John Prehora, Mike Risko, Metro Demyan, George Demyan and Mike Drabb. Three years later, Frank Mehalko purchased the property for $350.00 from Cambria County Sheriff, Carl Steurer, at a Sheriff's Sale. The Mehalkos lived in the upper house. The lower house was later bought by John Rubish. Some of the members returned to the Orthodox Church; others remained Russellites, who later became known as Jehovah's Witnesses.

There was a winner in all of this strife! It was not the faithful of Sts. Peter and Paul; it was not the Russellites; it was not those who dropped affiliation to either group. The winners were the lawyers for both sides.

Sts. Peter and Paul had to begin rebuilding its parish. The treasury was empty; morale was low, but the will of the people allowed the little church on the hill to survive this crisis and others to come. By 1930, the church was again functioning and sponsoring ethnic dances as fundraisers.

SAINTS PETER AND PAUL RUSSIAN ORTHODOX GREEK CATHOLIC CHURCH CEMETERIES

In October 1906, the parishoners of Sts. Peter and Paul, through Archbishop Tilchon, purchased a .45 acre tract in Buffington Township from R.W. and Eliza Mack. Purchase price was $200. This land, which was designated as a cemetery, is located on a township road leading up over the hill from the top of Broadway in Wehrum. A dedication monument, dated October 1, 1906, still stands in the center of the cemetery. The site was used for burials until the late 1920's. Two of the last buried there were Mike Babich and Helen Rosacha in 1927. Reasons given for the discontinued use of the Wehrum cemetery were the lack of grave sites and the rocky soil. Most graves are unmarked, but can be distinguished by the stones surrounding them.

The Wehrum cemetery was abandoned and neglected until the 1980's. The author, through an Academic Excellence Grant from the Indiana Area School District, was able for several years to take students to the cemetery each spring to try to keep ahead of the brush. Albert Parsley's Scout Troop also cleaned the cemetery as a service project. In the spring of 1990, the men of the church worked several days to clear the brush as the cemetery was on the tour route of a bus of dignitaries from the America's Industrial Heritage Project. This is hopefully the beginning of an annual community spring cleanup for the Wehrum Cemetery.

Unfortunately, there are no church records of locations of the graves. Through the same grant, the author was able to have the names on the tombstones translated. This work was done by Mrs. Helen Balog Oravec in 1984. In addition, another eighty-eight names of persons buried in Wehrum have been located in the Birth and Death Registers found in John Huth's attic. Both of these lists have been computerized and sent to area historical societies and Orthodox churches.

Vintondale Cemetery

In 1922, the parishoners of Sts. Peter and Paul approached the Vinton Colliery Company for a piece of ground in or near Vintondale which was suitable for a cemetery. On October 24, 1922, the company conveyed a 2.14 acre plot to the church for $1.00. Location of the lot was on the extension of Maple Street, facing the road leading to Goughnour's farm. A map of early burial sites is also not available for the new cemetery. The older members, like Mrs. Toton, knew who was buried where, but unfortunately no one wrote down the information. In both cemeteries, graves once marked with wooden crosses are now unmarked. Due to the dampness of the soil, the crosses rotted at the ground level. Even the large cross marking at the cemetery suffered the same fate and was replaced three times. A new granite cross was installed in 1983.

Outside the main cemetery, separated by a ditch, were several graves on a grassy knoll. A small bridge crossed the stream and flagstone walk led to the graves. The ravine was recently filled in and the two cemeteries were combined. Following years of controversy over who had the right to be buried in Vintondale cemetery, the committee decided to set aside a section of lots for non-members. The first Vintondaler buried in this area was Inga Shestak, German-born wife of Joe Shestak and tireless worker for the betterment of Vintondale.

Each year on Memorial Day, the pastor blesses the graves of former church members buried in the Wehrum, Vintondale, Twin Rocks, Lloyd, and Johnstown cemeteries. Several times a year, the names of the deceased members of the parish are remembered in a special liturgy. Included in these prayers are the miners killed in the Wehrum explosion.

PRESBYTERIAN CHURCH

In 1912, there was a short-lived effort to organize a Presbyterian church in Vintondale. Leading this drive was John Burgan, company shipping clerk, who applied to the Blairsville Presbytery for permission to set up a church. In September, 1912, he was elected the ruling elder. The congregation, which never had a church building, dissolved in 1922 because of the coal strike and a changing population in Vintondale. Most members joined the Baptist Church. Mary Shaffer, treasurer of the Vintondale Presbyterian Church, gave Baptist Church treasurer, Paul Graham, a check for $154.51. The Baptist Church was "sorely in need of our help." Mr. Burgan joined the Baptist Church and upon leaving the employment of the Vinton Colliery Company rejoined a Presbyterian church in Johnstown.

HUNGARIAN REFORMED CHURCH OF VINTONDALE

To meet the religious needs of recent Hungarian immigrants of the Reformed persuasion, an organizational meeting was held on February 27, 1916 in Vintondale to discuss the possibility of forming a branch church of the Johnstown Hungarian Reformed Church, located in Cambria City. Rev. Ernest Porzsolt, Johnstown pastor, accepted the unanimous vote of those attending to initiate the branch church.

Elected elders for the new church were: Janos (John) Gyorgy, Karol Toth, Peter Toth, Vince Hadar, Istvan Toth, Ferencz (Francis) Simon, Joszef Simon and Zigmond Yobbagy. Rev. Porzsolt agreed to conduct services on the second and fourth Sundays; his stipend for three months was $25 with travel expenses of $2.00 per trip. His salary was increased to $50 per quarter in 1919.

Following morning services, the new congregation held its annual general meeting on January 28, 1917. Additional council members elected were: Beni Markus, Antal Bagu, Baloz Bagu, and Joseph Gero.

The financial report for the first year was as follows:

Income: $410.55
Expenditures: $242.83
Balance: $167.72
        included January collection of $26.95

For several years, services were held in the town social hall and/or the Baptist Church, and a building fund was initiated. In 1921, the congregation began finalizing plans to construct a church. For $1.00, the Vinton Colliery Company deeded them Lot 13, Block F, last lot on the north side of Main Street adjacent to the cement bridge. The expressed purpose of the deed transfer, which was not recorded at the courthouse until January 30, 1923, was to erect a church. In addition to the lot, the VCC donated $100 to the building fund. Contributions also came from the 61st Branch of the **Verhovai Association**, from churches and/or Reformed members in Mosscreek, Emerich, Barnesboro, Clymer, Charles, Colver, and Windber.

The pre-construction stage of the building fund drive was not without its adversities. A general meeting was held in February, 1921. Attendance at that meeting was small. Church minutes imply that the meeting was rocky and that the elders were harrassed by those attending.

Actual construction began on September 22, 1921; contractor was C.R. Dilling of Nanty Glo; cost of the new church was $2,750. On September 25, 1921, the foundation stone was blessed. The white frame church was completed that year. Mr. and Mrs. Otto Hoffman attended the bell dedication ceremony on May 30, 1923; the Hoffmans were among the sponsors.

The Vintondale church became a self-governing mission in November, 1925. At that same meeting, the elders decided to hold services on Saturday evenings at 7:00 PM, in addition to the Sunday service. Also planned was a Hungarian dance for December 26th, the church's most successful type of fundraiser.

Vintondale's first full-time pastor was Zoltan Csorba, a native of Budapest, who was elected in January, 1926. His contract was from May 1, 1926 to May 1, 1927 at an annual salary of $500. Rev. Csorba was willing to serve as assisting minister for $60 per month until his contract started. The minister of the Rossiter congregation, Rev. Istvan Baktornyai, who had served Vintondale, "would not do" in the eyes of the council.

At that same January meeting, elders from neighboring towns were chosen. Wehrum was represented by Andras Markus, Balint Koszta, Gabor Toth and Josef Toth; Twin Rocks; Istvan Bok and Janos Bornyik; Nanty Glo; Josef Gyorgy and Ferencz Somogyi; and Cardiff; Istvan Pumeczki and Pal Somogyi.

At the January, 1926 annual meeting, the church had a balance of $380.53 in their account at the Vintondale State Bank. An inventory of church possessions was included in the minutes. Itemized were: one round table and five tablecloths, one mantle for the minister's use, one birth and marriage register, two money boxes, one bell, three ceiling lamps/chandeliers given by the Vintondale Reformed Women's Society, one standing lamp, two church bells, one marble wall tablet, one slate tablet, one piano, and two armchairs.

In March 1926, the congregation again contracted with Dilling to raise the church and build a basement underneath it. The Vintondale church advanced its status from a mission church in 1916 to a mother church in 1926. A branch church was formally organized on June 27, 1926 in Barnesboro. Vintondale's minister was released every fourth Sunday, and for the Lord's Supper on the second day of Christmas, Easter, and Whitsun to serve the Barnesboro church. Representing the Vintondale church at the ceremony were Josef Gero, Warden/Adminsitrator; Karoly Simon, Secretary; and John Toth, Secretary. The new Barnesboro church was to pay the mother church $100 per year.

The 1926-27 inventory added five church windows, six school windows, sixteen table legs, four short tables, two long tables, vessels used for christenings, one pitcher and plate, one slate tablet in school, two church hymnals, one sofa, twenty-one benches for church, sixteen benches for school and vessels used for the Lord's Supper.

In June 1927, a committee of Karoly Toth, Istvan Toth, Balint Izso, Zigmund Yobbagy and Karoly Simon were instructed to consult with craftsmen concerning repairs to the roof and gutters.

Rev. Csorba tendered his resignation in August, 1927 to return to Hungary to engage in YMCA work. Rev. Csorba was formally installed for this work in November, 1927 by Rev. Julius Melig of McKeesport, President of the Hungarian Synod; Rev. Alexander Kallassy, Director of the Hungarian Reformed Orphanage in Ligonier; and Rev. Ernest Persolz of Johnstown.

Replacing Csorba as pastor was Bela Csontas, also from Budapest, who continued the annual summer schools, the popular church bazaars, and the dances. One such social was held by the Women's Association on February 11 and 12, 1928 and featured fancy and handiworks, toys, games, cakewalks, music, dancing, refreshments and a beauty contest. Admission to the bazaar was 25 cents and 50 cents for both the bazaar and the dance. In the 1920's, a dance was a profitable fund raiser. In January 1927, gross profit was $365.90 with expenditures of $108.10. (See Frances Pluchinsky's "Memories of Vintondale" for a description of the grape dances.)

Rev. Csontas left Vintondale to assume a pastorate in Lorraine, Ohio. In July 1935, he was killed in an automobile accident while returning home from a church convention in Ligonier.

Replacing Rev. Csontas in January, 1929 was Rev. Ladislav Hunyady, a recent graduate of Franklin and Marshall College. In 1929, his schedule of services were as follows: Sunday School at 9:00 AM, Morning Worship at 10:30, Afternoon Service at 3:30, Star Club at 4:30, Christian Endeavor Meeting at 7:00 PM. A Midweek Service was held at 7:00 PM on Thursdays. Sunday School teachers in 1929 were G. Antal, P. Ure, H. Ure, E. Vargo, H. Bargo, E. Dulenszky, A. Antal and J. Soos.

Church Fire

The Hungarian Reformed Church of Vintondale was tragically destroyed by fire on New Years' Day, January 1, 1930. A service had been held that evening, and around 10:00 PM, a fire started in an overhead flue and entirely destroyed the church in spite of assistance from the Nanty Glo Fire Company. The church, valued at $7,000 in the newspapers and $9,500 in the church minutes, was only partially destroyed, but ruled a total loss. The adjacent company house, occupied by George Medwig, was damaged. Additional losses included Rev. Hunyady's books worth $300 and a violin, belonging to the late Vilmos Gero, which was used in the evening's service. The church council reimbursed Mrs. Gero $25.00 for the violin.

The congregation immediately resolved to rebuild and initiated various fund drives, such as dances, canvassing the Blacklick Valley, and a letter campaign to other Hungarian Reformed congregations. The congregation received permission to hold services at the Baptist Church; for the use of the church, a donation of $10 per month was offered.

During reconstruction, the schedule of services were: Thursday at 7:00 PM, Sunday at 2:00 PM, Sunday School at 8:00 AM, and Christian Endeavor at 7:30 PM on Sundays. Saturday School was temporarily suspended.

The church council also authorized renting a safety deposit box at the Vintondale Bank for safekeeping of church documents, sending out 500 fund-raising circulars, and ordering 2,500 offering envelopes to replace those burned in the fire.

At a meeting held on February 28th, the council, in response to Rev. Hunyady's request that the congregation inspect other newly-built Reformed churches, authorized him to visit churches in Manville, New Jersey; Middletown, Ohio; and Donora, Pennsylvania at the church's expense. Also canvassers were to be sent to churches in Homestead and McKeesport to raise money for the church. In addition to providing letters of recommendation, the church paid the canvassers $5.00 per day for expenses. The council also recommended that some members drive to Clymer to visit Ferenc Kes, curator of the disbanded church there; purpose of the visit was to see if there were any church possessions in Clymer which could be used in the new Vintondale church.

At the April 4th meeting, the council authorized payment for a $25 funeral wreath which had been sent by the church for Otto Hoffman's funeral on March 30.

By April 1930, the congregation was studying whether to build or purchase a manse. Owning their own land was vital for receiving a loan from the Building Fund of the Mission Church. By April 6th, the fund-raising circulars had been very successful. $751.45 had been received, and church members had paid $293.50 of their pledges. Five cent raffle tickets (brick tickets) also had been printed to aid the building fund.

Rev. Hunyady was authorized on April 26th to talk personally to Clarence Schwerin in New York City concerning the purchase of the house and lot adjacent to the church Lot. The Vinton Colliery Company made an offer to Rev. Hunyady that was acceptable to the congregation. Lot 11 in Block F, containing a two-story frame house was offered for $300; the old church's lot was valued at $200. Following the adoption of this measure, the session was authorized to put out construction bids after the deeds were received. The formal purchase and recording of the deed took place on May 27, 1930. Prior to the purchase of a parsonage, the ministers either commuted from Johnstown, Windber, or Ligonier; stayed with the Antal or Bagu families; or rented an apartment in town. Rev. Hunyady had an apartment in the Farabaugh Building until the church obtained the parsonage.

Summer school was held as usual on June 10th with Rev. Hunyady teaching the classes. Mrs. Hunyady taught the summer school classes at the Barnesboro church.

Bids were received from Harry Greenwood of Johnstown, $6,700; C.R. Dilling of Nanty Glo, $7,800; and Louis Shearer & Sons of South Fork, $6,200. On June 14, 1930, Louis Shearer was granted the contract to build the church of granite-faced concrete block. Shearer offered this type of block rather than the limestone-faced block at no extra charge, absorbing the extra $150 cost himself. Included in the project was the construction of a basement for the manse next door. The payment schedule was set as follows:

Completion of foundation $2,000
Completion of walls $1,200
Completion of roof $1,000
Completion of all outside work $1,000
Completion of entire project $1,000

Rev. Hunyady and Jack Huth, VCC assistant superintendent, were asked to supervise the project free of charge.

In July, the church purchased 100 chairs from the Royal Manufacturing Company at $1.94 each. Rev. Hunyady, the caretaker, and the treasurer recommended that a 1,300 pound bell be purchased for $105 from Sears, Roebuck and Company. A stove for the manse was also ordered for $85 from Sears. Additional purchases made in August were a linoleum rug for the manse's kitchen and a $15 plumbing bill for connecting hot water in the manse. The Women's Club purchased the curtains and also paid $147 for the electrical work done in the manse and the church. The Club also stated their intention of paying for the altar of the new church. The refurbished manse was consecrated at 4:00 PM on July 27, 1930. Following the service, a dinner was held in the newly completed church hall.

Visiting Vintondale that same summer and assisting at the church when needed was Frank Erdey, a ministerial student at Franklin and Marshall. He and Steve Dusza spent several years together at the Hungarian Reformed Orphanage in Ligonier.

Dedication of the new church was held on Sunday, August 31st at 4:00 PM. Invited were pastors from Youngstown, Ohio; Pittsburgh; Homestead; McKeesport; Uniontown; Johntown; and Windber; and Farrell. Also invited were the professors of Franklin and Marshall College. Rev. J. Meligh of McKeesport was in charge of opening services; sermon was given by Rev. B. Dienes of Homestead; dedication service in Hungarian was led by Rev. Drs. D.A. Toth and T.F. Herman of Franklin and Marshall. The dedication service in English was conducted by Rev. R.L. Folk of Esterly. The Lord's Supper was administered by Rev. Sigmund Laky of Youngstown and Rev. B. Kerekes of Windber.

A banquet was held in the church basement following the dedication service. Toastmaster was Rev. M. Daroczy of Farrell. Parish children enrolled in summer school presented a play called "Janos Vitez" ("John the Hero"). The entire community was invited to the service.

Following the dedication of the church and manse, Rev. Hunyady requested a vacation from September 8th to October 4th. In his absence, Rev. Bela Kerekes of Windber preached on the afternoons of September 14th and 21st. Presbyter Baloz Bagu volunteered to drive to Windber to get Rev. Kerekes and to take him back after services.

Consecration of the bell was postponed until November 16th when the Fall County Church Meeting was held. The presbyterium of the church decided to fully insure the church and manse. The church and its furniture were insured for $8,000, the house for $3,500, and its contents at $200; the agency underwriting the policies was George Kinkead of Ebensburg.

In October, the new church basement was put to good use. The Delano school building had burned, and acting on the school board's request, the council opened the basement free of charge as a temporary classroom.

Following the bell dedication on November 16th and the diocese meeting held on November 28th, dinners were prepared and served by the Women's Guild.

Christmas, 1930 was the scene of an innovation in services. At the request of certain parishioners, both Hungarian and English were used at the service. The year 1930 closed out with the acceptance of forty new hymn books and forty "Halleluija" books from the children, a Christmas party for the children, and a dance on December 27th. At a council meeting on December 31st, held in the new church, the group decided to pay the bell ringer $10 per month in the winter and $5 per month in the summer.

Summer School

Each summer, the Vintondale Hungarian Church provided a summer Bible school with a two-fold purpose in mind: provide religious education for the children and to teach the intricacies of the Hungarian language to the second generation. Classes lasted the full day, and discipline was strict. However, there were organized recreational activities, such as hikes, for the children.

The school, usually taught by the pastor, lasted six weeks. In 1925, eighty-five children attended, including several from Twin Rocks; closing services were conducted by Rev. Charles Dobos on August 23rd. Summer School fees in 1927 were $2.00 per local child; $3.50 for two children; $4.50 for four or more. Students from out of town paid $1.50 per child. The cost of summer school averaged $200; the church treasury made up the difference if there was a deficit. In 1928, school was expanded to eight weeks; seven Nanty Glo students and several from Twin Rocks participated. Rev. Bela Csontos was the director in 1928. The fee per student in 1930 was dropped to $1.50 per child, or $4.50 for four children. To help defray expenses, a picnic, play, and dances were held on July 12th and at the end of summer school.

A resident summer school was also held at the Hungarian Reformed Orphanage in Ligonier. In addition to intensive study in the Hungarian language, Hungarian folk dancing lessons were also given. Vintondale students who attended these classes were Margaret Antal O'Hara, Zolton and Arthur Antal and their niece, Ida Bagu.

Church Holidays

The Hungarian Reformed Church had three special church celebrations in addition to Christmas, Easter, and Whitsun. On March 15th, the church commemorated Hungarian Independence Day with entertainment for the children, included was an ice cream treat. Children's Day was on June 15th; the Vintondale church usually combined with the Windber and Johnstown churches for the program. October 31st marked the memorial service of the Reformation; this was usually followed by a Halloween party.

Ladies' Guild

The **Loranlfi Zsuzsanna** Women's Guild was formed on the second day of Easter, April 5, 1926 at the request of Rev. Csorba. A five member organizing committee included: Mesdames Baloz Bagu, Gyorgy Medwick, Karol Toth, Peter Toth, Gizella Antal.

The first general meeting was held on May 16, 1926 with twenty-eight members attending. The organizing committee signed up forty-five members and collected $9.50. First officers were: president, Mrs. Balint Izso; vice-president, Mrs. Peter Toth; secretary, Mrs. Karol Toth; cash-clerk, Mrs. Ignacz Barate, supervisor, Mrs. Gyorgne Medvik; treasurer, Mrs. Baloz Bagu; entertainment committee, Mesdames Karlo Marton, Ferenz Magyar, Josef Magyar, Ignacz Myeste, Istvanne Toth; committee for visiting the sick, Mesdames Pal Gulyas and Istvan Varga.

The second monthly meeting was held on the second Sunday of June. At the meeting, dues were set at ten cents per month. The first service project of the guild was to prepare and serve a luncheon to the congregation. Over the years, the Women's Guild was responsible for numerous dinners and for the church bazaars, for which townspeople waited in anticipation.

The charter members were: Mrs. Adam Antal, Gizella Antal, Mrs. Balasz Bagu, Mrs. Balasz Bajusz, Mrs. Ignatz Barate, Mrs. John Barna, Caroline Beregsozaszi, Rose Beregsozaszi, Mrs. Michael Beres, Mary Brozina, Mrs. George Drabbant, Mrs. Joseph Dulenszkci (Delaney), Mrs. Nicholas Farkas, Mrs. Joseph Gero, Mrs. Paul Gulyas, Marget Harnot, Mrs. Stephen Kaszas, Mrs. John Kovacs, Mrs. Frank Magyar, Helen Magyar, Mrs. Joseph Magyar, Mrs. Kalman Marcus, Karoline Marton, Mrs. George Med-Vig, Mrs. John Morey, Mrs. Louis Munkos, June Neyste, June Orbon, Mrs. Joseph Plisko, Mrs. Paul Sandor, Mrs. Peter Soos, Mrs. Sandor Szegedi, Mrs. John Toth, Mrs. Stephen Toth, Mrs. Charles Toth, Mrs. Peter Toth, Rosa Toth, Helen Ure, Mrs. Stephen Ure, Mrs. Stephen Vargo, and Mrs. Joseph Zoltanszki.

Other members were: Mrs. Gabriel Toth, Julia Gero, Mrs. Peter Szabo, Mrs. Balint Koszta, Mrs. Michael Koszta, Mrs. Stephen Dusza, Mrs. Frank Markus, Mrs. Frank Tar, Mrs. Stephen Kedves, Mrs. John Glidi, Mrs. Andrew Markus, Mrs. John Nagy, Mrs. Joseph Jendrick.

Through the use of dances, dinners, and bazaars, the Hungarian Reformed Church of Vintondale was able to convey some of the unique aspects of the Hungarian culture to townspeople who otherwise would not have had contact with those of Eastern European backgrounds. The impact of the work of the Hungarian Reformed Church accomplished in Vintondale cannot be understated.

CHRISTIAN AND MISSIONARY ALLIANCE CHURCH

In the 1920's and the 1930's, a branch church of the Nanty Glo Christian and Missionary Alliance Church was organized in Vintondale. The effort began with a tent revival at the ball field in July, 1923. Presiding at the nightly meetings was Reverend J.C. Gleen. The new members began meeting at the old Shaffer store adjacent to Jack and Mary Shaffer's house. Later, services were held in a small building beside that store. The first evangelical meetings were held from January 20 to February 10, 1924 by Reverend Henry, minister of the Nanty Glo church. A Sunday School was also started that January. The new church promised "oldtime Gospel preaching and good singing". Reverend Henry conducted a second tent revival in August, 1924.

The church offered Sunday School at 2:00 PM with John Wise serving as Sunday School Superintendent. A preaching service was held at 7:00 each Sunday evening. On Thursday nights, Prayer and Praise were conducted.

Rev. and Mrs. L.R. Carter were conducting revival meetings with "good reports" in January, 1925. In March and May, Reverend Henry held evangelical services at the Hungarian Lutheran Church (Hungarian Reformed). In June of that year, Esther Younkin of Coalport offered Bible School at the Gospel Tabernacle and the Baptist Church under the auspices of the Christian and Missionary Alliance Church. Another tent revival was set up in June by Reverend Henry, who was assisted by Rev. F. Sherman of Cresson and Rev. R. Gray of Portage. The revival was followed by a missionary convention in September. Guest speakers were Carrie Garrison of South China, Rev. M.B. Birrel of South China, and Rev. W.A. Tenney of Williamsport.

In January, 1926, the pastorate of the Vintondale church changed hands. Rev. F.D. Sherman of Cresson assumed the post. The annual missionary rally of 1927 yielded pledges of $300 to the cause. In 1928, the rally in Nanty Glo and Vintondale was led by Rev. Guy Gooderham. Ruth Wise took the pulpit for the evening services in 1929 and served as assistant pastor in 1930. Rev. Arthur Williams assumed the Nanty Glo pastorate and served the Vintondale church as a supply minister. A two-week revival was held in January, and a street meeting was announced for August 23, 1930.

In 1934, Mary Shaffer, Rev. Homer Baughman, his son Alvin, and Anna Watson of Nanty Glo drove to Birmingham, Alabama to attend the annual convention of the Christian and Missionary Alliance Church.

The date and the reasons for the closing of the Vintondale Christian and Missionary Alliance church are unknown.

Main Street, c. 1907.
Courtesy of Pauline Bostic Smith.

Vintondale, Plank Road, c. 1920.

Hotel Morris, later known as the Red Hotel, Lower Main Street, c. 1906.
Courtesy of Pauline Bostick Smith.

Chapter X
Potpourri

AUTOMOBILES

For the first three decades of its existence, Vintondale's main forms of transportation were the train and/or the horse and buggy. With the development of the internal combustion engine even Vintondale, snuggled in its remote little valley, succumbed to the lure of the automobile. The first cars were strictly luxury items for the town's more prosperous citizens. In February 1909, an Indiana paper reported that L.H. Davis, superintendent of the Blacklick and Yellow Creek Railroad, purchased a Model 10 Buick runabout from the newly-opened Indiana Motor Company. The same clipping mentioned that a similar car was sold to a Dr. Cummers. Was that Vintondale's Dr. Comerer? Very definitely! Ella Lantzy remembered watching the two new cars round the curve by her home on Rexis hill.

In spite of the price reductions due to Henry Ford's assembly line and the introduction of Model T, the automobile still remained out of reach of the average miner. A **Nanty Glo Journal** reporter thought it significant enough to include the automobile purchases in her weekly town summary. New owners in 1921 were:

News. Date	Purchaser	Automobile
May 5	Dr. MacFarlane	Chevy Runabout
May 12	Herb Daly	Five Passenger Buick
May 12	Sam Feldman	Five Passenger Buick
May 19	Lloyd Arbogast	Buick Roadster
May 19	Otto Hoffman	Buick Touring Car
June 9	Bill Roberts	Seven Passenger Buick (took it to Scranton a week later)
June 23	M. M. Hunter	Nash
June 30	Walter Morris	Dodge Truck

Note that all of the buyers were company personnel except Walter Morris.

Jack Huth purchased a car in 1916, and true to form, kept meticulous records of his expenses, down to the cost of some cotter pins. A gallon of gasoline averaged between twenty-five and thirty cents. In 1916, he bought gasoline from Arbogast-company store, Chickory, Nipps - near Cardiff?, and Joe Bennett. Mobiloil sold for sixty cents per gallon. "Dad" also purchased: chains, $2.80; 30" x 32" chamois, $0.98; inner tube, $2.52; tire paste, $0.23; and blow-out sleeve, $0.40. A license was $5.00 per year. In 1916, Jack Huth drove the car 725 miles.

In 1917, gasoline remained about thirty cents, despite World War I shortages. Tires were relatively expensive during the war. On September 1, 1917, Mr. Huth purchased a Goodyear tire and tube for $25.00. The price for a tire jumped to $26.40 in 1919; a battery sold for $26.40. His car payment appears to have been $25.00 per month. In his income and expenses ledger, Dad listed a monthly payment of $25.00 to American Motors Corporation. The speedometer reading in 1919 was 3,340 miles. If Mr. Huth made any long trips with the automobile, they would have been to Frackville to visit his family.

The automobile may have been a novelty, but manuvering it on the country roads was not a thrill, but a nightmare. Until the back roads were paved under Governor Pinchot, the average driver had to practically be an expert mechanic.

Company Teamsters. Borough office and jail in rear. John Morey's house on right. Steve Lantzy, second from the right. Jack Butala on right.

Courtesy of George Lantzy.

Gas Stations

In the 1920's several businesses in town applied for gasoline tank permits from the borough. Unlike today, where there is no gasoline available in town, the automobile owner had his choice of at least seven gas pumps. Davey Nevy received his permit on April 18, 1922; Andy and John Jacobs, who operated a Studebaker garage on Plank Road had their application granted on November 11, 1922. Baloz Bagu also received his permit at the same meeting. On July 15, 1924, R.W. Krumbine was granted permission to install two 500 gallon gas tanks in front of his garage. Ida Cooke was given a permit on September 24, 1929 to install a gas tank at her home on lower Main Street. Lee Cresswell applied for a permit to put a tank in front of the Farabaugh Building; this was granted on May 17, 1932. He obtained permission to install two gas pumps there in October and then that December asked for a third pump. John Rosacha applied in 1933 and was also granted permission to place a tank and pump in front of his store on lower Main Street.

Repair Garages

The first known repair garage was that of Philip Wayne who took over part of T.J. Callet's wholesale liquor building in May, 1921. The Jacobs brothers also repaired automobiles at their garage on Plank Road.

BRIEF BIOGRAPHIES

Lloyd I. Arbogast

Lloyd Arbogast was hired as chief clerk of the Vinton Colliery Company on August 4, 1912 and promoted to comptroller in 1916. Born February 16, 1884 in Mount Pleasant Hills in Mifflin County, Arbogast was educated at the Lewistown Academy. Before coming to Vintondale, he worked as a clerk for the Pennsylvania Railroad. After arriving in Vintondale, Arbogast also served three terms on borough council, giving himself credit in a Cambria County history for helping to make Vintondale a borough. In reality, Vintondale was a borough four years before he arrived. Arbogast also served as vice-president of the Vintondale State Bank.

For several years, he dated Amanda "Peg" Arble, a cashier in the company store. According to Mable Updike, Amanda was able to tell him what to do but she did not marry him because he was not a Catholic. Amanda eventually returned to Carrolltown and married her childhood sweetheart, Oliver Stoltz. Following their marriage, the couple lived in an apartment at the Vintondale Inn. Arbogast was also a member of the Sunnehanna Country Club in Johnstown. His employment with Vinton Colliery ended in 1929 due to a breakdown.

W.N. Lockard

For many years, the Pennsylvania Railroad station agent for Vintondale was William Lockard. He was born in Cookport on November 13, 1884 and educated at Marion Center. Beginning as a telegrapher at Glen Campbell and then Cherry Tree, Lockard was assigned to Frugality and Hastings as station agent before arriving in Vintondale in 1914. While in Vintondale, Mr. Lockard served as president of the school board. He married Nellie Anthony of Indiana in 1911; their only daughter, Lavon, was born October 28, 1912. Lavon graduated in the top three of her class at both Vintondale High School and Indiana Normal School. The Lockards lived across the street from the station.

John Burgan

Mr. Burgan was the shipping clerk for the Vinton Colliery Company until he resigned in May, 1929. The Burgans lived on upper Third Street in the house now owned by Lloyd Williams. While in Vintondale, Burgan served as secretary of borough council for nineteen years. He also was the proprietor of the movie theater before the company built the new one. Through his efforts, a Presbyterian Church was organized in Vintondale; it lasted from 1912-1922. After leaving Vintondale, the Burgans first moved to Ferndale and later to New York State where Mr. Burgan worked for the New York Electric and Gas Corporation. (See the chapter on Vintondale schools for biography of son Jack.)

Harry Ling

Harry Ling was the Prudential Insurance Company agent in Vintondale for twenty-seven years from 1923 to 1950. While in town, he took up residence in Paul Graham's house, then Mary Nancarvis', at Krumbine's, and finally next to Abe Solomon in the duplex above the post office. Mr. Ling also was a member of the Oakmont Hunting Club, another hunting club with Vintondale ties. He was a much welcomed guest at the Vintondale Homecomings each fall. Harry Ling died in 1988 at the age of 99.

Mike Farkas

Mikolas Farkas was born in Hungary in 1877 and at age 23 migrated to the United States. For several years, he huckstered fruits, vegetables and farm animals, especially pigs, to the immigrants in the mining towns of Cambria County and also ran a boarding house in Bakerton for seven years. In 1908, he settled permanently in Vintondale and purchased a store in Block B, Lot 3 from Sam Brett for $4,500 in December, 1908. On one side of the building, he opened a grocery store; the other side was leased to Joe Shoemaker, who operated a barroom there. The second floor was used for living quarters for Farkas and his first wife Annie and their boarders.

Over the years, Farkas purchased several trucks and continued to make trips to the stock yards and wholesale markets in Pittsburgh and brought back goods to sell in his store and door-to-door. In the spring, he took orders for pigs from the Hungarian families, who fattened them up through the summer and then butchered them in the fall. The pigs were shipped from Pittsburgh by stockcar; Farkas then drove the pigs up the street from the station, allowing the families to pick out the pig(s) they had ordered. He kept track of the orders and payments in a little black book.

To deliver groceries, Mike bought a 1923 Reo Speed Wagon. Farkas also used his trucks to move miners. This was a thriving business because of evictions and because miners tended to move frequently. Jim Wray often drove truck for him. In 1927, Farkas purchased a new Reo moving van which he kept busy in 1929-30 moving Wehrum families. The Wehrum shutdown proved to be lucrative to Mike. Not only were there families to be moved, but he also obtained building materials cheaply by purchasing several of the Wehrum houses. Step-son Steve Dusza and Frank Erdey, a classmate from the Hungarian Reformed Orphanage in Ligonier, spent the summer of 1930 using materials from the Wehrum houses to repair Mike's houses in Clymer, Nanty Glo, and Vintondale. Some of these houses had been obtained by loaning money to fellow Hungarians; when they could not pay the debt, Farkas confiscated the houses.

For his personal automobile, Mike Farkas liked a large luxury car. He had a 1918 Maxwell, which he bought at Leitenburger's in Johnstown; next was a 1923 seven passenger Reo Touring Car complete with curtains; in 1930, he bought a Wolverine; followed by a 1937 Packard, whose transmission Steve Dusza burned up on the Wehrum bridge. Farkas' last car was a 1940 Buick. Mike drove very, very slowly and put few miles on his automobiles, about 3,000 miles per year. However, when his daughters began driving, the mileage mysteriously jumped ten-fold within a year.

Farkas and his wife, Annie, began to accumulate lots in Vintondale, one on Dinkey Street, several on Plank Road and several on Maple Street. Some of these houses were picked up in his notorious card games. There were mirrors on the walls in strategic places in the dining room where the games were held. Also, as Mrs. Farkas was serving food and drinks, she was singing in code, alerting Mike as to which player had the best hand.

The company also found a useful asset in Mike Farkas. Because he could speak German and Hungarian, Farkas was often used during periods of labor shortages, such as 1917, to encourage Hungarian miners to move to Vintondale. He also served as one of the first directors of the Vintondale State Bank. His deposits during the 1930's helped keep the bank afloat.

Due to his many business enterprises, which included making and selling moonshine during Prohibition, Farkas was able to build up a sizeable estate. Mike Mihalik, cashier of the Vintondale State Bank, also gave him excellent investment advice, so during the Depression it was estimated that Farkas made $100,000. After the end of Prohibition, Farkas applied for a liquor license and re-opened the barroom in the Farkas Hotel. Because times were lean in the 1930's, many of his boarders could not afford to drink, so Mike sold them fifty cents worth of whiskey or beer for a $1.00 piece of scrip. Farkas then cashed in the accumulated scrip at the office. When the company store unexpectedly closed in March, 1940, he was stuck with hundreds of pieces of worthless scrip.

While Mike was on a trip to Europe in January, 1923, Annie Farkas died and was buried in St. Charles Cemetery, Twin Rocks. That same year, Mike's future wife, Helen Horvath Dusza, the author's paternal grandmother, was granted a divorce from Rudolph Dusza. Mrs. Dusza moved to Vintondale and eventually married Farkas.

Mike Farkas died in 1943 from injuries sustained in a fall down the cellar stairs of the hotel. His estate of approximately $250,000 was shared by his widow and their six children: Elizabeth, Helen, Mary, Margaret "Toots", Theresa, and Nicholas "Mick".

Paul Graham

Paul Graham was born in Lecontes, Clearfield County on August 18, 1892. He went to school in Clearfield County and then took a job in the lumber industry. In 1904, he moved to Vintondale and was employed by Sherman Ruffner. In May, 1909, Graham went into business for himself, purchasing Sam Brett's general merchandise store. In 1904, he married Ivy Rairigh of Indiana. The couple had four children: Chester, Viola, Geraldine and Dorothy. Mr. Graham, a Republican, served as borough treasurer for eleven years and was also a member of the school board.

Peter Toth

Peter Toth, like John Galer, was another one of those unknown, unsung inventors. Toth arrived in 1907 from Austria-Hungary; his brother Charles came about the same time. In January 24, 1911, Mr. Toth was granted a patent by the United States Patent Office for a double lock system now utilized in safety deposit boxes. The wintesses signing the document were Daniel Bajusz and Aladar Kuzdeny, bachelor friends or boarders of Mr. Toth. The patent was good for seventeen years, but Mr. Toth never realized any royalties. Part of the problem was that he was not an American citizen, but a "subject of the King of Hungary". Obviously, his lawyers did little to help Mr. Toth market his product. According to his son Albert, Mr. Toth was not a locksmith by trade in the old country, but simply puttered around with locks in his spare time. The Toths lived in the house behind the Farkas Hotel; Mrs. Toth cleaned for Mrs. Cresswell, the company store, and Mrs. Farkas. In fact, the clothes line in the Toth yard were frequently strung with the laundry of the Farkas Hotel.

Charles Bennett

Charlie Bennett, a Strongstown native, moved to Vintondale around 1900 and opened a livery stable and blacksmith shop across the street from the Morris (Red) Hotel. Bennett puchased the first two houses on Block AC. In the first building, he operated a meat market. At the store he occasionally sold fresh oysters which he brought from Baltimore in large buckets via the train.

Because the first house was large and had extra rooms, Mrs. Bennett, the former Katherine Peddicord, rented rooms to railroaders who had nightly layovers in Vintondale. However, she did not provide meals for them. Their son Oliver, a PRR employee, married the former Minnie Engle of Strongstown and moved to Rexis; shortly after the birth of their daughter Rhoda, they moved into the house beside his father. Oliver gave up his railroad job and moved to Indiana when Dr. MacFarlane warned the family that Rhoda's bouts with pneumonia would only get worse from the smoke and cinders from steam engines. The trains dumped their ashes from their fireboxes into a pit directly across the street from the Bennett houses. Soon, both father and son moved to Indiana. The building housing the store was sold to Mary Dodson Nancarvis, and the second was sold to the Bepplers.

Charlie Bennett's eldest daughter Lilly married Thomas Altemus and died in 1911 of complications following childbirth. Mrs. Altemus was one of about six or seven young mothers who died that year because of post-partum infections; others included Gertie Engle Graham and Cora Bennett Graham, wife of Allan Graham.

Bennett's younger daughter Margaret married Walter Schroth, who was employed by the Vinton Colliery Company. When Margaret was expecting their first child, her mother insisted that she go to a Johnstown hospital for the delivery. The Schroths also settled in Indiana where he was involved in lumbering and pine tree nurseries.

In Indiana, Charlie Bennett served as assistant sheriff and revenue agent and also had a slaughterhouse on the Clymer Road. Oliver Bennett got a job with the Miller-Sutton car dealership and later became a painter and sold mine props. His son Charles opened a gas station on the corner of Third and Philadelphia Streets. This business expanded into the home heating oil business and remains in family hands today.

DANCE YOUR TROUBLES AWAY

In a company town, much of the recreation and entertainment was orchestrated (Pun intended!) by the company. In Vintondale, to keep the masses happy, Saturday night dances at the company social hall on Plank Road were held on a regular basis. Local bands, such as Kaltreiter's of Wehrum, were hired. On other occasions, there was a pick-up group of local musicians who kept the music flowing. When Samuel DiFrancesco of Twin Rocks was fourteen or fifteen, he played in this group along with Jack Butala. Jim Haley of Nanty Glo tickled the ivories, and Doc MacFarlane strummed the banjo. The money was good, sometimes $40 a piece. Guaranteed to appear at every dance was "King Otto"; the superintendent wanted to check on those attending.

Dances were also held as fundraisers by churches and civic groups. Square dances were popular, but the most financially successful were the "Hungarian" dances. These were not the exclusive domain of the Hungarian Reformed Church. Between 1921 and 1940, the **Nanty Glo Journal** listed the following groups as sponsors of Hungarian dances: Sts. Peter and Paul Russian Orthodox Church, Hungarian Reformed Church, Immaculate Conception Church, the Athletic Fund, the Volunteer Fire Company, United Mine Worker's Local 621, and the Vintondale Baseball Club.

In spite of the hard times in the 1920's and the 1930's, attendance at dances was high, allowing residents to forget their personal problems for an evening. Admissions were reasonable. For men, the cost was 45 to 50 cents; for women, the prices were cheaper, 25 to 35 cents. Occasionally, there was free admission for the women.

Efforts were made to book "good" orchestras. In November 1929, because of his acquaintance with the manager, Bill Roberts was instrumental in securing the "Sunset Serenaders" for a masquerade dance. Two of its members had formerly played with Fred Waring's orchestra. For a dance in 1933, the Homestead Gypsies were engaged. Local Hungarian musicians who played for the dances, weddings, and funerals in the 1930's were John Morey and Joe Silagyi both of who played violin. Louis Kuhar of Cardiff and one other man rounded out Morey's gypsy orchestra. John Kuhar organized a band with his brothers Joe and Frank; they played at local dances well into the 1950's.

SPORTS

Sports and winning teams were synonymous in Vintondale from 1894 to 1940. Company-sponsored baseball teams in summer and high school basketball teams in winter provided some excitement and civic pride for the townspeople.

Baseball

Those who displayed exceptional talent on the diamond were rewarded with lighter tasks inside the mine, but most often were given light outside jobs. If a man had good pitching skills, he could find work even in the worst of times. During the Depression the late George Medlock of Waterman moved to Vintondale about the same time as Andy Chekan. He became a pitcher on the baseball team and found a house on Chickaree Hill. Dissatisfied with both living and working conditions in Vintondale, the Medlocks settled in Indiana.

Other ballplayers were college students, friends of Ken Hoffman. To earn the right to play ball, the boys were given summer jobs painting the company houses.

Baseball games were held at the ballfield behind Blair Shaffer's house on the lower end of town. Following the game, players showered at Village Hotel. Games were usually held on Wednesday, Saturday, and Sunday afternoons. National holidays, such as Memorial Day and Fourth of July were celebrated with town picnics which almost always included a baseball game with local rival Wehrum. Otto Hoffman attended every game possible and even drove to most away games.

Vintondale's competition included teams from mining and steel towns, some of whom were thirty to forty miles away. Over the years, Vintondale had games scheduled with teams from: Twin Rocks, Nanty Glo, Wehrum, Colver, Revloc, St. Michael, Conemaugh, Portage, Lucernemines, Mineral Point, Beaverdale, New Florence, Graceton, Jerome, and Punxsutawney.

Vintondale YMCA Baseball team, c. 1910. Row 1: Seated left to right, Ben Baron, Blair Hollsworth, John Huth, Ernie Coucill, Bill Shaffer, mascot: Sher. Row 2: Standing left to right, George Rodgers, Max Fleitzer, Pat Lynch, Joe Fleitzer, Irv Williams, Sam Sher.

Courtesy of John Huth Family Collection.

A post-1906/pre-1912 team consisted of George Rodgers, Max Fleitzer, Pat Lynch, Joe Fleitzer, Irv Williams, Sam Sher, Ben Baron, Blair Hollsworth, Jack Huth, Ernie Councill and Bill Shaffer.

The 1921 team was managed by John Wagner and included: Carlson, first base; John Wagner, right field; Findley, left field; Johnson, third base; James Wagner, short stop, Wiscovitch, second base; Gongloff, catcher; Williamson and Mort, pitchers.

In September 1930, the Vintondale baseball team challenged Cardiff, Twin Rocks, and Heisley (Nanty Glo) to some games; the managers of those teams were to contact Fred Michelbacher. The late 1930's team included: George Larish, Mike Grosik, Joe Lonetti, Andy Cheslo, Carl Michelbacher, Ross Anthony, Mike Surik, Albert "Cats" Oravec, Enoch Hallos, John Risko, Ed Felton, Pip Fulton, Andy Chekan, and Bill Cooke. Batboy was Nick Hozik, and Mr. Brown was the team manager.

World War II brought the end to the company baseball teams. Even before the United States took part in the war, the Vintondale miners were involved in a battle of their own with the Vinton Colliery Company over bankruptcy proceedings in 1940.

Back Row: George Larish, Mike Grosik, Joe Lonetti, -------- Brown, mgr, Andy Chesla, Carl Michelbacher, Ross Anthony. Kneeling: Mike Surik, Albert "Cats" Oravec, Enoch Hallas, John Risko, Ed Felton, Pip Fulton, Andy Checkan. Seated: Bat Boy - Nick Hozik, Bill Cooke, c. 1935.

Courtesy of Aileen Michelbacher Ure.

BASKETBALL

High school basketball was the sport of interest in the winter. As early as 1923, teams from neighboring schools competed against the Vintondale teams. In January of that year, two sleighloads of students made the trip from Nanty Glo. However, the driver of the sleigh which brought the seventh and eighth grade girls left about 11:30, stranding fourteen girls. Fortunately, Vintondale families generously put up the girls for the night. Wehrum students who attended the Vintondale High School and played basketball, usually spent the night in town with a friend. The boys often slept on the science tables at the high school.

Basketball reached its pinnacle of popularity between 1929 and 1935 when Vintondale teams broke record after record. Today, the younger generation of Vintondalers and Vintondalers-at-heart should be surprised to discover that it was the girls' basketball, not boys', that put Vintondale on the sports map in the 1930's. With a total school population of 100 or less, over half of the students participated on the two teams.

Superbly coached by supervising principal, Charles Mower, who had never played basketball, the girls' team won game after game. Between 1928 and 1931, the girls only lost seven games, but still lost the county championship to Gallitzin two years in a row. The 1932 team won the county championship and sported a record of 26-1. That June, the boys' and girls' teams were honored with a banquet at the Vintondale Inn. Sponsors were Dr. MacFarlane for the girls and Mr. Brandon for the boys. Each girl was presented with a vanity case. Speakers were the coaches: Charles Mower and Barry Morris. Bill Roberts loaned a victrola and radio for the dance.

Members of the 1932 girl's team were: Aileen Huth, Evans, Suprak, Nagy, A. Lynch, Davalli, Skvarcek, Anna Hrabor, Olga Kanich, Alice Sileck, Mary Margaret Morris, Sue Sileck, Virginia Daly, Balog, Hozik, E. Lynch, Lynetta Daly, E. Lynch.

1928-29 Vintondale High School. Front Row left to right: George Lybarger, Roy Kuckenbrod, Lloyd Williams, Russell Kuckenbrod, Jack Frazier. Second Row: Brad Lybarger (manager), George Hozik, Mike Hozik, Ed Nevy, Geno Avali, Clarence Huey, Rick Bennett, Jack Burgan, Bill Jendrick, Mario Biancotti.

Courtesy of Agnes Huth Dusza.

1934-35 Vintondale High School Girl's Team. Row 1: Claire Wise, Anna Sermeg, Elizabeth Hugya, Beula Ling, Margaret Vargo, Dorothy Whinnie, Elizabeth Berish. Row 2: Helena Shestak, Margaret Kuhar, Thelma Brown, Claire "Tood" Huth, Helen Kerekish.

Courtesy of Vintondale Homecoming Committee.

After their sole loss to Blacklick Township High School (Twin Rocks) in 1932, the Vintondale girls initiated a stunning winning streak, compiling sixty-three games without a loss. Blacklick's win in 1932 was their only victory against Vintondale in eight years.

The 1934 team included: Alice Sileck, guard; Anna Sermeg, guard; Anna Hrabor, forward; Olga Kanich, captain and center; Mary Skvarcek, forward; Margaret Huth, side center; Sara McPherson, guard; Dorothy Whinnie, center; Julie Yelenovsky, side center; Claire "Tood" Huth, forward.

The average points per game by the Vintondale girls' teams were 30.48. They outscored their opponents 1,524 points to 446. The team records were as follows:

Year	Wins	Losses	Ties
1926-27	8	8	2
1927-28	13	8	
1928-29	20	2	
1929-30	23	4	
1930-31	20	3	
1931-32	26	1	
1932-33	20	0	
1933-34	20	0	

The fantastic win streak was broken in January, 1935 by Blacklick Township by a score of 13-12. That loss did not break the spirit of the team. The next year, Vintondale again captured the Cambria County championship.

In addition to good coaching, Vintondale students in the 1920's and 1930's had a good practice area at the company social hall, the first building on the right on Plank Road. The hall was remodeled with locker rooms to accomodate the teams. Attendance was high for the girls' games, especially when they were on their winning streak. Often about half of the fans left the hall before the boys started their game.

The Vintondale boys' team, coached by 1920 graduate S. Barry Morris, also had a remarkable record in the 1930's, but were overshadowed by the girls. The 1930-31 boys' team consisted of: Hozik, Steve Dusza, Zoltan Antal, Ed Nevy, and Gary. Substitutes were Victor DeBona, Boyd Brandon and Gulyas.

The 1934 squad was made up of George, guard; Arthur Antal, center; Steve Oblackovich, forward; Charles Vaskovich, forward; Peter Kutasi, forward; Stephen Podlusky, forward; John Rosacha, guard; Mike Velesko, forward; Charles Balog, center; Jack Butala, Jr., center; Steve Ezo, forward; John Babich, forward. The team was undefeated.

In 1935, the squad won the Elmer Daly Trophy for good sportsmanship at the annual St. Francis Tournament in Loretto.

By 1941, the tide had turned, and Vintondale lost all games but one. The coal company bankruptcy had changed the makeup of the town, and World War II was raging in Europe and soon to engulf the United States. In 1942, the baseball season was cancelled due to lack of heat in the hall and transportation to away games.

Tennis

A clay tennis court was constructed in the superintendent's yard after Mr. Mower arrived in Vintondale. Since the court was directly across the street from the high school, there was a lot of student interest in forming a tennis club. In June 1930, a club was initiated with Eugene Nevy as president and Agnes Huth as secretary-treasurer. Other club members were: Mr. Mower, Marg Thomas, Mario Biancotti, Lawrence Nevy, Robert Greenwood, Tom Brandon, Lloyd Williams, Theodore Melcher, Jim MacFarlane, and Boyd Brandon. Boyd's dog Charlie was trained to retrieve the tennis balls.

The club held a tournament on August 4 and 5, 1930. The official for the event was Elmer Smith of Twin Rocks. Mower defeated Biancotti and Merle Nevy in the doubles round.

Boxing

Boxing and wrestling were also encouraged in Vintondale. In May, 1921, the social hall was the scene of a combined boxing and wrestling matches which benefitted the Scouts. Participants and victors in the boxing matches were:

Robert Mott & Mike Brozina (W)
Dave Griffith & Dwight Clyde (W)
Leslie Kempfer & Charles Duncan (W)
Don Blewitt (W) & Joe Sileck
Dewey Esaias (W) & Jim McGuire

Taking part in the wrestling matches were:

William Peterson & Merle Lybarger (W)
Gomer Evans (W) & Leslie Kempfer
Gomer Evans & Robert Mott (W)
Gordon Thomas (W) & Dwight Clyde
Jesse Bobenscak (W) & John Smidja

In a February 19, 1930 fundraiser, the Volunteer Fire Company staged eight three-round-bout boxing shows to raise money for its treasury. Participants came from Twin Rocks, Nanty Glo, Dunlo, Barnesboro and Johnstown. Charlie Butala, brother of fire chief and town policeman Jack Butala, represented Vintondale; he weighed in at 142 pounds. A bus brought spectators from Nanty Glo and Twin Rocks. Unfortunately, the results did not appear in the paper the following week.

Sports played a very important role in the coal mining towns as far as the company was concerned. It helped keep people in line and regulated their leisure time. It also gave them a diversion from thinking about hard work and low pay.

TOWN TRAGEDIES

Over the years, the community has been shaken by accidents and mishaps unrelated to the mines. These tragic events have been compiled below.

A severe storm caused the death of John Pazewell on November 20, 1900. Pazewell, Vinton Colliery Company's farm superintendent, started out for Max Trafford's farm that Sunday. When Pazewell failed to arrive, Trafford sent out a search party which did not find him. Trafford continued on alone and found Pazewell lying dead under a large tree limb. His head was crushed, and the gun he was carrying was broken. Pazewell, who had only been in Vinondale a few months, was survived by a wife and two children.

John Burke of Nolo, father of Mrs. John Cresswell and Mrs. George Kerr, was killed instantly by a train as he was entering Vintondale from the west in March, 1902. Because of the lumber stacked up at the Vinton Lumber Company yard, trains could not be seen at the crossing. Burke, who was 75 and hard of hearing, had operated a hotel for many years in Strongstown. His death aroused the anger of townspeople who had objected previously to railroad about the hazardous crossing and had seen no improvements. Squire J.B. Graham was authorized by Indiana County Coroner M.M. Davis to hold an inquest.

In May 1922, four year old Robert Dishong, son of Oscar Dishong, was killed instantly when he fell off the company store truck and was run over. The driver, David Bracken, was carrying a load of slate from the mine, and the truck stalled. Robert and some playmates climbed onto the truck without Bracken's knowledge. When he put the truck into reverse to get another start, he did not notice that the child had been thrown under the truck. Dr. MacFarlane and Mr. Dishong, who absolved Bracken of any blame, were nearby. The funeral was held at the Dishong home with burial at Heilwood. A coroner's inquest on June 5th exonerated Bracken from any blame in the tragic death. In another vehicular accident, Peter Hudja, aged 5, received a broken leg when he was rundown by a car in May, 1923.

On April 29, 1923 thirteen year old Mike Kushmusk was on the way home from Sunday mass at Immaculate Conception. He reached into his pocket, and the .22 caliber pistol, loaded with blank cartridges, that he was carrying went off accidentally. A week later on May 6, Mike died at 1:00 AM of lockjaw. Classmates paid their respects at the Kushmusk home on Chickaree and then attended the funeral as a group.

In September 1923, Robert McKeel and his brother, sons of Walter McKeel, were playing on a pile of railroad ties. Robert died of a broken neck when a railroad tie, pushed by his brother, fell on him.

Joe Vargo, 38, married with two children, was thrown out of an automobile and killed instantly at Singer Hill. With Vargo was Mike Julenosky and Andy Fetko, who was caught beneath the car. The men, under the influence of alcohol, were speeding in a Jewett automobile and sideswiped a truck driven by Chester Graham and Gordon Thomas of Vintondale.

A gruesome suicide took place in February, 1925.

George Bonezenture, a 42 year old unmarried miner, slashed his throat from ear to ear on a Sunday afternoon. Found along Plank Road, Bonezenture was taken to Dr. MacFarlane's office where he died two hours later. Bonezenture has arrived in Vintondale from Coupon just four months earlier. Burial was in Ashville. Authorities could give no justification for the suicide.

A freak accident caused the death of Joseph Tworcha in November, 1925. Tworcha has just drawn his pay and was standing at the counter of the company store when he was hit on the right side with a bullet which had pierced the ceiling from the second floor. Harry McCardle was unloading his gun when it accidently discharged, and the bullet went through the floor hitting Tworcha. He was survived by his widow and five children.

Five year old Helen Rosocha, daughter of cokeman John Rosocha, died of burns in January, 1927 when she fell into a tub of boiling water. She was buried in Wehrum Cemetery; her grave is marked by a stone inscribed in English.

Two railroad fatalities happened within a month of each other. Frank Danko, age 50, had been on a drunken spree for several days. He was found on November 11th near the Rexis Switch. Police believed that he had been struck the evening before by the PRR evening train. In early December, Paul Leroy Bratton, a PRR hostler from Cresson, attempted to cross the tracks ahead of the morning westbound train. Bratton slipped and fell on the tracks and was run over by the train. Survived by a wife and one child, his burial took place in Cresson.

Later that month, Mr. and Mrs. Arthur Kempfer were informed by the Navy that the S-4 submarine to which their son Paul was assigned had sunk off Cape Cod. In January 1928, divers brought the bodies to the surface, and Paul Kempfer was buried with honors at Arlington National Cemetery.

1929 saw a series of tragic accidents. Lawrence Dodson, aged 62, lived with his daughter, Mary Nancarvis. He had been an employee of the Vinton Lumber Company and remained in the area as an outside laborer for various employers after VLC moved to Kentucky. On April 11, Dodson, who had been despondent, left his daughter's home early in the morning. He was found dead later that morning in the old school house behind the iron furnace. Dodson had slashed his throat at the foot of the stairs, and struggled to the landing, leaving a trail of blood on the stairs and the wall. The razor that he had used had been a Christmas present. His suicide may have contributed to some of the legends which have grown up around the iron furnace. Mr. Dodson was a widower for twenty years; he was survived by five sons and a daughter. Sons Joseph and Russell remained in Vintondale and retired from the mines.

In August, Andrew Kerecman, 17, tried to hop a

ride home on the dinkey. He fell beneath the engine, sustaining a fractured ankle which sent him to Memorial Hospital. Nine days later, Ralph Nevy's three year old daughter Gloria died suddenly at Davy Nevy's home. She had eaten some ice cream and was put to bed for a nap. While sleeping, she became ill and choked on her vomit. Burial was in Cumberland.

In November, 1929 Frank Schmidt, age 76, was crossing a street about 8:00 PM and was killed instantly when he walked into the path of a car driven by E.C. Stiles. It was ruled an unavoidable accident. Schmidt was buried in Mundy's Corner cemetery.

There was another fatal car accident in December. Tom Hoffman, 25, and John Kerr, 26, were seriously injured in a wreck near the old streetcar crossing between Mundy's Corner and Ebensburg. Their car appeared to have gone off the road into a ditch and rolled up against a bank, trapping the men inside. A bus driver from the Southern Cambria Company found them and took them to Mundy's Corner where they were transported to Memorial Hospital by Ondrizek Ambulance. Kerr, the driver, died of his injuries a week later. His parents, Mr. and Mrs. George Kerr, owned the first farm on the road to Red Mill.

In October 1930, Steve Ure was in serious condition in the hospital with a gunshot wound of the thigh. Ure had been gathering wild grapes and was mistaken for a crow and shot by Dave Pesci, who said the shooting was accidental. Mr. Ure would have bled to death if he had not been found by Carl Boyer who summoned help. An ambulance drove as far as it could up Goat Hill and Mr. Ure had to be carried out of the woods on a stretcher.

In November 1930, an unidentified body was found beside the C&I tracks between Rexis and Vintondale. The man, about fifty years old and weighing 200 pounds, was six foot tall and had two fingers missing from the left hand. In his pocket was a slip of paper with the name Jacob Madrich scribbled on it. The label in his coat was from a Butler tailoring firm. Cause of death was believed to be heart disease.

William Lybarger, track boss for the PRR, was found seriously injured near the railroad bridge at the end of town. He was hit in the back of the head with a pick. He died in Indiana Hospital on April 27, 1931. No charges were ever pressed in the case.

In August 1931, Jim Dempsey, age 49, died at Memorial Hospital after a freak accident. Dempsey, coke boss at the mines, sat up late one evening listening to radio. He fell asleep in his chair and woke around 3:00 AM. Stepping out on the porch, he stumbled on the steps and fell off the porch, striking his back. Dempsey's vertebra were fractured, and his spinal cord was severed, leaving him paralyzed. He was survived by a wife and seven children, all at home. His funeral was held at the Catholic church with burial in Lloyd Cemetery.

The same week, John Vargo, son of Mrs. Joe Kerekish, died of injuries received in a motorcycle accident in Milwaukee. He was an engineering students at Wisconsin School of Engineering. Burial was in Grandview Cemetery following a funeral at the Hungarian church.

VINTONDALE HUNTING AND FISHING CLUB

A longstanding, but scarcely known, Vintondale institution is the Vintondale Hunting Camp, in existence since 1922. The campsite, located along Penn's Creek near Weikert in Union County, was chosen in 1922 after some Vintondale men accompanied powerhouse engineer John Galer to his family homestead on a hunting trip. They found the area to their liking and decided to invest in a permanent camp.

A charter was drawn up on April 1, 1922. The original membership fee was $15 with monthly dues of $1. Job descriptions for each officer were included in the club rules. Each member had a right to entertain friends at the camp oustide of the hunting season. If dues were not paid up within sixty days, the lost membership could be sold by the club. Original members were: Benton Williams, Tom Hoffman, William Dodd, Del George, Joe Bennett, George Rodgers, S.S. Krumbine, Murdick Findley, George McCreery, Thomas McCreery, Sam Longnecker, Roy Wetzel, L.B. LaBalle, Paul Graham, W.C. ?, and W.L. Long. The first officers were: President, William Dodd; Vice-President, George Rodgers; Secretary, Benton Williams; Treasurer, Joe Bennett; Purchasing Agent, Tom Hoffman; Gun Inspector, Del George.

A committee, consisting of Joe Bennett, Sam Krumbine, and George Rodgers, located a site at Pardee and bought one-fifth of an acre for the club. Sam Longnecker, an insurance man and resident of Blacklick, was a woodsman and carpenter by background. He and Joe Bennett planned the blueprint of the camp in their minds and then ordered the building materials, which were shipped to Weikert by boxcar. A trunk line, which paralleled Penn's Creek, ran from Bellefonte to the Sheffield Farms; milk cans were carried in the caboose. Near the club's lot was a siding, and the box car was moved onto the siding and used for storage and sleeping quarters until the camp was built. Bennett and Longnecker did the carpentry work; Bennie Williams served as a gofer. The finished two story building was 36' x 22' x 14' and consisted of a single large room on each floor. Total cost of the project was $215.19 for the lot and expenses and $675.80 for lumber.

In hunting season, a group of about twenty men and a full-time cook trekked to camp for a two-week outing. Since it was during Prohibition, the members made rules concerning the use of alcohol. Some were non-drinkers, but a few liked a drink or two during the day. Longnecker was a Methodist and a religious man, but he still liked a drink before and after supper. Some of the men, especially the cook, liked an "eyeopener" in the morning; so the camp rules limited the men to one drink in the morning.

Membership fluctuated during the 1920's and especially during the Depression. The following men belonged to the club at one time or another during the 1920's and 1930's: Walter Morris, Paul Graham, Leon Wenrick, L.C. McCreary, Doc Waltz, Percy Hoffman, Fred Bittorf, Horace Frampton, S.J. Shaw, Lawrence Long, Benton Edwards, Baloz Bagu, Jerry Lear, George Shomo, Lloyd Williams, Dr. C.S. Berkey, and John Wagner. Members had difficulty paying dues during the Depression. For instance, Joe Bennett was given credit on his dues in exchange for a load of wood for the camp.

In the 1940's new members were Arthur Mash, Ralph Wagner, Ed Felton, George Seigh, and Steve Dusza. Del George and George Shomo were reinstated, but in 1948, Shomo's share was repurchased by the club. Harry Brickner joined in 1955.

There was one tragedy in the club's history; Sam Longnecker of Blacklick, age 74, died suddenly at camp on November 26, 1953.

The club's 1971 by-laws and house rules limited membership to thirteen. New members were admitted when a member resigned or died. A new membership share cost $200 and annual dues were $10.00. By the early 1980's, taxes on the camp had risen from the original $9.00 per year to $80.00.

Improvements have been made over the years. The members requested in 1946 that electric lines be strung to the camp. This was accomplished in 1948. Water used to be carried, then pumped from a neighbor's well, but a few years ago, because of a drought, a well was dug to supply the camp. Indoor plumbing was installed by Ed Felton's son-in-law.

Years ago, there was no problem in closing up the camp for the winter, but now because of the indoor plumbing, the pipes have to be drained. To pay for these improvements and keep money in the treasury, the club now charges a daily fee of $1.00 a day for each member; 50 cents daily for each family member; and $1.50 for each guest. All in all, it makes for a very cheap vacation. The title of the camp probably should be changed because few people go to the camp to hunt anymore. It has become a fishing and vacation camp, especially during trout season.

The camp has experienced no problems with vandalism over the years. Lloyd Williams credits it to a long standing friendship with the people of the area. During the Depression, the area families were very poor and were glad to get any of the food left over after the men left camp. The camp members were happy to share with their neighbors.

The camp holds an annual meeting in August. For years, it was a corn roast held at Jerry Lear's home in Robindale. Later, it was held in Vinco. All members and their familes were invited. As a child, the author can remember dreading the corn roast ordeal, but later looked forward to them. The corn roast was discontinued in 1981, but the annual August meetings is still held. Today, Lloyd Williams is the only Vintondale resident who is a member of the Vintondale Hunting and Fishing Club.

BRACKEN

In December 1904, the Commercial Coal Mining Company of Expedit (Twin Rocks) began shipping coal from its new #4 mine which was located along the south branch of the Blacklick about half-way between Vintondale and Expedit. The mine and its patch town, called Bracken, could be reached by train or by a dirt road leading from the Twin Rocks-Vintondale road at Bracken Dip.

Bracken Mine; Commercial Coal Mining Company.

Courtesy of Gloria Risko.

Superintendent of the Commercial mines was William C. Smith, who resided in Twin Rocks. Other mines operated in the area by Commercial were #3, which eventually connected with #4 and #5; and #16 which had a portal along the Cambria & Indiana Railroad on the north branch of the Blacklick. #4 headed west toward Red Mill, and the #16 mine was probably an extension of #4.

Much of the coal mined in #3 and #4 was part of the Moore Syndicate Tract, of which the Griffith Family of Ebensburg owned a large share. Part of this tract was also leased to the Cambridge Bituminous Coal Company and later the Vinton Colliery Company.

The Bracken mine was a non-gaseous slope; its two air splits were ventilated by an eight-foot Robinson fan. Until 1906, the coal in #4 was machine-mined using compressed air purchased from the neighboring Cambridge Bituminous Coal Company. Haulage was done by as many as fifteen mules. Later, Commercial switched to electric haulage and began to purchase electricity from the Penn Central Power and Light Company. By 1913, Commercial's #3 and #4 mines were united, and production figures were often combined into a single account in the mine inspector's reports. Following an inspection in 1917, #4 was rated very good for ventilation, drainage, and timbering. By 1917, the slope in #4 extended 7,980 feet.

Bracken, c, 1910. George "Clate" Straw on left and Clarence Stephens on right.
Courtesy of Hazel Cramer Dill

Russell Dodson said he pulled stumps in the third water level of the Bracken mine when he was thirteen or fourteen years old. His partner was Mike Yelenosky, who had recently arrived from Europe. Mr. Dodson said that there was no coal for miles where the stumps had already been pulled. Neither he or nor Yelenosky realized how dangerous the work was.

The Commercial miners were unionized, at least in 1905, when Vintondale Superintendent, Charles Hower, informed Warren Delano that the Commercial miners living in Vintondale had been evicted, and he did not expect any trouble from them again. During the 1922 strike, the Commercial miners went out on strike. If there were any miners living in Bracken at the time, the Vinton Colliery was going to make sure that they had no contact with the Vintondale miners. On one of the guard sheds was a spotlight which could be directed toward Bracken. The guards could quickly spot anyone trying to come into Vintondale via the railroad tracks.

Michael and Anna (Skiba) Juba lived in Bracken from about 1910 to 1919. Three of their children: Steve, George, and Mary were born in there. Mrs. Juba's brother, Frank Skiba, who boarded with them, was killed in a mine accident in Bracken on January 17, 1917. The Jubas were active members of the Russian Church where Mr. Juba served as an officer on the church committee. The family moved to Plank Road and then on to Twin Rocks during the 1924 strike. The Jubas permanently settled in Colver.

Sam Kish and his family lived in Bracken from 1917 to 1920. Before that, they had lived in Vintondale. Sam's mother kept boarders, mainly Hungarians, on the Kaiser farm above the old Cambridge mine (#5). To cross the Blacklick, there was a plank swinging bridge near the PRR water tower at the Y switch.

George Kreashko and his family lived in Bracken in 1917. Daughters Mary and Pat decided to explore the Bracken mine and sneaked onto an empty trip. Once inside, the motorman discovered the girls and drove them back outside, warning them how dangerous their expedition was.

Pete Cramer, who cut lumber and props for the Vinton Colliery Company, lived in Bracken for about a year. He and his family, including daughter Hazel, lived near the hotel, which was run by Adam Barr. A future Buffington Township resident who resided in Bracken was John Parsley. The Palkos of Nanty Glo were also one-time residents of Bracken.

The village had three streets. Near the creek was a row of four single houses. There were eight double houses on the next two streets. A company store occupied half of the first double house; the store manager lived in the other half. In 1917, the company store manager was Russell Wagner. The company doctor, Dr. William Prideaux, drove to Bracken daily from Twin Rocks in a horse and buggy. He hoped the fences and yelled, "All right?" into the kitchens. If there were no problems, he proceeded to the next house.

In Bracken was a one room schoolhouse which Steve Kish, Mary Kreashko, Hazel Cramer and their brothers and sisters attended. The teacher, who was paid by the Blacklick Township School District, arrived daily on the 8:00 AM train and departed on the 3:00 PM. Ethel Kessler was one of the Bracken teachers.

In 1920, the Kishes moved to Twin Rocks, which many have coincided with the closing of the town. The last recorded Bracken birth and/or death certificate was in 1920. It appeared that the town was abandoned by 1923 because an entry in Hoffman's mine diary reported a fire in the "old tipple and sand house" in Bracken. He sent several men to investigate the fire and report back to him.

The patch of Bracken made very vivid impressions on Malcolm Cowley, poet and essayist. As a child, Malcolm rode the train between Pittsburgh and Vintondale or Twin Rocks. His memories of Bracken were included in an anthology of poetry entitled, **Blue Juniata**. In an undated letter to the author, Malcolm Cowley described Bracken in these words:

Bracken was the God-awfulest mining camp in Pennsylvania; all of the trees gone, all the houses squalid, and not even a company store.

In another letter dated August 28, 1982, Mr. Cowley relayed these feelings about Bracken:

When I was 12 years old Bracken -- I think it was Mine NO. Four, not Six -- hadn't a tree left standing. If you wanted shade, you could crawl under a blackberry bush Desolation and cinders and a huddle of two-family houses. The creek was orange . . .

Bracken may have been a desolate area, but the residents were still able to express their feelings in song. Mary Monyak Drabbant, who lived in the last house on Dinkey Street, enjoyed listening to the singing and accordian music which drifted down the valley toward Vintondale in the evenings.

Like many patch towns, Bracken's lifespan was very short. Once the two Commercial mines were connected, the coal could be brought out and processed at the Twin Rocks plant. Today, there are few traces of the village, even the foundations are difficult to find amid the brush.

"Mine No. 6"

They scoured the hill with steel and living brooms
of fire, that none else living might persist;
here crouch the cabins, here the tipple looms
uncompromising, black against the mist.

All day their wagons lumber past, the wide
squat wheels hub deep, the horses strained and still;
a headlong rain pours down all day to hide
the blackened stumps, the ulcerated hill.

Beauty, perfection, I have loved you fiercely
--even in this windy slum, where fear
drips from the eaves with April rain, and scarcely
brings forth its monstrous children--even here
   your long white cruel fingers at my brain.

The Hill Above the Mine

Nobody comes to the graveyard on the hill,
sprawled on the ash-gray slope above the mine,
where coke-oven fumes drift heavily by day
and creeping fires by night. Nobody stirs
here by the crumbled wall where headstones loom
among the blackberry vines. Nobody walks
in the blue starlight under cedar branches,
twisted and black against the moon, or speaks
except the mustered company of the dead,

and one who calls the roll.

   "Eziekel Cowley?"

Dead.

   "Laban and Uriah Evans?"

Dead.

"Jasper McCullough, your three wives, your thirty acknowledged children?"

   Dead to the last child,
most of them buried here on the hillside, hidden
under the brambles, waiting with the others
above the unpainted cabins and the mine.

What have you seen, O dead?

   "We saw our woods
butchered, flames curling in the maple tops,
white ashes drifting, a railroad in the valley
bridging the creek, and mine shafts under the hill.
We saw our farms lie fallow and houses grow
all summer in the flowerless meadows. Rats
all winter gnawed the last husks in the barn.
In spring the waters rose, crept through the fields
and stripped them bare of soil, while on the hill
we waited and slept firm."

   Wait on, O dead!
The waters still shall rise, the hills fold in,
the graves open to heaven, and you shall ride
eastward on a rain wind, wrapped in thunder,
your white bones drifting like herons across the moon.

From:

Blue Juniata: Collected Poems, copyright 1968 by Malcolm Cowley.
Reprinted by permission of the author.

Chapter XI
The Notorious Side

ROUGH, ROWDY AND READY TO FIGHT

From its earliest days, Vintondale resembled a western boomtown with houses and stores sprouting up all over. Accompanying this growth was violence caused by labor agitation, moonshine, domestic disputes, and extortion. Vintondale gained a reputation for rowdiness which still lingers on today. In the 1920's, Sixth Street had a well-earned label as the most unruly street in town. Included in this chapter is a chronicle of violent crimes, brawls, and Black Hand Activity in Vintondale before 1930.

Murders

In the 1920's Vintondale's wild reputation was enhanced by murders, four alone in 1923. Domestic problems was the cause of Vintondale's most tragic crime. The actual murder took place in Twin Rock, but involved a Vintondale couple. On July 28, 1923, Michael Soos, aged 24, approached the home of Mr. and Mrs. Joseph Vargo in Twin Rocks with the purpose of persuading his wife Mary, aged 15, to return home with him. The couple had been married April 29th, but separated several weeks later. Mary Soos moved to Twin Rocks to stay with the Vargos. Mary's father, John Meiser, had died earlier that spring in Vintondale; her mother lived in the hard coal town of Kulpmont. Mary's friends claimed that her parents persuaded her to marry Soos.

When Mary refused to return, claiming that she did not love him, Soos shot her five times with a .32 caliber revolver, twice in the chest, twice in the back and once in the arm. He ran to the mine flats adjoining the Twin Rocks ball diamond, put the gun muzzle to his head and fired. The bullet went through his right ear, paralyzing Soos. Recovering consciousness, he was sent to Memorial Hospital, where his condition was listed as critical.

Three weeks after the shooting, he was discharged from the hospital and transferred to the county jail to await trial. In September 1923, the court appointed William Dill of Barnesboro and J. Harrison Westover of Spangler as his defense attorneys. At his trial before Judge Evans in March 1924, the jury, after deliberating only twenty minutes, found Soos guilty of first degree murder. Soos was given the death penalty. On April 27th, 1925, Michael Soos was electrocuted at Rockview State Penitentiary in Bellefonte.

With Vintondale reeling from the Soos shooting, a second gruesome murder was committed the next day. Illegal liquor was a contributing factor in this murder. Late Sunday night on July 29th, a drunken "orgy" attended by "a crowd of aliens" led to a charge of murder levied on Frank Woydak (Wayda) of Sixth Street. He was accused on pouring oil or kerosene on Nick Marsenka (Mozinko) and setting him on fire. Marsenka, a 48 year old miner who worked at #6 and boarded on Sixth Street, died at 9:45 PM the next day in Memorial Hospital. County Coroner Smith and county detectives paid a visit to town on July 31 to gather evidence. Marsenko was buried on Wednesday, August 1st, and many miners took off work to attend the funeral. On Thursday August 30th, Attorneys Little and James paid Hoffman a call to obtain evidence for the murder hearing. A grand jury was called on Monday, December 2nd; Vintondale's police officers were called to testify. Woydak's trial began on Tuesday, December 11th and continued the next day. The verdict reached was not guilty due to lack of evidence. When the verdict was read, Woyak became violent and had to be restrained because he did not speak English and did not understand that he was found not guilty.

A third murder attributed to moonshine took place at the Anderson shaft in November, 1923. The details of that murder are treated in the Vintondale Mine Operations chapter in the section on mine shafts.

The fourth murder in 1923, which happened on Christmas night, was the result of another drunken quarrel. In Pappy Hoffman's diary, he first reported the crime as a shooting. His entry said, "Mrs. Joe Valiant shot John Stankovich on 6th St. at 5:00 PM." On December 27th, he noted that Joe Valent was in the Ebensburg jail charged with murder. John Stankovich died of a blood clot at 2:00 AM. Valent confessed to hitting Stankovich in the head with a pole ax. The **Nanty Glo Journal** reported that Stankovich was married, but not living with his wife and that his ten-year old son lived with him.

Legal and Illegal Liquor

Alcoholic beverages have contributed much to shaping Vintondale's reputation. Very early in her existence, Vintondale had a wholesale liquor outlet. In 1895, Francis Farabaugh purchased Lot 1, Block B at the corner of Second and Barker. A large three story building with apartments on the upper stories and two storerooms on the main level was constructed. On one side, Farabaugh conducted his liquor business; on the other side, Krumbine stored paint and wallpapering supplies and perhaps coffins. Mr. Farabaugh sold the structure to the company in 1912. Later occupants were Bill Roberts' Confectionery and Cresswell Electric.

The large hotels in Vintondale also served alcoholic beverages. The Vintondale Inn had a barroom in the basement; the bartender was John Cresswell. Shoemaker's, later Mike Farkas' Hotel, also served liquor. Patrons at the bar were men only. Women used the Ladies' Entrance, and then only if accompanied by a gentleman. This entrance, near the bar, led to a room equipped with tables and chairs. This custom was followed in Vintondale into the 1950's.

However, most consumption of alcohol was done at home, especially on the weekends after paydays. Orders were taken ahead of time by the dealer; then, then the kegs of beer were delivered to the doorstep by the dealer's driver and team.

After the departure of Farabaugh in 1912, Tobias J. Callet of Johnstown obtained a license, through the assistance of Judge Stephens, to operate a wholesale liquor business in the former Jacob Brett store on Third and Main. Callet, in partnership with Harry Silverstone, sold kegs of beer and whiskey rebottled from the large casks on the premises. Because he did not drive, Mr. Silverstone stayed in Vintondale during the week to run the business. Working with Silverstone at the distributorship was his nephew Henry Burger. Boarding at the Vintondale Inn, Silverstone took his meals from Mrs. Mary Nevy, who rented the third floor rooms above the distributorship.

Children in town were able to make a few pennies by redeeming empty bottles at Callet's. A quart bottle brought a penny; a pint was worth half a penny. Each family had a trash heap in the corner of the yard, so it was easy for the children to sift through the trash to find bottles. In addition, many were strewn along the streets and alleys by men staggering home from the bars or card parties. Callet washed the bottles, refilled them, and sold the whiskey. Often several brands came from the same cask, so one never knew if he was really getting Kesslers, Golden Wedding or some other brand.

The territory listed on Callet's license included Colver, but not Twin Rocks, Nanty Glo, or towns in Indiana County. Twice a week in Rexis, a boxcar was loaded with kegs of beer and whiskey. The next morning, the boxcar was hooked on to the Cambria & Indiana train and shipped to Colver. Callet sent his team to Colver to meet the train and make the deliveries. Clair Bearer, a C&I employee remembered beer being delivered almost daily in 1917. Callet and Silverstone were forced out of business when Prohibition came in. However as early as 1917, there was a remonstrance filed against the firm. The mine diaries blamed Mr. Krumbine for circulating a petition to close the business; the attempt failed, but Seymour Silverstone, son of Harry Silverstone, believed that the company was behind the petition. Callet sold the building after Prohibition closed him down.

Fights

So payday was a happy time in Vintondale. The bills were paid; the kids usually got a little candy; Sunday was a day of rest; and the booze flowed freely. Accompanying the drinking were card parties, bacon roasts, singing and fighting. As early as 1904, skirmishes were reported in the papers. The **Indiana County Gazette** reported on June 8, 1904 that six "foreigners" were in the Cambria County jail awaiting trial for shooting three "Americans" following a drunken spree on Memorial Day. The "Americans" were standing on a street corner when the shots were fired around 10:30 PM. Wounded were Philip Jones, who had the skin clipped from his nose; Isaac Michaels, who was hit by five bullets; and Robert Lynch, who was wounded in the leg. Names and nationalities of the "foreigners" were not revealed.

Alcohol played a role in the 1917 Charlie Daok case. Doak was charged carrying a concealed weapon; police officers Ed Timmons and Jack Butala were charged by Daok with assault and battery. According to Harry Ling, the police broke Daok's eardrums. On May 1st, Timmons, Dr. MacFarlane, and Lloyd Arbogast went to Ebensburg on the Daok assault suit. Hoffman drove to Ebensburg on May 31st to discuss the case with Judge Reed. Butala and Timmons appeared before a grand jury on June 6th; on the 8th, the jury found enough evidence to hold Timmons and Butala for trial. Dr. MacFarlane was discharged of any wrong-doing. The trial started on June 11th; attending for the Vinton Colliery Company were MacFarlane, Arbogast, Abrams, Timmons, Butala, and Shaffer. The next day, Arbogast and Abrams took the witness stand in the case. On the 13th, Daok was found guilty of carrying a concealed weapon and sentenced the next day by Judge Stephens to sixty days and 1/3 of the costs. Butala was found guilty of assault and battery, received thirty day sentence and 1/3 of costs. Timmons got ninety days and 1/3 of the costs. With both policemen scheduled to go to prison, John Cresswell as appointed temporary police officer. Hoffman took Mrs. Timmons and Mrs. Butala to visit their husbands at the County Jail on June 16th and July 6th. Butala served his thirty days and was back on the job on the 15th. Timmons was permitted to spend Sunday July 20th at home; Slutter was sent to get him. On July 25th, Timmons was back on duty, having served less than half of his sentence. Hoffman had a long talk with Judge Stephens on the 1st of August, and on the 5th, Timmons and Butala took a no-liquor pledges.

Not all residents relished the weekend drinking. Some early Vintondalers from Finland realized the problems created by excessive drinking and established a chapter of their temperance society in town. The Barkers also put their Prohibition Party philosophy into practice with the refusal to allow the sale of alcohol on the lots which they sold in town.

Vintondale entered the era of illegal use in 1920 when the 18th Amendment went into effect; accompanying the Amendment was the Volstead Act which legislated enforcement of Prohibition. Callet's closed, as did the barrooms of the hotels. Deprived of a legal source of beer and whiskey, some began to manufacture their own or stock up and hoard the real stuff.

Mike Farkas, who had converted his bar to a store, had a basement full of "good" booze.

Families made their own wine and beer at home. It was legal to sell blocks of compressed grapes which happened to include instructions for their use. Beer-making kits and ingredients were readily available in the stores, the most popular being Red Top. The illegality came in making, consuming, and selling the products.

A legal way of obtaining alcohol was to purchase a high alcohol content patent medicine, especially cough syrups. Most popular brand was Gingee, which sold for fifty cents for a small bottle. Users drank it straight or mixed it with a lemon pop bottled by the Goss Company in Twin Rocks. Vintondale merchants kept an ample supply of Gingee on hand and profited from brisk sales.

Speakeasies were also numerous, especially on Sixth Street. In fact, money made from sales of illegal liquor had a direct impact on the organization of the Vintondale State Bank. (See chapter on bank.)

Barrels of mash, fermenting at different stages, and stills could be found in homes and in the woods around town. Occasionally the federal revenuers, assisted by Jack Butala, appeared and staged raids in town. Many long-time Vintondale residents are of the opinion that those who paid Butala a bribe were tipped off before a raid. For the most part, the company and police turned their heads the other way when it came to making beer and whiskey. Goldie Hubner Brisky said that her mother hid the family still down the old #4 mine shaft near their house at Fourth and Main. During one impending raid, someone dumped the mash into the alley. A cow, wandering freely through the town as was the custom, happened across the mash and ate it. The cow died of a burst stomach.

The **Nanty Glo Journal** reported a few raids in the 1920's. Balus Bagu was arrested for bootlegging three times, 1922, 1924, and 1927. In 1922, he pled guilty to a charge of selling liquor, was fined $200, and sentenced to a year in jail. Bagu was immediately paroled. In 1924, Butala, in conjunction with county detective, John Gross, arrested Joe Vargo, Mrs. Charles Fenchalk and Bagu for selling moonshine. State police under the command of Sgt. Dahlstrom led the 1927 raid and seized a ten gallon keg, one gallon of whiskey and six barrels of mash at Alex Szegadis'. At Bagu's store, police found a jug of whiskey, twenty-four barrels of mash, and the coils and tops of two stills. Seven cases of beer were found and destroyed at Joe Kerekish's refreshment store. All three were arrested, brought before Justice of the Peace James T. Young, and bound over for court. Vintondale was a stop on a large county raid by prohibition officers in December 1931, but no arrests were listed.

Probably the most notorious bootlegger in the county was Harry Reitler. In the woods near his farm at the top of Chickaree Hill, he produced the famous or infamous "Chickaree Shine", with which politicians between Pittsburgh and Harrisburg were well-acquainted. Jim Wray acted as a driver for Reitler and had some narrow escapes with the law. On one occasion, they hid the whiskey barrels in a load of manure.

Moonshine use also led to serious fights and even murder, as mentioned above. Some were reported in the **Journal**; the mine diaries also gave accounts of the major altercations. Weddings among the immigrants were reasons for breaking out the booze. At the wedding of Nick Bobar and Lora Lonesky in February, 1924 in Twin Rocks, the moonshine flowed freely; tempers flared; and a free-for-all resulted. About one half of the guests were from Vintondale and Wehrum. In the course of the melee, Joe Butala of Vintondale was severely stabbed. Other participants listed in the paper were John Butala, John Wallie, and Mike and John Kerekes. The **Journal** remarked that, "The foreign population of Twin Rocks is usually well-behaved and trouble of this sort is rare among them."

Knives also played a role in many of the altercations. On August 28, 1923, Mike Demkevich, George Sisler, Mike Hassen and a Mr. Switch were involved in a "cutting affray" on Sixth Street. Hassen and Sisler, who were severely cut around the face, were taken to Memorial Hospital.

On October 29th of the same year, the police arrested a VCC cokeman and a Seward miner for carrying four gallons of moonshine on the streets. The Seward miner was lodged in the county jail, and the cokeman was out on $500.00 bail.

On Thursday, December 13th, the police arrested a "Horwatt" from West Virginia for carrying a large knife. Hoffman acted as town burgess on the 14th and fined George Nardella of Great Turn, West Virginia $20.00 for carrying a concealed weapon and for public drunkenness. Hoffman gave him one hour to leave town.

In March 1927, a shooting in House 160 on Sixth Street sent Stanley Placik, 44, and Nick Oblocovick (Oblackovich), 42, to Memorial Hospital. The two men were shot by Milan Bockrich, a guest of Placik. After consuming his own liquor, Bockrich ordered his host to find him some more. When Placik refused, an argument ensued, and Bockrich fired two shots at Placik. One missed, and the other hit Placik in the arm. He escaped into another room. When Oblackovich entered the room, he was hit twice, once in the thigh and once near the heart. Bockrich escaped from the house, and was tracked for a mile by Officer Harry McCardle before being captured. In April, Bockrich pleased guilty to felonious assault and battery and was sentenced to 2 and 1/2 to 5 years in Western Penitentiary.

In April 1930, Jack Butala's brother, Charlie, was hospitalized for a gunshot wound of the left arm. No explanation was given to the paper as the the cause. Charlie Butala was a baseball player and boxer.

Other Crimes Inside and Outside of Vintondale

Vintondale's police kept as much a lid on the town as they wanted, but occasionally they were called on to aid in cases which did not directly involve them. In September 1922, Mike Farkas pressed charges of larceny and fraudulent conversion against Fred Arnot. Farkas claimed that Arnot locked up his cow, used the milk, and refused to release it until Farkas paid him money. Arnot claimed that the cow had wandered into his cornfield and damaged the corn. So, Arnot was holding the cow for ransom. When the case came to court, testimony was given that Arnot had locked up the cow at least three times previously and did not notify Farkas why he was holding the cow. Additional testimony revealed that Arnot had locked up and ransomed cows of other townspeople. The court ruled that Arnot had no right to take the law into his own hands and should have proceeded with a civil action against Farkas for damages. Arnot was held under $800 bond, but defaulted and was sent to the county jail.

Butala helped solve several crimes outside of Vintondale. On November 22, 1927, the Wehrum Company Store was robbed of goods worth $175.00. About one month later, Butala arrested Raymond Holmes, a black. Holmes was suspected by the police because of a black hand letter he wrote. Under cross-examination, Holmes named John Doren as his accomplice. The men were tried in Indiana County Court because the crime happened in Wehrum, even though the men were living in Vintondale at the time. Holmes also robbed the Armorford Company Store in Dilltown and was sentenced to 2 and 1/2 years in the Allegheny Workhouse. Doren got six months in jail.

Butala solved a car theft on January 22, 1928. Someone reported to him that there was an abandoned Buick sedan on Martin Rager's farm on Chickaree. Upon making an investigation, Butala found footsteps leading from the car. Following the footsteps, he located John Henry Berry, a black from Johnstown. Berry admitted that he had helped steal the car from J.D. Sheasley of Johnstown. The mastermind of the crime was Francis Richards of Wehrum, who along with Berry had a criminal record. Richards stole the car to get home to Wehrum, and took Berry along to drive it back to Johnstown. However, Berry lost his way on Chickaree and abandoned the car.

In May 1928, Peter Rusko and Andrew Markas robbed the train station, using tools stolen from the company tool house to force open the station door and cash drawer. Their take amounted to $14.80. The money was buried near a tree, and the tools were found at their homes. Because they could not meet bail of $500., both men were sent to the county jail.

Thanks to Butala, a theft ring operating in the valley was broken up. In October 1928, the gas station at Rager's Grove near Gas Center was robbed of $30.00. The getaway car was described as a blue touring car without running boards. Upon hearing the description, Butala identified the getaway vehicle as one belonging to Howard Rutledge of Twin Rocks. Butala and Twin Rocks constable James Ross arrested Rutledge, Edward Ambrose, and Mike Semgo. The three were taken to the Vintondale jail to await a hearing before Justice of the Peace Dan Thomas. The men were held without bail, but Smego broke out and was not recaptured. Andy Smego and Frank Kendreski were charged with aiding in the escape. Blacklick Valley residents were hopeful that the arrests broke up a theft ring.

Mable Hampson Huth, c. 1906.

Agnes Huth at ball diamond behind Red Hotel, c. 1930.

Superintendent's house, 1906.
Courtesy of Denise Weber Collection,
Pennsylvania State Archives, Harrisburg

Olive McConnell, shipping clerk for Vinton Colliery and Vinton Coal and Coke. Edwin "Pop" McConnell, Ollie's brother, worked in the shop. Taken c. 1978 at the Stephen Dusza home. Nan and Pop were home for the Vintondale Reunion.

Courtesy of Agnes Dusza.

BLACK HAND

Practically every mining or industrial town with an Italian population had stories circulating about Black Hand extortions and/or murder. Vintondale was no exception. David Nevy, "Dealer in General Merchandise, Fruits, Tobacco, and Imported Macaroni", was the victim of at least two Black Hand extortions. A letter, dated December 31, 1909, was mailed by Nevy to Giovanni Lega, Box 17, Hackett, Washington County, Pennsylvania. In Italian, Nevy wrote that he was responding to "Johnny's" letter and that he (Nevy) was tired of the threat. If Lega did not stop demanding money, Nevy was going to sue and "take the shoes off your feet". (Translated from the letter.) The addressee never received the letter, and it was returned to Nevy by the post office. It is not known if Mr. Lega stopped his extortion attempts.

The big extortion case, which had Vintondale talking for months, included a Wild West chase and shootout. This cloak and dagger affair was executed on the night of March 11, 1914. An extremely elusive story to trace, the riddle of exactly when the event took place was solved thanks to the 1914 mine diary. Russell Dodson, Martha Abrams Young, Mable Davis Updike, and Paulina Nevy Pioli all relayed their knowledge of the story, but no one could pinpoint the date. Following the discovery of the date in the diary, local newspapers were consulted to back up their memories. This story appeared in the newspapers,

Tom Whinnie, Wes and Sarah Whinnie Extrom of Big Run, Ruth Whinnie.
Courtesy of Nancy Whinnie Grace

which offered conflicting names of both the conspirators and police involved in the case.

Paulina Nevy Pioli said she was, in a way, to blame for the episode. She had just arrived with her mother from Italy that year. Calling herself a greenhorn, Mrs. Pioli said that she smiled at all the customers who came into her brothers' store. One day, as Paulina was bagging carbide into five pound sacks, she smiled at a customer who was from southern Italy. The Nevys were from the province of Parma in northern Italy. Several days later, the man returned and told Davey Nevy that he was going to marry Paulina. Because Davey refused, the man planned to extort money from him. His accomplice was a southern Italian shoemaker who had a small shop behind the company hotel.

David Nevy received two Black Hand letters demanding money; the second arrived on March 6th. Estimates for the extortion amount ranged from $500 to $2,000. The mine diary listed it as $1,500; the **Indiana Evening Gazette** set it at $1,000. If the money was not paid, the men threatened to kill Nevy and his entire family. Nevy took the second letter to a postal detective in Altoona who advised him on how to respond to the threat. Knowing his every move was watched, Nevy went to the bank and withdrew a large sum of money. The letter instructed him to package the money and place it at Lackawanna's abandoned #3 mine in a boiler identified with three chalk marks. The detective showed Mr. Nevy how to package the money. Shredded paper filled the bottom of the box, and only a small amount of bills were placed on the top. Nevy made the drop as scheduled on Marcy 11.

In the meantime, two state policemen arrived at the Vintondale Inn that evening, and the innkeeper's wife fixed sandwiches for them. Mable Davis and Leila Packer, who boarded at the Inn, saw the police arrive. When they inquired about the police presence, the desk clerk said he knew nothing.

Postal inspector Densmore (or Detective Treat of Altoona), Sgt. C.T. Dent and Trooper B.C. Snyder from Troop A of Greensburg hid on the top of the boilers and in the washery. Others troopers were situated around #3. When the extortionists arrived to make the pickup, they were surprised by the troopers. A revolver battle ensued, followed by a chase in the dark through thick underbrush to the Blacklick, across the ice-covered creek and into hip deep snow. A steep cliff on the opposite bank prevented the extortioners from getting away.

Arrested were Lomaree Cisimero (Lonardo Tizinero, Leonardo Piscuneri), 24; Domenick Rasso (Domenico Raffa, Domeco Garrega), 23; and Asdmole Telibalo, (Pascanmole Pelicano), 40. (Will the real extortionists please stand up?) Raffa and Tizinero are given credit for the gun battle in which Tizinero was wounded in the hand. They threw the money in the snow during their attempted escape. Pelicano, the shoemaker, was arrested in Vintondale and accused of writing the letter. The men were brought back to town, thawed out and kept overnight in the jail. The next morning, they were taken to Johnstown for arraignment and held for court on charges of using mails to defraud and blackmail. The bail of $1,500 each was not met, and the men remained locked up until their trial began in Pittsburgh on May 4th, 1914. A U.S. Marshall arrived in Vintondale on May 1st to subpeona witnesses for the trial. Otto Hoffman left on the 4th to act as a witness; he returned on the 8:10 AM train the next morning. The local papers failed to carry any information about the trial, but Paulina Pioli said that the shoemaker received a sentence of five years in jail, and his family moved to Nanty Glo. So, Vintondale maintained its reputation as a Wild West town in the East.

About the same time Nevys were being terrorized, the town was shaken up by an explosion. Immediately, the kids and many adults thought the Black Hand was in action. Site of the blast was Mike Farkas' storeroom, next to the Farabaugh Building. Before dynamite came into common use in the mines, the miners had to buy black powder and make their own squibbs or purchase squibbs at the company store in order to shoot coal. The state then ordered miners to use permissible explosives, so stores kept dynamite sticks and detonator caps in stock.

In addition to operating a store, Farkas kept about fourteen boarders. Several of Farkas' boarders were showing another how to put up the dynamite for a shot. One man had about six sticks of dynamite; another had the cap; a third was fooling around with the old-fashioned unprotected battery wires. The dynamite went off, and blew out all the windows in both buildings. Fortunately no one was killed, but the force of the blast threw Louis Soos out of the window. Between the two buildings was an empty lot on which a clothes line was strung. Soos made a three-point landing into a basket of clothes which was conveniently sitting on the ground. In the company reading room of the Farabaugh Building next door, Russ Dodson and Will Findley were playing checkers. The force of the explosion blew the checkers off the board. The boys, thinking that it was the Black Hand, ran out of the building to see what was going on. When they got outside, glass was still showering down on the street. Years later when Dodson was working in #6, he noticed that Soos carried milk instead of water in the bottom of his bucket. Upon inquiring, Soos told Dodson that ever since the explosion, he carried milk in his bucket.

At about this time, Mike Farkas remodeled his building, adding a mansard roof which gave him extra rooms on the third floor and a porch roof which extended over the sidewalk. Steve Dusza said that the roof on the hotel had been damaged and hence the remodeling. There are two possible explanations for the roof damage: this dynamite explosion, or the tornado which swept through the town in 1920.

KU KLUX KLAN

In the early 1920's, the Ku Klux Klan staged a resurrection in the United States. Whereas the Klan of the 1860's and 1870's was geared mainly to harassing southern blacks, the Klan of the Twenties vented its hatred on blacks, Jews, Catholics, and Southern and Eastern European immigrants. This was done under the guise of patriotism. Xenophobia, hard economic times in the coal fields, and the fear of losing jobs to foreigners, led to the rapid spread of the KKK cancer in Pennsylvania.

In Cambria County, the Klan was especially active in 1924. Locally, there was a klaven (meeting place) near Belsano for Blacklick Township members. On April 5, 1924, Klan members from Cambria, Indiana, and Blair Counties converged on Lilly in special trains from Johnstown and Altoona. Because Lilly was a predominantly Catholic community of the Pennsy mainline between Cresson and South Fork, it was chosen by the Klan as a central location for a demonstration. Klan members, secretly armed, marched through Lilly to a field above the town where they burned a cross. When the Klansmembers paraded to the station after the cross burning, words were exchanged between the marchers and the townspeople. Fire hoses were turned on the marchers. A riot ensued; three people died from gunshot wounds. Many of the guns, carried by the Klansmen under their cloaks, were later found thrown along the railroad tracks between Altoona and Johnstown. Both Klansmembers and Lilly residents were arrested and tried for rioting. Vintondaler Jim Wray said he and Russ Hoffman went to Lilly that evening to see what was going on and were witnesses to the riot. At the trial held in Cambria County, John Huth, Vinton Colliery's engineer, was called as a prospective juror; but his name was rejected. There were several possible reasons why he was turned down. John Huth was Roman Catholic, but the most logical reason was a conflict of interest. One of the defense lawyers, Percy Rose, was also a lawyer for the Vinton Colliery Company.

The first serious indication that the Klan was active in Vintondale came on November 8, 1923. A window-rattling explosion shook Vintondale. Immediately afterward, a cross burst into flames on #6 hill. Several other crosses were burned between November and April, 1924. The Vintondale Baptist Church was the site of a Klan funeral. A member's son died, and the group, wearing their white hoods, marched in the funeral procession. Russ Dodson said that just about every non-Catholic male in town was a member of the Klan except Jimmy Dabbs and himself. Harry Ling was asked by a Wehrum resident why he did not belong to the Klan. Ling replied that if he had to cover his face to belong to any order, he would not join.

Otto Hoffman was furious when the crosses were burned and vowed to discover who was involved. Ruth Roberts Anderson said one of her neighbors said, "He'll have a hell of a time doing that." However, Mr. Hoffman did find out who was involved, and that neighbor was soon one of the many evictees. Perhaps one reason that the company reacted so strongly to the KKK was that Mr. Schwerin, company President, was Jewish. The large number of evictions was also tied into the strike going on at the same time. (See appendix.)

Olive McConnell, Mr. Burgan's clerk at the time, said all materials for the crosses and the oil came from the mine warehouse. The ringleader of the Klan was fired and told to leave town. At a basketball game that night, he was refused admittance to the game. As pre-arranged, the Klansmen and their families got up and walked out en masse. The next morning, none of them reported for work. The sheriff evicted these people the following day.

The Klan's influence greatly diminished following the Lilly riot. In Vintondale, the strike was lost, and many of the evicted Klansmen and/or strikers relocated. The burning crosses left a lasting impact on many residents of Vintondale, especially the children. People today still vividly remember the fear that the burning crosses invoked in them.

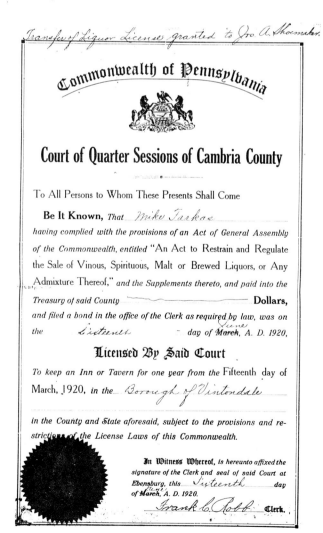

Chapter XII
Wehrum

LACKAWANNA COAL AND COKE COMPANY: 1901-1905

On December 4, 1900, a charter application for a new coal company was recorded at the Indiana County Courthouse. The new organization was titled the Lackawanna Coal and Coke Company. Purpose of the company was to: dig, mine, and sell coal; prepare and manufacture coke and other by-products; and transport the coal, coke and other minerals. Company business was to be transacted in Indiana and Cambria Counties with principal office in Vintondale. The new corporation was to have a lifespan of ninety-nine years.

Subscribers and their number of shares in the new company were:

Fred Barker, son of A.V. Barker, Ebensburg	100
Clarence R. Claghorn, Vintondale	100
H.B. Douglas, Vintondale	100
John Scott, Indiana, PA	100
Warren Delano, East Orange, New Jersey	100

The number of corporate directors was fixed at five. Serving on the board for the first year were:

Fred D. Barker - Ebensburg
Clarence Claghorn, General Superintendent - Vintondale
H.B. Douglas, Superintendent - Vintondale
Moses Taylor - New York City
Henry Wehrum, Elmhurst - Lackawanna County

The amount of capital stock was $500,000, divided into 5,000 shares with a par value of $100. Ten percent, or $50,000 of the capital stock, had been paid in cash to the treasurer of the new coporation, J.P. Higginson, 100 Broadway, New York City.

Lackawanna Iron and Steel Company

The parent company of the new corporation was Lackawanna Iron and Steel Company of Scranton, Lackawanna County, Pennsylvania. LI&SC and its predecessors were originally created in 1848 in the new settlement of Scranton by Colonel George Scranton for the purpose of manufacturing steel rails for the Erie Railroad. Eventually, New York financiers, the most influential of whom was Moses Taylor, acquired the firm. Other capitalists who wielded power in Lackawanna were: D.O. Mills; C. Ledyard Blair; Adrian Iselin, Jr.; H. McK. Trombley; Robert B. Van Cortlandt; and Cornelius Vanderbilt.

In the spring of 1899, several prominent businessmen of western New York State met at the exclusive Buffalo Club to discuss plans to entice Lackawanna Iron and Steel to move to the shores of Lake Erie. Local businessmen involved in raising the necessary capital to attract Lackawanna were John J. Albright, John Milburn, and Edmund Hayes. To make the move profitable, Lackawanna demanded capital, land, and dockage facilities. Albright provided fifteen hundred acres south of Buffalo through the Stony Point Land Company. As designated agent for Lackawanna, he acquired even more land from the city of Buffalo and sold it to Stony Point for a personal profit of $644,154.

The efforts of the Buffalo capitalists and Lackawanna's general manager, Henry Wehrum, were successful. The steel works of the Lackawanna Iron and Steel Company were transferred to Buffalo in 1901 to be closer to the iron ore supply. This move economically devastated the city of Scranton. The Scranton mills were demolished before the new mills were built in the new town of Lackawanna near Buffalo. At the time of the move, Walter Scranton held the presidency of the company, but not much of the power. Vice-President was Captain B.H. Buckingham of Cornwall; secretary was George Pierpont. Following the move to Buffalo, Lackawanna Iron and Steel Company formally became the Lackawanna Steel Company on February 14, 1902. Lackawanna Steel sought high level banking and railroad officials on its board of directors. Its first board represented thirty-three banks and thirty-six railroads. Featured prominently in decision-making was Moses Taylor, director of National City Bank, which was founded by his grandfather, Moses Taylor. The top executives maintained offices in company headquarters in New York City; the day-to-day matters were handled by the plant managers. The board, rather than the local managers, had authority to set labor policies, wages, production targets, products, etc.

Lackawanna scouted for coal and iron ore for their new mill and decided to develop their own captive mines and coking plants. Some of the properties acquired were the Tilly Foster Iron Mines in New York, the Scranton Mining Company in the Mesabi Range, the Franklin Iron Company in New Jersey. In addition to developing Wehrum, Lackawanna bought 15,000 acres of coking coal in Washington County, Pennsylvania in 1907. The Ellsworth Collieries Division had a capability of producing 2,000,000 tons per year from four mines.

A.V. Barker and Warren Delano were able to convince Lackawanna to survey the resources in the Blacklick Valley. Delano was convinced that the coal in his section of the valley could be made into marketable coke. In 1899 and 1900, these two men and Lackawanna agents began purchasing coal rights and surface rights in Buffington and East Wheatfield

Townships.

Vinton Colliery spokesman, Clarence Claghorn, predicted that the railroad would be extended to the village of Buffington by 1900. As the work on the railroad progressed, Claghorn kept the Scranton office updated with blueprints and other information. In the meantime, VCC purchased several farms, complete with livestock and farm equipment, with the intention of farming them. William T. Duncan and family moved to Morrellville after selling their farm in November, 1899. A.J. Croft of VCC's surveying corps moved to the farm. Others farms purchased included those of Burt Matthews in Buffington and James Cauffield, who moved to Armagh. Farms purchased in East Wheatfield Township included those of James Wallace, John Hill Heirs, James Bracken, and the Miller farm. In January 1900, Lackawanna Iron and Steel purchased the James Wallace farm near Buffington. Claghorn then had a road graded to the farm. Wallace acted as farm superintendent for LI&SC until October, 1900 when he was replaced by an Englishman. Later, Wehrum residents were permitted to graze their cows in the pastures for $2.00 per month.

A deed, dated July 10, 1900, saw Lackawanna I&S grant thirteen tracts to Walter Scranton for $23,632. On March 11, 1901, Scranton conveyed these properties for the same price to Lackawanna C&C. In a deed dated July 21, 1901, Lackawanna Iron and Steel Company sold ninety-five tracts of land and/or coal to the new Lackawanna Coal and Coke Company for $170,000.

On July 28, 1902, a deed transferred another twenty-eight tract in East Wheatfield, Buffington and Brush Valley Townships to Lackawanna C&C. These were called the "Hesbon Lands" and were sold by A.V. Barker for $141,717. Barker had a previous buyer for these tracts; the original offer was from John Patterson, an industrialist from Hamilton, Ontario. The deal fell through when Patterson could not arrange financing, supposedly $570,000, or $20 per acre. By checking the price Barker paid for each deed, the author calculated that Barker's costs were $66,331.81. Judge Barker realized a very handsome profit for his efforts. He was defeated in a re-election bid that fall and soon retired to Florida. One newspaper reproted that it thought this would be Barker's last major land transaction. In 1917, the "Heshbon Lands" were sold to Vinton Colliery Company. (See chapters on Vinton Land Company and on Claghorn.)

The purpose of the purchases between 1899 and 1902 was to provide surface and/or coal rights for the development of the two new mines, #3 and #4, which would supply coking coal to the Lackawanna mills. Tests were run on the coal as to its coking ability. Results proved that the coal was satisfactory for coking, but had to be washed first. In March 1901, the plans were to construct 600 coke ovens somewhere between Vintondale and Buffington. By that fall, the plans changed; coal was now to be shipped to Buffalo for coking.

Lackawanna #4 was to be built beside the newly planned town of Wehrum; #3, located at Edwards' Flats, was halfway between Wehrum and Vintondale. The choice of name for the new town was very appropriate; it was named for Henry Wehrum, general manager of the Lackawanna Iron and Steel works.

At the same time, LC&C purchased most of the Vinton Colliery holdings and the properties of the Blacklick Land and Improvement Company on February 5, 1901 for $325,000. The sale included seventeen tracts of land, excluding the lots already sold to private owners by BL&IC. Delano's lots in Vintondale were not included in the deal. With the sale of its lands to Lackawanna finalized, the Blacklick Land and Improvement Company disappeared after a short life of nine years.

The transaction with Vinton Colliery included #1 and #2 mines. An unnamed newspaper clipping listed the sale price as $175,000. Coal mined in Vintondale was to be shipped to Wehrum via the narrow gauge railroad completed July, 1901. Office messenger was Bill Roberts, son of #1 powerhouse worker John Roberts. Bill, who was crippled, rode the dinkey back and forth from #3 and #4 to #1 delivering memos.

Wehrum, c. 1912. Office of Lackawanna Coal and Coke Company. Superintendent Meehan is one of those on the porch.
Courtesy of Thelma Kirker.

Lackawanna #3 and #4

By 1902, #3 and #4 mines were sufficiently developed to be listed in the **Pennsylvania Department of Mines Annual Report**. Both mines were described as non-gaseous slopes; neither had fans installed in 1902. #3 employed thirty-one men, and #4 had fifty. The coal at both mines was machine-mined, using either Harrison or Ingersol compressed air

198

machines.

Lackawanna, #3 had two shafts and a slope opening. The machine-mined coal, using Claghorn's longwall method, was loaded by conveyors, dumped from a revolving tipple into two holding bins at the bottom of the shaft, and then hoisted at the top in a skiff.

#4's shaft and slope were sunk in October, 1901 on Miller's Flats. #4 operated on the same principal, except that conveyors took the coal directly from the bins to the washery, which also was to wash coal from #3. Clarence Claghorn even sent Walter Scranton a framed photograph of the Wehrum washery in January, 1902.

Lackawanna #4 Slope c. 1907 after double tracking.

Courtesy of Gloria Risko.

General superintendent of Lackawanna Coal and Coke Company was Clarence Claghorn. Superintendent of Lackawanna #3 and #4 was G.R. Delamater who was sent from Buffalo to assist Claghorn; his transfer included a salary increase from $105 to $125 per month. Managing the mines in Vintondale was H.B. Douglas. The Vinton Colliery Company remained in business and retained its #3 mine near the Big Curve.

The 1902 letter book of Walter Scranton has survived and included references to contract bids at Wehrum. In February, Scranton gave approval to Claghorn to award a contract for grading and laying a side track at the Wehrum washery to McMenamen & Sons for $9,500. In March, the Lackawanna building committee, which met every Friday, awarded contracts for six puncher or pick machines to Ingersoll-Sergeant Drill Company for $190 each. Each linen fibre wire wrapped air hose cost $20; each machine pick was $1.00. Several days later, the committee agreed to accept Edgar Kay's bid of $10,432.30 for the Wehrum water system.

The storehouse bid went to J.F. Cox for $2,380 and the mechanical equipment for the shafts and slopes was awarded to Webster Camp and Lane Machine Company at a bid of $23,000. These were awarded at the end of March. In mid-April, the building committee agreed to award the contract for 100 houses to J.P. Kennedy. His bid was $47,150. Blair Shaffer had the original contract, but died of a stroke in 1902. At the same meeting, the committee also approved the purchase of two head frames for #3 and #4 from the Wellman-Seaver Engineering Company for $15,640. Fort Pitt Bridge Company received the contracts to build the engine and boiler houses at #3 and #4 at a cost of $16,044. For these various contracts, two copies were sent to the Wehrum office for reference. The company store bid went to J.F. Cox for $6,070 in mid-May.

As the construction bids continued to be awarded, Scranton and Delano met with the railroad owners, Iselin and Yates, concerning construction of railroads into the Indiana County coal fields. Iselin and Yates had interests in the fledgling Buffalo, Rochester & Pittsburgh Railroad. Both the PRR and the BR&P were interested in extending their lines eastward from the Blacklick Station. In 1902, there was discussion about building a separate BR&P line which would parallel the PRR.

Within a year, work at Lackawanna's #1 and #2 had slacked off. The two mines worked 156 days from January to September and were idle for the remainder of the year. In 1903, a slump in the market led to a decline in the number of workers in Wehrum. Work on the new town of Claghorn halted. In the spring of 1904, work picked up some, and construction continued on the #3 washery. In October 1904, the Vinton Colliery Company leased its former #1 and #2 mines from Lackawanna.

The new Lackawanna mines were also beset with fatal accidents. In 1902, Ernest Monroe, a twenty-one year old black laborer, was killed while sinking the slope at #4. Mikeal Cowash and Charles Remish were killed by a rockfall at the bottom of the hoisting shaft in October. One month later, Paul Hertele, a former hard coal miner, was killed by an exploding charge. In 1903, Hungarian miners, Simon Boldi and Emerie Schmidt, were killed in two separate accidents in #4.

1904 was the worst year for accidents. In early May, a gruesome mishap occurred at the coal crusher in Wehrum. Some of the machinery on the crusher had been malfunctioning for several days, and William Diehl and Bert Noel, experienced machinists, were sent up to repair it. The crusher was grinding 150 tons of coal when the other workmen heard a cry. Looking up, they saw that the two men had disappeared and immediately signaled the engineer to stop the machine. The men found Noel's head sticking out of the coal directly above the grinders. Unconscious when they dug him out, Noel was taken to a nearby house, and the doctor called. Diehl's fate was the worst imaginable. His body went through the grinding wheels.

Finding his remains took four hours. Diehl, a native of Centre County, was 24 and unmarried; Noel, who was given no hope of recovery, was also unmarried. He was a native of Munster in Cambria County. Both men had moved to Wehrum when it opened.

On May 11, 1904, three foreign miners were killed instantly and a fourth later died of injuries when they crossed a danger board in #3, and their open flame lamps ignited a pocket of methane. Killed instantly were John Vantroga, George Shipley, and Andrew Drubant. Dying later of injuries was Frantz Gresico.

The 1904 Shutdown

The 1901 move to Buffalo, which involved transferring many of the Scranton personnel, may have been a major financial mistake. The Board of Directors made a serious error in ordering the dismantling of the Scranton works before those in Buffalo were completed. The Lackawanna mills were not ready for firing until 1904. Investments were estimated at $40,000,000., and the returns were small. By late 1903, Scranton lost the presidency of Lackawanna Steel Company to E.A.S. Clarke. Henry Wehrum was also forced into retirement in 1903 by the New York controlled Board of Directors.

On September 10, 1904, Clarence Claghorn posted closing notices, which were sent from Buffalo. Affected were over 700 workmen and 2,000 others. All residents were ordered to vacate the houses, and sixty men were sent from Vintondale to keep the pumps going, work in the rail yard and lumber yard and construct the #3 washer. The Vintondale men were to relocate in Wehrum, and only those working in the Vintondale mines were permitted to remain as renters. Reason given for the closure was the great expense of extracting the coal and moving to the tipples via the shafts. At the same time, the Vintondale mines were working at full capacity, and rumors zipped up and down the valley that the Vinton Colliery Company was going to repurchase its former mines, which were up for sale. Even though Lackawanna Steel had spent $3,000,000 in the valley, the company denied that it was hard pressed financially, but was open to negotiations on its Wehrum properties. However, another reason for the shutdown may have been that the company realized that they had made a bad judgment by investing in this property. The proposed coke ovens were never built by Lackawanna. By 1905, local newspaper articles hinted that the large washeries at #3 and #4 would be abandoned because the coal had too high a sulphur level even after washing it and that Lackawanna was going to market Wehrum's coal as steam coal.

1905 Sale

The **Johnstown Weekly Tribune** boldly announced in its December 8, 1905 edition that prosperity would be smiling on Vintondale following Vinton Colliery Company's repurchase of Lackawanna's #1 and #2 miners and its acquisition of the #5 mine of the Cambridge Bituminous Coal Company.

The Lackawanna purchase was discussed in a series of letters exchanged between Warren Delano and Charles Hower, superintendent of VCC in 1905. Delano relayed on November 2nd that he had proposed on October 20th to purchase all of the properties which had formerly belonged to Blacklick Land and Improvement Company, Vinton Colliery, and Vintondale Water Company. Delano stated:

I will take the property as I find it - absorbing the expenditure of over $10,000 which I have made during the past year of my tenancy (of which $3,000 has been for securing additional water supply and the balance for repairs and additional plant absolutely required for operation - see lease.)

Mr. Delano went on to state the favorable position of E.A.S. Clarke, LC&C president, for the sale except for the following reservations:

1. **Lackawanna wanted to reserve a fifty foot wide right of way on either side of the North Branch of the Blacklick Creek to the northern boundary of the land being sold.**

2. **Only water company and rights necessary to Vintondale property would be sold. Water rights which were mutually necessary would be adjusted.**

3. **Lackawanna also desired to reserve all telephone and telegraph wires and poles and the right of way from Lackawanna lands to the pike (Route 22).**

4. **Lackawanna also requested to keep control of the dinkey which ran between Lackawanna #3 and #1. A schedule was to be worked out between the two parties.**

5. **Lackawanna wanted to continue to furnish electricity to Vintondale until VCC was able to furnish its own.**

6. **Lackawanna requested that VCC pay for machinery and supplies sent from Lackawanna #3 and #4 to #1 and #2 when the lease was effective on November 1, 1904. These items included:**

a) ten puncher machines, fifty foot length of hose, foot block.

b) "Sullivan Baby" rock drills, with hose, one with drills, one without.

c) seven #5 "Little Giant" air drills for longwall faces, each with hose; four were old; three were new.

d) two side dump cars for rock.

e) twenty-two mine cars hitchings, such as used at Vintondale but not at Wehrum.

f) one rail bonding machine.

Lackawanna's Mr. Blanchard and Mr. Hower agreed on the list, and LCC set the value at $3,227.

7. All taxes, insurance premiums etc. had to be adjusted.

8. If Lackawanna accepted Delano's offer, it would not include any property with which Lackawanna had parted since 1901. (For instance, the Catholic Church lots on Fourth Street.)

In the same letter to Hower, Delano added his own comments to Clarke's recommendations.

1. On the fifty foot right of way, Mr. Delano asked for more information because the recommendation was ambiguous. He presumed that Lackawanna wanted the privilege of putting the railroad on either side of the creek.

2. According to Delano, Blacklick Land & Improvement developed the water companies. VCC took out the charter for the water company, sold the property, and the Vinton Water Company went out of business. Judge Barker advised the purchasers to organize four different water companies, one in each township. The property of the company was then transferred to them for $15,000: $12,750 paid by Black Lick Water Company, $1,750 paid by Jackson Water Company, and $500 paid by Buffington Water Company. Mr. Delano thought that VCC would only need the first two companies and desired a township map from Hower.

3. Regarding the phone lines, Mr. Delano did not want Lackawanna to keep them since most of the lines lay on the property Vinton Colliery was buying. He proposed a 50-50 compromise.

4. Mr. Delano had no objections.

5. Mr. Delano thought it reasonable.

6. Mr. Delano had intended that these items should have been included with the purchase of the entire plant.

7. and 8. were all right, just matters of detail.

In a return letter dated November 6, 1905, Hower also made his recommendations on Mr. Clarke's proposals. Hower sent Delano a geological map of Cambria County so that he could study the water situation.

Hower's recommedations for points 1-8 were as follows:

1. Hower had no objection for reserving a right of way, but the language must be clearer. He suggested:

a) that Lackawanna survey and locate the rights of way to be reserved within ten years or their rights cease.

b) that if a railroad is constructed, Vinton Colliery should have trackage rights as far as their land extends.

c) that Lackawanna's right of way should not interfer with any railroad the VCC should construct on the North Branch.

(This right of way desired by Lackawanna may have been for two possible purposes, extension of the Blacklick and Yellow Creek Railroad, or a branch of the Buffalo, Rochester and Pittsburgh Railroad to parallel the Ebensburg and Blacklick. The BR&P hypothesis is the more logical one because of the Buffalo connection, and Wehrum's coal was to be shipped to Buffalo. Eventually, the BR&P gained trackage rights on the E&BL.)

2. Regarding the water companies, Mr. Hower wrote that he had no access to maps or to a deed of the Vinton Water Company so he did not know which sources the companies controlled. He advised that if the Blacklick Water Company controlled the North Fork, it should grant 3/4 of the flow to Lackawanna for their #3. If the Buffington Water Company controlled the creek, then 1/4 of the water should go to the Blacklick Water Company. Hower wanted to reserve the right to take an additional 300,000 gallons per day (1/4 flow) from the North Fork during drought and also to serve customers in Rexis and the Vinton Lumber Company. Mr. Hower believed that the Jackson Water Company aid the $12,500, not the Blacklick Water Company. (This 1905 agreement was signed several years before the opening of the Colver mine and the subsequent polluting of the North Branch. Water in the North Branch at this time could be used for drinking. Jackson Water Company would be the logical of the four new companies to pay the $12,500. Vintondale's two resevoirs were on Shuman Run and Bracken Run in Jackson Township.)

3. Mr. Hower agreed with Mr. Delano that VCC should not allow LCC to keep the poles and wires which ran from Vintondale to the top of Chickaree. He suggested that Lackawanna be granted permanent right of way for existing poles from Wehrum to Vintondale through the property VCC was purchasing from them. There were two lines: a telephone and telegraph line which paralleled the Dinkey Track, and an electric line to Wehrum which went over the hill behind the Vintondale schools. Hower recommended granting LCC the right to put additional wires on the Vintondale pole line and constructing their own parallel lines. This would guarantee an independent outlet to the pike from Wehrum and at the same time reserve for VCC the electric copper power lines necessary for lighting Vintondale. (Vinton Colliery did manufacture electricity for the mine and the town, but some electricity was purchased

from the Penn Central Lighting Company.)

4. Regarding the Dinkey Track, Hower had no objections, but suggested that VCC have the right to lay and construct switches and sidings on their property.

5. All right.

6. Mr. Hower did not think the supplies transferred from Wehrum should be included in the deal. If they were to be paid for, Hower suggested that Lackawanna should bill them, and then the bill would pass through regular channels for complete records.

7. No information on the matter.

8. Regarding Lackawanna's previous sale of properties, Hower suggested a deed be drawn up with the rights of both parties listed on it. He wrote that it was impossible to get any maps or records from Wehrum and suggested a joint survey to establish boundary lines and reservations. Clarence Claghorn's notes on the above matters were lost. Mr. Hower suggested that the notes might be in Lackawanna's New York office. He also stated his willingness to oversee the survey.

Delano sent a return letter on November 8, 1905 which concentrated on the lack of power for VCC's #3. He assumed that the new hoisting engine had solved the problem. He thanked Hower for his suggestions concerning the Lackawanna deal, and said that he was writing to Claghorn concerning his "notes on the survey." (By 1905, Claghorn had moved to Tacoma, Washington.) Delano did not want to repeat the survey, and he felt that Lackawanna would agree with him. In the same letter, Delano asked about Hower's son, who had undergone a throat operation in Johnstown.

VCC's superintendents sent a weekly letter concerning the mine operations to the main office. Hower's letter of November 10, 1905 explained that the shortage of power in #3 was in the boilers only. He said they sometimes had to wait an hour for steam. He suggested that installing a heater would overcome the difficulty.

Claghorn's notes were also desired by Hower since he was anxious to have the property ready for the triannual assessment in the fall. He felt that the company was paying too much in taxes, and that Lackawanna could not identify the properties on which VCC had to pay taxes. This particular letter also contained a statement on union activity in town. (See chapter on union activity.)

A finalized deed was handed to Delano on June 8, 1906; selling price was $251,850, considerably less than the $350,000 Lackawanna paid in 1901. Included was the transfer of 2,500 shares of stock in Lackawanna Steel, new name for Lackawanna Iron and Steel. It was obvious that the new Lackawanna management in Buffalo saw the adventure in the Blacklick Valley as an expensive failure and were willing to unload the Vintondale properties as a loss.

On June 8, 1906, Delano announced the acquisition of the deed in a letter to Hower. The insurance premium on the properties was also settled. In the letter, Delano also recommended that Hower order cast iron pipe from Warren Foundry and Machine Company of Phillipsburg, New Jersey. He told Hower to mention his name in the order.

However, problems with the water companies were not easily resolved. Lackawanna gave Delano the qualifying stock to the Jackson and Black Lick companies. The Jackson company owned the plant, and the Blacklick owned nothing but its charter. The balance of the stock was held in trust under the Lackawanna Steel mortgage. Delano wrote that it would take a 3/4 vote of the Lackawanna Board of Directors to get the stock, which was in Delano's opinion almost impossible to get.

President Clarke of Lackawanna Steel was "worked up" and had contacted John Scott, Lackawanna's attorney and a stockholder in LCC, about condemning the water of the North Branch. Scott recommended that the Buffington Water Company condemn the entire flow and sell Blacklick 1/3. Delano stated that this was not satisfactory because it would be by lease and not by charter. He suggested that the companies work together on a 2/3, 1/3, basis according to the original agreement. He planned to contact his attorney concerning this.

The various problems were worked out, and the short-lived Lackawanna venture in Vintondale ended in 1905. Warren Delano, anxious to prove to Lackawanna's Board that marketable coke could be produced in the Blacklick Valley, proceeded with the construction of #6 and 152 coke ovens and took out the first of three large mortgages for VCC.

HENRY WEHRUM

Who was this man? What was his claim to fame? Was he important enough to have a town named after him? Henry Wehrum was born in Alsace-Lorraine in eastern France in 1843. At the age of sixteen, he was employed in an iron works in Alsace, working his way up from the lowest job. In his early career, Wehrum became acquainted with the Bessemer process and became one of its proponents. When the Prussians seized Alsace in 1871 after their stunning victory over the French in the Franco-Prussian War, Wehrum migrated to the United States and accepted a position with the struggling Lackawanna Iron and Coal Company, a predecessor of the Lackawanna Iron and Steel Company. Advancing to chief engineer, he resigned in 1881 and joined the Scranton Steel Company, aiding in the construction of its new works. These two companies merged, and Henry Wehrum emerged as chief engineer and general superintendent. In 1896, he was promoted to general manager. Wehrum was personally involved in the decision to move the steel works to Buffalo where Lackawanna would have the first fully-integrated iron and steel plant. He also played a role in the move to acquire bituminous coal holdings in Western Pennsylvania, believing that the coke made from soft coal was much more superior than that made from anthracite. At the new plant, Wehrum planned to use the latest technology to reduce waste and economize on the amount of fuel used in the production of steel.

Although Wehrum was involved in the move to Buffalo, the top executive managers, who were capitalists not steelmakers, entrenched themselves at company headquarters and made the decisions on capital expansion, marketing, labor policies, wages, and product lines. In the fall of 1903, there was a shakeup on the Board of Directors of Lackawanna Iron and Steel which led to Walter Scranton losing the presidency, and Henry Wehrum was fired for failing to delegate authority. Within a year, all plans to make Wehrum the coking center for Lackawanna Iron and Steel were scrapped, and signs were posted for the shut-down. Lackawanna's Board made a serious mistake in firing Wehrum, who was often seen as Buffalo's answer to Charlie Schwab. In the long run, the separation of the production and the Board had negative consequences for the Lackawanna Steel Company. It is ironic that Charlie Schwab eventually acquired the company and the town in 1922.

Following his dismissal, Wehrum retired to his home in Elmhurst, Lackawanna County, outside of Scranton. Henry Wehrum died in Buffalo on November 23, 1906 of the effects of a stroke he suffered on November 18. He was survived by his widow, three daughters and a son.

JOHN A. SCOTT

John A. Scott, attorney for Lackawanna Coal and Coke Company, was born in Clarksburg, Indiana County, on September 2, 1858. He attended local schools and the Elder's Ridge Classical Academy and gradated from Washington and Jefferson College. After teaching at the Elder's Ridge Academy and in Johntown, Scott entered the law firm of Honorable Silas M. Clark until Clark became a Justice of the Supreme Court of Pennsylvania. Working in the office of state senator, George Hood, was his next rung on the ladder to a law career. In December, 1884, Scott was admitted to the Indiana County Bar. He also served two three-year terms as Indiana County Prothonotary from 1888 to 1894. Shortly afterward, Scott left Indiana for Allegheny County where he practiced law. He returned to Indiana following the death of his father.

Scott married Edith Young (1876-1954) on September 7, 1899 at the home of her parents, Mr. and Mrs. J.H. Young, on Church Street in Indiana. For a honeymoon, the newlyweds took a train trip to the Great Lakes and returned via New York. The Scotts had three children: John Y., Albert H., and Florence. John Y. also pursued a law career in Harrisburg where he served as deputy attorney general and later opened a private practice there in 1937. John Y. Scott died in Harrisburg in 1964.

John A. Scott's speciality was corporate law and served as counsel for the Buffalo, Rochester and Pittsburgh Railroad Company, Rochester and Pittsburgh Coal and Iron Company, McKinney Steel of Ohio, and Lackawanna Coal and Coke Company. He was responsible for much of the title work for Judge A.V. Barker's coal purchases in Buffington, East Wheatfield, West Wheatfield, and Brush Valley Townships. In addition, Scott was president of Savings and Trust Company of Indiana. In 1914, he was elected president of the Central Banker's Association. In addition to the title searches, his expertise was used by Lackawanna Coal and Coke in June, 1909 after the Wehrum mine explosion and in 1914 when the company was being sued by a Dilltown gristmill owner for damages caused by pollution from Lackawanna's mines and washery.

John Scott died at the age of 77 on February 11, 1936 at his home on Seventh and Water Streets in Indiana.

WEHRUM'S SECOND CHANCE

The blast furnaces at Lackawanna Steel Company's new $40 million Buffalo plant were fired on April 15, 1904, but that good news did not have a positive impact on Wehrum. Her future was uncertain. After the shutdown notices were posted in the fall of 1904, the miners were asked to leave, but the completion of the #3 washery ironically proceeded on schedule. The proposed Wehrum coke ovens were never built, instead a coking plant was constructed at Lackawanna. Perhaps the Lackawanna Board of Directors had second thoughts and decided to recoup part of their losses by washing and selling Wehrum coal as steam coal rather than coking coal.

The Wehrum mines closed for a year, but were given a reprieve by Lackawanna Steel in the fall of 1905. Production for the remainder of that year was limited, and the work force only numbered 162. Serving as superintendent in 1905 was C.M. Blanchard. When Wehrum was given the go-ahead to re-open, #3 mine was not included in the order. In spite of the expense invested in the washery, #3 was permitted to fill up with water, and the buildings were abandoned. This remained in effect until 1914 when the Wehrum washery burned to the ground, and Lackawanna decided to repair the #3 washery rather than build a new one in Wehrum.

The status and improvements to the Wehrum mine can be found in that treasure trove of mining trivia, the **Annual Report** of the Pennsylvania Department of Mines. In 1905, the state mine inspector reported that Lackawanna's #4 had one pump with a capacity of 1,000 gallons per minute, sixteen tubular boilers with 4,000 horsepower, six electric dynamos, six air compressors, one steam and two electric locomotives, and thirteen engines of all classes. One major revision put into effect co-incided with the departure of Clarence Claghorn; the longwall method was abandoned and replaced with room and pillar mining.

There was a change in the superintendency in 1906. W.N. Johnson replaced Mr. Blanchard. Johnson had worked at the Claghorn mine in Bernice, Sullivan County with Harry Hampson and George Blewitt, two employees of the Vinton Colliery Company.

That year, the #4 slope was double-tracked, and a chain hoist was installed, allowing for the coal to be removed much more cheaply than through the shafts. The mine still only had one pump which had a capacity of 3,000 gallons per minute. Air was provided in three splits by a Capell steam fan. Thirty mining machines run by compressed air accounted for almost all of the annual production of coal.

There were no major changes in 1907, except that Mr. Johnson's son A.L. arrived from Bernice and started as a foreman. Six additional mining machines were added; cages were replaced in the shafts and used by the miners to enter and leave the mine. Some walked down the steps of the slope to reach their places.

In 1908, the Wehrum mine expanded enough to require five air splits. In that year, the miners used 505 kegs of black powder and 2,600 pounds of dynamite. Production in 1909 was down considerably due to the fatal mine explosion on June 23. (See following section.)

Mr. Johnson remained on in Wehrum as superintendent in 1910. In spite of the damage stemming from the explosion, ventilation and drainage in #4 were listed as good. The roadway was in good condition and improvements were made on the air courses.

THE WEHRUM MINE EXPLOSION*
BY
DENISE DUSZA WEBER

In 1909, Lackawanna's #4 mine employed a total of 142 inside workers, including miners, foremen, firebosses, drivers, and runners. Fifty-five were employed outside.

Wehrum's #4 mine consisted of a shaft opening for loading and unloading the coal; the main opening was a slope of 35 percent. The Miller (B) coal seam was about one hundred eighty feet below the surface. A second shaft opening nearby provided ventilation as a fan was placed on top of this opening. Many miners and inspectors stated that this mine was one of the best ventilated mines in the area.

On April 4, 1909, the **Johnstown Weekly Tribune** reported that Wehrum was operating three days a week. Due to depression in the market, a wage reduction of five percent for officials, monthly men, engineers, and pumpmen was posted on April 5. By June, the mine was working Tuesdays and Fridays only. On that fateful day of June 23, 1909, the Wehrum mine was not officially working, and no check was kept on the number of men entering the mine. Miners who went in were blasting down coal and getting things ready for work on Friday. Also in the mine was the surveying crew. Company officials and time keepers estimated that eighty to one hundred men were in the mine on the morning of June 23.

At 7:30 AM, a rumble shook the town. A cloud of dust and debris was blown out of the slope. The cage of the shaft, which was sitting at the bottom, was driven about one hundred feet to the top by the force of the explosion. Miners standing outside were thrown to the ground. The noise of the explosion was so loud that Russell Dodson heard it two miles away at #3.

Five year old Emma Huey and her younger brother Merle were eating breakfast at their house on First Street. The children were eagerly looking forward to the next week because they were to move to Broadway. There, the house had electricity. Mrs. Huey, seven and one half months pregnant, was upstairs when she heard screaming out on the streets. Coming downstairs to investigate, she looked out the back door. The kitchen table faced the door, so the children could also see out. The sky was the most gorgeous Emma had ever seen. Suddenly a large black cloud drifted up the street and over the hill. Mrs. Huey cried, "Oh my God, what's that awful cloud?" She ran around to the front of the house where neighbors helped her back into the living room. Her husband, Clarence, entered the mine that morning to assist the surveying crew.

John and Mike Orris, five year old twins, lived in the second house near the mine. They said the explosion sounded like a cannon firing and that the women ran screaming down to the mine. Their father, who had been given a permanent job after losing a leg in an accident in 1905, walked out about 11:30 without a scratch.

Rescue operation began immediately. Superintendent Johnson was assisted by Charles Hower, former superintendent of the Vinton Colliery Company. Joseph Williams, state mine inspector for the Tenth Bituminous District, arrived the same day to assist in the rescue and recovery of bodies. Nine other mine inspectors were ordered to report at once to assist Mr. Williams. Their job was to determine how many miners were killed and/or entombed in the mine and to inspect the mine thoroughly to locate the cause of the explosion. The 9:51 train from Ebensburg had to put on extra cars to handle the rescuers and the curious. About three hundred people gathered at the mine awaiting news. The Wehrum women provided baskets of sandwiches and buckets of coffee for the rescuers. The local hotel, the Blacklick Inn, also provided excellent service for up to two hundred people during the crisis.

By 1:30 PM, most of the dead and injured had been removed from the mine. Seventeen men were killed instantly, and sixteen others were injured, critically. Dr. Yearick, company physician, was assisted by Drs. Comerer of Vintondale, Barr and Stricker of Nanty Glo, and Grubb of Armagh.

Most of the injured were unconscious when brought from the mine and were revived by oxygen. The Johnstown **Weekly Tribune** credited Charles Hower, Frank Cloud of Cresson, and the doctors for heroically reviving almost all the miners who were not killed instantly. Four tanks of oxygen had been rushed to Wehrum in a car driven by Frank Cook of the Johnstown Automobile Company.

A special train provided by Cresson Trainmaster Henry Taylor left Wehrum at 3:00 PM for Miners' Hospital in Spangler. The following men were transported to Spangler: Patrick F. Burns, William Burns, Clarence Huey, Christopher Frazier, Sam Koncha, Louis Koncha, Frank Delegram, Tony Martin, Fred Thomas, Nick Spelli, Tom Batest and Joe Orwat.

Treated at home by Dr. Yearick were John Tobin, John Kessler, and Lee Johnson, the mine foreman and son of the superintendent. Johnson was critically injured with burns and lacerations, and at first it was thought that he would not survive. Rose Akers, a private duty nurse from Johnstown, was hired to care for him.

Killed instantly were:

Lovey Louis, Italian, miner, 22 years old, married, two children.

Ernest Barrochi, Italian, miner, 41, single.

Domenick Lilton, Italian, miner, 21, single.

Tony Batesta, Italian, miner, 20, single.

Tony Totena, Italian, miner, 22, single.

Charles Foldy, Slavonian, miner, 32, married, four children.

A.D. Raymer, American pumpman, 31, married, four children.

George Kovac, Slavonian, trackman, 23, single.

Simon Rominski, Russian, miner, 36, single.

Steve Base, Polish, miner, 36, single.

Kosti Sevick, Lithuanian, miner, 31, single.

George Lenn, Lithuanian, miner, 34, married, three children.

Joe Meniott, Italian, miner, 25, single.

Mike Lilton, Italian, miner, 23, single.

Alex Shaftock, Slavonian, miner, 46, married, two children.

Charles Georda, Italian, miner, 22, married, one child.

Charley Loray, Italian, miner, 20, single.

The spellings of the names of the miners vary from one source to another as often happened, for immigration officials and employers tended to spell the names as they sounded.

The bodies of the dead were placed on the machine shop floor and then moved to the livery stable, which became a temporary morgue. The late Russell Dodson, age 11 at the time, recalled the rescue operation. He said that bodies were hosed off at the stable and that the face of one Italian miner he knew was red as an apple. This was due to the exposure to the blackdamp, a gas mixture remaining after an explosion of firedamp (combustible gas). It is not explosive and will not support life. J.H. Krumbine, Vintondale undertaker, prepared the bodies for burial; the bodies were then removed to the victims' homes. A wagon, making numerous trips, brought coffins from Johnstown which were then delivered to the black crepe-marked houses. These victims, who according to the Indiana **Gazette**, were "all of the better class of working men," were buried for the most part by nationality. The Italian miners were buried in the St. Charles Cemetery, Twin Rocks. A marker in memory of the dead miners has been erected there by the people of Vintondale. The rest of the victims were listed as "Hungarians" and were buried in the Russian Orthodox Cemetery on the hill above Wehrum. A.D. Raymer, the only "American" killed instantly, was a pumpman and also a pitcher for the local baseball team. He was buried in Pittsburgh. Another miner, Alex Sevecik (Kosti Sevick) was buried in Windber.

Four of the injured died later at the Spangler Hospital, raising the death toll to 21. Frank Delegram died on June 25. His left arm was broken in two places, and he had severe burns on his hands, face and neck. He had also breathed afterdamp, a toxic mixture of carbon dioxide, carbon monoxide, and nitrogen, which remains in the mine after a firedamp explosion. William Burns, Patrick Burns and Clarence Huey died from severe burns and exposure to afterdamp. According to the June 25 **Weekly Tribune**, the injured miners in the hospital were not permitted any visitors, not even wives. However, the reporter did not receive the facts correctly. Emma Huey Stinson said that her mother was permitted to visit her husband. When the doctors realized that there was no hope for Mr. Huey, Emma and her brother, Merle, traveled by train to Spangler to visit their father. Clarence Huey survived for nine days. His funeral, under the direction of Ebensburg undertaker, Jeffrey Evans, was held at the Huey home on July 4. Burial followed in Lloyd Cemetery in Ebensburg. The funeral party traveled by train to Ebensburg. Evans had a hearse and a surrey there to meet the train. Mrs. Huey, her two children, and her mother, Emma Shaffer, rode in the surrey to the cemetery. The rest of the group walked. On July 18th, Zorah Huey gave birth to a son, Clarence.

Wehrum - Clarence and Zorah (Shaffer) Huey. Mr. Huey died of injuries received in the 1909 Wehrum explosion.
Courtesy of Emma Huey Stinson.

Rescue Operation

The slope entrance of the mine greatly aided in the rescue operation. Several miners escaped through that entrance, and some lives were saved because rescuers were able to enter the mine immediately. David Stutzman, a Vintondale miner aiding in the rescue, decided not to wait for a mine car to descend the slope. He walked down and encountered Clarence Huey, who was exiting the mine when the explosion took place. Huey was lying face down in one of the north headings. He was conscious, but unable to help himself. Other miners whom Stutzman found were George Penderd, Fred Thomas, and Lee Johnson; he helped these men to the foot of the slope. Fireboss Thomas Hogarth entered the mine with a rescue team about one hour after the explosion. His group was equipped with safety lamps. They encountered both living and dead miners at the foot of the slope. Those alive were unconscious. The position of the bodies indicated that the men had tried to crawl to safey, some as far as 100 feet. Eleven bodies were recovered between 1:00 and 3:00 PM. Eight hours after the explosion, Pat Batist was recovered in a fire and gas-filled entry; he recovered from his ordeal. Half of the persons in the mine escaped unhurt; many were unaware of an explosion until the rescue parties happened upon them. These miners were even wearing open flame lamps. They thought the noise of the explosion was just someone shooting coal.

Although the roof of the fan house was blown off, the fan, a steam-operated Capell, was not seriously damaged; the pressure gauge chart showed that it had only stopped for a few seconds. Consequently it was able to continue to clear the air in the five air splits in the mine, circulating 51,805 cubic feet of air per minute.

State mine inspectors examined the mine on June 24, 25, and 26. As they entered the mine by the main slope, they found evidence of great force and flame. At the foot of the slope was the main entry which was at a right angle to the slope and ran north and south. The force of the explosion seemed to extend no further than one thousand feet in any direction except in #1 south entry, where it reached the heading. The inspector found that concrete overcasts had been blown apart, and that brick and wooden stoppings and doors had been shattered. Mine cars were derailed, and hoses for compressed mining machines and electric wires were scattered all over. The force of the explosion seemed to weaken in areas where the mine was damp.

In some areas of the mine, the water was too deep to conduct the investigation. This was due to several of the pumps having been knocked out. One pumpman, A.D. Raymer, died, but two others survived. According to Russell Dodson, "old man Frazier," the night pumpman, was coming off his twelve-hour shift. Mr. Wurm, also a pitcher on the baseball team, was coming on. They were on the stairs of the slope when the mine exploded. Mr. Dodson said that both Frazier's and Wurm's pants were burned off. They recovered from their injuries, but Chris Frazier did not return to the mine, deciding rather to run a boarding house in Wehrum.

The inspectors found evidence that dynamite had been used in the first north entry left to bring down the coal; unused dynamite and burnt fuses were found there. They examined as much of the mine as possible and also interviewed mine personnel and survivors. They came to an agreement as to the cause of the explosion and presented it at the formal inquest.

The Indiana **Gazette** reported that the mine was inspected by a group of "competent miners." Included in this group were John Roberts, George Blewitt, and John Daly, all of Vintondale; other miners in the group were from Cardiff. They told the **Gazette** that they believed that the explosion was a firedamp explosion. (Firedamp is a combustible gas, mainly methane, created by the decomposition of coal.)

The disaster attracted attention from all over the country. The Indiana, Johnstown, and Pittsburgh daily newspapers kept staff correspondents on the scene. Telegraphs requesting information came in from all over, and a long distance telephone call was received from Toronto.

As the cleanup and mine inspections continued, rumors flew. Many people believed that the explosion spelled the end of the mine and of Wehrum. To clear the air, Lackawanna Coal and Coke Company released a news bulletin stating that the company which constructed the washery would be developing a process to rid the Wehrum coal of its four percent sulfur handicap which hindered its coking qualities. Also, because much of the mine was not seriously damaged in the explosion, it reopened on July 2.

Nine women were left widows and twenty-three children were orphans. For their benefit, a large picnic was scheduled for July 5. Even though Wehrum was a non-union mine, the Executive Board of the United Mine Workers of America donated $500.00 to the widow and orphan fund. The **Miner Workers' Journal** also wrote a blazing editorial about the Wehrum tragedy criticizing Wehrum and other mines for hiring unskilled foreign miners who were ignorant of the dangers of mining and of the use of dynamite and black powder in particular.

The December 8, 1909 **Indiana Evening Gazette** reported that the Lackawanna Coal and Coke Company made liberal settlements with the widows. Emma Stinson said that the company gave $500 to each family for hospital and other expenses and $100 for each child, but that the victims' families had to fight to get anything out of the company. Zorah Huey was permitted to move to the house on Broadway. To make ends meet, she cleaned for the company and was a seamstress. She also became one of Wehrum's midwives, joining Mrs. Crandall and Mrs. Sevilla Hanlon. Emma Stinson said that her mother was always busy; Zorah Huey did not remarry, but not for lack of offers.

Some victims' families were not satisfied with the company offer and filed suit against Lackawanna in the United States Circuit Court in Pittsburgh. Representing Tomaso and Pasquale DiBattista and the widow and son of Carmine Giodamo (Charles Georda) was Attorney Lawrence B. Cook. Each sued for $25,000. The outcome of the suit is unknown.

Preliminary and Formal Inquests

The Indiana County Coroner, Dr. J.S. Hammers, was at the scene by Wednesday evening and conducted a preliminary inquest. Jurors viewed the bodies and rendered a verdict that the men died from injuries caused by the explosion and by suffocation from blackdamp. A formal inquest was held on July 15 in Wehrum. Conducting the inquest according to strict state guidelines, Dr. Hammers convened it at 8:30 PM in the mine offices. The jury, by law, had to have a majority of experienced mine men. These men were chosen early by the coroner, some having visited the scene the day after the explosion. The jury was composed of Franklin Sansom, Indiana; Thomas Doberty, Graceton; Henry Kallaway, Ed McConville, Harry Dowler, all of Heilwood; and J. Dalton Johnson, Blacklick.

Six of the ten mine inspectors who assisted in the investigation were present. They were: Joseph Williams, Altoona; R.R. Blower, Scottdale; P.J. Walsh, Greensburg; E. Phillips, DuBois; N. Evans, Somerset; and I.S. Roby, Uniontown. Dr. Hammers was assisted in the questioning by Mr. Roby, whose stenographer recorded the testimony. Lackawanna Coal and Coke was represented by ex-judge Harry White in the absence of its lawyer, John Scott of Indiana. W.A. James of Buffalo, Lackawanna's chief engineer was also present.

Testimony was given by approximately twenty-five mine officials, inspectors, and survivors. Lee Johnson's testimony was taken at his home due to his injuries. One survivor, Mike Seafra, testified that he was blown one hundred feet through an open doorway from where he was working. In the course of the testimony, mention was made of a small methane leak that had been discovered several months before. S.N. Hazlett, engineer, said that it was unimportant and that the gas had dispersed before the fire bosses whose job it was to check for gas, had reached a reading on a safety lamp. Superintendent Johnson said that five years earlier there had been a discovery of methane when the fans had been shut down for thirteen hours. Mine inspectors and company personnel were in agreement that this was a well-inspected mine.

The most important testimony came from Tom Batist who made a statement explaining how he and two other workers on June 22 tried to shoot down the coal with black powder. The seam of coal was covered with fire clay, and the blast loosened a large piece of this. Mr. Batist used two and one-half sticks of dynamite in the same hole under the fire clay and inserted a six-foot fuse. He, his cousin, and another miner then went fifteen to twenty yards away from the blast area to a crosscut. When the shot went off, the room filled with flame. Batist and his cousin somehow survived, though severely injured. Tom Batist's original statement was taken in the hospital by Alexander Montheith, mine inspector. The third miner was killed instantly.

The inquest was adjourned at 11:30 PM and resumed the next morning. The jury retired at 11:30 AM and returned with the following verdict at 2:30 PM:

We the jury impanel (sic) to determine the cause of death of the seventeen (sic) miners or employees of the Lackawanna Coal and Coke Co., find that their death (sic) was caused by an explosion, presumably dust in Mine #4, owned and operated by said company located in Wehrum, Indiana County, Pennsylvania on June twenty third, one thousand nine hundred and nine. Said explosion was caused by the carelessness of Thomas Bestesta (Batist), a miner firing a dynamite shot, not tamped at the face of first left heading off north main heading.

In his annual report to the state, Inspector Williams stressed that there was a lack of knowledge on the part of the miners in the proper use of explosives. By 1909, inspectors were advocating the use of what they called permissable explosives instead of using black powder and dynamite. Just three weeks before the explosion, the **UMWA Journal** reported that the Geological Survey had just released a list of permissable explosives. Acceptable were: Aetna Coal Powder A&B, Aetna Powder Company, Chicago; Coal Special No. 1 & No. 2, Keystone Powder Company, Emporium, PA; Carbonite No. 1, No. 2, No. 3, No. 1 L.F., No. 2 L.F. Meteor Dynamite & Monobel, Dupont company, Wilmington, Delaware; Coalite No. 1 & No. 2 D, Potts Powder Company, New York City; Collier Dynamite No. 2, No. 4, No. 5, Sinnamahoning Powder Company, Emporium; and Masurite MLF, Masurite Explosives Company, Sharon, PA. However, most miners were reluctant to change and continued to use the black powder and dynamite.

Although the inspectors deplored the Wehrum disaster and Batist's way of blasting the coal, they also commended him for his honesty in his evidence. The inspectors believed that he was truly ignorant of what could happen after a blast like that; otherwise, he probably would not have confessed.

The mine inspectors made a list of recommendations for the Wehrum mine and any other mine in the state. These were published on the July 19, 1909 front page of the **Indiana Evening Gazette**. Some of these recommendations follow:

1. Use only permissable explosives.

2. Keep mine wet and/or dusted with calcium chloride; coal dust be removed at least once a week.

3. Non-combustible materials to be used in stemming shot holes.

4. Extreme caution should be used when handling and shooting explosives.

5. No shot should be laid deeper than the undercutting.

6. Safety lamps should be used when and where directed by law.

7. Rigid discipline should be enforced and maintained.

8. Sufficient fire bosses should be employed.

This explosion and the loss of twenty-one lives continued to support the inspectors' theory that a mine is not safe to work in "when black powder is used by ignorant men who know nothing of the dangers of coal dust."

Although the official verdict was a dust explosion caused by the dynamite blast, there were many who believed that there was methane in that heading which was ignited by the blast. The dust in the mine was then touched off by the methane exploding.

The Wehrum miners have not been forgotten. A memorial service was held in June, 1984 at Sts. Peter and Paul Russian Orthodox Church in Vintondale; the occasion was the seventy-fifth anniversary of the disaster. In attendance was Emma Huey Stinson, daughter of victim Clarence Huey and Nick Molnar, official of District 2, UMWA. To mark the eightieth anniversary, a liturgy was celebrated prior to the 1989 Wehrum Reunion.

Indiana County Heritage, Summer, 1984, pp. 20-27. Revised and reprinted with the written permission of the Executive Board of the Historical and Genealogical Society of Indiana County.

LACKAWANNA AFTER THE EXPLOSION

In 1911, H.J. Meehan became the superintendent. During the year, Lackawanna made some safety improvements at the bottom of the shaft and slope. A large brick stopping was installed so that the openings were separated and could be used in an emergency. The ventilation was good, but drainage in several places was defective.

The 1912 mine inspector's **Report** showed that #4 now had seven air splits and forty-eight mining machines run by electricity or compressed air. Two pumps with a capacity of 4,000 gallons were running; 1,500 gallons per minute reached the surface. Ventilation drainage and general conditions were much better than in 1911.

The big improvement for 1913 was the sinking of a 13-foot shaft close to the face of main south heading. An escape stairway was built inside the shaft. Overall ventilation was good, drainage was bad in a few places and the number of mining machines increased to fifty-five.

In 1914, disaster struck Wehrum on August 8; the washery and tipple burned to the ground. A hot afternoon, John and Mike Orris, aged 10, were swimming in the dam behind the boiler house. The water in the Blacklick was dirty, but kids stayed away from the Wehrum resevoir because snakes swam in it. The boys heard the mine whistles go off around 4:00 PM and scurried home because they had to be there when their father got home from work. As they were running home, they saw smoke pouring out of the tower where the coal exited the washery, and soon it exploded into flames. The fire, fanned by a steady wind, actually threatened to destroy the town. Only the hard work of the fire fighters prevented an even bigger catastrophe. Vintondale sent its fire company hose and a dozen of its employees to help fight the blaze. By midnight, all that remained of the three hundred foot washery was the steel skeleton, which remained an eyesore for years. In addition to the washer, thirty carloads of coal were destroyed. Damages were estimated at $100,000. A temporary tipple was built, and arrangements were made with Vinton Colliery to clean and ship some of Wehrum's coal until #3 was drained and put into working condition. The first shipment of washed Wehrum coal left VCC on August 13th. Eleven cars were loaded on August 27th and twelve more on September 3rd. There were several meetings between Vintondale and Wehrum officials over the shipment of coal. Superintendent Frank Dunbar and Mr. Burnett paid a business call to Vintondale on September 23rd. LCC officials spent the afternoon in Vintondale on the 29th. On November 10th, General Manager Luce of Pittsburgh paid a visit to see President Schwerin. The next day, Schwerin, Huth and Hoffman returned the visit to study mine maps.

Wehrum #4 Washery before fire in 1914.
Courtesy of Wehrum Reunion Committee.

Wehrum - Lackawanna Coal and Coke's #4 plant, taken after 1914, the year the washery burned.
Courtesy of Mabel Davis Updike.

However, the mine inspector's **Annual Report** stated that the Wehrum mine was out of production for the rest of that year. Whatever the case, Vintondale and Wehrum co-operated on washing coal, and Lackawanna proceeded to remedy the loss of the washer. Rather than rebuild another washery, the company decided in February, 1915 to rehabilitate the #3 washery, which had been completed during the 1904 shutdown and then abandoned. Lackawanna's #3 and #4 mines were to be connected, and the company had a full crew driving a heading toward the #3 mine. When connected, the coal extracted from #4 was taken out and cleaned at #3. Before that happened, the #3 mine had to be drained, and its plant put in working order. At the time that the decision was made, only 75 to 100 men were working in Wehrum. Local newspapers predicted that 400-500 would be employed by that April.

Some major improvements were made in 1915. Five concrete overcasts and fifty permanent brick stoppings were built. The main haulage road was widened. A 2,000 gallon per minute turbine electric-driven pump was installed along with a new pump house. A large sump for water storage was also made. Mining machines were switched from compressed air to electricity. A fifteen kilowatt generator and two-three hundred kilowatt motor-generator sets were installed. To supply electricity for the revitalized #3 plant, a high power transmission line was laid out; four transformers were also installed at the #3 powerhouse; one electric hoist was erected at #3 shaft. The hardly-used #3 washer, built ten years earlier by Heyl and Patterson, was updated with equipment to wash 2,000 tons per day. With the expansion of #4 into #3, providing air in the gaseous mines became even more difficult; fifteen airsplits circulated 60,000 cubic feet per minute.

Wehrum Yard Engine, c. 1915. Mr. McGough second from the right.
Courtesy of Thelma Kirker.

However, in 1916, the state mine inspector reported that ventilation was not satisfactory in some places and very good in others. Drainage was fair. Six new shortwall mining machines were put into operation, bringing the total to seventeen. Also added were ten five-horsepower pumps, six five-horsepower room hoists, and 400 electric lamps. Five telephones were installed in the mine along with a rescue station equipped with five sets of Fluess-Proto breathing apparatus and a complete lung motor. An additional thirty-nine concrete stoppings and three overcasts were built. The shaft at #3 was 161 feet, #4's, 184 feet. The coal seam in #3 was forty-two inches and forty-six inches in #4.

In 1917, the ventilation and drainage conditions were listed as only fair. An old air shaft at #3, which had been permitted to close, was cleaned out and lined with concrete. A new ten-foot Jeffrey fan was installed there.

In 1918, there was yet another change in superintendents. Dan Galbreath, who had previously worked at Vintondale, became superintendent in 1917, but soon departed for Kentucky. He was replaced by George Lindsey. Most of Wehrum's later superintendents had seen service at Ellsworth, Slickville, and/or Heilwood.

The Wehrum mine in 1918 was ventilated by two fans, a ten-foot Jeffrey and the sixteen-foot Capell, circulating 117,500 cubic feet of air in ten splits. A little bit more than half of the coal production was done by seventeen electric mining machines. Twenty pumps had a capacity of 8,535 gallons per minute. Two pumps brought 1,941 gallons per minute to the surface.

The Department of Mines terminated their detailed reports of every mine in the state after 1918. After that year, information on the situation in the mine at Wehrum is sketchy at the very least.

Like the Vinton Colliery Company, the Lackawanna Coal and Coke Company also advertised in the **Keystone Coal Catalog** in 1920. The company general office was Lackawanna, New York; company president was C.H. McCullough of Lackawanna. Moses Taylor of New York City maintained his influence on the company through a vice-presidency. Treasurer of LC&C was J.P. Higginson of Lackawanna. Purchasing agent was F.H. Burnett. General manager for the company was W.A. Luce, who kept his office in the Farmers' Bank Building in Pittsburgh. The electrical engineers, W.A. James and T.E. Tynes were both out of Lackawanna, NY. The purchasing agent for the company store was J.W. Hugus of Ellsworth, another Lackawanna Steel subsidiary. In Wehrum itself, the superintendent in 1920 was George Lindsay, and W.P. Francis was the mining engineer. The system of haulage for their #3 and #4 mines was ten trolley pole-type locomotives, using thirty-six inch track. Seventeen electric punchers mined the coal. In the power plant, there were seven tube boilers, two 300 kilowatt and one 5,000 kilowat turbine generating units.

The Lackawanna Steel Company was bought out by the Bethlehem Steel Company in May, 1922; at the time of the merger, Moses Taylor, Jr. was the Chairman of the Board of Lackawanna Steel. In 1923, Bethlehem Steel, through subsidiaries, began acquiring mines in the Indiana County area. The local papers announced that Frank Horton was appointed as division superintendent for all Bethlehem mines in the Johnstown area, including Wehrum and Slickville. In December, 1923, their West Virginia based Bethlehem-Cuba Mining Company acquired the Penn-Mary Coal Company in Heilwood for $1,291,621.40. Heilwood was originally a Coleman-Weaver operation. The month before, Bethlehem Mines, headquartered in Delaware, transferred their quarries near Bethlehem, Lebanon, Steelton, and York to Bethlehem-Cuba. The stage was set for the merger of Lackawanna Coal and Coke Company and the Ellsworth Collieries of Washington County in November, 1924. The Ellsworth Colliery Company had been organized in January 1907 with capital stock of 10,000 shares at $100.00 each. In 1924, Ellsworth had outstanding debts of $1.916 million, whereas, Wehrum had no outstanding indebtedness. The name was to remain as the Lackawanna Coal and Coke Company. The newly merged company had only 5,000 shares of capital stock. Those holding stock in the old companies were to deliver them to the new company office in Bethlehem. The new board of directors were all from Bethlehem. President was C.A. Buck; Vice-President, H.E. Lewis; Vice-President, H.S. Snyder; Secretary, R.E. McMath, who also served as the firm's attorney; and F.A. Shick as auditor. Charlie Schwab, as president of Bethlehem Steel, made his presence felt in Wehrum, especially for the first aid meets. Wehrum teams competed at national contests, and Schwab footed the bill.

Lackawanna #4 Boiler House on left. Dam on Blacklick Creek, c. 1916-20.

Courtesy of Mable Davis Updike.

For the time being, Wehrum's future seemed secure in spite of a lessening demand for coal caused by competition from cleaner fossil fuels. But by reading the merger agreement, one can see the handwriting on the wall. Wehrum was being used to pay off Ellsworth indebtedness.

According to Russell Dodson, Behtlehem "hogged it up" and ruined the mine. He claimed that the company spent no money on improvements for two years. Bethlehem did not take up the bottom or put in roads to advance. Abe Rager, who worked in Wehrum before moving to Vintondale, told Dodson that there was a 900 foot heist to his working place. That meant that Bethlehem was working 900 feet beyond the air. Locomotives could not have been run that far without the air breakthroughs, which were to be every ninety feet. Dodson, probably the knowledgable person in Vintondale on mine ventilation, said that the state inspectors told Bethlehem that the ventilation system had to be revamped or shut the mine. Several new air shafts were needed, and Bethlehem decided to close the mine. Notices were posted unexpectedly in May, 1929. Residents were given a year to relocate. That closing, which was listed as temporary in the 1930 mine inspectors' reports, has lasted over sixty years. Wehrum's fate was sealed when the houses were put up for sale for lumber.

Wehrum #4, c. 1920.
Courtesy of Wehrum Reunion Committee.

LABOR ACTIVITY

Wehrum was strictly a company town; all houses were owned by the company. Even the church buildings remained in company hands, for no deeds were ever granted for the Catholic or Methodist churches. The bank was one of the few buildings on a private lot. In this way, company control of the miners was easy to maintain.

As in Vintondale, company police kept close watch on those coming and going. Harry Ling, a Prudential agent for twenty-seven years, had Wehrum as part of his territory. He got off the train in Wehrum and heard someone whistling. Officer Crane told him, "I'm the policeman here." Mr. Ling, with a quick comeback, told him, "Well, I'm the insurance man here." Because he had a license to collect insurance any place, company police usually accompanied him to make sure he was not organizing his customers.

Union organizers occasionally attempted to get the Wehrum miners to come out on strike. But one organizer, Harvard graduate Powers Hapgood, actually felt that the foreign miners were difficult to organize because of the landlord system under which the immigrants had lived in Europe; they were used to taking orders. Hapgood had stopped in Wehrum in 1921 while on an organizing excursion down the Blacklick Valley. In an article in the March, 1922 **Survey**, Hapgood had this to say about Wehrum miners. "*They seemed, like most other non-union men I've met, too inactive mentally to consider the question.*" It would be interesting to have seen him react to his own words in 1922 when the Wehrum miners joined the national strike called by the UMWA.

The Lackawanna Coal and Coke Company had been expecting labor problems in 1922 and made plans in case there was trouble. The company requested in advance for state police protection as can be seen in the State Police Files in the archives of the State Historical and Museum Commission. On March 7, Mr. James of Lackawanna conferenced in Harrisburg with Lynn Adams, Superintendent of State Police. Immediately after the meeting, the company contacted Indiana County Sheriff, J.R. Richards, to secure a promise of protection. Richards promised a force of deputies throughout the coal region. In a March 27 letter to Adams, the company president reported that they had already deputized "*twenty of our own and trusted employees.*" "No Trespassing" signs were posted around the operating plant and buildings. The situation was the same at Lackawanna's Ellsworth mines except that Ellsworth and Cokeburg were organized boroughs. There trusted employees were sworn in as borough police. Wehrum expected state police protection if their deputies could not handle the situation.

In the spring of 1922, Wehrum was working one to two days per week; the men were paid seventy-two cents per ton. The pay of the companymen was also reduced. Wehrum miners were in the mood to make a statement. The UMWA strike call went out on April 1, and Wehrum continued to work. On Tuesday, April 18th, UMWA board member William Broad and organizer Charles Dias got into Wehrum, and Board announced that he was there to organize the town. They were promptly arrested by State Constabulary at 2:00 PM, charged with trespassing and taken to the company offices. Throughout the day, under the pretenses of wanting to see the superintendent, miners came to the office to see what an organizer looked like. Broad and Dias were released and ordered to leave town after the company waived a hearing and granted bail.

Charles Dias' statement of the affair was found in the UMWA District 2 files at IUP. According to Dias' affadavit, when he, Broad, and a third organizer, John Lucas, arrived, they were met by Officer McMullen and his men and told that they could not get supper in town and to get out of town and go to Wheatfield. As the organizers headed to Wheatfield, via the railroad tracks, two state police and three coal and irons rode up to them, questioned them and roughed them up. They were arrested for trespassing on private property-the railroad line. Taken to the company office, Justice of the Peace McCabe wanted to fine the organizers $10.00 and costs for trespassing. The men refused to pay. That night, the watchman told the organizers he would let them out about 11:00. Again the organizers refused because they heard voices in nearby rooms and figured it was a set-up. The watchman returned after midnight and asked them why they did not leave. They said that they had not done anything. The next morning, the justice of the peace came in, gave them a hearing, wanted them to pay the $10.00. The organizers again refused, and the company had to let them go anyway.

The Wehrum miners were so agitated at the arrests that they struck the next day. John Brophy, president of District 2, notified local newspapers that a mass meeting was held on Thursday, and that a union local had been established. State troopers called in by the company had been interfering with the organizing, according to Brophy.

The Wehrum miners also attempted to organize Vintondale, marching en masse up to the iron bridge, where they were stopped by a barricade of six or seven trucks. The strikers said that they wanted to go through town to Nanty Glo. They were told that they had to walk single file. There were also six mounted police at the scene. Frank Mehalko said that he fell on the bridge and Butala hit him on the head and told him to go back to Wehrum. Otto Hoffman must have feared the Wehrum miners enough that he went around to the boarding houses to convince the boarding house "misters" to get their boarders out to work.

Striking miners in Wehrum were evicted. The only formal evictions listed in the Indiana County Eviction Docket were requested in September, 1922. Only twenty-four names were on the docket, which seems unusual for a town where 350 miners went out on strike in April. Evictions probably started in April and continued throughout the summer. There was a tent colony set up on Steve Mihalik's farm in Wheatfield. Brophy had ordered 1,000 tents from Governor Sprowls as early as April, 1922.

Many families, rather than live in a tent colony, moved to other towns. Harry Stinson, husband of Emma Huey, went out on strike and found work in Ebensburg vulcanizing tires. Sam Roberti was a young boy when he arrived in the United States from Italy; he soon changed his last name to Roberts. Many of the Roberts family and their relatives lived and worked in Wehrum. One of Sam's sons worked on the dinkey-which hauled refuse to the rock dump. Another relative, Romy Delagram, could translate Italian and was used by the company in instruct the miners how to vote. He lost his job when he told the miners to vote a different way and the company lost the election. Frank Delegram was also a relative; he died of injuries received in the explosion. Sam Roberts, a miner who had survived the explosion, was evicted by the coal and irons during a rainstorm. Most of the family's furniture was ruined by the rain, so they left it behind. The Roberts family lived in the tent colony for a while, but decided to leave. The family walked to Cresson with Daisy, the family cow. There they lived in an old schoolhouse until relatives were able to find them a house in Sankertown. Sam's son, Joseph, was born in Wehrum; the attending doctor was Twin Rocks' Dr. Prideaux. Joe Roberts has served as Cambria County Commissioner for over twenty-five years.

John and Mike Orris, twins who had experienced the 1909 explosion and the 1914 washer fire, said that their father was not evicted in the 1922 strike because he had been given a life-long job with the company. Mr. Orris had lost a leg in the Wehrum mines; miners on strike accused Mr. Orris of being in with the superintendent.

For those who remained in the tent colony, life was hard. Bread was transported in large wooden boxes. John Belitza's father provided cabbage for the strikers. Victor Spongross' father decided he did not want to spend the winter in a tent either and bought a farm.

The striking miners held out for several months, but when John L. Lewis settled with his union mines before the non-union mines, the strike was lost in Wehrum.

WEHRUM: THE TOWN, 1901-1904

Buffington

Lackawanna Coal and Coke Company's new town was built near Buffington, but what and where was Buffington, a village or just a post office? By checking maps, local histories, old newspapers and using the process of elimination, the location of Buffington was pinpointed. It was a smattering of houses and a church on the Dilltown road near the Blacklick Furnace. When the post office was located on the west side of the Blacklick Creek, the settlement was named Buffington.

In 1845, the middle section of the Blacklick Creek came alive with the construction of several iron furnaces. David Stewart built the Blacklick Furnace against a bluff on the Blacklick about a mile or so from the Pittsburgh Pike and close to the Rager's Hollow-Pleasant Valley road. That year, another road was built to connect the furnace and Armagh. Like Ritter's furnace, Stewart's was a failure, and Attorney Edward Shoemaker of Ebensburg offered it for sale in 1848. Purchasing the furnace, timber, and ore deposits were George King and Peter Shoenberger, future developers of the Cambria Iron Company of Johnstown. It is not known whether they smelted iron there, but Cambria Iron Company retained the timberlands and mineral rights for the next fifty years.

All traces of the furnace have disappeared, but several families who had settled near the furnace farmed the land for several generations. These early settlers were listed as residents of the Buffington area in the 1886 **Indiana County Directory**: Jas. L. Bracken, Fletcher Bracken, Henderson Bracken, John Blakeley, Robert Blakeley, Twinan Bracken, Thomas Jefferson Bracken, Anthony Blackburn "Bun" Bracken, Watson Bracken, Robert Best, W. Cameron, Wm. Cameron, David Coho, David Cramer, W.M. Craig, John Dodson, Wm. Davis, T.J. Davis, Thomas Dodd, and John Hill, all Republicans. The lone Democrat was Wm. Gordon.

The church in Buffington was the Blacklick Methodist Episcopal Church, founded around the time the Blacklick Furnace was in operation, and it remained open to serve the spiritual needs of the residents. Much of the land surrounding Buffington was part of the land settled by William Clark in the late 1700's; parts of it were occupied in 1900 by the Bests and the Buterbaughs, descendents of Mr. Clark.

There was also a store in Buffington. Known proprietors were William Wilson, Thompson Clark, Morris Buterbaugh, and in 1913, J.M. Mack. The post office, named Buffington, was located in Buterbaugh's store. This store was located on the second sharp curve in the road after crossing the Wheatfield Bridge. In October 1900, Michael Misner of Rexis hauled lumber to Buffington for Morris Buterbaugh who was building a new house and storeroom. Isaac Skiles of Dilltown sold D.M. Buterbaugh an acre and a half in January, 1901. A month later, the new buildings burned; local residents thought it was the work of an arsonist. D.M. sold the property to his brother, another D.M., in 1903 for $3,000.

The post office was moved across the Blacklick and renamed Wheatfield. Located at the foot of Gas Center hill and Rager's Hollow-Pleasant Valley Road, the settlement also boasted a store run by Thomas Dodds. A single-lane iron bridge, built in 1882, connected Buffington and Wheatfield. Often called the Wheatfield or the Dilltown Bridge, it was a single lane, ninety-foot span. Original cost was $1,360. Later repairs cost $723.06. The bridge was painted by P.S. Allen in November, 1924 at a cost of $120.00.

Lackawanna #3

When Lackawanna Iron and Steel Company was scouting over its Blacklick holdings for suitable locations for its mines, coke ovens, and towns, they chose a site called Edwards' Flats for their #3 mine. It was located about one mile west of Vintondale. Here they built thirteen houses beside a small stream which flowed through the mine flat. Six houses were on one side and seven on the other. One house was a little larger, with interior sliding doors, a large furnace and radiators. This was to be for the superintendent, but was never used as such. Lawrence Dodson, whose wife died in 1907 at #3, had a boarding house there for several years; his daughter, Mary Duncan Nancarvis, kept house for the family. At one time, the Dodsons also lived on a farm above #3, next to Hoffman's, where Mr. Dodson cared for Burt Rogers' race horses.

Lackawanna #3, c. 1912. Two years before being reopened.
Courtesy of Wehrum Reunion Committee.

There was a school house at #3 for students from #3 and Rexis, it burned in 1910 and was reopened in the building behind the Eliza Furnace. Number 3 was closed in 1904 and remained idle for ten years. When the #3 plant was reopened in 1914, a large boarding house was built by the company; Anna Petrilla Dancha, who was the "missus" for several years, said at times, she had up to 200 boarders. She told the superintendent not to send her any more boarders, so he sent seven men and two truckloads of lumber to build a big dining hall and two trestle tables. The boarders bought the tablecloths.

One of Anna's boarders was homesick for Russia and wanted to return. He developed stomach problems, and Annie told him to get better because they would not let him return if he were ill. One day, when she sent someone up to wake him, he was dead. He left behind $2,000 and a note written in Russian that the money was to be sent to his wife in the old country. Mrs. Dancha hired a lawyer to send the money to the widow.

Also boarding with Mrs. Dancha was Charles Kovash, whose father and brother were killed in the 1909 explosion. Mr. Kovash stayed there until his wife arrived from Europe. Enroute, their child died of pneumonia, which had set in after catching measles. Mrs. Kovash was quarantined at Ellis Island, and her husband could not find her when he went to meet her. He had to return to Wehrum without her. When the quarantine was lifted, Mrs. Risko, a neighbor, went to meet her. The Kovash family eventually settled in Central City after the 1929 shutdown.

Lackawanna Coal and Coke Company's #3 mine, c. 1916, facing west.
Courtesy of Gloria Risko.

The Arminini Family also lived in #3 where Louie was born in 1917. Later, his father moved to the company farm above #3 and then to Vintondale in 1925.

In 1901, Lackawanna planned to make #3 the center of their coking facilities, which were to supply their new steel mill in Lackawanna, New York. By 1905, the coking plans were scrapped, and the ovens were never built. The #3 houses and mine buildings disappeared between 1930 and 1934, just like Wehrum's. The only traces of the mine today are the large washery pits and the rock dump which has largely been recycled.

Wehrum

Searching further downstream for an area large enough for a mine, a washery, and a full scale town, Lackawanna chose Miller's Flats, a large flood plain between Dilltown and Vintondale.

County Surveyor, D.L. Moorhead of Indiana, surveyed the tract and laid out a town plan of six streets, each of which was sixty feet wide. The streets were: "A", Broadway, First, Second, Third, and Fourth. Intersecting these streets were Walnut Street, Chestnut Street, and the Bowery, which separated the office and the hotel. The tract was cleared, and the construction contracts let. An electric line from Vintondale to Wehrum was surveyed in February, 1901; Nicholas Cameron was one of the locals who secured employment on that crew. A company sawmill, located at the mouth of Laurel Run near #3, was put into operation in March, 1901; Wellington Cameron, a local landowner near the mill, sold his timber rights to Lackawanna. A brickworks with an output of 20,000 bricks per day was erected.

The contract for construction of some of the houses was awarded to Blair Shaffer of Vintondale, whose large planing mill was to supply the materials. However, Shaffer had built only three houses when he died of a stroke in 1902. Shaffer's son Logan, a carpenter, continued to work on the construction crew; in one house, he left his signature on the rafters in the attic. His brother, Bert, hauled lumber to Wehrum and also helped wire some of the houses for electricity in 1909. Shaffer's daughter, Zorah, and husband, Clarence Huey, lived in one of her father's houses. Their daughter, Emma Huey Stinson, was born there in 1902, the third child born in Wehrum.

After Shaffer's death, two Blairsville contractors finished the houses, a Mr. Buchanon and J.P. Kennedy. In his letter book, Walter Scranton authorized Henry Wehrum on May 9, 1902 to accept J.P. Kennedy's offer to roof fifty houses with #2 American Black slate instead of hemlock shingles. Kennedy's bid was $25 per house. Scranton also accepted Kennedy's bid of $4,750 to build the office building on Broadway. A bridge was completed by the company across the Blacklick, and a train station was built in the summer of 1902. Eventually there were about 225 houses in Wehrum, each equipped with running water. The water was supplied by a resevoir on Rummel's run. In the 1920's, a larger dam was built above the earlier resevoir; a 50,000 gallon water storage tank set above the town adjacent to the Vintondale road. Before the 1909 explosion, only a select few of the houses had electricity.

Other locals who worked in Wehrum when she was in her infancy were Ben Williams, J.C. Nicholson and G.T. McCrea. Williams, who was a carpenter, moved from Martintown to Wehrum in November, 1901. Nicholson, a laborer at the dump site of the shaft, left in March, 1902 to return to farming. McCrea, who did some painting for Claghorn, contracted with Rochester and Pittsburgh Railroad for timber that same month.

Broadway

Wehrum, being a company town, had few businesses, and these were found on Broadway, the main street. At the foot of Broadway were the company office, doctor's office, post office, bank, barber shop, pool hall, company store and hotel. All of these businesses were connected by board sidewalks.

Wehrum Bank, c. 1913, A Street in background.
Courtesy of Thelma Kirker.

Harry Graham, whose father Allen Graham was a Ford agent and also had a farm near Strongstown, operated an auto repair garage until 1921 when he left Wehrum. During the winter months, he worked in the machine shop at the flat and did a lot of chores for Superintendent Frank Dunbar.

Wehrum, Chief Clerk's and later Superintendent's House on upper Broadway, burned in 1923 (originally built for Dr. Gile). Photograph taken c. 1918-1920.
Courtesy of Mable Davis Updike.

At the top of Broadway was a large Dutch colonial house which was occupied at various times by the company doctor, chief clerk, and later the superintendent. Originally built for Dr. Gile in 1901, the large cut stone retaining wall is all that remains of this house. Chief Clerk Philip Updike and his wife, former office secretary, Mable Davis, resided there until he was transferred to Buffalo in the early 1920's. Superintendent Richard Abrams and his family occupied the house in October, 1922. Abrams had moved to Wehrum from Heilwood the previous October as part of the Lackawanna-Bethlehem Steel merger. In February, 1923, the house burned to the ground with the loss estimated at $10,000. Only a few household items were saved. A smaller house, costing $20,000 was built on the same site. Several other large houses were located along Broadway at the top of the hill. These were occupied by companymen, such as yard boss Charles Kirker. Houses on First Street were also reserved for companymen.

Superintendent's house at the top of Broadway, built to replace the one that burned.
Courtesy of Mable Davis Updike.

Charles Kirker and Family, c. 1913. Upper Broadway. Left to right: Charles W., Thelma, Ida Kirker, Carl, Charles Kirker, Robert E.
Courtesy of Thelma Kirker.

Wehrum's Banks

The bank was chartered in May, 1903 as the First National Bank of Wehrum. The charter application was filed by Clarence Claghorn, M.S. Scott and E.J. Blackley of Wehrum; and D.L. Moorhead and John Scott, Indiana. The capital stock was valued at $25,000. In September, 1903, the bank purchased Lot #9 on Broadway for $300 from the company. The transfer permitted the coal company to repurchase the property if it were not being used for banking purposes.

Due to adverse economic conditions in 1904, Wehrum closed for a year. Anticipating a run on the bank, Cashier Charles Cunningham requested $10,000 from Indiana banks. The bank, by a vote of the stockholders, went into voluntary liquidation in October and closed its doors on November 1, 1904. Office fixtures and bank property were moved from the premises. Vintondale unsuccessfully tried to obtain the bank. Banking gossip also hinted that the Wehrum directors were to meet with some interested people in Marion Center concerning opening a bank there. The Marion Center National Bank opened on August 5, 1904, but its first board of directors did not include any Wehrum directors. As late as 1909, Nanty Glo parties were interested in obtaining the Wehrum charter.

The bank property was then sold for $2,400 to Christ Brixner in 1904 because LCC did not want to purchase the lot and building. In 1907, Brixner sold the site to M.B. Estep of Ebensburg for $2,500; Jennie Brixner bought the property from him in March of the same year for $1.00. This was one of the few privately owned lots in Wehrum.

A second bank opened in Wehrum in the mid-1920's. The photograph of the Blacklick Inn shows the bank next door; the caption stated that the small brick building was being remodeled for the bank. In its annual report, the National Bank of Wehrum as of December 31, 1928, had assets and liabilities of $238,390.29. In June 1929, the bank announced closure as soon as the patrons' accounts were adjusted. Many customers, like Mrs. Ann Longazel, did not receive the full amount that they had deposited.

M.W. Smith was cashier and postmaster at the time Wehrum closed. He accepted a position with Miners and Merchants Bank in Nanty Glo and became cashier in 1931. Smith was very unhappy in Nanty Glo, and after that bank failed, he committed suicide.

Wehrum Supply Company

The company store, known as the Wehrum Supply Company was chartered in July, 1906 for the purpose of "Buying, selling, trading in dry goods, groceries, hardware, boots, shoes, notions, clothings and furnishings, furniture and other similar articles." Shareholders were I.E. Lewis of Indiana with 47 shares. Three other Indiana residents each held a single share. They were: J. Wood Clark, Alexander Stewart, and D.L. Moorhead, county surveyor. The capital stock of the company was $5,000 divided into 50 shares at $10 per share.

**Wehrum Company Store, 1904.
Courtesy of Helen Bretzin Neumayer.**

Blacklick Inn

The company hotel, often called the Blacklick Inn, was a large two-story wooden structure with forty rooms. When it opened on March 18, 1903, Christ Brixner, a graduate of Cambria-Rowe business school and a former employee of Swank's Hardware in Johnstown, was the leasee. His application for a liquor license was approved in March, 1903, so the Blacklick Inn had the only legal bar in Wehrum. In April, 1909, he sold the lease, furniture and liquor license to Mr. Scrandi of Pittsburgh and moved to Johnstown. Barney Murray was another proprietor of the Inn.

The Blacklick Inn served the many salesmen who paid business calls at each mining town along the Ebensburg and Blacklick Railroad. Many company employees also boarded there. Harry Graham stayed there before his marriage to Maude Decker on January 29, 1918.

Known employees who worked at one time or other at the Wehrum hotel were Stella Wojtowicz, Anna Jacobs Sileck, and Jean Keim Colbert, all of whom became permanent residents of Vintondale. Sarah Jones McNaulty whose father, Alvan Jones worked in the office, worked as a waitress at the hotel before her marriage. Pay for hotel employees was $3.00 per week when Murray was the proprietor.

In April 1929, the hotel was operated by C.C. Crissey, who then bought the La Finke Hotel in Duncansville. Between April and the shutdown, John Galer, who had worked in the Vintondale powerhouse until the 1924 strike, became the last proprietor. When he took out the lease, he had been assured that the mines were going to stay open. Galer, originally from

Pardee in Union County, was an electrical engineer and invented a way to purify sulphur water to make it drinkable. After losing his job in the 1924 strike, money from his patent enabled him to live in Rexis several years without working and to purchase the Blacklick Inn. His family lived in the upstairs front rooms of the hotel. The hotel had several regular boarders, including two wealthy old ladies who dressed in their best clothes and pearls every night for dinner. Employees included upstairs girl, Phyllis Jones, and waitresses, Ann and Mary Shook. Galer leased the hotel, but owned the furnishings. He could have opened a speakeasy, but chose not to because of his family. When the Galers moved, they took everything they could and left a lot of things behind. Galer eventually settled in Nanty Glo where he died of a stroke two years later.

Helen Best Bretzin, Assistant Postmistress, Wehrum 1902-1909.
Courtesy of Helen Bretzin Neumayer.

Blacklick Inn, Wehrum, PA, c. 1925.
Courtesy of Gloria Risko.

Post Office

Wehrum's post office was originally chartered on February 14, 1902 with Clarence Claghorn as the first postmaster, just as he had been in Vintondale. James Cunningham, Claghorn's assistant, served as temporary postmaster until 1902 when Helen Best Bretzin of Buffington was granted a commission as assistant postmistress. Employed from about 1902 to 1909, Mrs. Bretzin boarded in Wehrum. She left Wehrum to take a position with Bell Telephone in Johnstown. Other known postmasters were E.J. Blackney and company chief clerks. Known post office clerks was Irv Burkhart and Mae Cornman. The post office was first located on the second floor of the company store and then moved to a small building to the north of the company store. The mail came in on the evening train and was locked in the company safe until time to sort it the next morning. The Wehrum post office was officially abandoned on May 31, 1930, and rural route mail service was provided from Vintondale.

Assistant Postmistress Helen Best, Clerk Ira Burkhart, 1904.
Courtesy of Helen Bretzin Neumayer.

Claghorn's Mansion

Dominating the entire town was the superintendent's house built by Clarence Claghorn in 1901 on top of the hill on the opposite side of the Blacklick, also known as "Super's Hill." Again, Claghorn built an English-style mansion which gave him a full view of the town. This large house reportedly had twenty-nine rooms, beautiful chandeliers and window seats and was surrounded by a large porch. The house could be approached by a winding drive or by steps which met a swinging bridge which crossed the Blacklick. There also was a coach house near the mansion which was used to store Claghorn's English style open carriage, his three bob-tail horses, and five greyhound coach dogs. Mike Warsko, his coachman, had a four-room apartment on the second floor of the carriage house. Located above the house was a concrete covered resevoir which provided running water for the mansion.

When Claghorn left for Tacoma in 1904, the house sat empty for a year or more. Later superintendents Johnson, Dunbar, Meehan lived in it. When Dan Galbreath was superintendent in 1917, he chose to occupy a double house on Broadway. George Lindsey resided in the Claghorn mansion, but Richard Abrams lived in the chief clerk's house on Broadway. The Claghorn mansion may have been rented out, but remained neglected during the 1920's and burned sometime between 1929 and 1930. It was eventually demolished. Super's Hill was also the scene of Ku Klux Klan cross burnings in the early 1920's.

Wehrum in Her Early Years

Buffington Township was a hotbed of activity between 1900 and 1903 with the sinking of the shafts and slopes for two mines, the extension of the railroad to Blacklick Station, the planning and construction of a town.

Before the 1904 shutdown, Wehrum was building a reputation as a violent village, much the same as Vintondale's. Between 1902 and 1904, the Indiana weekly newspapers regularly carried stories about the problems in Wehrum. As early as May, 1902, three Italian miners, Frank Berryman, Pellegrim Lassino, and Tony Domengo were charged with assault with intent to kill a fellow Italian. The men were arrested after a drunken brawl in which knives "figured prominently." The three were sent to the Indiana County Jail.

In July 1902, Wehrum was the scene of a cold-blooded murder at the boarding house of Mrs. Sarah Powers. The victim was Mrs. Powers' son, George Clinton, who worked at the washery. The accused murderer was John Dubie, a French-Canadian, whom the papers described as a "desperate character with a criminal record." Dubie ran a speakeasy at the boarding house. Dubie had boarded with Mrs. Powers at various times in other towns.

According to evidence given at the coroner's inquest the next day, Dubie and Clinton went to Vintondale to drink and returned around 11:00 PM. Dubie went into the kitchen, a separate building, and began abusing Mrs. Powers. She called for help; when Clinton arrived, he ordered Dubie out of the house. The men argued outside. Dubie was heard to say, *"George, you wouldn't hurt me, would you?"* before he shot Clinton in the chest. Dubie escaped through the sleeping arpartment. The two buildings were connected by a boardwalk. Other boarders had been awakened by the noise, but refused to track him unarmed. Vintondale's doctor, B.C. Gile was summoned, but it was too late.

The next day, a Vintondale policeman located Dubie at a big lumber camp near Vintondale, but he escaped. Even though word of the murder had reached the camp, no one attempted to detain him. A coronor's jury consisting of Bert Hoorhead, H.H. Green, G.R. Delamater, L.G. Gorsuch, Walter McLean, and E.J. Blackley concluded that Dubie had fired with intent to murder. According to witnesses, Dubie had shot a man in West Virginia, broke out of the Addison Prison, was wanted in Windber for selling illicit liquor, and was reportedly a member of a horse theft ring in Michigan. Mrs. Powers testified that she had objected to Dubie running the speakeasy, but he had threatened the lives of several people if he were turned in.

Since Wehrum had only a Justice of the Peace, E.J. Blackley, and no sworn law officer, the constables in Vintondale and Buffington Township were charged with locating Dubie. Jacob Whettling was then sworn in as constable. For a while, rumors circulated the Valley that Dubie was sighted in various places, but he was not apprehended and escaped to commit other crimes elsewhere.

There were two more shootings in February, 1903. J. Walton Thomas, 26, bookkeeper for the Vindale Supply Company was shot by Sterling Aiken, a black teamster, who had been fired two weeks earlier by Thomas for drunkeness and disorderly conduct. On a Saturday night, Aikens waited for Thomas along an unlit section of the boardwalk and shot him in the hip with a .38 caliber revolver; the motive was revenge. Found by those alerted by the shots, Thomas was taken to Memorial Hospital in Johnstown where the bullet was extracted.

About fifty white residents decided to take the law in their own hands and force the blacks to leave Wehrum. About twenty blacks who lived in a shanty on the Bowery refused to leave. So the self-proclaimed posse, on February 20th, tied heavy ropes to the structure and pulled it down. If the blacks did not peacefully leave, the posse planned "more vigorous measures."

That same weekend, there was a second shooting that was caused by a three-day old fight. Thomas Phillips, an Italian, shot fellow Italian Frank Beringer in the back and leg on Saturday. Phillips was taken to Blairsville by Squire Blackley on Saturday evening where he was met by Sheriff Neal and taken to the Indiana County Jail. Brought to trial in March on charges of assault and battery with intent to kill, Phillips was found guilty and fined $25 and court costs.

Because of three shootings in eight months, with two men still at large, law-abiding citizens were getting anxious and telephoned Sheriff Neal in Indiana to send someone to Wehrum. They believed that the presence of a law officer would "quiet the threatening element." The sheriff paid a visit to Wehrum and told the **Indiana Progress** that the town was abused by the public, that the lawlessness was exaggerated, and that Wehrum was not quarantined.

Sheriff Neal also said that the stories about the

whites driving out the blacks was untrue, that he found many industrious blacks working in Wehrum. Several undesirable persons who arrived in Wehrum were ordered to leave. Was the story of driving out the blacks really true? Most likely! Bill McMullen said that while he lived there, 1914-1924, Lackawanna had a policy of hiring no blacks.

In March, 1903, there was a case of threatened harm. Jules Walles, Italian miner, brandished a revolver and knife in the face of a Slavish girl identified as Miss Wolfinsky. Walles had been trying to court her for some time, but she refused his proposals partly because he was Italian. Wehrum's new police officers, John Smith and Officer Davis followed the same routine of taking the prisoner to Blairsville to meet Sheriff Neal.

Both the **Indiana Progress** and the **Indiana Weekly Messenger** carried glowing articles on Wehrum in their March 4, 1904 editions. Most likely, the reporters were invited by the company to take a tour of the town in order to improve the town's image.

Crime continued to plague early Wehrum. As the town was in the throws of a major shutdown, the town was rocked by an explosion. In November 1904, the post office was robbed of some stamps and money; the thieves blew open the safe with nitroglycerine. Edward Blackley was postmaster at the time. Helen Best, assistant postmistress, boarded near the post office and noticed that evening that someone carrying a light was roaming around in the post office. By the time it was investigated, the thieves had completed their work and escaped. The thieves were not apprehended, but it was rumored that they resided nearby.

ROADS AND BRIDGES IN WEHRUM

When the town plan of Wehrum was laid out in 1901, the road connecting Wehrum to Buffington paralleled the north bank of the Blacklick Creek. This road, which existed from at least 1845, led from the Pittsburgh pike to the Ritter Furnace via the Blacklick Furnace. In 1904, a petition was filed to lay out a road which connected R.W. Mack's farm and Dr. Gile's house in Wehrum. Dr. Gile, who had formerly lived in Vintondale, occupied the large house later known as the chief clerk's house at the top of Broadway. The road in question leads up the hill and passes the Russian cemetery. (That ground was purchased by the church in 1906.) The court-appointed road viewers were D.L. Moorhead, county surveyor; Bruce Wagner of Heshbon; and J.D. Drips of Armagh. Judge Harry White confirmed the petition in September.

On August 20, 1917, two petitions were filed in Indiana County Court; one to construct a road on the south side of the Blacklick and one to vacate the road on the north side. Judge Langham appointed Thomas Sutton, John Elrich, and D.C. Doty to view the petition to vacate the road between Wehrum and the Dilltown road. On December 3, 1917, the judge rejected the vacation of the road because it was the only road available to get to school for students who lived along it.

The petition to build a new road on the south side of the creek was approved because of public demand. The new road was to skirt the Wehrum ballfield, follow the bank of the creek, and reconnect with the old road at the Wheatfield Bridge. Viewers for the petition were Thomas Sutton, G.J. McCrae, and John Carson. A public bridge in Wehrum had to be constructed; in the meantime, Wehrum's superintendent Galbreath and solicitor John Scott agreed that the public could use their private bridge and that the company would make no claims for damages. Somewhere around 1917 or 1918, the wooden bridge built by Lackawanna washed out in a flood. The county agreed to put out bids for a two span concrete arch bridge to replace the 1901 wooden one. The arches on the new bridge were each sixty-eight feet long; the roadway was eighteen feet wide. However, due to the curves on the approaches, caution had to be taken when crossing the bridge. The construction bid went to M. Bennett of Indiana for $26,611.60. The county replaced this bridge and rerouted the road through the ballfield in the 1970's. Practically all traces of the old road on the north bank of the Blacklick have disappeared.

Wehrum - First bridge across the Blacklick, built by Lackawanna Coal and Coke Company. #4 mine in the background, c. 1905-06.
Courtesy of Gloria Risko.

Old bridge, after 1914. Note skeleton of washery in background. Bridge washed out in a flood around 1914 and 1917.
Courtesy of Wehrum Reunion Committee.

Wehrum: #4 slope and air shaft in the foreground, office at center right, freight station on left, concrete arch bridge, built 1918, Wehrum ballfield in upper left.
Courtesy of Mabel Davis Updike.

WEHRUM SCHOOLS

When the village of Wehrum was in the construction phase, the Buffington Township Board of Education was also initiating a plan to educate all the new students who would be arriving in Wehrum, Rexis and #3 mine. In 1901, six experienced teachers were hired at a salary of $35.00 per month. To handle this influx of students, a two-room schoolhouse was constructed at Lackawanna's #3; Rexis students walked to #3 for class.

Each year, the parents cleaned the #3 building the weekend before school started. On the Friday before school started in 1910, the schoolhouse burned to the ground. However, the extended summer vacation was short-lived as a school was hastily set up in the white frame building behind the Iron Furnace. The first teacher at the iron furnace school was Maude Altemus Graham; the second was Sue Altemus; another was Margaret Barclay. Later, when the building was converted into badly needed housing for Vintondale miners, students from Rexis and #3 walked or traveled the train to the Wehrum school. A special train pass was used by most students. After the loss of passenger service, students were transported by bus.

On May 21, 1904, the Lackawanna Coal and Coke Company granted a plot of land for the purpose of building a four-room schoolhouse to the Buffington Township School District. Instead, a four room wooden schoolhouse was built above Wehrum on the north side of the Vintondale road. Others have suggested that a boarding house was converted into a schoolhouse. The classrooms on the upper level were reached by a wooden ramp. Known teachers were Bertha Wagner, Miss Troxele, Mr. Jessie Hadden, Arthur Bracken, Ray Mardis, and Esther Kuchenbrod.

When Vintondale opened the first high school in the valley in 1916, some Wehrum students opted to continue their education there. They either walked to school or rode the train. Later, they could take a bus which dropped them off at the iron bridge, meaning that the students still had to talk over a mile to the school. The hardest part of that trek was the climb up Third Street in the winter.

The Wehrum grade school outgrew itself, and on November 21, 1925, the Buffington School Board accepted $10,000 from LCC for the plot of ground on which the old school stood. A new school had already been started on the lower part of A Street. The land for the new school was leased to the school district for $1.00 by Bethlehem-Cuba Mines.

The new yellow sixteen room school building was completed by contractor L.F. Swinter of Ebensburg in 1925; cost of the school was $50,000. Even though Buffington Township students were bussed in, the school was not filled to capacity. After Wehrum closed in 1929, the school remained open, and all students were bussed in from Rexis and the farms near the former village. Only three rooms were used for the fall term in 1930.

The school remained opened throughout the 1930's and 1940's, even initiating a hot lunch program in 1948. The cafeteria was located in the brightly illuminated basement; the walls were painted green and white. Flowered plastic curtains (very much in vogue at the time) were hung in the twelve windows. Mrs. Charles Madill was the supervisor; cooks were Frances Pluchinsky and Zalla George; helpers were Ann Adams and Jean Davis. Custodian was Sam Mack.

Students paid twenty cents a day for lunch; a weekly ticket was available. The cost of preparing the food was subsidized by federal surplus food and state aid of nine cents per meal. Sanitary Diary of Johnstown supplied the milk. In 1951, 46,892 meals were served by the staff. The cafeteria was also open to the public. Vintondale teachers often drove to Wehrum to buy hot, nourishing lunches. The author was also a patron several times while in grade school. In a child's eyes, it was a special treat to experience the choices of the cafeteria line.

In 1952, the Wehrum school had classes for grades 1-5. Arthur Bracken was the principal. Education was not Mr. Bracken's first choice of occupations. He worked in the Wehrum mines until the day his brother Elmer was killed in a rockfall. Arthur vowed if he made it out safely, he would go back to school. His first teaching jb was at the Wagnel School in Rager's Hollow. When he was hired by the Buffington Township School Board to teach in the new Wehrum school, he asked for an upstairs front classroom so he could keep an eye on the children at recess. Mr. Bracken's faculty included:

First Grade - Mrs. Ina McKnight and Mrs. Mary McFeaters.

Second Grade - Miss Mildred Findley and Miss Ethel Stewart.

Third Grade - Mrs. Patricia Kittka and Mrs. Florence Stiles.

Fourth Grade - Mrs. Viola Empfield and Mrs. Dorothy Strong.

Fifth Grade - Mrs. Ruth Findley and Mrs. Bernice Dixon.

Wehrum School, 1928
ROW 1: Marion Kuckenbrod, Freda Wagner, Myrtle Simmons, Doris Hunter, Charlotte McCullough, Lois Ondrezack, Nellie Peddicord Madill, Pearl Mack, Mary Anderson Smith, Peggy McGinnis, Bertha Anderson.
ROW 2: Richard Starr, Stella Cunningham, Margaret Widmar, Alverta Kinter, Anne Dutko McGaughey, J. Arthur Bracken-Teacher, Josephine Verba Marcus, Irma Boring, Matilda Davis Hughes, Nora Stutzman, Eva Starr Misner, Richard Mazey.
ROW 3: James George, Raymond Empfield, Roy Duncan, John Dutko, Steve Rosko, Boyd Boring, John Boychuck, John Rosko, Lester Clouser, Bernard Hunter.
Courtesy of Anne Dutko McGaughey.

One of the teachers, Katherine Jones of Armagh wrote a history of Wehrum for her Masters of Education Degree from Indiana State College in 1963. Six teachers who taught at the Wehrum school in the mid-1920's sent each other a round robin letter and held a reunion when possible. Included in this group were: Mrs. Hazel Peters, Ila Diehl Brett, Mrs. Anna Rehn Kettles, Mrs. Ysable Bostick Frazier, Elizabeth Beechey, and Mrs. B.W. Cassett.

The Wehrum Elementary School was abandoned in the 1960's when the United School District consolidated its elementary schools into a new building across from the high school on Route 56 north of Armagh. The Wehrum school was sold and dismantled. On the site today is a sawmill.

WEHRUM CHURCHES

At one time or another, Wehrum could claim three churches: Russian Orthodox, Roman Catholic, and Methodist Episcopal. Sts. Peter and Paul Russian Orthodox Church was organized in Wehrum in 1902 or 1903, its location unknown, but probably was on A Street near the hotel. The church, however, did not have a charter. When Wehrum closed in 1904 with little hope for reopening, the congregation transferred to Vintondale. No evidence thus far has surfaced as to whether the church remained opened while Wehrum was in shutdown.

In October, 1906, the congregation purchased three quarters of an acre from R.W. and Eliza Mack for $200.00. This property was used as a cemetery until 1930 and is accessible by taking the township road at the top of Broadway. It is about one-half mile up over the hill from the ghost town. Efforts have been made by the author's history club and local Boy Scout Troops to keep the overgrowth to a minimum. In the appendix is a translation of the tombstones and a partial list of those known to be buried in the cemetery.

In Vintondale, parishioners of Sts. Peter and Paul purchased a lot from Barker Brothers and built the present church, which was dedicated in 1907. After Wehrum reopened, residents of the Orthodox faith traveled to Vintondale to attend church.

Methodist Episcopal Church - Wehrum.
Courtesy of Wehrum Reunion Committee.

Methodist Episcopal

A Luthern church was constructed in 1903 on A Street, but it is not known whether this church reopened in 1906. In 1913, this building was sold to the Methodist Christian Endeavor for $75.00. The church was then turned over to the Methodist congregation. Serving the church as pastors up to 1913 were: Dillon, N.H. Nevins, Maddocks Andrews, Carrol, and Samuel Hill. One pastor, Rev. Shaffer, traveled from Homer City for services. He usually had dinner at Huey's home before traveling back.

Trustees in 1913 were Dr. W.H. Nix, Charles Kirker, and Jesse Craig. When Alban Jones moved to Wehrum from Heilwood, he joined the church and also gave music lessons to Wehrum students. Eleven residents claimed membership in the church. Sunday school superintendent was Mr. Smead; enrollment was fifty students. Wehrum was part of the Belsano Charge, which also included the Methodist churches in Strongstown, Belsano and the Blacklick Church at Buffington.

Reverend Clarence Bennett was present at the last services in September, 1929 and helped finalize the closing of the church. Pastor at that time was Rev. W.E. Seiss. To fulfill the spiritual needs of Methodists in the surrounding area, the Blacklick Methodist church was reopened after being closed for many years. Before that, it was opened for special services, such as a reunion held in August, 1909. Attendance at that event was estimated at 1,000.

Rev. Bennett said that some of the church furnishings were given to the Belsano Methodist Church, which also purchased several of the Wehrum houses for $50 each. These were rebuilt at the church camp in Belsano. The piano and chairs were donated to the camp for use in the tabernacle. Zora Huey received the church's organ; she later traded it in for a larger one.

St. Fidelis

According to the records of the Diocese of Pittsburgh, the parish of St. Fidelis was not officially formed until 1912 when the first mass was celebrated in the company recreation hall. At that time, giving a church the name St. Fidelis was unusual because it was a name closely associated with the Germans and the Franciscans Friars. Perhaps a church was consecrated by Bishop Boyle, who frequently named a church after the saint whose feastday was being honored that day.

Eventually, St. Fidelis occupied a church across the side street from the hotel; this building may have been used earlier as Sts. Peter and Paul Russian Orthodox Church. In any case, the property belonged to Lackawanna Coal and Coke Company and not to the Diocese of Pittsburgh. Prior to 1912, pastors of parishes in Vintondale and Nanty Glo reportedly said mass in Wehrum, but hard evidence supporting this contention is scanty. The new parish, located in IndianaCounty, was part of the Pittsburgh Diocese and later the Greensburg Diocese and encompassed Buffington, East Wheatfield and the eastern half of Brush Valley Townships.

In spite of being located in the Pittsburgh Diocese, baptismal and marriage records at Holy Family Roman Catholic Church in Colver confirm that Fr. Ignatius Herkel of the Johnstown Diocese, first pastor in Colver and its mission in Vintondale, administered the Wehrum church from 1912 to July 31, 1915. He was succeeded by Fr. Edward B. Daly (d. 1970) who served Colver, Vintondale, and Wehrum until August 25, 1918. During the tenure of these two priests, 144 baptisms were performed at St. Fidelis.

After Fr. Daly was transferred in 1918, St. Fidelis returned to the jurisdiction of the Pittsburgh Diocese and became a mission of St. Bernard's in Indiana until 1923. The pastor in Indiana at the time was Fr. Neil McNelis; his assistants were usually sent to say mass at these mission churches. Parishes formed at about the same time as Wehrum's were Assumption in Ernest, St. Louis in Lucernemines, and St. Francis in Graceton. Indiana priests known to have offered mass in Wehrum were Fr. Michael Barry in 1919, and Fr. Amelio Farri, who eventually became pastor of Assumption Church in Ernest and served there for over thirty years.

Sts. Peter and Paul Russian Orthodox Greek Catholic Cemetery, c. 1922.
Courtesy of Wehrum Reunion Committee.

In February, 1923, Bishop Boyle of the Pittsburgh Diocese sent informational questionaires to all churches under his jurisdiction. The response sent to Bishop Boyle indicated that St. Fidelis was unattended in 1923 and that the parish records from 1912 were in Indiana. However, a check of the records show that the Wehrum baptisms were not registered in Indiana until 1919.

From 1923-29, the parishioners of St. Fidelis were ministered to by the Franciscans priests of Loretto. Merle Hunter, who had moved to Wehrum when he was eight, said he, Mike Grosik, and John Smith, Jr. served as acolytes at Sunday mass at the church. John's father sang in the choir. Mr. Hunter said that two of the Franciscans who traveled by train to Wehrum on Sundays were Fr. Paschal Lauffle and Fr. Bernardine Dillon. Fr. Paschal had also occasionally served as a substitute in Colver for Fr. Daly.

Sts. Peter and Paul Russian Orthodox Greek Catholic Church Cemetery, Wehrum, PA, 1986. Courtesy of Diane Dusza.

When Wehrum closed, St. Fidelis was torn down and the lumber used to build several small structures near the Ebensburg airport, one of which may still be standing.

Constructing an accurate history of St. Fidelis has been difficult. The Pittsburgh Diocese has sketchy records, based mainly on those sent from St. Bernard's in Indiana. The Greensburg Diocese now has jurisdiction over the territory which was once St. Fidelis Parish, and the source of its records on St. Fidelis also come from St. Bernard's. Records from the Altoona-Johnstown Diocese, if they ever existed, must have been discarded. Local parish records have aided in piecing together an incomplete and sometimes conflicting history of St. Fidelis.

LEISURE TIME IN WEHRUM

Wehrum, like any typical company mining town offered a variety of leisure activities for its residents. As early as 1903, the Lackawanna Coal and Coke Company was represented in the June 23rd Industrial Parade of Indiana. Displayed were mining machines followed by a contingent of Wehrum miners.

A very popular activity were the dances, held almost every Saturday night at the Blacklick Inn. Tables in the dining room were moved back against the wall to provide a dance floor. Usually the music was provided by Wehrum's own Kaltreiter's Orchestra. These dances usually attracted the young men from Vintondale.

The Blacklick Inn had the only liquor license in Wehrum, but alcoholic beverages were available in many speakeasies in town. With large Italian and Hungarian minorities living in Wehrum, the demand for beer and wine was high, especially on payday weekends. Many times, the stakes for the card games were drinks of illegal liquor. For weddings and baptisms, the families often celebrated with liquor flavored with anisette or peppermint. However, Lackawanna and later Bethlehem Mines must have regulated either the drinking or the press a little more effectly that Vintondale because after 1904, one does not read stories or brawls in Wehrum caused by drinking. During Prohibition, there were many families who manufactured alcoholic beverages and made a lot of money running speakeasies in their houses. See the Bethlehem police reports in the appendix for arrests for violating the Volstead Act.

Corn roasts, Labor Day and Fourth of July picnics, and church socials were other forms of entertainment. The pie social was a popular fundraiser at the Methodist Episcopal Church. For the town picnics, the company provided a picnic ground and playground near the hotel and office. A moonlight picnic in September, 1909 was described in the Indiana papers. Included was a corn roast, racing and climbing apple trees. Guests at the picnic included Annie Riley, Winifred McDermott, May Cornman, Edna Casset, Anna Duncan, Mary Hewlett, Twilight Gaster, Mae Tobin, Mable and Olive Piper, John Tobin, Tom Gaster, J.S. McQuesney, Alvah Brickley, Charlie Hogue, Ira Ben and Lee Casset.

The company also built a recreation hall on a side street behind the hotel. Its greatest use was for basketball practices and games, but volleyball and boxing was also available there for the townspeople.

Baseball

A standard feature of any company town was the baseball team. Many team members were recruited for the baseball skills, not their mining abilities. In Wehrum, the ball diamond, complete with bleachers and backstop, was across the Blacklick near the road which led to the superintendent's house. Baseball games were a vital part of the town picnics on holidays. On Labor Day 1909, the Wehrum Bulldogs played Nanty Glo in a double header. Wehrum's pitcher, Grow, had eighteen strikeouts in the first game. Nanty Glo took the first game, 7-2. The second game was called so that the Nanty Glo players could catch the train home. Score for the second game was 4-4. In 1909, the Wehrum team consisted of: Watson, shortstop; Anderson, centerfield; Grow, pitcher; Gaster, first base; Smart, second base; Noke, centerfield; Littleworth, right field; Stutzman, left field; and Johnson, third base. Several members of the team, Johnson, a foreman, and Wurm, a pumpman, were injured in the 1909 explosion. A.D. Raymer, a pitcher, was killed.

Wehrum also had a poolroom for the entertainment of the gentlemen. John Smith, mine foreman, operated it in the old bank building until the mid-1920's when the bank reopened.

Boccie, or Italian lawn bowling, was also a popular sport among Wehrum's large Italian population. Many of the yards on Third and Fourth Streets had boccie courts. Dorothy and Chincy Milazzo said that when they moved into Wehrum in 1920, there was a court already constructed between their house and the one above them.

Wehrum's Bands

Another form of company-sponsored entertainment was the band. During its three decades of existence, Wehrum had several bands, and the company provided a bandstand for concerts.

An organizational charter for the Wehrum Citizens Band Association was filed at the Indiana County Courthouse in 1908. Purpose of the Association was to promote advancement of instrumental music and furnish music to the general public. Its playing members were to maintain a high professional standard. Subscribers to the Association were S.N. Hewlett, E.P. Goodwin, J. Beale, G. Pendred and T. Hogarth. The first year officers were S.N. Hewlett, president; J. Beale, vice-president; J.H. Dyer, secretary; and G. Pendred, treasurer. Trustees were J.E. Elder, William Burns, and John Kessler. Band leader was J. Hogarth. The Association issued no capital stock nor assessed any annual dues. Its yearly income was not to exceed $2,000. Later, because the band had many Italian members, including band director, Patsy Codispoti, it was nicknamed the "Italian Band." Codispoti also composed and arranged music. While in Wehrum he wrote the "Richard Abrams March." After the Wehrum shutdown, he moved to Vintondale where he organized another band.

Musical instruments were at a premium in the 1920's. John Orris purchased a trombone from Ann Piper Wolf's brother. When John left Wehrum, his twin Mike learned to play it. Mike joined the Wehrum band, and after the shutdown moved to Pittsburgh where he played a tuba in a Hungarian band.

**Lackawanna Band, before 1924, Wehrum.
Courtesy of Tom Shook.**

Bethlehem Mines Corp Band

BACK ROW: Fourth from left, Joseph Shook.
NEXT TO LAST ROW: Third from left (holding saxophone), John Shook. Fourth from left, Frank Shook. Fifth from left, Andrew Shook (died July 31, 1985). Band Director: Patsy Codispoti. Picture taken around 1928.

Courtesy of Thomas Shook.

Also playing the Wehrum band were several of the Shook brothers. Joseph played the trombone; John was on the saxophone; Frank and Andy played trumpet in the twenty-five member band.

When the coal companies changed hands, so did the name of the band. The Lackawanna Band became the Bethlehem Mines Corporation Band.

WEHRUM ANECDOTES

In the course of conducting interviews, the author heard several stories which were worth repeating. Thelma Kirker, who lived in Wehrum from 1912 to 1915, vividly remembered life in Wehrum. Her father, Charles Kirker, was yard boss, and the family lived in one of the large houses at the top of Broadway. Thelma remembered the time Dr. Nix, company doctor, decided to wash his buggy in the Blacklick by driving the horse and buggy into the creek. The horse stepped into a deep hole and drowned. Dr. Nix was unable to save the horse because it was still hitched to the buggy, and there was no one nearby to assist him. Until he bought a new horse, Dr. Nix borrowed Kirker's horse, Dan, to make house calls. This occurred the same summer that the washer burned.

On the afternoon of August 14, Dr. Nix was enroute to Vintondale to make a house call. When he heard the whistle, he rushed back to Wehrum and tied the horse at the hotel. Mrs. Kirker, who had been raised on a farm, knew that the horse would bolt into the fire. She told the children to stay put and headed down the boardwalk carrying a blanket and a towel. She dried off the horse and put the blanket over his head, talking softly to him the entire time. Then she led him home and put him in the barn. When the fire was extinguished, Dr. Nix and Mr. Kirker searched for the horse and could not find it. They feared that Dan had run into the flames. It was not until Kirker reached home that he discovered that his wife had saved the horse from certain death.

Annual Gypsy Migration

Each spring, a caravan of gypsies passed through Wehrum enroute to parts unknown. In 1912, they were permitted to stop in Wehrum to buy groceries, etc. Mrs. Kirker had her doors and windows locked for fear of the gypsies. A large gypsy woman pounded on the door, but Mrs. Kirker would not open it and asked her what she wanted. The woman said that she wanted the piece of yard goods that she had in the lower bureau drawer of the guest bedroom. Mrs. Kirker replied that she did not any material, and the woman replied that Mrs. Kirker had it and she wanted it. When Mrs. Kirker looked in the drawer, she discovered that she did indeed have a five-yard piece of cloth which she purchased at the company store and had forgotten about. She handed it to the woman and locked the door again. When her husband came home from work that evening, she told him what happened. He said that he would make sure that the gypsies would not be allowed to stop in Wehrum in the future.

In 1913, when the gypsies came through, Mrs. Kirker and Mrs. Donahue were out of their houses, fetching water from the spring, about three blocks from Broadway. Mrs. Donahue left her three month old baby in the crib. The oldest Kirker boy saw the gypsies coming up Broadway and hid the younger two siblings in the bedroom under three blankets and then ran to get the women. Because of the episode of 1912, the women dashed back. They thought that gypsies were clairvoyant and would steal the baby. However, their fears were unjustified, and all the children were safe. The next year, the gypsies passed through Wehrum without stopping.

Actually, one method the gypsies used to obtain information was by talking to the children about the town and their families. Mary Kreashko Averi Smay said tht she was offered a ride in a gypsy wagon. The gypsies did not harm her, but asked her all sorts of questions about the town and its residents.

**Wehrum washery after August 1914 fire.
Courtesy of Lovell Mitchell Esaias.**

One gypsy stopped at the house of Joseph Milazzo and said that she would, for a price, tell Mrs. Milazzo, who was sick at the time, if and when she was going to die. Mr. Milazzo's reaction was swift; he pulled out a shotgun and told the gypsy he knew when she was going to die if she did not leave immediately.

WEHRUM TOWN TRAGEDIES

In addition to the accidents which occured in the mines, Wehrum was also party to several tragic incidents over the course of her three decades. Reports of the following accidents and fatalities were found in various local newspapers.

In March 1902, Hiram Kemry was fatally injured

when he was hit by a steel rail while helping to unload a car of rails. Kemry, 37 years old, was survived by a wife and five children. He was buried at Wesley Chapel.

In September 1913, twelve year old Raymond Congnara, son of Mr. and Mrs. Daniel Congnara, died in Mercy Hospital in Johnstown. A .38 caliber revolver discharged accidentally, fatally wounding the boy.

Mrs. Elizabeth (Howatt) Lindsey, wife of superintendent George Lindsey, burned to death in the family car following the annual Halloween dance in 1919. The car stalled going up the hill to the house. Mr. Lindsey went to the house to get some gas. The gas tank was located beneath the front seat, and Mrs. Lindsey held a lantern while he poured the gasoline into the tank. The fumes ignited and caught her clothes on fire. Mrs. Lindsey died November 2, 1919 at Memorial Hospital.

On Friday, January 4, 1923, Cella de Guiseppi died at home of burns of the head and arms. Her mother, a widow, was washing the children's heads with kerosene. She left the room, and Cella put a match to her head. Mrs. de Guiseppi was burned on the arms when she attempted to extinguish the fire and was in serious condition in Mercy Hospital.

In September 1926, Mike Popp, while drunk, attempted to board a train and was thrown under the wheels.

As early as 1926, Route 22 had the reputation of being a killer highway. In November, a bus carrying twenty-seven men from New York City crashed at Thomas Rager's garage. Killed outright was Martin Regan, 23, of Tarentum; James Murphy of Jamaica Plains, Massachusetts died later at Memorial Hospital. The men had been recruited by Bethlehem Mines to work in Wehrum. These two fatalities raised to seven the total killed in nine days on the William Penn Highway.

Pete Kerekish, 31, died in a hunting accident on November 22, 1928. He was hunting rabbits near Vintondale and shot himself in the thigh when he stumbled, and the gun discharged accidentally. His dog ran to a nearby farm house while Kerekish dragged himself 1,000 feet. The accident happened about 9:00 AM, and Kerekish was not found until noon. Taken to the doctor's office and then by Ondriezek ambulance to the hospital, Kerekish's leg had to be amputated. He died in the operating room about 8:50 PM. Kerekish was survived by his wife and four small children. Funeral services were at the Russian Church in Vintondale. Just the night before, Kerekish had a narrow escape in a wreck while on hootowl. He had escaped with his bucket and lamp smashed.

Wehrum continued to be jinxed even after the closing. Just as workemn were killed constructing the shafts and mine buildings, one was killed demolishing it. At the #3 power plant, Walter Furman of Johnstown, 25 years old, had been cutting steel girders with a blow torch. He was killed instantly when he fell twenty feet off a wall; Furman's neck broke when his chin hit a piece of metal.

Diptheria stalked the Wehrum area between 1937 and 1947, tragically affecting the Mardis family. In December, 1937, two year old Wesley Mardis, son of Mr. and Mrs. Ray Mardis, died of diptheria. Because of the rash of diptheria cases in the area, the Wehrum school closed for the week. In February 1947, Gerald, Richard, and Philip Mardis, sons of Robert Mardis caught diptheria. Ruth Gibbons of the Wehrum area also had the disease. Gerald Mardis, 14, died at home on February 6th. His funeral was private, with only adult members of the family attending.

YA DONE GOOD BOY!

As with every small town, Wehrum produced her hometown heroes who became successful businessmen, teachers, executives, inventors, etc. The following are just a few of Wehrum's successful native sons, several of whom were very proud to have lived in Wehrum and saw the town as a very special place in which to grow up. In addition to producing a county commissioner, Joe Roberts, the town produced other notables.

George Lesak

George Lesak moved to Wehrum around 1919. His father had been killed in the mines at Denisonville in 1917, and George's mother, a widow with four children, married Mr. Datsko, a widower with three sons. While in Wehrum, George delivered the Sunday newspapers, which arrived by train. He peddled the Pittsburgh, Johnstown, and Philadelphia papers, in addition to the **Grit**. The newspapers arrived in bundles, and George loaded them into a wagon and walked up and down the streets selling them. According to Mr. Lesak, the foreign miners rarely bought weekly papers, but did purchase the Sunday edition. Later in the week, the housewives, after scrubbing their kitchens, spread layers of newspapers on the floors to keep them cleaner longer. Newspaper was also used as shelf liner. Very few daily papers were sold in Wehrum, but George did drop off a few at the hotel.

While in Wehrum, George participated on the first aid team, but was too young to be on the baseball team; he was permitted to warm up the players. Later, he played ball for the Garden City industrial team in Michigan.

At the age of sixteen, George set out on his own. First, he worked as water boy on the Twin Rocks/Vintondale road paving project. Then he lived in Colver where he worked in the mines for six months. Unhappy as a miner, George borrowed $20.00 and headed for Detroit. On the way, he stopped in Youngstown, Ohio and worked briefly in a grocery store. While in Detroit, he worked at several jobs at the same time and still found time to attend the Henry Ford Trade School. Mr. Lesak said it took him eleven years to secure the education that he wanted. He acquired a car dealership, sold it and "retired" in 1944. When he returned to the Ebensburg area, he opened another Ford dealership which he later sold to Frank Castelli. In the meantime, he also organized the Stevens Manufacturing Company, located on the New Germany Road in Ebensburg, which built government trailers. When he sold the firm to Penn Metals, there were 300 employees. Today, he is actively involved in Kasel Manufacturing Company which fabricates trailers in the old state highway shed on the eastern end of High Street in Ebensburg. Mr. Lesak was a first time guest at the Wehrum Reunion in 1989.

Stevens Manufacturing Company, March 10, 1967.
Courtesy of George Lesak.

Tom Longazel

Tom Longazel was the son of Mike and Ann (Prebola) Longazel, longtime Wehrum residents. Ann's family arrived in Wehrum in 1909, shortly after the explosion. Her father was a pumpman. Before her marriage in 1913, Mrs. Longazel worked for chief clerk Chase, the superintendent's wife, and at the hotel. Mike Longazel, a motorman and later an electrician, played baseball for the Wehrum Bulldogs. In 1922, he went out on strike and found work in Pittsburgh. The family remained behind in Wehrum, and Mr. Longazel returned home on weekends. The Longazels remained after the shutdown in 1929, but had to move out after the water and electricity were turned off. Mr. McMullen gave them a couple of days notice to leave. The family later lived in Beaverdale, Pittsburgh, and Nanty Glo. After Mr. Longazel's death in 1945, Ann Longazel worked in the kitchen at Memorial Hospital in Johnstown and at the State School at Ebensburg.

Their son, Thomas, served in World War II and the Korean War and worked as a marketing representative in the insurance business. But through the years, he remembered all the folk tales and mining lore that his father had told him. In 1984, his first folklore book, **The Magic Reed of the Woodpecker**, was published by Dorrance & Company of Bryn Mawr. The story is about a magical blade of grass which can be used to open doors and the man Juddac who discovered the secret of the blade of grass. Many references to Cambria and Indiana Counties can be found in the book. Longazel's book is available in area public libraries and at the Cambria County Historical Society.

Emmett Averi

The author is finally fulfilling a promise to Emmett to put his story in the book even though it is a bit too late for Emmett to personally read it. Emmett was born and raised in Wehrum, son of Guerino and Virginia Averi. When Emmett was nine, his mother gave birth to a baby boy who died the same day. Mrs. Averi died two days later. Between 1918 and 1923, the Averis lost four boys and a girl, several to the Spanish influenza. In 1926, Guerino Averi moved to Vintondale and opened a store, first in the Shaffer building on lower Main Street and then in the former Cupp Store uptown. Emmett claimed that his father had first class information that Wehrum was going to close. From 1926 to March 4, 1930, Emmett lived in Vintondale. then he chose to go out on his own, even though he had only gone as far as eighth grade in school. He worked for N. Cavalle & Brother, a wholesale grocer, then for J.P. Kingston, the largest beer distributor in the area. On July 1, 1938, he was to marry Myrtle Albarano, but the wedding was postponed because Emmett's best man, his brother Natalie Averi, had been killed in the mines in Beaverdale two days earlier. Emmett and Myrtle were married on November 20, 1938.

In 1940, he bought out Whistle-Vess Distributing Company in Johnstown. The soda pop was shipped by rail from Connellsville, and the beer came from Connecticut. Emmett had to close the distributorship during the war because of the draft. He was classified 4F and was sent to the Fairless shipyard in Baltimore for a year. Then he returned to Johnstown to Bethlehem Steel where he worked as a painter, armature cleaner, and crane operator. After the war, Emmett applied for a beer distributor's license and ran a distributorship in Lilly until he retired in 1976. Until his death in April, 1986, Emmett was an annual visitor at the Vintondale Reunion.

Christopher Frazier

Christopher "Lefty" Frazier was the son of Chris Frazier, who operated a boarding house in Wehrum after suffering burns on his legs in the 1909 explosion. Lefty supposedly earned his nickname from his baseball playing days for the Wehrum Bulldogs. Lefty, who worked in the office, married Ysabel Bostick, granddaughter of Edward Findley of Vintondale and niece of Wehrum office secretary, Verna Findley. Before her marriage, Mrs. Frazier taught for two years at the Wehrum elementary school. The Fraziers settled in Williamsport where Lefty invented the plastic shoe polish top. After receiving a patent for his invention, Frazier opened the Plastovac factory. Eventually, he sold out for two million dollars; his son, Thomas, still works as an engineer at the factory.

Angelo Milazzo

There were two Milazzo families in Wehrum, Bennie's and Joseph's. Each family had a son Angelo. Like many of the mining families, the Joseph Milazzos lived in many of the mining towns of the Blacklick Valley, including Clover and Wehrum. Their sojourn in Wehrum was from 1920 to 1924 where they lived on either Third or Fourth Street above the Shooks. Across the street was their uncle Bennie and the Marpellis. As a youngster, Angelo traveled by train to Blacklick to learn barbering. After the family left Wehrum, they settled in Old Conemaugh Borough in Johnstown, but Angelo went to Pittsburgh where he continued in the barbering business. He had unsuccessfully attempted to obtain a permit to open a barbershop at the Pittsburgh Airport when it first opened, and at the time of his death was part owner of the Pittsburgh School of Barbering.

AND THE BOMB DROPPED!

In the 1920's, the sale of the Lackawanna Steel Company to Bethlehem Steel Company topped the interest list of the residents of Wehrum. Life went on as usual, except for the interruption of the 1922 strike. The baseball teams matched their skills against the neighboring mining town rivals. The community band entertained at concerts and accompanied funerals corteges to the cemeteries. First aid meets became more competitive as Bethlehem mining teams competed in state and national contests. Personnel changes were made; most of the new companymen in town had worked at either Heilwood or Slickville. and on rare occasions, even Charlie Schwab paid a visit when he was staying at his summer home in Loretto, opening the Cambria County Fair, or attending the first aid meets.

Then on June 1, 1929, Bethlehem dropped the bombshell. The Wehrum mines were closing indefinitely. The previous day of operation, May 31, was the last day for Wehrum mines; employes were advised to seek employment elsewhere. Some were offered jobs at other Bethlehem mines at Ellsworth, Heilwood, or Slickville. For instance, Jesse Craig, engineer at #3, was offered a job in West Virginia. He went there to check on working conditions and declined the offer. Three hundred men were immediately out of work. Since Vintondale was working fairly steady at the time, many showed up at the Vinton Colliery employment window the very next day. Families were permitted to remain temporarily as long as they paid the rent.

Wehrum's few businesses wrapped up their operations and closed the doors. The bank wrapped up its affairs and closed its doors forever. James Beechey, who had operated a barber shop in Vintondale for ten years before moving to Wehrum, moved his shop to Nanty Glo. He died there in 1932. His daughters were teachers; Mary taught in Vintondale, and Elizabeth in Wehrum and then in Nanty Glo. John Smith obtained a job with the Vinton Colliery Company and moved his wife and ten children to Vintondale, occupying the house now owned by the author's father. Mr. Smith died in a freak mine accident in 1935.

Some families, including the Craigs, remained in Wehrum to allow their children to finish their senior year of high school in Vintondale; final eviction notices were given to the last fifteen families in April. All houses were to be vacated by May 15. Jesse Craig and his family moved to Indiana in August. At that time, the houses were still furnished with running water. By September 39, 1930, only five families remained. One was Bill McMillen's father who remained on as watchman. His house on Broadway was one of 4 or 5 houses out of the 225 in the once thriving village called Wehrum. The **Nanty Glo Journal** reported, "*The once happy village presents a deserted and pathetic appearance today.*"

Bethlehem Mines offered the houses for sale for $50.00; the purchaser had to demolish the structure for the lumber. Many houses and garages in the area have been built from Wehrum houses. Mike Farkas, Vintondale merchant, had a moving van and did a brisk business in Wehrum. He also bought some houses and used them for building materials to repair his various properties in Clymer, Nanty Glo, and Vintondale. Garfield McGinnis demolished Wehrum houses and used the lumber to build three buildings in Gas Center. He also helped tear down the new superintendent's house at the top of Broadway. Claghorn's old mansion burned down after the town closed; the remains were torn down. McGinnis continued to live in Wehrum during the 1930's. Bill Conrad, who served as bus dirver and janitor at the school, occupied what became Wehrum's last house. After he moved to a farm above Wehrum, Robert Mardis of Belsano moved into the house on January 1, 1943. Wehrum's last house is now owned by John and Gloria Mardis Risko.

In addition, Bethlehem contracted salvage companies to tear down the various mine buildings. Harry Peddicord, now of Indiana, helped take the pumps and rails out of #3 and #4. By 1934, almost all traces of mines #3 ad #4 and the village of Wehrum were gone. Some of the coal on the south side of Wehrum's holdings were supposedly mined by Bethlehem's Rosedale mine. In 1940, Bethlehem Mines sold the Wehrum property to Meade Cauffield.

Wehrum's foundations slowly disappeared from view. Brush and then trees began to sprout among the stones. Within several years, all traces of the town, except Mardis' house and the school, disappeared. For years, the mine shafts were closed off only by wooden fences, but with the last fifteen years, the shaft and slope openings have been sealed with concrete. But the ruins continue to attract visitors. Within three months in 1957-58, two deer fell into the washery pits at #3. In early November, the author and her sister, Diane, were riding bikes near Three Springs and noticed a deer desperately trying to jump out of the pit. After summoning help, Raymond "Frenchy" Hugar and Dennis Toth climbed down and roped the six-point buck. Lloyd Williams and Bill Toth pulled it up and freed it. The following February, a 125 pound doe was found in the pit by Charley Lynch and Joe Pluchinsky. It was rescued by Everett Miller and Andy Jacobs, Jr. who climbed down and tied the deer's feet. Helping to pull out the deer were Red Frantz, Emil Delosh, Ed Shestak, George Pytash, Jr.

The physical aspects of Wehrum are gone, but the spirit of the town has lived on in its former residents, enough spirit for a group of them to meet each year on the first Sunday of August. Early reunions were held at Jondora Park in Munday's Corner; in Youngstown, Ohio; Cleveland, Ohio; Mack Park in Indiana; Rager's Grove in Gas Center; and the Wehrum school. For the last ten or so years, the Wehrum group has met on the first Sunday in August at the Vintondale Community Park. Although the number attending understandably dwindled each year, the children, grandchildren and even great-grandchildren of former residents kept the reunion going for forty-three years. At the 1990 reunion, it was decided that the number attending the reunion did not warrant continuing it. After a lengthy discussion, it was decided, not by a unanimous vote, to meet at the Vintondale Homecoming during the Labor Day Weekend.

Wehrum - Foundations for machine shop and blacksmith shops, 1934.
Courtesy of Gloria Risko.

Wehrum flat after the removal of the powerhouse and head frame. Company store is in the background, August, 1934.
Courtesy of Gloria Risko.

Steve Dusza, Wehrum, 1932. Train station on left, school in background.
Courtesy of Steve Dusza.

Wehrum after the shutdown, 1934 - General Storehouse.
Courtesy of Gloria Risko.

Wehrum Jail, 1986.
Courtesy of Diane Dusza.

Wehrum Flat, 1986.
Courtesy of Diane Dusza.

Wehrum after the shutdown, 1934 - Powerhouse with boiler house in background.
Courtesy of Gloria Risko.

Wehrum Lamphouse, 1986.
Courtesy of Diane Dusza.

Wehrum - House of John and Gloria (Mardis) Risko, 1986.
Courtesy of Diane Dusza.

Rockdump of Wehrum's #4 mine, 1986, partially recycled in the 1970's. Michael Weber, age 7, is the dot on the extreme right top of the dump.
Courtesy of Diane Dusza.

Wehrum - Retaining wall for Chief Clerk's House, 1986. Upper Broadway.
Courtesy of Diane Dusza.

Foundations of Clarence Claghorn's mansion on hill above Wehrum, 1986.
Courtesy of Diane Dusza.

Wehrum - Slope or shaft entrance to #4, sealed with concrete, 1986.
Courtesy of Diane Dusza.

Present bridge crossing the Blacklick at Wehrum, 1986.
Courtesy of Diane Dusza.

Wehrum - One of the foundations on Second Street, 1986.
Courtesy of Diane Dusza.

Site of Wehrum's former baseball field, the highway now cuts through the center of the field.
Courtesy of Diane Dusza.

Wehrum School, 1986.
Courtesy of Diane Dusza.

Wehrum: Ruins of Lackawanna Coal and Coke Company's #3 washery, April, 1986.
Courtesy of Diane Dusza.

Wehrum: Spring located on the other side of the bridge near the ball field. Used frequently for Hungarian bacon roasts - kept the beer cold, 1986.
Courtesy of Diane Dusza.

Rockdump at Lackawanna Coal and Coke Company's #3 mine, 1986, largely recycled in 1970's and early 1980's.
Courtesy of Diane Dusza.

Chapter XIII
Delano's Other Towns

CLAGHORN - THE TOWN

Claghorn, was this new town to be Clarence Claghorn's crowning achievement, or was it a salve for his bruised ego? Blacklick Valley hearsay has long hinted that Claghorn left Wehrum in a huff because that town was not named for him. Alas, the facts do not support the rumors. The town of Claghorn was actually in the construction stage in 1902-03, a full year before an economic downturn forced the closing of Wehrum. It was this closing, not hurt feelings, which forced Clarence Claghorn to depart for Tacoma, Washington. However, hints have surfaced that Claghorn overspent the Lackawanna budget when the Wehrum was built.

The first sign of development in the Claghorn area was Ash's sawmill at the mouth of Brush Creek in 1891. The nearest town was Heshbon, one and one half miles west of the sawmill. Heshbon's population in 1890 was 36.

In 1900, Clarence Claghorn was conducting coking experiments in Vintondale with coal from the Heshbon area. In the fall of 1900, the Indiana papers reported that coal lands ranging from 20,000 to 25,000 acres in East Wheatfield and Brush Valley Townships had been sold to John Patterson of Hamilton, Canada. The negotiators for the sale were Clarence Claghorn and A.V. Barker. Indiana attorney, John Scott, prepared abstracts on sixty tracts of land. Patterson's lawyer, G.H. Levy, arrived in November to pay out $100,000 to the landowners for the coal rights. The Patterson deal fell through due to lack of funding. It is not known if Patterson had any connection to the newly formed Lackawanna Coal and Coke Company.

In 1902, the Lackawanna Iron and Steel Company, formerly of Scranton, and now of Buffalo, New York, bought the surface and coal rights around Claghorn from none other than Judge A.V. Barker, who had been purchasing these twenty-eight properties since 1890. (See appendix for the list of same properties conveyed to the Vinton Colliery Company in 1916.)

With the extension of the Pennsylvania Railroad competed to Wehrum and Dilltown by 1902, rumors began to fly about extending the tracks to Blacklick Station. Land owners in the lower Blacklick Valley began to receive offers for their coal. In June of 1902, Lackawanna Coal and Coke was conducting test drilling on the Dixon Tomb farm. Barker had purchased the property, and Lackawanna's attorney, John Scott, did the abstracting at $25 per acre. County surveyor D.L. Morehead and civil engineers were surveying all the tracts. The total size of these coal lands was 3,000 acres and included **Buena Vista**, Dr. Stephen Johnston's iron furnace in Dias. Clarence Claghorn, representing Lackawanna, also made some purchases of coal. By October, 1902, the site for the mine was chosen, and test drilling continued on the tract. A new iron bridge was completed at Heshbon, making access to the site easier.

By March of 1903, Lackawanna had constructed a large two story sawmill on Brush Creek; a spur line connected the mill with the PRR. Capacity was 50,000 board feet per day. Manager was Mr. Crise, who rented a house from John Dripp of Armagh. In addition to the sawmill, Lackawanna's plans in 1903 were to open several drifts, build a tipple as soon as possible and construct a coal washery later.

Plans for the new town, which was to be built on the H.E. Wagner farm, were made public. Company offices were being built, and J.P. Kennedy of Blairsville, who also had several contracts in Wehrum, received one to build a three-story hotel for $5,000. County surveyor Morehead was appointed superintendent of plant construction and relocated to Wehrum to oversee constuction of the mine buildings and houses.

Because the railroad had not been completed by the end of September, 1903, work on Claghorn was suspended until spring. All materials had to be hauled overland, a slow and expensive proposition. Even though the laying of the rails was to be completed by December 1, work on the town and the proposed 500 coke ovens was put on hold. Due to an economic downturn in 1904, work on Claghorn was not resumed with the spring thaw. Lakawanna's plans for a large coking center in the Blacklick Valley were never realized. Most of its earlier pioneer steel-making personnel, like Henry Wehrum, were sacked by 1906. Warren Delano, determined to prove that he could make a profit in coke, repurchased #1 and #2 in Vintondale and proceeded to construct the #6 plant complete with 152 coke ovens.

Sale to the Vinton Colliery Company

The town of Claghorn remained undeveloped and abandoned until the fall of 1916. With the US entry into World War I eminent, the coal market was booming; Delano decided to gamble further and purchase additional coal lands. At this time, Lackawanna transferred its 5,800 acres, three drifts and town to Delano's new company, the Vinton Land Company, for $100,000. To finance this purchase, plus twenty five additional houses for Claghorn and more houses for Vintondale, the new firm took out a mortgage for $225,000 with The Pennsylvania Company for Insurances on Lives and Granting Annuities. The date of the mortgage was August 1, 1916. The previous day, the same property was leased to the Vinton Colliery Company. (See related chapter on the Vinton Land

Company.)

On October 9, 1917, Vinton Land Company purchased a small sawmill near Claghorn from Sherman Smith and Mark Engle at a cost of $3,015. Included in the sale were: one pair of grey horses and harnesses; one sorrel horse and one grey mare and harness; one pair of bay and brown horses with harness; two Cramer wagons; one set mainsleds; one twenty-horsepower Frick boiler and engine; one #1 Frick sawmill; one sawdust blower; one log turner; one cutoff saw; one edger; one bull wheel; belt and saw chains; spreaders; axes; crosscut saws; sledges; mauls; wedges; seven hundred feet of 3/4 inch pipe and other small tools.

All houses in Claghorn were five room houses, similar to those in Vintondale and Wehrum. There were no double houses. Known construction workers were John Daly and Mr. Davis. Many of the houses were surrounded with wooden fences; another feature of Claghorn were wooden sidewalks. Former residents described it as a nice town in which to live.

Main Street - Claghorn, c. 1923. Houses for companymen.

Courtesy Robert Cresswell.

Claghorn was laid out on the southside of a bend in the Blacklick across from where Brush Creek emptied into the Blacklick. There was a main street and three side streets. Part of the town lay beside the creek and the rest of the houses were on a bluff overlooking the Blacklick. The Pennsylvania Railroad paralleled the north bank of the creek. A passenger station stood about thirty feet east of the bridge. Since the nearest bridge crossing the Blacklick was a mile away in Heshbon, a concrete arch (rainbow) bridge of three sixty-foot spans, similar to those in Wehrum and Vintondale, was constructed in 1917-1918. The fourteen-foot wide bridge was erected by M. Bennett and Sons for $19,315.98. To get the county to build the bridge, VCC requested that Curtis Mack circulate a petition. Then, a grand jury concurred that there was a need for this bridge. Today, it is the only standing reminder that a town once existed in this rugged, beautiful valley.

Claghorn - Three Span Concrete Bridge, built 1917-1918 - October, 1989.

Courtesy Denise Weber.

According to tax records, there were seventy-three houses. Each house, renting for $7.00 per month, was equipped with running cold water and electricity, $1.00 extra. Water for Claghorn was supplied from Brush Creek at a site 3,650 feet from where the creek entered the Blacklick. The company-owned Claghorn Water Company was chartered on October 8, 1917. Its capital stock was listed as $5,000; each share was worth $100. According to the charter, stock subscribers were John Burgan, thirty shares; Otto Hoffman, ten shares; Lloyd Arbogast, ten shares. According to Joe Anderson, former resident, Claghorn had an adequate water supply, originating from several streams. There was no large resevoir, but small concrete dams which used gravity flow to supply the houses.

There also was a twenty-two room hotel where room and board cost $35 per month. For entertainment, there was a theatre which showed movies every two weeks; patrons came from nearby settlements like

Clyde and heshbon. Outside the theatre was a dancehall/pavilion. The theatre also doubled as a schoolhouse. A curtain could be pulled to create two classrooms. One teacheer taught all the grades and drew a salary of $100 per month. Mrs. Howard Cunningham taught three years at the Claghorn school. The Vinton Land Company entered an agreement with the West Wheatfield School District to build a two-room schoolhouse on a 1.7 acres plot bordering a stream which flowed into the Blacklick. The land company was to pay one-third and the school two-thirds. This probably was not built, because all evidence points to the school sharing the theatre.

Company doctors were Dr. Curtise in 1918 and Dr. Francis Reilly of Blairsville. According to the newspapers and tax records, Dr. Reilly resided in Claghorn until 1923. Michael Wagner, a car inspector for the company, lived across the street from Dr. Rilley, who delivered Merle Wagner on November 14, 1922. Michael and Bruce Wagner also ran the Wagner Lumber Company in Heshbon.

The company store was run as part of the Vinton Supply Company and was located in one of the houses. The first known manager was Mr. Davidson; in 1923, the manager was Mr. N.L. Buck, who was listed as chief clerk in the tax records. Available financial records for 1923 valued store inventory at $7,439.74 and expenses were $2,027.83. The gross gain of the Claghorn store during that period was $1,238.10. By December, 1925 the store was not included in the balance sheet of the Vintondale Supply Company.

At its peak, the population of Claghorn numbered about 400, 150 of whom were miners. Appearing on a map of the Heshbon Lands were five drifts, although the **Department of Mines Annual Report** list production for only four mines. In 1917, the Vinton Colliery Company drained and reopened #11 and #12, the drifts started in 1903 by LCC. Both of these mines were located on the north side of the Blacklick with #11 on the west side of Brush Creek and #12 on the east side. These mines ceased operation in 1918. #13 was on the hill south of Claghorn; it was a drift mine with a railroad spur or narrow gauge connecting with the PRR and #16 about a mile upstream from Claghorn.

Trestle piers are still standing in the Blacklick, and blueprints for the trestle are in the author's possession. Exact date of its demolition is unknown. In 1918, #16 was opened about a mile east of Claghorn on the north side of the Blacklick. #17, opened in 1922, was directly across the Blacklick from #16; it also used the trestle. Both mines were idle by 1923 and declared abandoned by 1927.

Claghorn - Trestle between #16 and #17, c. 1920.

Courtesy Jack Huth Family Collection.

Claghorn - Trestle between #16 and #17, c. 1920.

Courtesy John Huth Family Photos.

Company movments in Claghorn have been preserved in the superintendent's diaries for 1917 and 1923, the first and last major years for operation. Day by day events for 1917 have been summarized and are as follows:

January 2: Less than one month after purchasing the property, Jack Huth, company engineer, took the new foreman, Mr. Olsen, to Claghorn to look over the property.

January 4: Schwerin, Huth, Arbogast and Hoffman spent the day in Claghorn.

January 5 & 6: Hoffman spent time on plans for #11, decided to start with mule haulage.

January 9: Hoffman spent the day at Claghorn, let Thomas of the engineering crew go and appointed Hannan in his place. A wagon bridge was needed across the Blacklick, so Mr. Hoffman got Curtis Mack to circulate a petition.

January 12: Hoffman again at Claghorn all day with Lamb and Hannan. A decision was made to use chain haulage on the #11 tipple.

January 15: Hannan spent the day with two chainmen.

January 16: Company visitors were Huth, Hannan and Lamb.

January 17: A busy day in Indiana County. Arbogast was in Indiana closing the Altemus land deal. Hannan, Hoffman and Azzarra spent the day in C. In the evening, back in the office, the plans for #11 and #12, made by former employee Thomas were studied by Huth, Hannan, Slutter, James, Abrams, Arbogast and Hoffman. All agreed that the coal should be diverted to the tipple by motor.

January 18: Hannan with two helpers was in Claghorn looking at tipple and railroad siding sites. Policeman Timmons picked up a mule in Johnstown and delivered it to Claghorn.

January 20: Saturday, Hoffman spent the day in C. Water had been pumped out of #11 Friday night, and track was laid in #11 Saturday. A car of ten-pound rails was unloaded, and Mr. Schwerin okayed the new plans for the #11 tipple.

January 22: Arbogast took Paul Williams, who was to serve as timekeeper, to C. The engineers were tallying costs for the tipple and the railroad siding.

January 23: Hoffman spent the day in C. with the Anderson Brothers.

January 27: Saturday, Hoffman spent the day again.

January 31: Schwerin, Willard, Town, Hoffman and Anderson Brothers paid a visit.

February 5: Mr. Lamb, the Hess Brothers construction representative, visited Hoffman at the Vintondale office to "talk Claghorn".

February 7: Arbogast in Claghorn for the day.

February 9: Hoffman and Lamb in C.

February 10: Hannan in Claghorn to give grading stakes to Anderson Brothers.

February 13: Hoffman in C.

February 15: Hoffman returned to meet with the board of viewers concerning the bridge. He used Bennett's automobile to get to C.

February 18: Sunday evening spent by Hoffman going over Claghorn situation.

February 19: The company store opened with Mr. Davidson as timekeeper and storekeeper. He was accompanied to Claghorn by Clement and Arbogast.

February 20: Hoffman returned to C. to go over tipple and motor track grading plans with the Anderson Brothers.

February 21: Arbogast traveled to Indiana to settle the Graham land deal.

February 22: A miner evicted for drunkenness and disorderly behavior in Vintondale was hired to dig coal at C. Mr. Slutter was planning installations in C.

February 23: Hoffman and Huth took V. miner and his family to C.; also covered all the workings. The fourth car was loaded today.

February 26: Poles for a telephone line to the hotel was being installed by Arthur Johns. The contract to build tipples and tramways at #11 and #12 was awarded to Anderson Brothers.

February 27: Hoffman was in C. as was Dr. MacFarlane, who was visiting the sick.

February 28: Coal mined at #11 between February 8th and 27th was loaded into five cars. Slutter spent day at C.

March 1: Hoffman spent the day in C. with A.L. Anderson.

March 2: Arbogast and Timmons in C. paying out.

March 3: Saturday, Hoffman in C. for day.

March 5: Huth and the engineering crew in C. to turn the heading. Slutter there to lay out the M.G. (motor generator) and shop building. Morton and his crew erecting #11's fan.

March 6: A National Tile Company representative called at the V'dale office and left with orders for shop fan and M.G. houses at #11.

March 7: Hoffman and Slutter spent the day in C. where concrete was poured for the foundation for the M.G. set.

March 8: Huth and Hoffman spent the day at the Indiana County Courthouse testifying before a grand jury on the proposed bridge at Claghorn. The jury reported favorably.

March 10: Hoffman at Claghorn with Clement and Arbogast.

March 13 and 14: Hoffman at C. with Delano, Schwerin and Clarence Claghorn. (C.R.'s first appearance in the 1917 diary.)

March 15: O.P. Thomas called to show Hoffman and others the Heshbon watershed.

March 17: Hoffman at C. with Arbogast who was paying off. A clean-up was being done at #12.

March 18: A car of wire, fan hoist, etc. was loaded for #12. That evening Hoffman met with Olsen and Edmondson in his office. He decided to put Olsen at #6 and Edmonson at Claghorn.

March 20: Hoffman at C.

March 22: Hoffman at C.; he decided to get a

hoist and boiler to load #11 coal.

March 23: Slutter on construction at #11 and #12. A hoist and boiler was obtained from Anderson Bros. to load coal at #11. The teams were eliminated.

March 24: Hoffman in C.

March 26: Hoffman in Johnstown with Mr. Campbell about coal land in Heshbon.

March 27: Hoffman at C.

March 28: Slutter at C.

March 29: Hoffman at C.

March 30: Arbogast to Claghorn, Pay Day. Slutter at Claghorn.

April 2: Slutter and Huth made estimates on Claghorn houses.

April 4: Hoffman at C.

April 5: Arbogast and Slutter at C.

April 11: Hoffman at C.

April 13: Huth at C.

April 14: Hoffman at C.

April 15: Meisner loaded car for Claghorn: cement, etc.

April 17: Hoffman, Schwerin, Arbogast, Thompson, Hewitt and Public Service Company in Claghorn. Mr. Mack called to talk about Vinton Colliery loading coal at Dias Wharf.

April 20: Huth and Slutter to C. National Tile Company Representative drew sketch of proposed company store for Claghorn.

April 21: Hoffman and Slutter to C.

April 23: Slutter at Claghorn. Hoffman drove to Wehrum to talk to Supt. Dunbar about the Dias siding.

April 24: Hoffman was at C. for day with Huth and A.L. Anderson.

April 25: The Dias loading dock problem continued to simmer. Mr. Wolk stopped to talk over the loading block. Mr. Blair of PRR telephoned and wanted more information about loading and unloading at Dias, threatening to stop loading and unloading at #11.

April 26: Arbogast drove to Ebensburg to consult Mr. Bird about Dias loading wharf.

April 28: Saturday, Hoffman at C. as was Shaffer who was getting acquainted with the plant.

April 30: Slutter at C. Arbogast and Timmons were also in C. to pay off.

May 1: Hoffman in C. for part of day trying to get some action from the Hess Brothers on the houses.

May 2: Slutter at C.

May 3: Arbogast at C.

May 5: Saturday, Hoffman at C. with Abrams.

May 8: Hoffman at C. with Abrams.

May 9: Slutter at C., also the #1 track layer was sent to put in main switches.

May 10: Hoffman at C.

May 12: Hoffman at C. Jackson Mardis and Bruce Wagner (future Indiana County Commissioner) estimating timber tract from Dias to Claghorn on the south side of the Blacklick.

May 14: Arbogast and Timmons to C. for payday.

May 16: G.E. promised substation meters for C. for June 1. Hoffman received their promise of temporary meters so that machines in #11 could be started.

May 17: Mr. Hoffman spent the day in C. and met with Mr. Campbell, East Wheatfield assessor, who was going to tell the county school board that there must be a school in Claghorn.

May 19: Hoffman spent part of the day at C. and then drove to Indiana with Burgan and Arbogast to incorporate the water company. Mr. Hess of Johnstown telephoned and offered to sell his coal in the Heshbon Tract.

May 22: The M-G set, motor and hoist were started. Slutter, Hoffman and G.E. representative, Mr. Skinner, spent the day there. Visiting Claghorn that day was BR&P president Noonan. He and his party arrived on his special train.

May 23: Slutter was at #12 getting the sub-station started.

May 24: Slutter and Hoffman at C. concerning power problems. Power was off in #11 for four hours.

May 25: Slutter was at Claghorn working on the M.G. sets. Delano and Hoffman also visited.

May 27: Hoffman spent most of the day in C. with the Hess Brothers.

May 29: Huth and Slutter made an estimate for housing at C. and then supervised estimate of Hess Brothers and Mr. Lamb.

May 31: Pay Day in Claghorn. Arbogast, Burgan and Timmons paid off. #11 and #12 were inspected by State Mine Inspector Crocker.

June 2: Hoffman spent the day in C., which was celebrating its first known wedding.

June 4: Slutter started up the puncher machine in #12.

June 6: Slutter wired up the meters on the M-G sets.

June 8: Slutter, Clement and Hoffman spent the day visiting both the mines and tipples at #11 and #12 and chose a house to be used as a store.

June 13: Arbogast and Hoffman were in Johnstown consulting with Mr. Grumbling concerning one brother's share of the land sale.

June 15: Arbogast and Clements were at #11 and #12 paying off.

June 16: Saturday, Hoffman at C. for the day.

June 20: Hoffman at Claghorn with Cameron.

June 21: Slutter at C.

June 23: Arbogast went to Indiana to close the Campbell Land Deal.

June 24: Sunday, Clarence Claghorn arrived in

Pittsburgh. Hoffman drove to pick him up.

June 25: Hoffman spent the day in Claghorn showing the property to Clarence Claghorn.

June 27: The Campbell land deal was closed. Vinton Colliery Company bought the Hess coal lands in the Heshbon Tract. Hoffman, Delano, Claghorn and Schwerin met Liggert and Mack in the afternoon at Dias.

June 28: Hoffman, Delano, Schwerin and Claghorn spent the day in C. In the evening Hoffman met the Hess Brothers in the office to discuss construction of the new houses in Vintondale and Claghorn.

June 30: Hoffman, along with Roth and C. Claghorn, spent the day at C., meeting Mr. Therlkeld of National Tile Roofing Company who brought sample tiles.

July 2: Pay day at Claghorn.

July 3: Clarence Claghorn and Hoffman spent the day together in the office.

July 7: Saturday, Hoffman drove to Johnstown to meet Claghorn and his family at the 7:50 train.

July 16: Arbogast in C. to pay off.

July 19: Arbogast in C. to teach Maloney the office routine.

(Could this be the same Joe Maloney who served as trustee in the 1940 bankruptcy?)

July 28: Slutter spent that Saturday installing low voltage releases in MG sets.

August 2: Inspector Abrams visited #11 and #12 mines.

August 3: Arbogast and Timmons drove to C. to settle boarding house accounts.

August 5: The Hoffmans and the Claghorns took a Sunday drive to Waspsanotti. (Wapsanonnick Mountain near Altoona.)

August 15: Delano, Schwerin, Claghorn and Hoffman drove to the top of the hill above Claghorn, surveyed the Claghorn tract on horseback and returned to Vintondale by "machine" by 3:00 PM.

August 16: A particularly busy morning! Delano, Schwerin, Claghorn and Hoffman read the mail until 9:30, then caught PRR Superintendent Smith's special train which was heading for Claghorn. There they went over plans for the railroad station. At 11:00, they met Abrams and the engineer and also purchased the Johnson and Weir coal tract near Dias. After a visit to #11, they drove to Dilltown to visit Superintendent Hewitt before returning to Vintondale.

August 17: Delano, Schwerin and Claghorn drove to Claghon and spent the day on the Heshbon Tract. Mr. Jack, Indiana attorney, came to Vintondale in the late afternoon to discuss the Claghorn Water Company.

August 22: Slutter drove to Claghorn to pick up coke braize, timber and odds and ends to fill car.

August 24: Timmons at Claghorn to shut down speakeasies. Two tenants were moved out. (This was three years before Prohibition went into effect.)

August 27: Abrams went to Claghorn to make a compensation inspection. A machine runner accompanied him to start the Goodman chain machine.

August 30: The lumber that was delivered to Claghorn for the houses that the Hess Brothers were building was placed under lien by Eihle Lumber Company.

August 31: M.H. Madigan of Cresson stopped at the Vintondale office; he was able to furnish laborers for Vintondale and Claghorn.

September 2: Claghorn, Schrock, Holt and Hoffman journeyed to Dias and walked to Claghorn (about two miles). While there, they inspected the prospect holes.

September 5: Hoffman drove to Claghorn at noon to pick up Mr. Claghorn and Mr. Dickinson of Goodman Manufacturing Company. The purpose of the visit was to discuss delivery of chain machines. Arbogast also drove to Claghorn to go over the hotel books.

September 11: Hoffman drove Mrs. Claghorn and her party to Claghorn in the afternoon. Mr. Claghorn returned to Vintondale with them. Mr. Maloney went to Claghorn for several days.

October 2: Hoffman drove to Johnstown to pick up Clarence Schwerin and sons, Clarence and Joseph. They then drove to Claghorn where they spent the day and then motored to Vintondale.

October 5: Mr. Schwerin and his sons spent the day at Claghorn, traveling both ways by train.

October 8: Arbogast and Blewitt were at Claghorn for the day to start up the card system.

October 10: Abrams made a compensation inspection at Claghorn.

October 15: Jack Huth traveled to Indiana to close the Johnston-Weir tract deal at Claghorn.

October 19: The company sent Mr. Cook to Claghorn with a barrel of gasoline.

October 22: Arbogast, Blewitt and Timmons to Claghorn for the day.

October 30: Hoffman spent the day at Claghorn with Clarence Claghorn, who was nursing an injured foot.

November 1: Mr. Jack of Indiana called in Vintondale on account of the Claghorn Water Company.

November 2: Schwerin spent the day at Claghorn.

November 3: Abrams made a compensation inspection at Claghorn.

November 4: Mr. Meisner unloaded a ten-ton G.E. motor for #6; plans were to ship it later to Claghorn.

November 16: The fan for Claghorn was ready for

shipping.

November 20: At 1:00 PM Arbogast, Timmons and Hoffman took money to Claghorn to pay off. Mrs. Hoffman delivered checks to Mr. Jack for the Grumbling land deal.

November 21: Abrams made another compensation inspection at Claghorn.

December 6: Hoffman spent the day at Claghorn with Clarence Claghorn, visiting all mines and real estate.

December 7: Delano, Schwerin and Hoffman spent the day at Claghorn with Clarence Claghorn. Schwerin met Mack and Bedill and talked coal contract.

Between 1904 and 1917, Lackawanna Coal and Coke continued to pay the county taxes on the Claghorn property. Tax notices were sent to Atty. John Scott in Indiana. Total assessed value in 1916 was $7,357, and the county tax was $73.57. When the Hesbon Lands were sold to the Vinton Land Company and then leased to the Vinton Colliery Company, the payment of taxes was handled by Atty. S.M. Jack of Indiana.

As can be seen by the diary entries, there was a lot of construction work done in Claghorn in 1917. Forty-seven new houses were built, but not without the same problems that the construction company, Hess Brothers, had in Vintondale. Tipples, trestles, and other mine buildings were erected. A motor-generator set was installed to provide power for the mines, which in 1917 were #11 and #12.

But there were hints in those diary entries which forecast doom for Claghorn. One was the turning of the heading in #11 in March. Martha Abrams Young, daughter of Abe Abrams, superintendent in 1918-1919, said that the coal was low, and the miners had to lie on their backs to dig the coal. This is supported by various mining books, such as the **Keystone Directory**, which lists the seam of coal at Claghorn at thirty inches. A comparison check of Rochester and Pittsburgh Coal Company's mining maps with those of the Vinton Land Company show an unforseen problem: rock channels, old stream beds which have solidified. The coal seams unexpectedly disappeared. Mines #11 and #12 were abandoned by 1920.

Also, the company had trouble keeping workable management in Claghorn. Clarence Claghorn stayed long enough to build himself a mansion away from the rest of the town. By 1918, he was gone to parts unknown and for unknown reasons. Abe Abrams, #6 foreman, was sent to Claghorn against his wishes. He did not live in the Claghorn mansion because it was too far away from the town, and the company did not open up the house as promised. Abrams was dismissed in 1920. Others filled the management role, the last being Allie Cresswell who was promoted to superintendent from foreman. When Claghorn closed, he returned to Vintondale as a companyman.

The wartime prosperity which led to the re-opening of Claghorn soon fizzled. Between 1917 and 1923, there was boom, postwar depression, the Red Scare, and two coal strikes (1919 and 1922) which the Vinton Colliery vehemently opposed.

The 1922 Strike

In Vintondale, the Vinton Colliery Company was able to prevent the miners from joining the strike, but in Claghorn, they were not successful. The miners walked out and were evicted. A tent colony went up near Heshbon, and the company brought in strikebreakers, housing them at the Heshbon Hotel. Problems between union organizers and company deputies started in early May when UMWA organizer, Charles Diaz of nearby Dias, tried to help the evicted miners. Diaz was met at the station on May 8th or 9th by Jay Jensen and Harry McArdle. When he returned to the station, the two deputies tried to goad Diaz into a fight. He refused to try to take on the two armed men. They tried to force Diaz off the train platform and make him walk to Dias so that they could take care of him out of sight of the townspeople. McArdle broke his club over Diaz's head; then Jensen blackjacked him over the head. The beating stopped when Harold Gray and his sister, Mrs. C.E. Evans, came along. Diaz managed to walk the two miles back to his home.

On July 8, there was a confrontation in Heshbon between the strikers and the coal and irons. The coal companies termed it a "riot." The UMWA announced that it was planning a rally at the Heshbon tent colony. Charles Diaz and twenty-seven miners boarded the train at Dias. Their train was met in Heshbon at 9:15 AM by a crowd of striking miners and families and by the coal and irons and several state policemen. The state police had been called in by Margaret Coal Company Superintendent, S.J. Davis. Pretext for calling in the state police was that violence was expected. To receive assistance from the state police, the coal companies had to pay to transport the troopers to the troubled area.

When the miners debarked, the coal and irons told the crowd to hurry up the hill; Diaz refused, claiming that they could not move until the train left and that he had a right to hold a meeting. Diaz was hit from behind by a club for not moving fast enough. His brother Thorn was talking to the police and was soon handcuffed. Mike Murray, who was with the Diaz brothers, started cussing at the coal and irons, but mainly after the Diaz boys were arrested. While this was happening, other coal and irons were chasing the strikers back to the tent colony. Justice of the Peace, A.W. Campbell of Heshbon, refused to hear the riot case, and the state police had to travel to Indiana to find a justice of the peace who would hear the case.

Witnesses to the incident included A.R. Wagner, postmaster of Heshbon, who gave testimony similar to Diaz's; he laid the blame for the situation on the coal and irons. Several other witnesses believed that the riot was another setup to try to get Diaz because of his success in organizing the mines in the area. The witnesses also agreed that the Diaz brothers were very capable of defending themselves, but did not actively seek trouble that day.

One of the state policemen at the scene, Andrew Yesavige, testified in March, 1923 that 300 men got off the train in Heshbon and were met by 150 strikers. His was the only testimony which was so far removed from the facts. This testimony and the fact that the coal comapnies could request state troopers any time makes one wonder as to the quality of officers and the role of the state police in 1922. Also, about one month before the strike, a questionnaire concerning potential problems during the upcoming strike was sent to each colliery in the state by Lynn Adams, Superintendent of the Department of State Police.

The John Anderson, Sr. family moved to Claghorn from New Jersey in 1922; their son Joe said that only about one third of the houses in Claghorn were occupied at the time. The Simon family was not evicted from their house in Claghorn because Mrs. Simon was pregnant. However, she was subjected to the taunts of the strikers' wives who greeted her with, "Hello, scab woman!" The strike was a failure, and soon the Claghorn mines were back in full production.

But by 1923 with the coal markets being challenged by new sources of energy such as natural gas and oil, marginal coal lands such as Claghorn could not compete. #11 and #12 were closed by 1920. #16 and #17, about a mile from the town, were being worked on an abbreviated work week. Otto Hoffman had unintentionally recorded the demise of the town in his mine diary, which is summarized below.

Daily production of #16 and #17 are listed in the diary. However, the author has chosen not to list these statistics. Yearly production of the Claghorn mines can be found in the appendix.

1923

January 11: Hoffman and Huth took the 8:30 AM train to Claghorn; they toured all planes in #16 with Riley and Cresswell. In the afternoon they visited the #17 tipple and the store, returning to Vintondale on the 4:04.

January 15: Blewitt was in Claghorn on account of pay day.

January 17: Huth and State Workman Insurance Rating Inspector Penrod spent the day in #16 and #17.

January 22: Cresswell, foreman at #16 and #17, spent the day in third slope, #6 to see how pillars were drawn. State Mine Inspector Crocker was conducting inspections in #16 and #17.

January 24: Hoffman went to Claghorn on the 8:30 to meet Mr. Mack to offer to buy his coal instead of paying a royalty. Mr. Mack immediately replied no. While in Claghorn, Hoffman arranged to double-shift #17.

January 25: Claghorn's mines suffered from a shortage of coal hoppers just like Vintondale. At 9:00 AM they received three cars and loaded out at 3:30 PM. The double shift with four men started in #17, and four more were to begin in a few days.

January 30: #16 and #17 idle, no cars.

February 8: Huth was at Claghorn and found only two motors in commission. There had been a collision the evening before which was not reported to Hoffman. Details had to be pieced together by Huth. Hoffman decided to personally look into matters the next day.

February 9: Hoffman took the 8:30 train to Claghorn to look into the collision and other matters. He was picked up at 1:00 PM by the jitney. His comment on the Claghorn matter: "The management at Claghorn is rotten."

February 14: Hoffman drove to Claghorn at 3:30 and put Allie Cresswell in charge of Claghorn, replacing Paul Riley. When he returned at 6:30, the roads were a sheet of ice.

February 19: UMWA organizers Diaz, Larson, and three others got off the 3:30 at Claghorn.

February 20: Hoffman went to Claghorn on the 8:30 to "find facts about fight Diaz and other union man started." (Unfortunately the newspapers do not carry an account of this fight.)

February 26: Huth at Claghorn for the day.

March 1: Jensen, policeman at Claghorn, was put on the tipple cleaning coal.

March 9: Arbogast visited the flats early and then took the 8:30 to Claghorn, spending the day in #17. He returned on the 4:04.

March 12: Huth at Claghorn.

March 22: Hoffman took the 8:30 train to Claghorn to visit Dr. Reilly and Mr. Buck, the store manager. He also visited the tipple and #16 and #17 mines. He made unspecified arrangements for April 1. (Vintondale was expecting a strike.)

April 3: Huth and Inspector Penrod worked on #16 and #17 ratings.

April 9: Hoffman picked up Mr. Schwerin in Johnstown in the morning, and in the afternoon they drove to Claghorn, returning at 4:30 PM.

April 19: Mr. Stiles called and wanted to cut props at Claghorn.

April 20: Huth spent the day at Clahgorn and then reported to Hoffman upon his return.

April 23: Huth, Hoffman and Mr. Mitchell of Link Belt Company drove to Claghorn at 8:15 to go over the question of a picking table for #16 and #17. They returned at noon.

April 27: At 7:15, Hoffman, Huth and the representative of Heyl and Patterson drove to Claghorn in the jitney go go over the picking

table for #16 and #17. Returned at noon.
May 8: Huth spent the entire day at Claghorn.
May 23: In the afternoon, Hoffman drove to Claghorn to talk over dumping coal in the tipple only three days a week. Back at 4:00 PM.
May 26: Cresswell in Vintondale to get the Claghorn pay.
May 29: Hoffman and Huth drove to Claghorn to watch coal preparation, which Hoffman found satisfactory. Back in Vintondale by 4:00 PM.
June 5: Heyl and Patterson representative was in Vintondale to go over the #16 picking table plans. In the afternoon, Hoffman drove to Claghorn to visit #16 and #17.
June 8: Huth at Claghorn for day.
June 11: Hoffman, C.M. Schwerin, and Hufford drove to Claghorn and visited the town and the mines. They returned at 4:30.
June 15: A Vintondale policeman was sent to Claghorn to investigate a candy-stealing case at the depot. The parents paid.
June 21: Hoffman spent the afternoon at Claghorn.
June 24: Lightning hit the #16 substation building and burnt off the roof and burnt up the entire switchboard and coils in both ends of the MG set. The only thing left standing was the frame, shaft, bearings and rotary.
June 25: At noon, Blewitt, Daly and Hoffman drove to Claghorn to meet the representative from G.E. about repairs to the MG set. Hoffman drove the rept. and Blewitt to Johnstown to talk with Mr. Buchanon about getting repairs done quickly. They returned at 6:00 PM.
June 28: #16 and #17 were idle, waiting for new MG set.
June 29: The new switchboard for #16 and #17 arrived at 9:00 AM and the MG set at 5:00 PM. Erection started at once. Daly and two men were sent from Vintondale as soon as the parts arrived. A vote of thanks was sent to the Pittsburgh G.E. office and W.W. Miller.
June 30: Four of the electric crew were in Claghorn working on the erection of the MG set. Hoffman spent two hours in Claghorn.
July 1: Hoffman drove to Claghorn to observe the MG progress.
July 2: Electrician Daly and Hoffman drove to Claghorn.
July 5: The new MG set was started at 7:05. Hoffman left for Claghorn at 5:50 and returned at 8:45.
July 9: Hoffman drove to Claghorn with C.M. and Joseph Schwerin. They drove on to Graceton where the Schwerins stayed until the next day.
July 12: Police were investigating the Grove (?) killing case at Claghorn. (There was no information about this incident in the local papers.)

July 23: Strike at #16 and #17 (?)
July 24: Hoffman drove to Claghorn for the balance of the day on pump arrangements.
August 4: Hoffman spent the afternoon at Claghorn with Huth on maps.
August 15: Hoffman made a trip to Claghorn while the police made a mounted trip there.
August 30: Hoffman went to Claghorn in the afternoon to see the new tipple, back at 3:07.
September 13: Policeman Bautala (sic) and Hoffman to Claghorn early to look over material at #12. Back at noon.
September 19: John Cresswell, #6 car repairman, died suddenly at 3:30 AM. Hoffman drove to Claghorn in the afternoon with the Miller-Owens representative concerning repairs to the burnt up MG set. He also made arrangements for work with the assistant foreman during Cresswell's absence due to his father's funeral.
September 21: Hoffman attended the Cresswell funeral at 12:00 and then at 1:30 drove to Claghorn to look over things while Cresswell was off. He returned at 4:50.
September 26: Hoffman spent PM in Claghorn and arranged to work water courses only on day shift.
October 4: Hoffman to Claghorn, back at 4:40.
October 11: Blewitt and Hoffman to Claghorn in PM.
October 13: President Schwerin sent orders to close #16 and #17 mines. Foreman Cresswell and electrician Steele were to keep the mines free of water and do the watching day and night. Dr. Riley was going to look for a practice in another location. Tenants were permitted to remain in the homes for 50 cents charge for rent and pay for lights they used.
October 16: The carpenters started working on the picking table at Claghorn.
November 5: Huth took Link Belt visiting engineer to Claghorn to see the picking table job.
November 8: Huth and Mr. Mitchell at Claghorn on picking tables.
November 9: Police trip to Claghorn.
November 14: Schwerin stopped in Claghorn on way from Graceton to look at the tipple, then went on to Vintondale.
November 22: Hoffman took Mr. Mitchell of Link Belt Company to Claghorn and returned at noon.
December 7: Picking table for Claghorn was completed.
December 17: In the afternoon, Garrity, Arbogast and Hoffman drove to Claghorn and were back at 4:00.
December 31: Foreman Cresswell was in Vintondale to arrange to start #17 on the fifteenth of January.
In spite of being closed during much of 1923, the

Vinton Colliery Company invested in a new picking table at the Claghorn tipple. Joe Anderson said that he worked on the table before the Claghorn mines closed permanently. The company's reasoning for investing in the table is unknown.

According to former resident, George Pytash, Vinton Colliery was forced to close #17 because Curt Mack would not sell his coal to VCC. His statement is supported by the Claghorn mine map which shows the Mack property directly south of the #17 workings which headed east and west of the drift mouth. The diary entry for January, 1923 also backs Pytash. Mr. Hoffman went to Claghorn to offer to buy Mack's coal rather than pay a royalty on it. Mr. Mack's immediate response was no.

When the Claghorn mines closed in 1924, the Andersons remained behind as watchmen. They moved the furniture from the hotel and the superintendent's house to the company store. Later, it was loaded into a box car and shipped to the Vintondale company store to be sold as second-hand furniture. The Andersons remained in Claghorn for about a year and one half and then transferred to the new #6 shaft near Red Mill.

In addition to the Andersons, Cresswells, and Steeles, other Vintondale residents who had lived in Claghorn at various times were the Pytashes and the Simons. Paul Pytash, Sr. leased a farm above Claghorn; the owners turned out to be moonshiners. When the family moved ine, there were thirteen barrels of mash in the house. Hooch from area stills was regularly flown out from a landing strip on a nearby farm. George Pytash also remembered picking apples in the Claghorn orchard and taking shortcuts across the trestle to get to town. He also said that he and his friends would sneak into the music hall to play the piano. After short stints in Vintondale, Nanty Glo, Scott Glen, Heshbon, Elkin, and Dilltown, the Pytashes settled in Vintondale on Chickaree Hill in 1932.

The Simon family lived in Claghorn several different times and located permanently in the town of Charles. Their daughter, Mary, occasionally spent part of her summer vacation in Claghorn visiting with the Tomsulas. Mr. Tomsula served as the company watchman after the Andersons. After their marriages, daughters, Mary Kerekish and Libby Biondo, became Vintondale residents.

The Claghorn mines were officially abandoned, but the Vinton Land Company continued to rent a few of the houses in the 1930's. Margaret Serene of Indiana, who was a social worker for the State Emergency Relief Board during the Depression, made trips into Claghorn to check on the health and welfare of the residents. Her caseload in Claghorn was about six families. By that time, to get into Claghorn, one had to park in Heshbon and walk up the sand road on the south side of the creek.

The abandoned buildings, in particular, Claghorn's mansion, supplied heating and plumbing equipment to several houses in Vintondale. In 1936, with Mr. Brandon's reluctant permission, Steve Oblackovich, Steve Kish, and Nick Palovich went to Claghorn to retrieve furnaces and radiators. Oblackovich took the steam furnace and fixtures out of the Claghorn mansion. There were fifteen radiators in the house, but he only took six. The furnace, which was blanketed with asbestos, was like new. After he removed the furnace, Mr. Tomsula approached Oblackovich and asked him if he had seen any copperheads while taking out the furnace. If he had known about the snakes beforehand, Mr. Oblackovich said that he might have thought twice about removing the furnace. John Biondo said that he and several other men had a letter from Mr. Brandon granting permission to remove items from the company store. Since the watchman could not read, the men took more items than were on the list. These materials were used to construct the Hillside Inn on Chickaree Hill, later known as the Citizen's Club.

Mr. Tomsula eventually moved to Homestead. Joe Kerekish of Vintondale was a friend of Tomsula and occasionally drove to Pittsburgh to visit him. The last Claghorn watchman was a bachelor who erected himself a shanty. Yearly, during hunting season, Paul Pytash, Steve Pytash, and Kalman Antol stayed with him for several days.

The Claghorn property was sold to the Indiana County Commissioners in 1934 for $2,420.11, the amount of the unpaid taxes for 1931. In 1948, the county sold the property for $3,274.28 to the Vinton Coal and Coke Company, the successor of the Vinton Colliery Company. The sale included twenty-seven houses, one boarding house, one theatre, one store, one garage, and one mule barn. The company then sold the buildings to Nick Kovalchick of Indiana for $4,300. The Kovalchick Salvage Company was given two years in which to remove all salvageable materials. Perhaps some of the Claghorn materials are still sitting at the bottom of Kovalchick's junk piles on Wayne Avenue in Indiana.

Claghorn disappeared via the same route as Wehrum. Today, all that remains are a few foundations, traces of mine drifts, a few rock dumps, and the cement bridge, built in 1917, which was strong enough to survive the 1977 flood.

For those who would like to visit Claghorn, there are several routes; all involve hiking at least one mile or more. One can park the car in Heshbon or Dias and follow the railroad tracks, eastward from Heshbon or westward from Dias. The Dias route is not recommended because of the difficulty of climbing down a steep embankment to reach the railroad tracks. This section of track from the Oneida Cleaning Plant at Dias to Blairsville is still operated by Conrail. A third route follows the old Claghorn road which parallels Brush Run.

Claghorn, c. 1923.
Courtesy Robert Cressell.

Johnsons visiting Halldins one Sunday afternoon at Claghorn about 1920. From left to right - Louisa, Harold and Lydia Johnson, Sylvia Halldin Lentz, Eric Johnson, Walter, Arthur, Gustaf Halldin, Astrid Halldin Akins and Edla M. Halldin.

Claghorn, c. 1923.
Courtesy Robert Cresswell.

At Claghorn about 1920. From left to right - Boy with rabbit - name unknown, Arthur G. Halldin*, Walter Halldin and Astrid Halldin Akins.
*Mr. Halldin is Founder and President of The A. G. Halldin Publishing Co., Inc. in Indiana, PA 15701.

Johnsons visiting Halldins at Claghorn. From left to right - Lydia Johnson Granlund, Louisa and Eric Johnson. Mr. Johnson is driving his new Model T Ford.

Claghorn trestle piers, October, 1989.
Courtesy Denise Weber.

GRACETON COAL AND COKE COMPANY

Shortly before his untimely death in 1920, Warren Delano purchased an established Indiana County mining and coking venture. That July, the local newspapers reported that Youngstown Steel Company had sold its Graceton Coke Company to Delano and Associates for $750,000. The sale included all coal lands, machinery, equipment and houses for one hundred thirty families.

The town of Graceton, located in southern Indiana County between Homer City and Black Lick, incuded sixty-seven double houses, several singles, nine shanties near the coke ovens, a company store, a pool hall, and two bowling alleys. When a new company store was built, the old one became a dance hall and basketball court. Youngstown Steel had built a community park with a bandshell where the company band practiced on Sundays. There were several privately owned stores on the edge of town. Aspers had a dry goods store, and Pearlsteins had a meat market and grocery. Mr. Pearlstein was killed during a robbery. The town justice of the peace had a shoe repair shop in Graceton so the hearings were often held there. The town also had its own local character, named Whitey, who, with Mr. Brandon's permission, lived in the last coke oven.

The Delano purchase was the fourth sale of the Graceton property since its opening in the 1880's. Graceton #1 was first opened by George Mikesell who built and operated the firm of J.W. Moore. John and Harry McCreary of Indiana built Graceton #2 and the town in 1890. Harry McCreary bought out his partners and later sold the property to Youngstown Steel Company.

Youngstown's general manager was Charles M. Lingle, originally from Osceola Mills. During his tenure, additional houses, a general store, and the Methodist Episcopal Church were built. The plant at #2 was also constructed. Since the population of Graceton increased by twenty-five percent, the school was also enlarged. Following the Delano purchase, Mr. Lingle transferred to Nemacolin where he worked for the Buckeye Coal Company, subsidiary of Youngstown Steel. The Graceton employees presented Mr. Lingle with a gold watch and chain and a diamond-studded Masonic emblem charm as a departing gift.

The deed for the Graceton property was dated December 21, 1920; the amount of the sale was $600,000. The purchase included forty-five parcels of coal and real estate in Center Township. The breakdown of the $600,000 was as follows:

Washery $100,000
202 coke ovens at $600 each 121,000
Coal and minerals with Rights 300,000
Houses and buildings 72,800
Surface lands 6,000

The company was renamed the Graceton Coal and Coke Company. To pay for this new purchase, capital stock worth $250,000 and 10,000 shares of common stock were issued. Graceton Coal and Coke's indebtedness increased from nothing to $450,000. It also authorized a mortgage for $450,000 plus six percent interest with the Gracetown Coal Company. Payments of $90,000 were to be made annually on July 1 from 1921-1925. This mortgage was marked satisfied on August 3, 1922 by the Indiana County Recorder of Deeds. This was because Graceton executed a mortgage for $500,000 with the Merchanics Trust Company of New Jersey. The bank remained a trustee until March 15, 1932 when it resigned its trusteeship. On March 16, 1932, Eugene F.E. Jung of New York was appointed trustee for the mortgage.

Company Store

On July 21, 1921, the Graceton Supply Company sold its building, lot, furniture and fixtures to the Graceton Coal and Coke Company for $24,838.52. The deed listed John Stambaugh as president and W.R. Merrick as secretary of the Graceton Supply Company. The company store, which was located beside a railroad siding along Rt. 119 in Graceton then became a branch of the Vinton Supply Company, which in 1924 became the Vintondale Supply Company. Financial records found at the FDR Library show that from June to December, 1923, the Graceton Company store grossed $6,780.27. Its inventory was worth $17,294.17, and operating expenses were $4,969.69. Other available records for December, 1925 show practically the same amounts.

For the purchase price, the Graceton Coal and Coke Company received two mines, 202 coke ovens, and the corresponding preparation plant. The mines were serviced by the Buffalo, Rochester and Pittsburgh Railroad and the Pennsylvania Railroad. Superintendent of mines between 1920 and 1930 was Milton Brandon, formerly of Homer City. Following the death of Otto Hoffman in 1930, Brandon became superintendent of both towns and moved into the superintendent's mansion in Vintondale. Graceton lacked a suitable superintendent's house; the super and the companymen lived in the double houses on the upper side of town. The Valent Varesak family settled in Graceton where Mr. Varesak was employed at the coke ovens as a mason. Frequently, he was sent to Vintondale on coke oven-related work.

(All coal and coke production, number of days worked, and number of Graceton employees from 1920-1936 can be found in the appendix. Also found in the appendix are names of miners evicted during the 1922 strike.)

In January, 1925, Graceton Coal and Coke also entered into a gas lease with the T.W. Phillips Gas and Oil Company of Butler. The lease covered 15,234 acres. In January 1934, the lease was declared null and void and was rewritten, covering the same land. As long as gas was produced in paying amounts, the lease was to last twenty years. The royalties were to be paid quarterly, and operations were to start by July 15, 1934. This agreement was released on April, 1942 to the new owners of Graceton.

Bankruptcy

In March, 1936, the company store lost a boxcar of company store items that was to be shipped to the Vintondale company store. The boxcar, which was sitting on a siding near the Blacklick Creek was washed away in the March, 1936 flood. Less than three months later, the Graceton mines closed with warning, and the Graceton Coal and Coke Company declared bankruptcy. Albert Oswalt, later postmaster of Graceton, worked in the company office and helped get the last payroll ready. Milton Brandon was appointed temporary receiver by the Indiana County Court of Common Pleas on June 25, 1936. That same day, he filed bond for $25,000. On July 13, 1936, the court authorized him to sell all the assets of the company at a public sale, which was then scheduled for 10:00 AM on September 10, 1936. For six weeks prior to the sale, advertisements were placed in the **Indiana Evening Gazette**, Indiana Progress, **Philadelphia Inquirer**, and the **New York Post**.

The largest purchaser at the sale was A.L. Light of Punxsutawney who bought most of the property for $37,400 and changed the name of the company to Coal Mining Company of Graceton. Nick Kovalchick of Kovalchick Salvage, Indiana bid $600 for the company store and .349 acres. However, A.L. Light bid an additional $250, which the court accepted. Also at the sale, 116 acres of surface were sold to John Benamati for $1,000, and 34.25 acres of coal were sold to Lisle Flickinger of Homer City for $1,300.

Coal continued to be mined in Graceton until the company was liquidated in December, 1970 at a stockholders' meeting. Oscar and Samuel Light each owned 31.25 per cent of the stock. The rest was owned by the Rochester and Pittsburgh Coal Company of Indiana.

Today Graceton is much like Vintondale and the other mining towns of the area, a residential-commuter town. Most of the traces of the coal mines and the coke ovens cannot be seen from Route 119, and a mine reclamation program was started in the spring of 1990. Much of the remaining coal of Graceton and neighboring Coral has been strip-mined.

1933-34 Vintondale High School Boy's Basketball Team
Row 1: Boyd Brandon, Mike Velisko, Arthur Antal, Charles Vaskovich, George Larish. Row 2: George Cooke, Pete Katasi, Steve Oblackovich, Mr. Barry Morris, Jack Butala, Jr., _____, _____, Charles Balog.
Courtesy of Vintondale Homecoming Committee.

1933-34 Vintondale High School Girls' Basketball Team
Row 1: Margaret Huth, Mary Skvarcek, Olga Kanich, Anna Hrabor, Anne Sermeg, Alice Sileck. Row 2: Claire Wise, Julia Yelonsky, Sara McPherson, Claire "Tood" Huth, Mary Rosacha, Elizabeth Hugya, Dorothy Whinnie.
Courtesy of Vintondale Homecoming Committee.

Additional Photographs

Charles Bennett Family. Standing in rear; Lilly Bennett Altemus, Oliver Bennett. Seated; Charles Bennett, Catherine Peddicord Bennett. Standing in front; Margaret Bennett Schroth.
Courtesy of Rhoda Bennett Garvin.

Otto Hoffman c. 1920. Dr. James P. MacFarlane in front on the superintendent's house.
Courtesy of John Huth Collection.

Coralee Craudell, Luke _____, and Twilight Gaster, Wehrum.

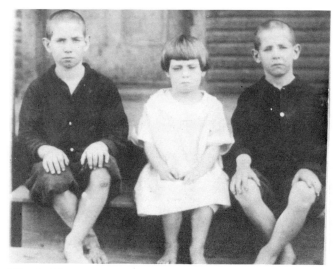

Merle Hunter, age 11, Doris Hunter, age 5, Bernard Hunter, age 7, c. 1923 in Wehrum.
  Courtesy of Winnie & Merle Hunter.

Amos Chick and Catherine, two daughters Alice and Cevilla taken in 1923 in Wehrum
  Courtesy of Catherine Chick.

Charles Hullihan on left, John Dutko on right c. 1918.
  Courtesy of Anne Dutko McGaughey.

John Dodd, Luke _____, A.M. Graffin, Wehrum Depot.

Annie Cilip Dutko holding daughter Anne, John Dutko with son John William, c. 1918.
  Courtesy of Anne Dutko McGaughey.

LACKAWANNA COAL & LUMBER CO.

701 Linden Street

SCRANTON, PA.

Courtesy of the late Robert Darr, retured executive of Rochester & Pittsburgh Coal Company.

**Lackawanna #3 c. 1922.
Courtesy of Wehrum Reunion Committee.**

**Blacklick Creek, narrow gauge between Wehrum and No. 3.
Courtesy of Wehrum Reunion Committee.**

**Lackawanna #3 c. 1922.
Courtesy of Wehrum Reunion Committee.**

**Freight station, Wehrum, 1914.
Courtesy of Wehrum Reunion Committee.**

Wehrum, doctor's office, between hotel and office on Broadway, 1934, Dr. Evans' name on window, Saint Fidelis Catholic Church on the right.
Courtesy of Gloria Mardis Risko.

Wehrum Company Store, 1926.
Courtesy of Glorida Risko.

Interior of chief clerk's house, c. 1918.
Courtesy of Mabel Davis Updike.

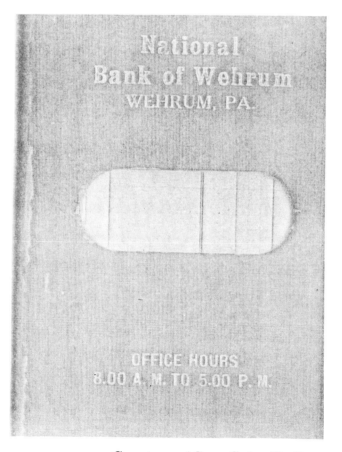

Courtesy of Sara Galer-Sholter.

Wehrum School, c. 1965.
Courtesy of Bernice Dixon, former teacher at the school.

Wehrum: Second, Third, and Fourth Streets, c. 1923.
Courtesy of Gloria Risko.

Broadway in Winter, c. 1918.
Courtesy of Mabel Davis Updike.

Wehrum. Philip Updike, Chief Clerk of Lackawanna Coal and Coke Company, inside chief clerk's house, c. 1918.
Courtesy of Mabel Davis Updike.

Charles Kirker, Wehrum yard boss, c. 1915.
Upper Broadway.

Courtesy of Thelma Kirker.

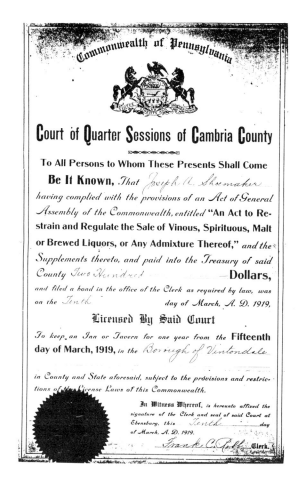

Vintondale. First Company Office of Vinton Colliery Company. Goat Hill houses to the right, 1906.
Courtesy of Denise Weber Collection
Pennsylvania State Archives, Harrisburg.

Wehrum, c. 1924. Star automobile won by Jim Tonagila - behind driver's seat, Domenick Monteleone in back seat, seated Alfred _____, and Dorothy Milazza Rok.

Courtesy of Dorothy Milazzo Rok.

Wehrum, 1926. Taken by Bethlehem Mines Corporation. Courtesy of Gloria Risko.

Town Plan: Wehrum
Revised January 20, 1925

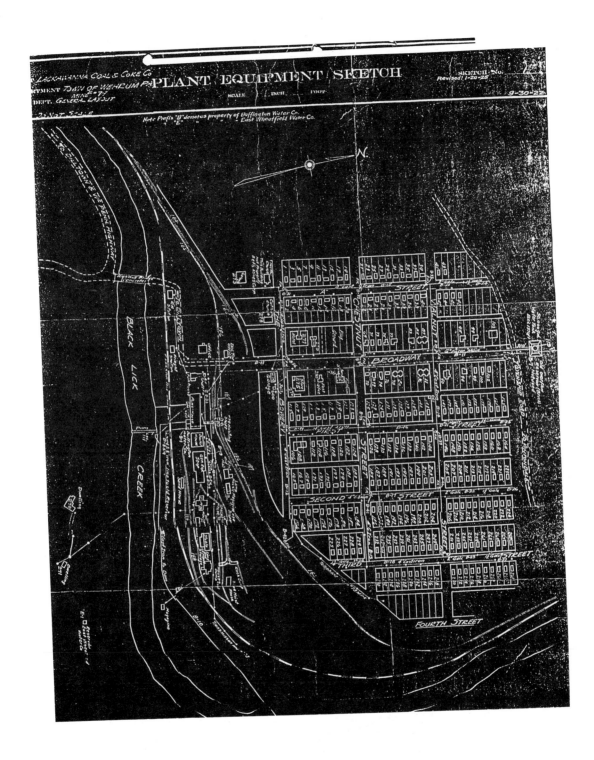

1928 Vinton Colliery Company

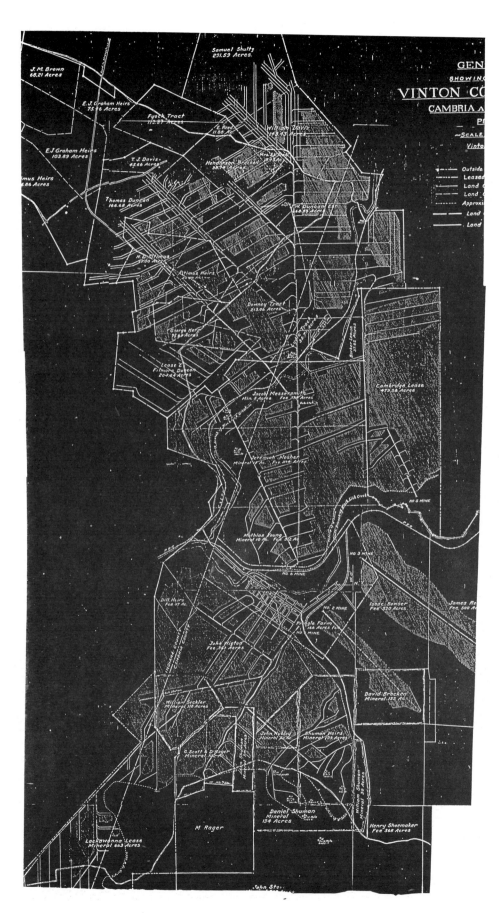

Lackawanna Coal and Coke Company
Indiana and Cambria Counties, 1901

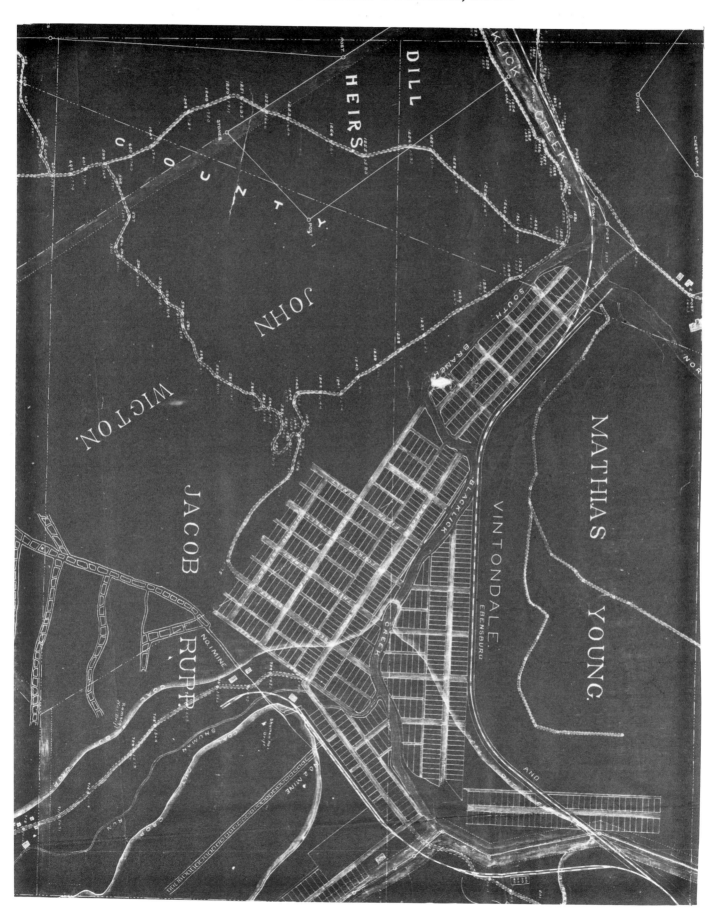

Appendix A
Birth Certificates: 1906-1921

AUTHOR'S NOTE: The following birth and death certificates were taken from Blacklick Valley register books found in the John Huth attic. The last two registrars were Harry Hampson, the author's great-grandfather, and John Huth, the author's grandfather. Other registrars were A.V. Caldwell, Samuel Williams, John Daly, Dr. Alvin Comerer, and L. Davis. These listings are incomplete as some of the registers are missing. Interpreting the various styles of handwriting caused the author some difficulty. Parenthesis next to a word indicates that the author was unable to correctly interpret the handwriting. Only the entries with a Vintondale address were included. Eventually, all entries for Nanty Glo, Twin Rocks and surrounding townships will be computerized and made available to the public.

BIRTH CERTIFICATES: 1906

Name of Child Dr. or Midwife Occupation	Date of Birth Father Mother	Time Age Age	Order of Birth Birthplace Birthplace	No. of Living Children
Laura Bracken Dr. J.A. Comerer Teamster See Death Certificate 1/8/1907	3/8/1906 Harry Bracken Bertha Walker	5:00AM 41 39	10 Indiana State Clinton Co.	5
Dorothy Beatrice Brett Dr. J.A. Comerer Merchant	3/11/1906 Abraham Jacob Brett Ida Sarah Block	11:00PM 29 24	3 Russia Russia	3
Frank Monar Mrs. M. Petrilla Miner	3/12/1906 Frank Monar Julie Krusman	7:30AM 28 26	1 Ouyar (Hungary?) Ouyar	1
Walter Arthur Lindros Dr. J.A. Comerer Miner	3/14/1906 Frank V. Lindros Alina Bitkavesi	6:00AM 30 31	4 Finland Finland	3
Joseph Gasko (Filed 1911) Agnes Such Toth Miner	4/5/1906 Joseph Gasko Margaret Neweonyik	 28 23	1 Hungary Hungary	1
Chester Harvey Graham Dr. J.A. Comerer Cook	4/14/1906 Paul Graham Iva Rairigh	4:00PM 23 18	1 Clearfield Co. Clearfield Co.	1
Michael Pisak (Filed 1/9/1911) Agnes Such Toth Miner See Death Certificate 9/2/1906	5/31/1906 Stephan Pisak Maria Kasras	 30 29	3 Hungary Hungary	3
Mary Ada Ruffner Dr. J.A. Comerer Merchant	6/5/1906 Sherman Ruffner Maggie Elkin	7:00AM 38 31	9 Indiana Co. Jefferson Co.	6

Name / Doctor / Occupation	Date	Time	Father / Mother / Ages	Birthplaces	
Elda M. Daughenbaugh Dr. J.A. Comerer Engineer	6/11/1906	1:00AM	Birt Daughenbaugh 35 Maggie Dearimin 34	7 Blair Co. Indiana Co.	6
Charles Edward Duncan Dr. J.A. Comerer Laborer	7/20/1906	6:30AM	George Duncan 30 Ella Stiles 23	2 Indiana Co. Indiana Co.	2
Donald Brown Blewitt Dr. J.A. Comerer Clerk	7/22/1906	1:00PM	George S. Blewitt 33 Jeanie Brown 34	2 England Pennsylvania	2
Stillborn - Marsh Dr. J.A. Comerer Miner See Death Certificate 7/23/1906	7/23/1906	2:00PM	Cyrus Marsh 24 Elsie Wagner 23	4 Johnstown Johnstown	1
Samuel Randolph Morris Dr. J.A. Comerer Merchant (Filed Twice)	8/3/1906	10:00AM	Thomas E. Morris 45 Ellen Burgoon 38	8 Cambria Co. Cambria Co.	6
No Name Given-Female Dr. G. Yearick, Wehrum Miner	9/22/1906		Albert Kreamer 26 Margaret Straw 22	2 Cambria Co. Clearfield Co.	1
Andrew Borcsik (Filed 1911) Agnes Toth Miner	10/14/1906		Joseph Borscik 38 Elizabeth Roh 26	2 Hungary Hungary	2
No Name Given-Male Dr. G. Yearick Miner	10/23/1906	11:40PM	George Lyle 30 Elizabeth Carins 21	1 Bradford Co. England	1
Anna Gray Dr. J.I. Pollum, V'Dale Miner	11/22/1906	4:30PM	Daniel Gray Nora Mulvihill 23	2 Loyalhanna, PA Broadtop, PA	2
Stillborn-Female Dr. J.I. Pollum Miner See Death Certificate 12/1/1906	12/1/1906	11:00PM	Frank Zorezo 35 Julia Corusut (?) 22	2 Hungary Hungary	1
Sheridan Rager Dr. J.I. Pollum Laborer	12/9/1906	2:00PM	S.H. Rager 28 F.E.M. Lefevere 25	5 Pennsylvania Pennsylvania	2
Paul O'Reilly Farabaugh Dr. J.A. Comerer Wholesale Liquior Dealer	12/9/1907 (Sic)	5:35PM	Frank S. Farabaugh 42 Felicitas Kane 36	3 Cambria Co. Cambria Co.	3
No Name Given-Male Dr. J.A. Comerer Miner	12/27/1906	7:00AM	Markus Heitanen 29 Areliima Ohrikuoma 28	6 Finland Finland	5

Birth Certificates: 1907

Lorrane Fitzgerald	1/17/1907	5:30AM	6	6
Dr. J.A. Comerer	Joseph Fitzgerald	37	Pennsylvania	
Miner	Cynthia Buchannon	29	Pennsylvania	
Maria Estok	2/1/1907		1	1
Agnes Toth	George Estok	27	Hungary	
Miner	Susan Varga	22	Hungary	
Mary Morin (Filed 1907)	2/6/1907		2	0
Dr. J.A. Comerer	Steven Morin	28	Hungary	
Miner	Mary Patiy	21	Hungary	
See Death Certificate 2/6/1907				
Andy Kes (Kish)	2/12/1907	5:00PM	1	1
Dr. J.A. Comerer	Samuel Kes	27	Hungary	
Miner	Maria Christian	19	Hungary	
Female, Name Not Sent	2/13/1907	10:00PM	6	6
Dr. J.A. Comerer	Anthony McMullen	31	Pennsylvania	
Miner	Emma Irwin	26	Pennsylvania	
Herrnui Marui, Stillborn-Female	2/15/1907		1, First Born Twin	0
Agnes Toth	Stephen Marui	29	Hungary	
Miner	Maria Peti	23	Hungary	
(Filed 1911)				
Margaret Marui (Filed 1911)	2/15/1907		2, Second Born Twin	1
See Above Information				
Helen May Kempher	3/18/1907	9:00AM	6	5
Dr. J.A. Comerer	D.H. Kempher	43	Pennsylvania	
Laborer	Anna Greoster (?)	32	Pennsylvania	
Jennie Rebo	4/1/1907	6:00AM	1	1
Dr. J.A. Comerer	Frank Rebo	27	Hungary	
Miner	Betha Urban	19	Hungary	
Fay Comerer	4/7/1907	9:00AM	2	1
Dr. F.C. Jones	Dr. J.A. Comerer	38	Pennsylvania	
Physician	Mary Lynch	21	Pennsylvania	
See Death Certificate 4/7/1907				
T.B. Hower	4/7/1907	10:00AM	5	5
Dr. J.A. Comerer	C.L. Hower	32	Pennsylvania	
Mine Superintendent	Ida Giddes	31	Pennsylvania	
Charles Donald Lewis	4/28/1907	8:00AM	5	
Dr. J.A. Comerer	William Lewis	32	Pennsylvania	
Miner	Bessie Comings	27	Pennsylvania	
Margaret Alma Riddle	5/1/1907	1:00AM	1	1
Dr. J.A. Comerer	Jno Riddle	25	Pennsylvania	
Engineer	Myrtle Brett (?)	21	Pennsylvania	

Name	Date	Time	Col5	Col6
Basilio L. Nicolazio Dr. J.A. Comerer Track Foreman	5/2/1907 Basilio Nicolazio Franeischina Bonadia	8:00PM 32 36	4 Italy Italy	3
No Name Given-Female Dr. J.A. Comerer Miner	5/16/1907 Charles McCoskey Elizabeth Younkins	9:30PM 29 25	5 Pennsylvania Pennsylvania	4
No Name Given-Male Dr. J.A. Comerer Miner	5/17/1907 Bert Teeter Bertha Singer	6:00PM 28 26	3 Pennsylvania Pennsylvania	3
Anne Helli Soffia Kivesta Dr. J.A. Comerer Miner See Death Certificate 8/9/1908	5/18/1907 Jacob Kivesta Soffia Ramsteam	9:00AM 31 31	3 Finland Finland	2
Marya Asplunb Dr. J.A. Comerer Miner	5/22/1907 Andy Asplunb Marya Valinoke	6:00AM 31 27	4 Finland Finland	4
Eino August Lippala Dr. J.A. Comerer Miner	5/22/1907 Matte Lippala Sonna Ranhala	11:00PM 22 27	1 Finland Finland	1
No Name Given-Male Dr. J.A. Comerer Miner	6/4/1907 E.S. Smith A. Carnahan (?)	3:00AM 26 26	1 Pennsylvania Pennsylvania	1
Stephen Szobonya (Filed 1911) Agnes Toth Miner	6/13/1907 John Szobonya Terezia	 39 32	5 Hungary Hungary	5
George C. Quelch Dr. J.A. Comerer Master Mechanic	6/25/1907 George C. Quelch Ella Hartpence	9:00AM 29 28	4 New York New Jersey	3
Unnamed Stillborn-Male Dr. J.A. Comerer Miner See Death Certificate 7/2/1907	7/2/1907 M. Labanitz Mary Risko	12:30AM 31 25	4 Hungary Hungary	0
Margaret Alice Dunmire Dr. J.A. Comerer Miner	7/10/1907 William Dunmire Clara Agnes Kelly	1:00PM 32 35	6 Pennsylvania Pennsylvania	6
No Name Given-Male Annie Petrilla Miner See Death Certificate 7/18/1S07	7/18/1907 Jno McKay Pearl Rager	5:00AM 33 18	2 Arnot, PA Jackson Twp., PA	1
Name Not Supplied to Reg. Dr. J.A. Comerer Miner	7/26/1907 E.L. Smith Aciustia (?)	7:30AM 26 26	1 Pennsylvania Pennsylvania	1

Name / Doctor / Occupation	Date	Time / Age	Col1	Col2
Fedora Kuhn Dr. J.A. Comerer Miner	7/27/1907 Walter Kuhn Harriet Hullihan	11:00AM 23 22	3 Pennsylvania Pennsylvania	2
James Penatzen Dr. J.A. Comerer Miner	8/9/1907 David Penatzen Millie Ashcroft	9:00AM 25 25	2 Pennsylvania Pennsylvania	2
Augustus Edward Coleman Dr. J.A. Comerer	8/14/1907 Peter Coleman Mary Rager	10:00PM 24 25	1 Pennsylvania Pennsylvania	1
No Name Given - Female Dr. J.A. Comerer Driver	8/17/1907 Paul Graham Iva Rairigh	5:00AM 25 19	2 Pennsylvania Pennsylvania	2
No Name Given - Male Dr. J.A. Comerer Merchant	8/17/1907 David Nevy Maria Racasi	7:00PM 28 22	1 Italy Italy	1
Annie Dalaney (Filed 1912) Mrs. M. Petrilla Blacksmith	8/29/1907 Joe Dalaney Mary Pator	1:30AM 38 26	4 Aboye (Hungary?) Bomere Madk	4
Mike Ribneziski Dr. J.A. Comerer Miner	9/6/1907 Mike Ribneziski Mari Gyacsok	10:00PM 27 21	1 Austria Austria	1 Austria
Grace Mary Watkins Dr. J.A. Comerer Miner	9/13/1907 David Watkins Carrie Hopkins	11:00PM 37 32	7 Pennsylvania Pennsylvania	7
Mary Kemecak Dr. J.A. Comerer Miner	9/21/1907 Jno Kemecak Lizzie Puios (?)	11:00PM 36 30	5 Hungary Hungary	4
Vaina Mikael Hakanen Dr. J.A. Comerer Miner	10/8/1907 Mikki Hakanen Maki Leakem	4:00PM 29 25	3 Finland Finland	3
Louise Magee Dr. J.A. Comerer Clerk	10/10/1907 Robert Magee Edith Rairigh	5:00AM 30 22	1 Pennsylvania Pennsylvania	1
Steve Hungo (Filed 1912) Mrs. M. Petrilla Storekeeper	10/22/1907 Steve Hungo Lizzie Kish	7:00AM 29 26	3 Hungary Hungary	3
John Barlogh Dr. J.A. Comerer Miner	10/26/1907 John Barlogh Bertha Nebes	7:00AM 22 21	2 Hungary Hungary	1
Reginna Isaacson Dr. J.A. Comerer Miner	11/3/1907 Jno Isaacson Alexandria Makki	10:00AM 29	5 Finland Vintondale (sic)	4

Chas. Douty	11/8/1907	2:00AM	5	5
Dr. J.A. Comerer	Chas. Douty	33	Pennsylvania	
Butcher	Mary Elands (?)	32	Pennsylvania	
Ellie Dausca	11/20/1907	9:00PM	3	2
Dr. J.A. Comerer	Demitus Daucsa	22	Hungary	
Miner	Annie Risko	24	Hungary	
Elizabeth Anna Saxon	11/21/1907	9:00PM	4	4
Dr. J.A. Comerer	William W. Saxon	27	Pennsylvania	
Machinist	May Bennett	26	Pennsylvania	
Mary Risko	11/23/1907	1:00PM	1	1
Dr. J.A. Comerer	Joseph Risko	29	Hungary	
Miner	Elislie Risko	26	Hungary	
See Death Certificate 1/30/1908				
Marshall L. Williams	12/1/1907	9:20PM	11	6
Mrs. Jno Condo	Samuel R. Williams	40	Pine Twp., PA	
Merchant	Margaret Sides	37	Green Twp., PA	
Leonard Archie Brett	12/6/1907	11:00PM	4	4
Dr. J.A. Comerer	Jacob Brett	31	Russia	
Merchant	Ida Block	26	Russia	
Unnamed Female	12/10/1907	3:00PM	4	1
Dr. J.A. Comerer	Paul Palone	31	Hungary	
Miner	Mary Balogh	28	Hungary	
See Death Certificate 12/10/1907				
Robert Rood Adams	12/27/1907	7:00PM	2	2
Dr. J.A. Comerer	Robert B. Adams	26	Pennsylvania	
Clerk	Nellie M. Rood	27	Pennsylvania	
Stif John Dobos	12/27/1907	3:00PM	1	1
Dr. J.A. Comerer	Stif Dobos	32	Hungary	
Miner	Juli Markovics	19	Hungary	
Jennings J. Schultz	12/29/1907	3:00PM	2	2
Dr. J.A. Comerer	Blair Shultz	34	Pennsylvania	
Mine	Linnie Foredline	30	Pennsylvania	

Birth Certificates: 1908

Ernest M. Meisner	1/4/1908	9:00AM	6	6
Dr. J.A. Comerer	Henry Meisner	30	Pennsylvania	
Laborer	Annie McMahon	27	Pennsylvania	
Laaja Mikael Hakanen	1/11/1908	6:00AM	1	1
Dr. J.A. Comerer	Jno Hakanen	26	Finland	
Miner	Mary Kivesta	19	Finland	
Kenneth F. Causer	1/16/1908	2:00AM	1	1
Dr. J.A. Comerer	Frederick Causer	22	England	
Miner	Edna Hunter	19	Pennsylvania	

Child	Date	Time		
Elmer Harclerode	1/29/1908	7:30AM	1	1
Dr. Geo. Yearick	J.E. Harclebrode	28	Pennsylvania	
Agent-Railroad	Margie A. Booth	25	Pennsylvania	
Robert Custer	2/3/1908	5:00PM	2	2
Dr. J.A. Comerer	Robert M. Custer		Pennsylvania	
Mining Engineer	Ida Filer	29	Pennsylvania	
Edward Elmer Dill	2/3/1908	2:00PM	1	1
Dr. J.A. Comerer	Edward L. Dill	25	Pennsylvania	
Miner	Mary Wright	17	Pennsylvania	
No Name Given-Male	2/17/1908	3:00AM	7	5
Dr. J.A. Comerer	Harry Bishop	36	Pennsylvania	
Miner	Lenie Jose	32	England	
Stillborn Twins-Female	2/24/1908	4:00PM	7	5
Dr. Geo. Yearick	Martin Rager	28	Pennsylvania	
Farmer	Harriet Whysong	32	Pennsylvania	
Mary Renik	2/26/1908	9:00AM	2	2
Dr. J.A. Comerer	Andy Renik	29	Austria-Hungary	
Miner	Annie Salog (?)	21	Austria-Hungary	
Mabel Clare Hotten	3/6/1908	4:00AM	1	1
Dr. J.A. Comerer	Joseph Hotten	30	Austria	
Miner	Margaret McQueeney	20	Pennsylvania	
Catherine D. Stopp	3/13/1908	2:00PM	2	2
Dr. J.A. Comerer	Jno Stopp	35	Pennsylvania	
Miner	Fannie Hullihan	28	Pennsylvania	
Fravey Crakus	3/15/1908	3:00AM	5	3
Dr. J.A. Cromerer	Jno Crakus	28	Poland	
Miner	Mary Bagan	22	Poland	
Malcolm P. Williams	3/17/1908	10:00PM	1	1
Dr. J.A. Comerer	Cloyd Williams	20	Pennsylvania	
Miner	Annie Madill	21	Pennsylvania	
May Davis	4/10/1908	1:00AM	9	7
Dr. J.A. Comerer	Arthur Davis	42	England	
Mine Foreman	Sarah A. Ward	40	England	
Anna Louise Duncan	4/19/1908	12:00PM	4	3
Dr. J.A. Comerer	David C. Duncan	32	Scotland	
Miner	Maggie E. Duncan	27	Scotland	
John Louma	4/20/1908	5:00AM	3	3
Dr. J.A. Comerer	John Louma	25	Finland	
Miner	Hulda Maki	22	Finland	
Ellie Kozer (Filed 1910)	5/3/1908		3	3
Dr. J.A. Comerer	Charley Kozer	26	Austria	
Miner	Ellie Balogh	21	Austria	

Name	Date	Time	Father / Mother	Age	Birthplace	Col
Annie Jeshe Dr. J.A. Comerer Miner	5/9/1908	2:00AM	Joseph Jeshe Mary Marta	23 25	1 Poland Poland	1
Jno Farkas Dr. J.A. Comerer Miner	5/13/1908	1:00PM	Jno Farkas Bessie Lessas	27 23	1 Hungary Hungary	1
Mary Helen George Dr. J.A. Comerer Miner	5/25/1908	2:00AM	Edward George Lulla Farwell	31 19	3 Pennsylvania Pennsylvania	2
Helen May Causer Dr. J.A. Comerer Miner	6/11/1908	3:00PM	Jno H. Causer Alice A. Lovatt	42 40	15 England England	7
Dorothy Virginia Williams Dr. J.A. Comerer Bookkeeper	6/22/1908	2:00PM	William G. Williams Mary Berkey	20 20	1 Pennsylvania Pennsylvania	1
Andrew Balog (Filed 1911) Agnes Toth Miner	6/28/1908		Vincent Balog Christina Kovacs	24 21	1 Hungary Hungary	1
Ora Douty Dr. J.A. Comerer Driver, Delivery Wagon	7/3/1908	2:00AM	Lester L. Douty Alice A. Millen	35 35	7 Pennsylvania Pennsylvania	5
Marion J. Smith Dr. J.A. Comerer Machinist	7/28/1908	1:00PM	Andrew C. Smith Sarah Replogle	27 27	3 Pennsylvania Pennsylvania	3
Stif Risko Dr. J.A. Comerer Miner	8/1/1908	1:00PM	Jno Risko Annie Kozab	55 40	5 Russia Russia	4
Mary Apsj Dr. J.A. Comerer Miner	9/5/1908	5:00AM	Jno Apsj Mary Yuhasz	28 18	1 Slav Slavish	1
Saul Smith Douty Dr. J.A. Comerer Teamster	10/6/1908	2:00AM	Sydney B. Douty Mary D. Snoak	32 29	4 Pennsylvania Pennsylvania	4
Joseph Kisch Dr. J.A. Comerer Miner	10/7/1908	1:00PM	Samuel Kish Mary Christian	28 20	2 Hungary Hungary	2
Joseph Borscik (Filed 1911) Agnes Toth Miner	10/23/1908		Joseph Borscik Elizabeth Roh	40 28	3 Hungary Hungary	3
Rena Commons Dr. J.A. Comerer Motorman	10/30/1908	7:00AM	George Commons Bertha Butterbaugh	34 27	4 Pennsylvania Pennsylvania	3

Child / Attendant / Father's Occupation	Date	Time	Father / Mother / Ages	Birthplace	
Vida May Lynch Dr. J.A. Comerer Laborer	11/5/1908	11:00AM	Eugene Lynch 20 Gretta Strausbaugh 18	1 Pennsylvania Pennsylvania	1
Nicoletta Bacilio Dr. J.A. Comerer Miner	11/7/1908	11:00AM	Bacilio Nicalozio 34 Francischina Bonadio 38	5 Italy Italy	4
Mildred I. Stephens Dr. J.A. Comerer Laborer	11/8/1908	4:00AM	Bailey S. Stephens 36 Laura Lynch 26	5 Pennsylvania Pennsylvania	5
Jno M. Mayer, Jr. Dr. J.A. Comerer Clerk	11/21/1908	3:00AM	Jno M. Mayer 28 Catherine M. Logan 26	2 Pennsylvania Pennsylvania	2
Iona Edwards Dr. J.A. Comerer Merchant	11/29/1908	5:00PM	Burton T. Edwards 26 Maggie Wilt 25	1 Pennsylvania Pennsylvania	1
Mary Monar (Filed 1912) Mrs. M. Petrilla Miner	12/1/1908	11:00AM	Frank Monar 30 Julie Krusnam 28	2 Ougur (Hungary?) Ougur	2
Margaret Gasko (Filed 1911) Agnes Toth Miner	12/3/1908		Joseph Gasko 30 Margaret Nemesnyik 25	2 Hungary Hungary	2
Selina Maria Kavna Dr. J.A. Comerer Miner See Death Certificate 12/8/1908	12/8/1908	11:00AM	Victor Kavna 28 Mary Aspland 21	1 Finland Finald	1
Samuel Griffiths (Filed 1911) Agnes Toth Miner	12/21/1908		Evan Griffiths 40 Marian Phillips 35	7 Wales Wales	7
Lucy Rager Dr. J.A. Comerer Machinist	12/28/1908	10:00AM	J. Henry Rager 31 M.M. Stayer 34	4 Pennsylvania Pennsylvania	4

Birth Certificates: 1909

Child / Attendant / Father's Occupation	Date	Time	Father / Mother / Ages	Birthplace	
Jno Peloni Dr. J.A. Comerer Miner See Death Certificate 2/14/1909	1/2/1909	4:00PM	Paul Polani 31 Mary Balogh 28	5 Hungary Hungary	2
Julie Monar (Filed 1912) Mrs. M. Petrilla Miner	1/31/1909	1:30AM	Frank Monar 31 Julie Krusnam 29	3 Ougar Ougar	3
Aili Amaliia Kivesta Dr. J.A. Comerer Miner	2/18/1909	5:00AM	Jacob Kivesta 33 Sofia Ranstam 33	4 Finland Finland	2

Francis R. McMullen Dr. J.A. Comerer Miner	3/5/1909 Anthony McMullen Emma Irwin	5:00PM 32 27	7 Pennsylvania Pennsylvania	6	
Unnamed Male Dr. J.A. Comerer Miner See Death Certificate 3/26/1909	3/26/1909 George Bracken Maggie Rhodes	8:00PM 38 36	9 Pennsylvania Pennsylvania	7	
Catherine Hower Dr. J.A. Comerer Mine Superintendent	3/31/1909 Charles H. Hower Ida Geddes	7:00AM 34 32	6 Pennsylvania Pennsylvania	6	
Ellie Resco Evi Petrilla Miner See Death Certificate 4/19/1909	4/19/1909 Joe Recko Ellie Recko	8:00PM 32 26	2 Austria Austria	1	
Laura Edith Patterson Dr. J.A. Comerer Laborer	4/21/1909 W.R. Patterson Josephine Showers	10:00PM 32 30	3 Pennsylvania Pennsylvania	3	
Stillborn-Male Dr. J.A. Comerer Miner See Death Certificate 5/7/1909	5/7/1909 Mike Rosie Katie Kalibell (?)	11:30PM 25 25	1 Austria Austria	0	
Jennie Alice Daniels Mrs. Lester Douty Miner See Death Certificate 8/15/1909	5/20/1909 Franklin Daniels Emma F. Daniels	4:00AM 40 37	10 Somerset Co., PA Illinois	8	
Regina Theresa Farabaugh Dr. J.A. Comerer Wholesale Liquoir Dealer	5/27/1909 Francis Farabaugh Felicitus Kane	6:00PM 44 38	4 Pennsylvania Pennsylvania		
Unnamed-Male Mrs. Evi Petrilla Miner See Death Certificate 6/9/1909	6/9/1909 Joseph Delaney Mary Pasigtor (?)	1:00AM 40 28	8 Austria Austria	4	
Annie Gaboda Dr. J.A. Comerer Miner	7/5/1909 Joseph Gaboda Annie Druga	3:00AM 28 21	3 Austria-Hungary Austria-Hungary	3	
Dora Catherine Graham Dr. J.A. Comerer Merchant	7/7/1909 Paul Graham Iva Rairigh	8:00PM 26 21	3 Pennsylvania Pennsylvania	3	
Peter Kmeeak (Filed 1911) Agnes Toth Laborer	9/7/1909 John Kmeeak No Name Given	 36 30	6 Hungary Hungary	6	
Lizzie Hango (Filed 1912) Mrs. M. Petrilla Store Keeper	9/12/1909 Steve Hongo Lizzie Kish	2:30AM 31 28	4 Hungary Hungary	4	

Harold Alorn Douty Dr. J.A. Comerer Butcher	10/4/1909 Charles Douty Ellen Condo	8:00PM 35 34	4 Pennsylvania Pennsylvania	4
Vincent Balog (Filed 1911) Agnes Toth Miner	10/13/1909 Vincent Balog Christina Kovacs	25 22	2 Hungary Hungary	2
Clarence Edward Beers Dr. J.A. Comerer Carpenter	10/17/1909 Porter N. Beers Susan Kough	7:00PM 30 26	4 Pennsylvania Pennsylvania	4
John Wilbur Dill Dr. J.A. Comerer Laborer	11/15/1909 Elmer L. Dill Mary Wright	12 Noon 27 19	2 Pennsylvania Pennsylvania	2
Raymond Samuel Irvin Dr. J.A. Comerer Miner	11/13/1909 Samuel Irvin Annie McQueeney	11:30AM 23 20	1 Pennsylvania Pennsylvania	1
Mary J. Meisner Dr. J.A. Comerer Laborer	11/22/1909 Henry H. Meisner Annie McMahon	4:00AM 30 28	7 Pennsylvania Pennsylvania	7
Giacomino Lentine Dr. J.A. Comerer Miner	12/9/1909 Giacorui Lentine Alfonsena Gravagna	11:00AM 29 22	3 Italy Italy	3
Mary Ribnicki None Miner See Death Certificate 2/9/1910	2/20/1909 Mike Ribnicki Mary Dahok	8:00AM 30 23	3 Austria-Hungary Austria-Hungary	2
Giovanni Righetti Dr. J.A. Comerer Miner	12/20/1909 Guiseppe Righetti Angela Peccinni	10:00PM 28 27	3 Italy Italy	3
Jerome Raymond Patterson Dr. J.A. Comerer Laborer	12/27/1909 Fred Patterson Nancy Robinson	3:00AM 22 22	1 Pennsylvania Pennsylvania	1
Florence Lewis Dr. J.A. Comerer Miner	12/29/1909 Wm. Lewis Bessie Cummins	2:00AM 34 28	Pennsylvania Pennsylvania	

Birth Certificates: 1910

Katie Jusfustia Dr. J.A. Comerer Miner See Death Certificate 3/22/1910	2/9/1910 Joseph Jusfustia Mary Marta	5:00AM 25 26	2 Austria-Hungary Austria-Hungary	2
Maria Pisok (Filed 1911) Agnes Toth Miner See Death Certificate 5/5/1911	2/21/1910 Stephan Pisok Maria Koszos	34 33	6 Hungary Hungary	6

Name / Physician / Occupation	Date / Parents / Ages	Time	Col1	Birthplace	Col2
Isabel Madill Dr. J.A. Comerer Miner See Death Certificate 7/22/1910	3/6/1910 Wm. J. Madill — 26 Mary Chambers — 20	11:00PM	1	Pennsylvania Pennsylvania	1
Stif Kisch (Steven Kish) Dr. J.A. Comerer Miner	3/10/1910 Samuel Kisch — 30 Mary Christian — 23	7:00PM	3	Hungary Hungary	3
Susie Arauka Kuzdenyi Dr. J.A. Comerer Clerk	3/11/1910 Allanar Kuzdenyi — 46 Mary Kizak — 28	8:00AM	3	Hungary Hungary	1
Andrew Gasko Agnes Toth Miner	3/25/1910 Joseph Gasko — 32 Margaret Neinesnyik — 27		3	Hungary Hungary	3
Robert Charles Morris Dr. J.A. Comerer Butcher	4/14/1910 Walter Morris — 29 Lottie Barry — 22	10:00PM	3	Pennsylvania Pennsylvania	3
Guisseppe Verderico Dr. J.A. Comerer Miner	4/17/1910 Gos. Verderico — 28 Giovoneoieo Gentui (?) — 20	2:00AM	1	Italy Italy	1
Frederick Adams Dr. J.A. Comerer Clerk	5/20/1910 R. Bruce Adams — 28 Nellie Rook — 29	10:00PM	3	Pennsylvania Pennsylvania	3
John Sodost Dr. J.A. Comerer Miner	5/20/1910 John Sodost — 36 Katie Orvat — 32	6:00AM	6	Connecticut Austria	5
Americo Bazzoli Dr. J.A. Comerer Miner	5/24/1910 Nicola Bazzoli — 32 Soffia Imistadi — 31	5:00AM	3	Austria Austria	3
John Sodost (Filed Twice) Dr. J.A. Comerer Miner	5/30/1910 John Sodost — 36 Katie Owat — 32	6:00AM	6	Connecticut Austria	5
Charley Cyrus Dr. J.A. Comerer Miner See Death Certificate 1/5/1911	6/7/1910 Albert Cyrus — 34 Lizzie Skura — 29	6:00PM	5	Austria Austria	5
Julia Belicsak Agnes Toth Miner	6/9/1910 John Belicsak — 38 Anna Yakuboowelie — 26		5	Hungary Hungary	5
Susan Estok Agnes Toth Miner	6/9/1910 George Estok — 30 Susan Varga — 25		2	Hungary Hungary	2
Mike Gaboda (Filed 1912) Mrs. M. Petrilla Miner	6/14/1910 Mike Gaboda — 26 Elama Famonik — 22	7:00PM	1	Mad Gocknak Mad Gocknak	1

Name / Doctor / Occupation	Date / Father / Mother	Time / Father Age / Mother Age	Child # / Father Birthplace / Mother Birthplace	Living
Edward Leonard Daniels Father Miner	6/18/1910 Frank P. Daniels Emma Balsinger	7:00AM 41 37	11 Pennsylvania Pennsylvania	8
Richard Blair Shilling Dr. M. Austin, Wehrum Blacksmith	6/21/1910 David R. Shilling Ida Manies (?)	11:00AM 27 20	1 Pennsylvania Pennsylvania	1
Frank Beregsasy (Filed 1912) Mrs. M. Petrilla Miner	6/29/1910 Charley Beregsasy Borok Borkey	4:00PM 47 38	11 Bosoke Stolet Bosoke Stolet	11
Francis Eudress Dr. J.A. Comerer Butcher	7/1/1910 Herman Eudress Nina Gaster	9:00PM 29 28	1 Pennsylvania Pennsylvania	1
Marian H. Thomshaw Dr. J.A. Comerer Miner	7/7/1910 Michael Thomshaw Agnes Pointon	11:00PM 19 20	1 Pennsylvania Pennsylvania	1
Lester Lloyd Williams Dr. J.A. Comerer Merchant	7/7/1910 Samuel R. Williams Margaret Sides	2:00PM 43 38	12 Pennsylvania Pennsylvania	7
Vincenty Luergynski Dr. J.A. Comerer Miner	7/19/1910 Pete Luerozynski Goanns Gontouske (?)	3:00AM 31 19	1 Poland Poland	1
John Klem (?) Dr. J.A. Comerer Miner	7/26/1910 Charley Klem (?) Annie Balogh	11:00PM 30 24	5 Austria Austria	3
Metro Balogh Dr. J.A. Comerer Miner	8/6/1910 Metro Balogh Mary Verbies	6:00PM 22 19	1 Austria Austria	1
Zoltan Babincsak Dr. J.A. Comerer Miner	8/27/1910 John Babincsak Barbala Balogh	1:00PM 29 26	5 Hungary Hungary	4
Mabel Elenora Long Dr. J.A. Comerer Barber	8/29/1910 Samuel L. Long Blanche George	6:30AM 28 27	1 Pennsylvania Pennsylvania	1
Julius Balog Agnes Toth Miner	9/4/1910 Vincent Balog Christine Kovacs	 26 23	3 Hungary Hungary	3
August Corti Dr. Austin, Wehrum Miner	9/4/1910 Oscar Corti Lavia Vitzelyon	1:45AM 30 30	1 Italy Italy	1
Elmer Douty Dr. J.A. Comerer Clerk	9/13/1910 Lester L. Douty Alice Miller	5:00PM 38 38	 Pennsylvania Pennsylvania	

Name / Attendant / Occupation	Date / Father / Mother	Time / Father Age / Mother Age	Children / Father Birthplace / Mother Birthplace	Living
Robert H. Stephens Dr. J.A. Comerer Engineer	9/14/1910 Bailey S. Stephens Laura Lynch	4:00PM 38 28	6 Pennsylvania Pennsylvania	6
Majorie Emma Travers Dr. J.A. Comerer Miner	9/23/1910 Joseph Travers Elsie M. Sowers	11:00AM 24 17	1 Maryland Pennsylvania	1
Pearl Palonyi Agnes Toth Miner	9/30/1910 Louis Palonyi Julia Pastor	 37 29	3 Hungary Hungary	3
Mary Bury Dr. J.A. Comerer Miner	10/3/1910 George Bury Pelouji Posulki	 22 18	1 Austria Austria	1
John Szabo Agnes Toth Miner	10/13/1910 George Szabo Julia Ceuk	 35 26	 Hungary Hungary	
Clorinda (?) Caliari Dr. J.A. Comerer Miner	10/28/1910 Carle Caliari Pregneti Fortunato	2:00AM 29 20	2 Italy Italy	2
Mike Sobo Dr. J.A. Comerer Miner	11/3/1910 George Sobo Julian Senkia	 33 26	3 Austria Austria	0
Erina Grace Nale Dr. J.A. Comerer Miner	11/5/1910 Marcus C. Nale Ella Yingling	6:00AM 50 43	12 Pennsylvania Pennsylvania	10
Charles Hayden Griffith Dr. J.A. Comerer Miner	11/9/1910 Evan Griffith Mary Jones	9:00PM 43 37	8 Wales Wales	8
George Russel Coleman Dr. J.A. Comerer Laborer	11/17/1910 Peter Coleman Mary Rager	8:00AM 27 28	3 Pennsylvania Pennsylvania	3
Anna Wesnicki Dr. J.A. Comerer Miner	11/28/1910 Villem Wesnicki Elizabeth Andrichik	1:00PM 24 17	1 Russia Russia	1
John Borzcek Dr. J.A. Comerer Drayman	11/30/1910 Joseph Borzcik Elizabeth Roh	9:00AM 41 28	5 Austria Austria	4
Albert Adams Dr. J.A. Comerer Teamster	12/10/1910 Harry A. Adams Ada P. Rowe	4:00PM 20 17	1 Pennsylvania Pennsylvania	1
Mary Margaret Burgan Dr. J.A. Comerer Shipping Clerk	12/12/1910 John L. Burgan Annie Davidson	1:00AM 21 24	1 Pennsylvania Pennsylvania	1

David H. Kempfer Dr. J.A. Comerer Station Engineer	12/14/1910 D.H. Kempfer Anna R. Treaster	2:00AM 47 37	7 Pennsylvania Pennsylvania	7
Mabel O. Meisner Dr. J.A. Comerer Laborer	12/18/1910 Henry H. Meisner Annie McMahon	6:00AM 31 29	8 Pennsylvania Pennsylvania	8
Catherine Bronzbry Dr. J.A. Comerer Miner	12/24/1910 Michael Bronzbvry Anna Burnett	3:00AM 23 19	2 Poland Pennsylvania	2
Rachel Parsley Dr. J.A. Comerer Pumpman	12/27/1910 John Parsley Margaret B. Warner	2:00AM 36 20	2 Pennsylvania Pennsylvania	2
Granville K. Eaton Dr. J.A. Comerer Fireboss	12/30/1910 William R. Eaton Isabel Butterbaugh	6:00AM 31 28	7 Scotland Pennsylvania	6

Birth Certificates: 1911

Tony Wojtawics Dr. J.A. Comerer Miner	1/12/1911 Andy Wojtawics Name Not Given	 24 	1 Poland 	1
Mary Babich Dr. J.A. Comerer Coal Miner	1/14/1911 George Babich Mary Risko	2:00AM 23 16	1 Austria Austria	1
Maria Sandor Harnocs Agnes S. Toth Miner	1/17/1911 John Harnocs Maria Sandor	3:00AM 28 27	2 Hungary Hungary	1
Bertha Gertrude Michaels Had No Physician Farmer	1/21/1911 Isaac Michaels Clara May Jones	8:00AM 32 25	6 Pine Twp., Indiana Co. Blacklick Twp., Camb. Co.	6
Helen Norgvist (Filed Twice) Dr. J.A. Comerer Coal Miner	1/30/1911 Jacob Norgvist Sofia Kainstr	2:00AM 31 32	8 Finland Finland	5
Andy Toht Dr. J.A. Comerer Coal Miner	2/8/1911 Andy Toht Annie Gabich	6:00PM 38 34	3 Austria Austria	2
Mary Mildred Beers Dr. J.A. Comerer Carpenter	2/10/1911 Porter N. Beers Susan E. Kough	6:00PM 31 28	5 Pennsylvania Pennsylvania	5
Joseph Brisko Agnes Such Toth Miner	2/10/1911 MikeBrisko Julia Szabo	5:00AM 33 30	1 Hungary Hungary	1
Amelia B. Kempfer Dr. J.A. Comerer Laborer	2/17/1911 Arthur S. Kempfer Gertrude Hassinger	3:00AM 40 35	5 Illinois Pennsylvania	5

Emma Nevy Dr. J.A. Comerer Merchant	2/18/1911 Davy Nevy Mary Reachasin	7:30AM 31 25	2 Italy Italy	2
Stillborn-Male Dr. J.A. Comerer Miner See Death Certificate 3/9/1911	3/9/1911 John Shurlock Katie Krivak	3:00PM 30 23	3 Russia Russia	0
Joseph Landon Agnes Such Toth Miner	3/17/1911 Paul Landon Elizabeth Mohnar	4:00AM 23 22	1 Hungary Hungary	1
Sarah Yete Myers Dr. J.A. Comerer Merchant	3/23/1911 Isaac M. Myers Rachel Brett	11:00AM 34 31	2 Russia Russia	1
Stillborn-Female Agnes Such Toth Miner See Death Certificate 4/1/1911	4/1/1911 Steve Pisak Maria Kaszat	6:00AM 35 34	7 Hungary Hungary	3
Stillborn-Female (Filed Twice) Dr. J.A. Comerer Miner	4/1/1911 Stif Pisak Mary Kosach	5:00AM 37 32	7 Austria Austria	3
Frederick Hollas Dr. J.A. Comerer Miner See Death Certificate 7/30/1911	4/8/1911 Enick Hollos Mary Lacoski	2:00PM 26 21	3 Poland Poland	3
Dominik Malpoli Dr. J.A. Comerer Miner	4/9/1911 Louis Malpili Josephine Consinati	4:00AM 26 20	2 Italy Italy	2
Katrina Delaney Mrs. M. Petrilla Blacksmith See Death Certificate 10/4/1912	5/2/1911 Joe Delaney Mary Pastor	5:00AM 41 29	6 Aleoye Borsm Madk	6
Unnamed-Female Dr. J.A. Comerer Machinist	5/3/1911 Thos. F. Stein Mary O'Colnan	3:00AM 38 39	10 Pennsylvania Pennsylvania	6
Lewey Marian Dr. J.A. Comerer Miner See Death Certificate 11/15/1911	5/20/1911 Steve Marian Mary Pate	8:00PM 32 26	4 Hungary Hungary	1
Margaret Hongo (Filed 1912) Mrs. M. Petrilla Store Keeper	5/20/1911 Steve Hongo Lizzie Kish	5:00AM 33 30	5 Hungary Hungary	5
Gino Avalli Dr. J.A. Comerer Miner	5/22/1911 Gont (?) Avalli Maria Nevi	4:00PM 25 24	4 Italy Italy	3

Name / Doctor / Occupation	Date / Father / Mother	Time / Father Age / Mother Age	Children / Father Birthplace / Mother Birthplace	Living
Mary Suray Dr. J.A. Comerer Miner	5/25/1911 John Suray Annie Andrew	 27 26	4 Austria Austria	3
Dominico Arbertelli Dr. J.A. Comerer Miner	5/29/1911 Angelo Albertelli Ninciato Covazzini	6:00AM 55 38	8 Italy Italy	6
Charley Risco Dr. J.A. Comerer Miner	5/31/1911 John Risco Ann Kozab (?)	2:00AM 55 42	12 Russia Russia	8
Margaret Ivans Agnes Toth Miner	6/18/1911 John Ivan Verona Aeil (?)	4:00AM 25 21	2 Hungary Hungary	2
Ella Jendrik Dr. J.A. Comerer Bartender	6/22/1911 Joseph Jendrick Maggie Kalina	4:00AM 36 23	1 Hungary Hungary	1
Stillborn Female Dr. J.A. Comerer Miner See Death Certificate 7/3/1911	7/3/1911 Metro Ballogh Mary Verbich	11:00AM 23 21	2 Russia Russia	1
Albert Alvin Cramer Dr. J.A. Comerer Lumberman	7/5/1911 Peter B. Cramer Mary Bracken	4:00AM 30 21	4 Pennsylvania Pennsylvania	3
Victor Sopina Mary Shisler Miner	7/18/1911 Mato Sopina Matilda Olgosky	11:30PM 30 25	2 Austria Austria	2
Joseph Ostafin Mary Shisler Miner	7/25/1911 August Ostafin Mary Taryanya	10:00PM 35 30	5 Austria Austria	5
Mary Griscisin Dr. J.A. Comerer Miner See Death Certificate 11/6/1911	8/16/1911 Joseph Griscisin Mary Marcyak	2:00PM 27 29	3 Russia Russia	2
Stillborn-Male Dr. J.A. Comerer Laborer See Death Certificate 8/21/1911	8/21/1911 Thos. Altemus Lilly M. Bennett	10:40PM 21 20	1 Pennsylvania Pennsylvania	0
Thelma Marie Travers Dr. J.A. Comerer Laborer	8/24/1911 Joseph F. Travers Elsie Sowers	9:00PM 24 18	2 Maryland Pennsylvania	2
John Monar (Filed 1912) Mrs. M. Petrilla	8/30/1911 Frank Monar Julia Krusnam	7:00AM 33 31	4 Ougar Ougar	4

Charles Occkan (?) Marie Sisler Miner See Death Certificate 2/17/1914	9/2/1911 Pete Occkan Marie Demyan	11:00PM 30 28	2 Austria Austria	2
Mary Catherine Clyde Dr. J.A. Comerer Motorman	9/22/1911 Robert Clyde Noona V. Hullihan	8:00AM 38 32	5 Pennsylvania Pennsylvania	5
Elizabeth Daly Dr. J.A. Comerer Engineer	10/3/1911 Herbert Daly Cora Ella Shaffer	2:00AM 25 23	1 Pennsylvania Pennsylvania	1
Daniel Mathias Kuzdenyi Dr. J.A. Comerer S.S. Agent	10/3/1911 Aladar Kuzdenyi Mary Kizak	2:30AM 49 32	4 Hungary Hungary	2
Ruth Abrams Dr. J.A. Comerer Miner Foreman	10/6/1911 Abram Abrams Mary Powell	1:00AM 38 29	4 Pennsylvania Pennsylvania	3
Elizabeth Gasko Agnes Such Toth Miner	10/14/1911 Joseph Gasko Margaret Memesnyik	8:00PM 27 23	3 Hungary Hungary	3
Mary Marko Mrs. M. Petrilla Miner	10/23/1911 John Marko Annie Podlake	8:00AM 32 26	3 Galetamm, Austria Do, Austria	3
Goldie Agnes McMullen Dr. J.A. Comerer Coal Miner	10/24/1911 Anthony McMullen Emma Irvin	5:00AM 36 30	8 Pennsylvania Pennsylvania	7
Sadie Lewis Mrs. N.A. Evans Miner	11/5/1911 Joseph Lewis Minnie Jenkins	11:30 34 26	3 Austria Pennsylvania	3
Elizabeth Madill Dr. J.A. Comerer Fireboss	11/7/1911 Wm. J. Madill Mary Chambers	2:00AM 27 20	2 Pennsylvania Pennsylvania	1
Helen Kisch Dr. J.A. Comerer Miner	11/15/1911 Samuel Kish Mary Christain	4:00PM 31 23	4 Hungary Hungary	4
William A. Schroyer Dr. J.A. Comerer Miner See Death Certificate 8/13/1912	11/20/1911 Wm. L. Schroyer Annie Pugh	8:00AM 36 25	5 Pennsylvania Pennsylvania	4
Norman Kenneth Stephens Dr. J.A. Comerer Engineman	12/2/1911 Bailey Stephens Laura Lynch	8:00PM 39 29	7 West Virginia Pennsylvania	7
Annie Babich Dr. J.A. Comerer Miner	12/15/1911 George Babich Mary Risco	7:00AM 25 18	2 Hungary Hungary	1

Mattie Berge Bell	12/17/1911	7:00PM	1		1
Dr. J.A. Comerer	George W. Bell	26	Virginia		
Staty. Engr.	Florence Butler	19	England		
Andy Gaboda	12/26/1911	7:00AM	2		2
Mrs. M. Petrilla	Mike Gaboda	27	Austria		
Miner	Elana Famomk	23	Austria		
Stillborn-Male	12/20/1911	2:15PM	2		1
Father	Daniel F. Rager	26	Pennsylvania		
	Nellie Selders	23	Pennsylvania		
See Death Certificate 12/20/1911					
Antonio Soponara	12/30/1911	10:00PM	1		1
Mary Sisler	Mariano Sopenora	26	Italy		
Watchman	Guissepina Pardullo	23	Italy		
Mary Toth		7:00PM	4		4
Mrs. M. Petrilla	Steve Toth	30	Austria		
Miner	Ester Walseson	28	Austria		

Birth Certificates: 1912

Marie Balog	1/7/1912	2:30PM	1		1
Marie Sisler	John Balog	25	Austria		
Miner	Marie Kakas	18	Austria		
Margaret Virginia Thomas	1/8/1912	3:00AM	7		5
Dr. J.A. Comerer	D.A. Thomas	29	Pennsylvania		
P.R.R. Agent	Mary A. Davis	30	Pennsylvania		
Kristena Gyardey	1/9/1912	6:45AM	1		1
Dr. J.A. Comerer	John Gyardey	24	Austria		
Miner	Anna Arsay	19	Austria		
Ruth E. McMullen	1/16/1912	3:00AM	4		4
Dr. J.A. Comerer	Hayden McMullen	34	Pennsylvania		
Laborer	Elsie Hare	26	Pennsylvania		
Unnamed-Female	1/30/1912	2:00PM	2		0
Dr. J.A. Comerer	Jos. D. Bennett	32	Pennsylvania		
Miller	Mary McCreerey	27	Pennsylvania		
See Death Certificate 1/30/1912					
James Clifford Boyer	2/25/1912	2:00AM	2		2
Dr. J.A. Comerer	Elmer H. Boyer	24	Pennsylvania		
Coal Miner	Edith Goughnour	23	Pennsylvania		
Edward Buzi	3/1/1912	1:00AM	1		1
Dr. J.A. Comerer	Tony Buzi	22	Italy		
Miner	Lizzie Leser	18	Italy		
Joseph Stoucy Konis	3/2/1912	6:00PM	4		3
Dr. J.A. Comerer	Martin S. Konis	33	Russia		
Miner	Martha Seilions	30	Russia		

Name / Attendant / Occupation	Date	Time	Father / Mother / Ages	Birthplace	
Mike Moniak Eva Jacobs Coal Miner	3/4/1912	10:00PM	Charley Moniak 37 Mary Papeleia 26	4 Russia Russia	4
Dowey Croyle Dr. J.A. Comerer Blacksmith	3/5/1912	4:00AM	Philip Croyle 23 Martha Evans 23	2 Pennsylvania Pennsylvania	2
Kenneth Woodring Dr. J.A. Comerer Miner	3/7/1912	5:00AM	Howard B. Woodring 23 Mary Jones 19	2 Pennsylvania Pennsylvania	2
Elmer W. Gallaher Dr. J.A. Comerer Miner	3/22/1912	1:00PM	Elmer F. Gallaher 22 Bertha Schroyer 25	4 Pennsylvania Kansas	4
Elmer Beers Dr. J.A. Comerer Carpenter	3/25/1912	11:00AM	Porter N. Beers 33 Susan Kough 30	6 Pennsylvania Pennsylvania	6
Mary Sabo Mrs. M. Petrilla Miner	3/28/1912	5:00AM	George Sabo 37 Julia Gavia 27	4 Sock Mayr Selogi Mayr	4
Frank Kerekes Marie Sisler Miner	4/24/1912	10:30PM	John Kerekes 22 Maggie Luika 21	2 Austria Austria	2
Martin Skubik Dr. Nix, Wehrum Miner	5/10/1912	4:00PM	Martin Skubik 23 Anna Boch 30	3 Russia Russia	1
Winiford Marie Williams Dr. J.A. Comerer Salesman Parents' Residence: Tyrone	5/12/1912	3:00AM	W.G. Williams 25 Mary Elizabeth Berkey 24	2 Pennsylvania Pennsylvania	1
William Shimonitz Marie Sisler Laborer	5/15/1912	7:00PM	Mike Shimonitz 30 Melie Belanich 22	 Austria Austria	2
Dezso Babinsak Dr. J.A. Comerer Miner	5/21/1912	8:00AM	John Babinsak 32 Barbula Bologh 29	6 Austria Austria	5
Jason Wallace Shaffer Dr. J.A. Comerer Car Inspector	5/26/1912	11:00AM	Wallace Shaffer 34 Alberta Frankhouser 33	1 Pennsylvania West Virginia	1
Howard S. Brady Dr. J.A. Comerer Clerk	5/26/1912	7:00PM	Howard H. Brady 23 Nellie Chambers 25	1 Pennsylvania Pennsylvania	1
Tony Defazio Dr. J.A. Comerer Miner	5/29/1912	10:00PM	Sam Defuts 38 ? Cicco 36	7 Italy Italy	5

Name / Attendant / Occupation	Date / Father / Mother	Time / Father Age / Mother Age	Children / Father Birthplace / Mother Birthplace	Living
Alvira Bucci Dr. J.A. Comerer Miner	5/30/1912 Alford Bucci Rosara Minkiona	11:00AM 28 23	2 Italy Italy	2
Stillborn-Female Dr. J.A. Comerer Miner	6/8/1912 John Bossolo Noka Angels	9:00AM 43 32	11 Italy Italy	1
Tony Bucci Dr. J.A. Comerer Miner	6/13/1912 John Bucci Angeline Giacoino	4:00PM 24 28	 Italy Italy	
Stephen Soos Dr. J.A. Comerer Miner	7/14/1912 Joseph Soos Frea Toth	 21 19	2 Hungary Hungary	1
Harry Clement Dr. J.A. Comerer Store Manager	7/19/1912 Wm. H. Clement Emma Hohhe (?)	4:30PM 37 35	5 Pennsylvania Pennsylvania	4
Ellen Stoyka Mrs. Mary Sisler Coal Miner	7/24/1912 John Stoyka Christine Kovvez	2:00PM 26 25	1 Austria-Hungary Austria-Hungary	1
Yoan Oistfine Marie Sisler Miner	8/24/1912 Gustavin Oistfine Marie Taleinie	11:00PM 32 31	2 Austria Austria	2
Margaret Schroyer Dr. J.A. Comerer Miner	9/7/1912 Joseph Schrayer Mary Grant	9:00AM 22 18	 Pennsylvania Pennsylvania	
Mary Baliscak Agnes Toth Miner	9/12/1912 John Baliscak Annie Jakiboni	2:00PM 34 29	6 Hungary Hungary	5
Julius John More Agnes Szues Miner	9/15/1912 John More Mary Nagy	3:00PM 40 35	7 Hungary Hungary	4
Yohan Smayda Marie Sisler Miner	9/15/1912 Frank Smayda Magie Roseline	3:00PM 42 29	5 Austria Austria	2
Mary Cecilia McConnell Dr. J.A. Comerer Car Inspector	9/23/1912 Stephen McConnell (?) Hoffman	 28 23	2 Pennsylvania Pennsylvania	2
Alvin Comerer, Jr. Dr. W.H. Nix Physician	10/15/1912 J.A. Comerer Mary J. Lynch	7:00AM 44 27	3 Pennsylvania Pennsylvania	2
Rosa Martha Williams Dr. J.A. Comerer Merchant	12/3/1912 Samuel R. Williams Margaret Sides	2:00AM 45 42	13 Pennsylvania Pennsylvania	8

Stillborn-Male Agnes Such Toth Coal Miner	12/16/1912 Paul Ignac Victoria Pink	5:30AM 37 21	2 Hungary Hungary	0
Bula Joseph Jendricke Dr. J.A. Comerer Bartender	12/16/1912 Joseph Jendricke Margaret Kalina	7:00PM 37 25	2 Hungary Hungary	2

Birth Certificates: 1913

Vincent Gastelione Dr. J.P. MacFarlane Miner	8/17/1913 Tony Gastelione Mary Cardulo	5:00PM 27 20	1 Italy Italy	1
Leonard Kass Dr. J.P. MacFarlane Merchant	8/26/1913 Wolf. Kass Ilda Brett	3:00AM 34 30	2 Russia Russia	2
Annie Uvega Dr. J.P. MacFarlane Miner	10/3/1913 Louis Uvega Mary Famosio	5:00PM 42 38	5 Austria Austria	5
William Thomas Dr. J.P. MacFarlane Station Agent See Death Certificate 9/14/1914	10/16/1913 Daniel Thomas Mary Ann Davis	11:00PM 31 31	8 Pennsylvania Pennsylvania	6
James Virgil Dempsey Dr. J.P. MacFarlane Machinist	10/26/1913 James Dempsey Zula Kinter	7:00AM 29 20	1 Michigan Pennsylvania	1
Jim Telesekometva None Miner See Death Certificate 11/30/1913	10/29/1913 Jim Telesko Sophia Volosin	1:00AM 24 19	 Russia Russia	
Gerald Leatherman Dr. J.P. MacFarlane Machinist	11/4/1913 Charles Leatherman Mary Hassinger	8:00AM 39 34	4 U.S.A. U.S.A.	2
James Walter Clyde Dr. J.P. MacFarlane Motorman	1/7/1913 Robert Clyde Nona Hullihen	6:00AM 41 35	7 U.S.A. U.S.A.	6
George Wilkner Dr. J.P. MacFarlane Miner	11/13/1913 Nick Wilkner Ruby Stradick	11:00PM 23 18	1 Hungary U.S.A.	1
Ethel Watkins Mrs. Evan Griffith Laborer	11/16/1913 Byron Watkins Sara Hanna Weaver	10:30AM 30 25	4 Scranton Yorkshire, Eng.	3
Verna Kish Dr. J.P. MacFarlane Miner	11/25/1914 (Sic) Steve Kish Mary Nodolanick	2:00PM 29 26	5 Austria Austria	5

Mary Jane Johns (Jackson Twp.) Dr. J.P. MacFarlane Blacksmith	11/26/1913 Edward E. Johns Clara Cross	5:00AM 44 40	14 Pennsylvania England	11
Andrew Heda Dr. J.P. MacFarlane Miner	11/30/1913 John Heda Julia Plisco	11:00PM 26 19	1 Austria Austria	1
John Jewhas Dr. J.P. MacFarlane Miner See Death Certificate 12/14/1913	12/14/1913 Andy Jewhas Mary Porule	11:00PM 29 26	2 Hungary Hungary	1
Stillborn-Female Dr. J.P. MacFarlane Blacksmith See Death Certificate 12/15/1913	12/15/1913 Louis Trevelano Annie Maritt	4:00PM 40 34	4 Italy Italy	3
Stillborn-Female Dr. J.P. MacFarlane Miner See Death Certificate 12/18/1913	12/18/1913 Mike Balock Lena Balock	1:00AM 27 19	1 Hungary Hungary	0
Mary Miller Dr. J.P. MacFarlane Miner	12/18/1913 Mike Miller Mary Lewack	2:00AM 28 19	1 Russia Russia	1
Victor Vincent Dr. J.P. MacFarlane Miner	12/23/1913 Ignotas Vincent Bosso Barbola	4:00AM 37 28	3 Hungary Hungary	3
Stillborn-Male Dr. J.P. MacFarlane Miner See Death Certificate 12/28/1913	12/28/1913 Joseph Sabo Rose Cassa	9:00AM 39 26	3 Hungary Hungary	1

Birth Certificates: 1914

Juliska Hador Mrs. Agnes Such Toth Miner	1/1/1914 Vinces Hador Doka. Tuliance	2:00AM 32 32	6 Hungary	6
Elizabeth Molnar Mrs. Agnes Such Toth Miner	1/2/1914 John Molnar Elizmadid Mary	5:00AM 26 22	3 Hungary Hungary	3
John George Mrs. Agnes Such Toth Miner	1/11/1914 Joe George Julian Garea	9:00PM 29 21	2 Hungary Hungary	2
Kalman Suto Mrs. Agnes Such Toth Miner See Death Certificate 2/23/1914	1/16/1914 Kalman Suto Elizabeth Toth	10:00AM 27 22	2 Hungary Hungary	2

Name / Informant / Occupation	Date / Father / Mother	Time / Father Age / Mother Age	Child # / Father Birthplace / Mother Birthplace	Living
Julian Gorcsos Mrs. Agnes Such Toth Miner	1/23/1914 Joseph Gorcsos Julian Tortnon (?)	3:30 PM 40 35	7 Hungary Hungary	5
Slojna (?) Myritich Dr. J.P. MacFarlane Miner	3/10/1914 Chas. Myritich Antonia Glogaski	9:00PM 38 27	1 Hungary Hungary	1
Gerald Stephens Dr. J.P. MacFarlane Mine Laborer	3/26/1914 Bailey Stephens Laura Lynch	6:00AM 41 32	8 West Virginia Pennsylvania	8
Mary Gubish Dr. J.P. MacFarlane Miner	3/27/1914 John Gubish Mary Toth	11:00AM 40 38	4 Hungary Hungary	4
Mary Kozma Mrs. Agnes Sencs Miner	3/27/1914 Mike Kozma Elizabeth Levay	7:00AM 23 19	2 Hungary Hungary	2
Not Named-Male Dr. J.P. MacFarlane Coal Miner	4/4/1914 George Contrez Mary Balog	11:00AM 38 36	5 Austria Austria	4
George Slevka Mrs. Agnes Toth Miner	4/5/1914 George Slevka Mary Tyctia	1:00AM 23 22	2 Hungary Hungary	2
Gallistine Bubnock Dr. J.P. MacFarlane Miner	4/6/1914 John Bubnock Mary Drabyack	5:00PM 28 20	1 Russia Russia	1
Steve Aristafin Mrs. Agnes Such Toth Miner	4/12/1914 Gusty Aristafin Mary Faryany	12:00PM 33 32	7 Hungary Hungary	3
Unnamed-Male Dr. J.P. MacFarlane Miner See Death Certificate 4/4/1914	4/14/1914 George Contrez Mary Balog	 38 36	5 Austria Austria	4
George Slevka Mrs. Agnes Toth Miner See Death Certificate 7/23/1914	4/15/1914 George Slevska Mary Tyctia (?)	1:00AM 23 22	2 Hungary Hungary	2
Anne Horriety Dr. J.P. MacFarlane Mine Laborer	4/20/1914 Chas. Horriety Mary Gamble	9:30PM 33 23	4 Russia Russia	3
Edith Viola Rager Dr. J.P. MacFarlane Miner	5/1/1914 Daniel F. Rager Nellie Selders	7:00PM 35 24	 Pennsylvania Pennsylvania	
Joe Barate Agnes Such Toth Miner	5/19/1914 Ignatz Barate Ester Demien	8:00PM 35 30	3 Hungary Hungary	3

Annie Balog Dr. J.P. MacFarlane Miner	5/10/1914 Mitro Balog Mary Verbish	2:00AM 26 24	4 Russia Russia	3
Mary Beres Agnes S. Toth Miner	5/14/1914 Mike Beres Etelka Orban	2:00PM 30 24	6 Hungary Hungary	6
Willard Rodney Feldman Dr. MacFarlane Machinist	5/16/1914 Sam F. Feldman Ivy Cooke	2:00AM 26 21	1 Michigan Pennsylvania	1
Piraska Munkosci Agnes Toth Miner	5/24/1914 Louis Munkasci Julia Sefrak	8:00AM 36 28	4 Hungary Hungary	3
Gravomina Sopanov Agnes Toth Night Watchman	6/12/1914 Mariana Sapanov Groseffina Cardulla	8:00AM 29 26	2 Italy italy	2
Mary Rosman Dr. Roy St. Clair, V'dale Laborer	6/15/1914 Charles Rosman Annie Danch	10:30AM 26 19	1 Austria Austria	1
Edward Retco Dr. Roy St. Clair Miner	6/17/1914 Matthew Retco Pearl Schroyer	6:00AM 24 22	2 Germany Kansas	2
Mike Rabola Agnes Toth Coal Miner	6/18/1914 Alex Rabola Helen Bilake	8:00AM 30 27	2 Hungary Hungary	2
Leo Esuardo Dr. Roy St. Clair Miner	6/19/1914 Donati Esuardo Bonyos Ferroa	12:00PM 34 29	3 Italy Italy	3
Jan Buovaventuro Agnes Toth Laborer	7/6/1914 Samuel Buovaventuro Mary Dispefacno	6:00PM 35 37	4 Italy Italy	3
Lucia Coletti Agnes Toth Miner	7/14/1914 Dominico Coletti Dominica Corti	8:00AM 32 29	1 Italy Italy	1
Altha Marie Hunter Dr. Roy St. Clair Laborer	7/20/1914 William Hunter Mary McCloskey	11:30PM 30 26	4 Pennsylvania Pennsylvania	4
Ward Davis Dr. Roy St. Clair Miner	7/25/1914 T.W. Davis, Rexis Ella Booterbaugh	10:00PM 36 35	7 North Carolina Pennsylvania	7
Margit George Agnes Toth Miner	7/28/1914 John George Mary Stankovich	11:00PM 26 20	2 Hungary Hungary	2

Name / Attendant / Occupation	Date / Father / Mother	Time / Ages	Father Birthplace / Mother Birthplace		
Joe Triboezky Agnes Toth Miner	7/29/1914 Louis Triboezky Lizzy Kovato	4:00PM 33 27	Hungary Hungary	1	1
Clarence Dwight Johns Dr. MacFarlane Laborer	8/16/1914 Charles Johns Elizabeth M. Miller	6:00AM 26 23	Pennsylvania Pennsylvania	1	1
Giyas Paul Agnes Toth Miner	8/20/1914 Paul Giyas Susie Harnocs	1:00PM 25 19	Hungary Hungary	1	1
Lizzy Bliska Agnes Toth Miner	8/27/1914 Joe Bliska Bilakorics Maria	12:00PM 28 22	Hungary Hungary	3	3
Paul Shandor Agnes Toth Miner	8/27/1914 Paul Shandor Lizzy Molnar	11:00PM 26 25	Hungary Hungary	2	2
Stillborn-Female Dr. Roy St. Clair Miner See Death Certificate 8/30/1914	8/30/1914 Joe Delasko Helen Yena	5:00AM 27 20	Austria Austria	2	1

Birth Certificates: 1915

Name / Attendant / Occupation	Date / Father / Mother	Time / Ages	Father Birthplace / Mother Birthplace		
Andrew J. Markus Dr. MacFarlane Miner	4/?/1915 Andy J. Markus 	6:00AM 	Austria Austria	2	2
Everitt Dempsey Dr. MacFarlane	4/26/1915 James Dempsey Zula	8:00PM 33 22	Michigan Pennsylvania		2
Hugh Thomas Williams Dr. MacFarlane Storekeeper See Death Certificate 1/5/1916	5/5/1915 S.R. Williams Margaret Sides	1:00AM 48 45	Pennsylvania Pennsylvania	14	9
Margaret Morris Dr. MacFarlane Butcher	5/20/1915 Walter Morris Lottie Barry	11:00AM 35 28	Pennsylvania Pennsylvania	5	4
Helen Dunch, First Twin Dr. MacFarlane Miner	5/24/1915 Charles Dunch Annie Sofilko	8:00PM 28 22	Austria Austria	2	2
Mary Dunch, Second Twin	Same As Above				
Rosie Pendur Dr. MacFarlane Miner	5/24/1915 Mike Pendur Mary Monchock	5:00PM 32 23	Austria Austria	3	2
Isabel Stopp Dr. MacFarlane Miner	5/31/1915 John Stopp Fannie Hullihan	4:00AM 42 35	England Pennsylvania	5	4

Name / Doctor / Occupation	Date	Time	Father / Mother / Ages	Birthplace	Notes
Evline Emma Wray Dr. MacFarlane Mine Laborer	6/16/1915	6:00AM	Frank Wray 24 Zetha M. Brown 18	1 Pennsylvania Pennsylvania	1
Peter Balog Dr. MacFarlane Miner See Death Certificate 8/8/1915	6/21/1915	6:00AM	Mike Balog 22 Annie Kogar 19	1 Austria Austria	1
Karlovitch-Female Dr. MacFarlane Miner	6/24/1915	10:40AM	John Karlovitch 29 Antoine Meloceuton 24	4 Austria Austria	4
Twinan Bracken Dr. MacFarlane Teamster	6/25/1915	5:45AM	Twinan Bracken 19 Sylvia Mardin 20	2 America America	2
Fay Deloris McMullen Dr. MacFarlane Miner	6/27/1915	12:05AM	Anthony McMullen 39 Emma Irvin 34	9 America America	7
Frank Delasco Dr. MacFarlane Miner	7/11/1915	3:00PM	Joseph Delasco 27 Helen Genu 21	3 Austria Austria	2
Isabelle Jendrick, Filed 7/23/1915 Dr. MacFarlane Bartender		11:30AM	Joseph Jendrick 41 Margaret Kalina 28	3 Hungry Hungary	3
Stillborn-Male Dr. MacFarlane Miner See Death Certificate 7/21/1915	7/21/1915	1:00AM	Joseph Soos 27 Teressa Cotar 24	4 Hungary Hungary	1
Elvera Marpelli Dr. MacFarlane Miner See Death Certificate 11/13/1915	7/22/1915	3:00AM	Levi Marpelli 30 Josaphine Camenatti 25	4 Italy Italy	4
Mary Yanuti Dr. MacFarlane Miner	7/30/1915	4:00PM	Chas. Yanuti 23 Mary Walosen 20	1 Austria Austria	1
Joseph Crinity Dr. MacFarlane Laborer	8/12/1915	5:00AM	Mike Crinity 30 Melea Markus 24	4 Austria Austria	4
Kosta-Male Dr. MacFarlane Miner	8/14/1915	12 Noon	Mike Kosta 28 Bertha Garyne 21	2 Hungary Hungary	1 3rd House, 6th St.
Alice LaFay Decker Dr. MacFarlane Mine Laborer	8/25/1915	1:00AM	Daniel Decker 21 Jennie M. Douty 23	1 Pennsylvania Pennsylvania	1

Annie Konack	8/26/1915	8:00AM	1	1	
Dr. MacFarlane	John Konack	24	Austria		
Miner	Annie Wallesin	19	Austria		
Metro Kerikish	8/31/1915	7:00AM	1	1	
Dr. MacFarlane	George Kerikish	21	Austria	House 59	
Miner	Annie Sefar	18	Austria		
See Death Certificate 9/1/1915					
Katherine Custer	9/4/1915	4:00AM	4	3	
Dr. MacFarlane	John A. Custer	26	Pennsylvania		
Restaurant Keeper	Beatrice Rairigh	23	Pennsylvania		
Stillborn-Male	9/7/1915	3:00PM	4	3	
Dr. MacFarlane	Adam Molnar	38	Hungary		
Coal Miner	Ida Klie	37	Hungary		
See Death Certificate 9/7/1915					
John Krisfolisy	9/11/1915	1:00AM	1	1	
Dr. MacFarlane	Chas. Krisfolisy	27	Austria	House 250C	
Miner	Paraska Todosia	20	Austria		
Mary Malhowky	9/16/1915	4:00AM	3	3	
Father	Teddy Malhowky	26	Austria		
Miner	Mary Strelosky	22	Austria		
Annie Siniczeti	9/24/1915	7:30PM	2	2	
Dr. MacFarlane	Charles Siniezeti	26	Austria-Hungary	241 Plank Rd.	
Miner	Bertha Kusma	19	Austria-Hungary		
Mike Skubick	10/3/1915	7:00AM	7	5	
Dr. MacFarlane	Andy Skubick	25	Austria		
Miner	Elizabeth Roche	34	Austria		
John Busnock	10/10/1915	11:00PM	2	2	
Dr. MacFarlane	John Busnock	28	Austria		
Miner	Mary Duyesk	22	Austria		
Frank Monyuk	10/10/1915	6:00AM	1	1	
Dr. MacFarlane	John Monyuk	26	Austria		
Miner	Mary Franks	21	Austria		
Annie Holat	10/11/1915		1	1	
Dr. MacFarlane	Metro Holat	26	Austria	158 Sixth St.	
Miner	Annie Lobic	20	Austria		
See Death Certificate 10/17/1915					
Gulodi-Male	10/17/1915	3:00AM	8	2	
Dr. MacFarlane	Andy Fulodi	40	Hungary	59 Main St.	
Miner	Ada Boraty	30	Hungary		
See Death Certificate 1/23/1916					
Stillborn-Male	10/20/1915	7:30 PM	3	2	
Dr. MacFarlane	Steve Borchick	36	Hungary		
Miner	Mary Franks	29	Hungary	36 Plank Rd.	
See Death Certificate 10/20/1915					

Birth Certificates 1916

Name / Doctor / Occupation	Date / Father / Mother	Time / Father Age / Mother Age	# / Father Birthplace / Mother Birthplace	#
Elen Lettica Esias Dr. MacFarlane Miner	8/4/1916 Thomas Esias Grace Carew	1:00PM 27 23	1 Pennsylvania Pennsylvania	1
Jay Charles Leatherman Dr. MacFarlane Carpenter	8/6/1916 Charles Leatherman Mary Hassinger	3:00AM 42 36	5 Pennsylvania Pennsylvania	3
Armando Sole Dr. MacFarlane Miner	8/7/1916 Alexander Sole Vita Standard	1:00PM 43 31	7 Italy Italy	5
Helen Podhor Dr. W.A. Prideaux Miner	8/11/1916 Adam Podhor Sofia Vineni	11:00PM 41 38	10 Austria Austria	9
Elizabeth Markus Dr. H.C. Thomas, V'dale Miner	8/13/1916 Carmine Markus Julia Feyu	11:30AM 38 28	4 Hungary Hungary	4
Helen Balko Dr. MacFarlane	11/11/1916 Not Listed Tona Vervorcy	4:00 28	 Austria	4
Roy Grant Gongloff Dr. MacFarlane Motorman	11/25/1916 Grant Gongloff Edith Gardiner	9:00PM 29 31	6 Pennsylvania Pennsylvania	4
Michael Hornity Dr. MacFarlane Miner	Nov. or Dec. Charles Hornity Mary Etceshery	6:00AM 24 22	 Austria-Hungary Austria-Hungary	
John Kovak, Jr. Dr. MacFarlane Miner	12/4/1916 John Kovak Martha Popp	11:00AM 28 24	2 Hungary Hungary	2
Unnamed-Female Dr. MacFarlane Miner	12/16/1916 Joe Novitsky Katie Kozolosky	2:00PM 22 17	1 Pennsylvania Pennsylvania	1
Steve Bocchi, Jr. Dr. MacFarlane Miner	12/27/1916 Steve Bocchi Mary Frank	11:00AM 36 32	4 Hungary Hungary	3

Birth Certificates: 1917

John Stopp, Jr. Dr. MacFarlane Miner	1/2/1917 John Stopp Anna Hullehan	1:45AM 43 37	4 Clearfield Co. Somerset Co.	4
Stillborn-Female Dr. MacFarlane Miner See Death Certificate 1/12/1917	1/12/1917 Joseph Sabo Rosa Zreasa	11:00PM 42 35	3 Hungary Hungary	2

Name	Date	Time	Col4	Col5	Col6
Bertha Florence Gongloff Dr. MacFarlane Miner	1/22/1917 Milton Gongloff Lulbie (?) Jenkins	9:15PM 21 27	3 Pennsylvania Pennsylvania	3	
Mary Ylenosky Dr. MacFarlane Miner	1/30/1917 Mike Ylenosky Annie Costa	2:00AM 21 18	1 Austria Austria	1	
Bossy Yobbagy Dr. MacFarlane Miner	1/30/19171 Bossy Yobbagy Elizabeth Hugya	2:40PM 31 24	3 Austria-Hungary Austria-Hungary	3	
Elizabeth Kisk Dr. MacFarlane Miner	2/3/1917 Sam Kisk Mary Christian	2:00AM 36 29	7 Hungary Hungary	7	
Paulina Selan Dr. MacFarlane Miner	2/3/1917 Frank Selan Mary Delaogramy	8:45AM	4 Austria-Hungary Austria-Hungary	3	
Shandor Molnar Dr. MacFarlane See Death Certificate 8/21/1917	2/4/1917 John Molnar Mary Chismadea	4:00PM 28 24	5 Hungary Hungary	5	
Balaryis-Female Dr. MacFarlane Miner	2/6/1917 Bajoni Balaryis Esther Helmecayr	3:50PM 28 24	4 Austria-Hungary Austria-Hungary	3	
Stillborn-Female Dr. MacFarlane Foreman See Death Certificate 2/10/1917	2/10/1917 Henry Misner Annie McMahon	6:00PM 38 35	11 Pennsylvania Pennsylvania	8	
Mary Gulash Dr. MacFarlane Miner	2/13/1917 Paul Gulash Susie Hornoty	2:00AM 27 22	3 Hungary Hungary	3 310 Chicora	
Julius Nyesta Dr. H. Paul Blake Miner	2/14/1917 Ignatz Nyesta Mary Filip	8:00PM 37 33	5 Hungary Hungary	5 290 Chicora	
Mike Vereb Dr. H. Paul Blake, V'dale Miner	2/15/1917 Mike Vereb Julia Olexa	24 20	1 Austria-Hungary Austria-Hungary	1 54 Plank Rd.	
Annie Volosin Dr. MacFarlane Motorman	2/27/1917 John Volosin Annie Berilla	3:00AM 21 20	1 Austria Austria	187 Plank Rd.	
Mary Holot Dr. H. Paul Blake Miner	2/28/1917 Metro Holot Annie Zabie	5:00AM 28 24	3 Austria Austria	1	
Helen Turoras Dr. MacFarlane Miner	3/9/1917 Peter Turoras Lizzie Pina (?)	3:00PM 32 38	Russia Russia	10 First St.	

Child	Date	Time	#	Birthplace	#
Mary Kriskilusa Dr. MacFarlane Miner	3/17/1917 Charles Kirskilusa No Name Given	3:00AM 29 21	2 Austria-Hungary Austria-Hungary	2 276 Sixth St.	
Lizzie Shandor Dr. MacFarlane Miner	3/18/1917 Paul Shandor Lizzie Molnar	4:00AM 29 27	3 Hungary Hungary	3	
Harry Edward Wise Dr. MacFarlane Fireboss	3/21/1917 John Wise Mable Hall	3:00AM 28 26	3 Scotland Pennsylvania	3	
James Findley Dr. MacFarlane Laborer	3/24/1917 Wm. Findley Ruth Kough	5:00PM 28 28	4 Pennsylvania Pennsylvania	4	
Edwin Long Dr. MacFarlane Laborer	3/27/1917 Clay M. Long Edna Jane Daly	1:00AM 26 24	2 Pennsylvania Pennsylvania	2	
Unnamed-Male Dr. MacFarlane Mine Laborer	3/30/1917 Metro Volosin Annie Wasnock	5:00PM 25 20	1 Austria Austria	1	
Steve Hudja Dr. MacFarlane Miner	4/1/1917 Steve Hudja Lizzie Hedill	11:00PM 27 22	2 Hungary Hungary	2	
No Name Given - Male Dr. MacFarlane Laborer	4/3/1917 Rudolf Pabst Caroline Lucorasish	4:00AM 23 20	1 Austria Austria	1	
Alfred John Harvey Dr. MacFarlane Miner	4/17/1917 John Harvey Grace Lovelace	2:00PM 23 20	1 Pennsylvania Pennsylvania	1	
Joseph Wargo Dr. MacFarlane Miner	4/23/1917 Steve Wargo Lizzie Christiace	11:00AM 27 23	3 Hungary Hungary	3	
Mary Bushock Dr. MacFarlane Miner	4/25/1917 John Bushnock Lizzie Parlie	6:00AM 36 32	7 Austria-Hungary Austria-Hungary	5	
Unnamed-Male Dr. MacFarlane Laborer	4/27/1917 Sam Gromley Rose Witskosky	1:20AM 25 20	 Pennsylvania Pennsylvania	2	
Ellen Wagner Dr. MacFarlane Clerk	4/27/1917 John M. Wagner Ellen Brown	1:00AM 25 25	3 Pennsylvania Pennsylvania	3	

Birth Certificates: 1918

Wilma Molnar Dr. MacFarlane Miner	12/21/1918 Paul Molnar Julia Szinko	8:00AM 30 23	3 Austria Austria	2	

Birth Certificates: 1919

Child / Doctor / Occupation	Date / Father / Mother	Time / Father Age / Mother Age	Child # / Father Birthplace / Mother Birthplace	Living Children / Address
Julia Yelenask Dr. MacFarlane Miner	1/7/1919 Mike Yelenask Annie Horst	3:00PM 23 22	2 Austria Austria	2
John Yowarski Dr. MacFarlane Miner	1/19/1919 Mike Yowarski Katie Lonoski	4:00PM 27 15	 Austria Austria	65 Goat Hill
Eileen Rariagh Dr. MacFarlane Teamster	1/21/1919 Dewey Raraigh Amanda Krumbine	5:00AM 21 23	1 Pennsylvania Pennsylvania	1
Mike Rubish Dr. MacFarlane Miner	1/22/1919 John Rubish Anna Anuta	3:00PM 34 30	 Austria Austria	
Joseph Yobbogy Dr. MacFarlane Miner	1/22/1919 Sigmond Yobbogy Elizabeth Huyga	4:00PM 33 26	4 Austria Austria	4 51 Plank Rd.
Stillborn Male Dr. MacFarlane Miner	1/26/1919 John Pipscak Elizabeth Salil	7:00PM	1 Hungary Hungary	287 Fourth St.
Irene Dr. MacFarlane Mine Foreman	1/28/1919 Harry Clark Ida May Walkins	 47 46	16 England Pennsylvania	10
Stillborn-Female Dr. MacFarlane Miner	2/3/1919 John Krisfolusirs Annie Kerekes	4:00PM 34 31	8 Austria Austria	3

Birth Certificates: 1920

Child / Doctor / Occupation	Date / Father / Mother	Time / Father Age / Mother Age	Child # / Father Birthplace / Mother Birthplace	Living Children / Address
Virginia Lovelace Dr. MacFarlane Mine Laborer	6/24/1920 Carl Lovelace Bessie Frampton	3:00PM 24 19	1 Pennsylvania Pennsylvania	1
Annie Simmons Dr. MacFarlane Railway Laborer	7/4/1920 James Simmons Mary Guth (?)	1:00PM 26 23	 Pennsylvania New York	
Florence Altemus Dr. MacFarlane Farmer, Buffington Twp.	7/18/1920 Hugh Altemus Carrie Stewart	10:00AM 25 25	1 Pennsylvania Pennsylvania	1
No Name Given-Female Dr. H.F. Gockley, Vintondale Miner	7/19/1920 Steve Holupka Mary Youhas	2:30AM 28 29	 Austria austria	116 Main St.
George Vaskovitch Dr. H.F. Gockley Miner	7/28/1920 George Vaskovich Anna Risko	1:50AM 27 24	 Austria U.S.	333 Main St.

Steve Pelechic Dr. MacFarlane Miner	8/3/1920 Mike Pelechic Helen Copenatz	5:00AM 44 35	 Austria Austria	 280 Sixth St.
John Rosko Dr. H.F. Gockley Miner	8/7/1920 George Rosko Helen Risko	6:30AM 45 28	7 Austria Austria	3 Plank Road
Annie Copenatz Dr. H.F. Gockley Miner	8/26/1920 John Copenatz Annie Dehoe	3:00AM 39 34	7 Austria Austria	3 275 Sixth St.
Mary Drubb Dr. H.F. Gockley Miner	8/26/1920 Mike Drubb Helen Silody	3:00AM 27 17	1 Austria Austria	1
Margaret Bodnor Dr. H.F. Gockley Miner	8/20/1920 Joe Bodnor Mary Matzos	8:30PM 39 26	 Hungary Hungary	 292 Chickoree
Russell Floyd Dodson Dr. H.F. Gockley Coal Operator	8/28/1920 Russell A. Dodson Lillian May Rosner	12:15AM 22 16	1 U.S. U.S.	1 Main St.
John Luzar Dr. H.F. Gockley Miner See Death Certificate 12/6/1920	8/29/1920 Paul Luzar Sophia Stanislaw	1:30PM 28 20	2 Austria Austria	2 327 Chickoree
Annie Zuba Dr. H.F. Gockley Miner	8/30/1920 Charlie Zuba Helen Hoka	4:00PM 35 25	3 Austria Austria	3 64 Goat Hill
Stillborn - Female Dr. MacFarlane Miner See Death Certificate 9/4/1920	9/4/1920 Metro Balog Mary Verbish	11:00AM 34 31	7 Austria Austria	4 341 Main St.
Charles William Hunter Dr. MacFarlane Miner	9/4/1920 William Hunter Mary McCloskey	10:00PM 36 32	7 Pennsylvania Pennsylvaina	7
Elmer Ezo Dr. H.F. Gockley Miner	9/5/1920 William Ezo Elizabeth Brisky	11:45PM 30 26	4 Austria U.S.	4 Main Street
Mike Bobinsak Dr. MacFarlane Miner	9/8/1920 John Bobinsak Barbara Balog	1:00AM 39 35	10 Hungary Hungary	8
Irene Hajdo Dr. Gockley Miner	9/8/1920 William Hajdo Sophie Sabo	7:15AM 30 24	2 Austria Austria	1
Robert James Sebulsky Dr. Gockley Miner	9/12/1920 Frank Sebulsky Edith Cook	12:30 22 19	2 U.S. U.S.	2 169 Main St.

Margaret Bagu Dr. MacFarlane Miner	9/21/1920 Baloz Bagu Ida Antol	3:00PM 30 18	2 Hungary Hungary	2 291 Chickaree	
Josephine Gerodo Dr. Gockley Miner	9/29/1920 Paul Gerodo Mary Georda	2:00PM 40 38	4 Italy Italy	4 61 Plank Rd.	
Eleanor Grace Clouser Dr. MacFarlane Miner	9/30/1920 Ivan Clouser Annie Kalon	9:00AM 27 25	2 Pennsylvania Austria	2 87 Main St.	
Annie Stroga Dr. MacFarlane Miner	9/30/1920 Harry Stroga Annie Kalon	10:00PM 29 25	4 Austria Austria	4 18 Second St.	
John Richard McGowan Dr. MacFarlane Miner	10/7/1920 Lawrence McGowan Lizzie Banfield	1:00AM 29 26	3 Pennsylvania Pennsylvania	3 149 Third St.	
Helen E. Wilson Dr. MacFarlane Painter	10/7/1920 Charles Wilson Mary Ann Brown	2:00PM 22 25	2 Pennsylvania Pennsylvania	2	
Margaret Custer Dr. MacFarlane Miner Laborer	10/11/1920 John Custer Beatrice Rarigh	3:00AM 31 28	6 Pennsylvania Pennsylvania	5 Res., Wehrum Res., V'dale	
John Gasko Dr. MacFarlane Miner	10/23/1920 Joseph Gasko Margaret Namunick	4:00PM 36 32	7 Hungary Hungary	6 111 Main St.	
Munie (?) Robanky Dr. Gockley Miner See Death Certificate 10/28/1920	10/27/1920 Ignots Robankey Annie Kochenko	7:00PM 41 38	 Russia Russia	101 Main St.	
John Katovich Dr. MacFarlane Miner	10/29/1920 John Kutovich Annie Statz	2:00PM 33 35	2 Austria Austria	2 343 Main St.	
Annie Maruska Dr. MacFarlane Watchmaker	11/11/1920 John Maruska Mary Lutza	1:00PM 28 23	4 Austria aUSTRIA	4	
Nellie E. Custer Dr. MacFarlane Miner	11/14/1920 Norman Custer Esther Rager	6:00AM 30 23	 Pennsylvania Pennsylvania	 306 Chickoree	
Mary Balog Dr. Gockley Miner	11/22/1920 Mike Balog Helen Balog	12:30PM 36 26	5 Austria Austria	4 300 Chickoree	
Stillborn-Female Dr. MacFarlane Miner See Death Certificate 11/28/1920	11/28/1920 John Kritzman Julie Pliska	11:00PM 34 26	5 Hungary Hungary	4 194 Plank Rd.	

Child	Date	Time	Col4	Col5	Col6	Address
Mary Secora Dr. Gockley Miner	12/6/1920 Wasko Secora Katie Tsizon (?)	7:45 23 26	4 Austria Austria		4 277 Sixth St.	
Margaret Kish Dr. Gockley Miner	12/7/1920 Sam Kish Mary Christian	9:45AM 40 33	9 Austria Austria		9 100 Plank Rd.	
Christina Bobich Dr. Gockley Miner	12/16/1920 Mike Bobich Annie Dohanich	6:00AM 36 35	8 Austria Austria		5	
John Youkuskin Dr. Gockley Miner	12/21/1920 Leo Youkuskin Mary Shank	10:00AM 24 20	1 Pennsylvania Indiana, PA		1 279 Sixth St.	
Mary Turoras Dr. MacFarlane Miner	12/22/1920 Peter Turoras Elizabeth Pinas	 35 42	10 Austria Austria		8 10 First St.	
Mary Blozolsky Dr. MacFarlane Miner	12/26/1920 Charles Blozolsky Mary Joseph	1:00AM 39 27	6 Austria Austria		4 154 Sixth St.	
Ruzina Bobnack Dr. MacFarlane Miner	12/31/1920 John Bobnack Mary Drynzak	6:00PM 33 28	5 Austria Austria		5 349 Main St.	

Birth Certificates: 1921

Child	Date	Time	Col4	Col5	Col6	Address
John Valosin Dr. J.P. MacFarlane Coal Miner	1/7/1921 Charles Valosin Mary Faber	6:00PM 28 23	5 Austria Austria		3	
Emma Lizzie Bobinscak Dr. MacFarlane Miner	1/17/1921 Joseph Bobinscak Lizzie Setic	10:00AM 35 25	3 Hungary Hungary		2	
John Nemetz, Jr. Dr. MacFarlane Miner	1/18/1921 John Nemetz, Jr. Mary Chester	7:00AM 28 26	3 Austria Austria		2 129 Fourth St.	
Barbara Palko Dr. MacFarlane Miner	1/21/1921 Joseph Palko Lizzie Dango	7:00PM 38 34	7 Hungary Hungary		5 5 First St.	
Katie Soos Dr. MacFarlane Miner	2/8/1921 Charles Soos Annie Soltes	4:00PM 34 22	3 Austria Austria		3 131/2 Second	
Charles Yosie, Jr. Dr. MacFarlane Miner	2/11/1921 Charles Yosie Mary Rapinitz	 30 26	 Austria Austria		345 Main St.	
Charles Edward Blair Dr. J.P. MacFarlane Mine Laborer	2/26/1921 John E. Blair Alice Misner	3:00PM 28 26	1 Pennsylvania Pennsylvania		1	

Name / Doctor / Occupation	Date / Father / Mother	Time / Father Age / Mother Age	Child # / Father Birthplace / Mother Birthplace	Living Children
Elmer Hezakiah Rager Dr. J.P. MacFarlane Teamster	3/3/1921 Sebastian Rager Malissa Moore	10:00PM 25 24	1 Pennsylvania Pennsylvania	1
Bertha Toth Dr. J.P. MacFarlane Coal Miner	3/6/1921 Joe Toth Pearl Boden	8:00AM 31 30	5 Hungary Hungary	4
Rosina Casseri Dr. J.P. MacFarlane Miner	3/10/1921 Joe Gasseri Lucy Innocenti	11:00PM 26 21	1 Italy Italy	1
Ruth Irene Mazey Dr. J.P. MacFarlane Blacksmith	3/15/1921 William Mazey Bessie Krukenbrod	6:00AM 31 30	7 Pennsylvania Pennsylvania	7
Irene Soos Dr. J.P. MacFarlane Coal Miner	3/16/1921 Peter Soos Mary Popp	3:00PM 30 26	3 Hungary Hungary	3
Ralph Carlson Dr. J.P. MacFarlane Miner	3/17/1921 Willard Carlson Pauline Hudzinski	9:00PM 21 21	2 Pennsylvania Pennsylvania	2
Stillborn-Female Dr. J.P. MacFarlane Coal Miner	3/23/1921 George Kotza Sophie Pelich	11:00AM 38 33	11 Austria Austria	6
John Edward Clark Dr. J.P. MacFarlane Coal Miner	3/23/1921 John Clark Irene Carranza	2:00PM 25 21	2 Maryland Costa Rica	2
Paul Kerchinski Dr. J.P. MacFarlane Coal Miner	3/29/1921 John Kerchinski Helen Dunlap	3:00PM 22 17	1 Pennsylvania Pennsylvania	1
John Yelenoski Dr. J.P. MacFarlane Coal Miner	4/1/1921 Mike Yelenoski Anna Kovta	6:00AM 25 23	3 Austria Austria	3
Helen Balog Dr. J.P. MacFarlane Coal Miner	4/13/1921 John Balog Annie Reish	2:00AM 25 23	3 Hungary Hungary	3

Appendix B
Birth Certificates: Bracken

Name of Child Dr. or Midwife Occupation	Date of Birth Father Mother	Time Age Age	Order of Birth Birthplace Birthplace	No. of Living Children
Steve Gregor Dr. W.A. Prideaux Miner	2/18/1910 Steve Gregor Mary Katur	8:00AM 34 25	1 Russia Russia	1
John Chartar, Legitimate Dr. W.A. Prideaux Miner	4/4/1910 Frank Lordon Eliza Chartar	9:00AM 33 30	1 Russia Russia	1
Mike Orsick Dr. W.A. Prideaux Miner	5/3/1910 Mike Orsick Susana Kom	10:00AM 25 26	2 Austria Austria	2
Demetrius Sumanek Dr. W.A. Prideaux Miner	5/18/1910 Stanco Sumanek Eva Misko	 30 25	2 Bracken (Sic) Austria	1
Helen Bernot Dr. W.A. Prideaux Miner	7/14/1910 Mike Bernot Mary Kulka	4:00PM 31 37	3 Russia Russia	3
Helen Martin Dr. W.A. Prideaux Miner	8/2/1910 Joseph Martin Mary Marlin	4:00PM 35 30	5 Russia Russia	5
Katie Martin Same As Above	8/2/1910	4:00PM	6	6
Joseph Kolesko Dr. W.A. Prideaux Miner	8/3/1910 Tony Kolesko Katie Poscovich	10:00AM 36 27	4 Austria Austria	3
Cornelius Bowman Dr. W.A. Prideaux Miner	8/28/1910 Lou Bowman Phoebie McDonald	8:00AM 39 22	2 Pennsylvania Pennsylvania	2
Andy Chopka Dr. W.A. Prideaux Miner	9/18/1910 Mike Chopka Petrona Roberts	10:00PM 34 26	3 Russia Russia	3
Ignots Toma Dr. W.A. Prideaux Miner	10/7/1910 Charles Toma Rosa Decan	2:00PM 31 20	1 Italy Italy	1
James Dessatorie Dr. W.A. Prideaux Miner	12/22/1910 Tony Dessatorie Lucy Parakina	6:00AM 40 30	6 Italy Italy	4

Name	Attending	Date	Father/Mother	Time	Age	Birthplace		
George Eaton	Dr. W.A. Prideaux	12/30/1912	Wm. Eaton	8:00PM	38	8	Scotland	7
	Mine Foreman		Della Buterbaugh		37		Pennsylvania	
Mary Gregor	Dr. W.A. Prideaux	1/1/1913	Steve Gregor	4:00AM	32	2	Austria	1
	Miner		Mary Zosse		27		Austria	
Mary Lorio	Dr. W.A. Prideaux	10/1/1913	Talix Loria		29	3	Italy	2
	Miner		Angelina Giranlo		26		Italy	
Josephine Kneda	Dr. W.A. Prideaux	4/14/1914	Mike Kneda	11:00PM	23	1	Austria	1
	Miner		Katea Grisdula		25		Austria	
Hedva Chopka	Dr. W.A. Prideaux	4/29/1914	Mike Chopka		29	5	Austria	5
	Miner		Petrona Roberts		30		Austria	
Girennie Lewis	Dr. W.A. Prideaux	5/15/1914	Joseph Lewis	4:00AM	37	4	Pennsylvania	4
	Miner		Minnie Jankins		29		Pennsylvania	
William James Musser	Dr. W.A. Prideaux	4/24/1915	James Musser	2:00AM	29	2	Pennsylvania	2
	Miner		Olive Reed		20		Pennsylvania	
Edward J. Little	Dr. W.A. Prideaux	4/25/1915	Edward Little	12:00AM	28	1	Pennsylvania	1
	Miner		Elsie Musser		23		Pennsylvania	
Rosy Lavitski	Dr. W.A. Prideaux	8/22/1916	John Lavitski	8:00PM	33	3	Russia	3
	Miner		Sophia Worbun		31		Russia	
Louis Muzer	Dr. W.A. Prideaux	8/23/1916	Tony Muzer	11:00AM	27	3	Austria	3
	Miner		Pauline Parakaoish		24		Austria	
Ladisla Yako	Dr. W.A. Prideaux	8/31/1916	Mike Yako	4:00AM	37	4	Austria	3
	Miner		Annie Vercoski		36		Austria	
Stanley Substanley	Dr. W.A. Prideaux	2/13/1917	Joseph Substanley	2:00AM	42	7	Pennsylvania	5
	Miner		Katie Kordor		35		Pennsylvania	
Frank Dolosko	Dr. W.A. Prideaux	4/21/1917	Frank Dolosko	7:00PM	33	3	Austria	3
			Verrna Yina		25		Austria	
John George	Dr. W.A. Prideaux	7/18/1918	Joseph George	9:00PM	34	4	Austria	4
	Miner		Julia Gassa		26		Austria	

Name	Date	Time	Col4	Birthplace	Col6
Rosie Bosika	7/21/1918	10:00PM	9		5
Dr. W.A. Prideaux	Birt Boskia	48		Italy	
Miner	Manetta Rosa	38		Italy	
James John Toth	9/7/1918	2:00PM	2		2
Dr. W.A. Prideaux	John Toth	31		Austria	
Miner	Lizzie Gregor	37		Austria	
Louellen Jarnas (Male)	9/16/1918	12:00	1		1
Dr. W.A. Prideaux	Louellen Jarnas	23		Pennsylvania	
Soldier	Regina Arnot	23		Pennsylvania	
Julia Sineski	10/3/1918	2:00PM	4		4
Dr. W.A. Prideaux	Charles Sineski	23		Austria	
Miner	Bertha Kusman	27		Austria	
Annie George	10/5/1918	8:00PM	4		4
Dr. W.A. Prideaux	John George	30		Austria	
Miner	Mary Stunkovich	24		Austria	
James Sowers	10/28/1918	10:00AM	5		2
Dr. W.A. Prideaux	George Sowers	37		Pennsylvania	
Miner	Edna Albright	25		Pennsylvania	
Joe Moskalaski	1/5/1919	9:00PM	6		6
Dr. W.A. Prideaux	Joe Moskolaski	33		Poland	
Miner	Eva Skeva	32		Poland	
No Name Given	1/18/1919	11:00AM	2		2
Dr. W.A. Prideaux	Russel H. Wagner	21		Pennsylvania	
Store Manager	Mable Bennett	21		Pennsylvania	
Mary Ann Wagner	7/28/1920		3		3
Dr. W.A. Prideaux	Russell H. Wagner	25		Pennsylvania	
Store Manager	Mable Bennett	23		Pennsylvania	
Steve Monar	8/13/1920	11:00PM	11		5
Dr. W.A. Prideaux	Steve Monar	46		Austria	
Miner	Annie Rosko	40		Austria	

Appendix C
Death Certificates: 1906-1921

Death Certificates: 1906

Name-Occupation / Doctor / Date of Birth / Place of Birth	Date of Death / Age / Father's Name / Undertaker	Time / Place of Birth / Place of Birth / Place of Burial	Cause / Marital Status / Mother's Name
Weina Mariia Ekman J.A. Comerer 2/1/1906 Finland	2/21/1906 21 Days Samuel Ekman Krumbine, Rexis	4:00AM Vintondale Finland Nanty Glo	La Grippe Sopua Hilma
Mike Croney, Miner J.A. Comerer 3/4/1857 Hungary	3/4/1906 49 Years Louis Crony Krumbine	1:00PM Hungary Hungary Ebensburg	Pulmonary TB Married Mary Brozina
John Adams J.A. Comerer 2/28/1900 England	3/5/1906 6 Yrs., 5 Da. Willie Adams Krumbine	12:30PM Bellwood, Blair Co. Johnstown Belsano	Diptheria, 6 Da. Maggie Townsend
Maggy Patrick J.A. Comerer 8/15/1905 Austria	4/5/1906 7 Mos., 15 Da. Geo. Patrick Krumbine	6:00AM Dayton, Ohio Austria Ebensburg	Pneumonia Mary Patrick
Gust Nikkola, Miner J.A. Comerer Finland	4/16/1906 36 Yrs. G. Nikkola Kumbine	 Finland Finland Nanty Glo	Pulmonary TB Single Unknown
Laczy Kis, Miner J.A. Comerer 8/25/1878 Hungary	5/5/1906 27 Yrs, 8 Mos., 10 Da. Luczy Kis Evan & Son, Ebens.	8:00AM Hungary Hungary Ebensburg	Toxalusia (?) Married Libeta Enzsikot
Sherman Buchanan, Miner J.A. Comerer 9/22/1886 Indiana	6/9/1906 19 Yrs., 8 Mos. S.G. Buchanan Krumbine	10:30PM Indiana Co., PA Indiana Co. Armagh	Accident: Internal Injuries Single Sadie Ruffner
John Hozak J.A. Comerer 6/17/1906 Hungary	6/17/1906 Geo. Hozak Krumbine	 Vintondale Hungary Ebensburg	Premature Labor Mary Crony

Name / Undertaker / DOB / Birthplace	Date of Death / Age / Father / Informant	Time / Place of Death / Father's Birthplace / Burial	Cause / Marital / Mother
Unnamed Female J.A. Comerer 7/6/1906 Pennsylvania	7/6/1906 One-Half Hour Chas. Hullihan	Pennsylvania Pennsylvania Blacklick Church	Miscarriage Verna Davis
Robert H. Doron J.A. Comerer 1/7/1906 England	7/10/1906 6 Mos. Herman Doron Krumbine	4:00AM Mt. Pleasant, PA Germany Mount Pleasant	Detention, Convulsions Naomia Hickley
Amanda R. Roberts, Mrs. J.A. Comerer 12/19/1868 Pennsylvania	7/17/1906 37 Yrs., 9 Mos., 6 Da. Ezra Krumbine Krumbine	10:00AM Pennsylvania Pennsylvania Strongstown	Pulmonary TB Married Catherin Steavenson
Stillborn Female J.A. Comerer 7/23/1906 Pennsylvania	7/23/1906 Cyrus S. Marsh	Vintondale Pennsylvania Vinco	 Elsie Wagner
George McClellan, Jr. J.A. Comerer Clearfield Co.	8/11/1906 (sic) 10 Mos., 11 Da. Aaron McClellan Krumbine	12:30AM Cambria Co. Clearfield Co. Belsano	Gastro-Intestinal Intox. Florence Williams
John Joseph Evans J.I. Pollum England	8/30/1906 3 Yrs., 6 Mos., 27 Da. John Evans Krumbine	8:20PM Wehrum Scranton Blacklick	Scalded by Accident L. Kerus
Mike Pisak J.A. Comerer 9/2/1906 (Sic) Hungary	9/2/1906 3 Mos. Stephan Pisak Krumbine	4:20PM Vintondale Hungary Blacklick	Marasunus Mary Kaosoos
Helen Butala W. Nedder, Expedit 4/11/1906 Austria	9/16/1906 7 Mos. Samuel Butala Krumbine	11:00AM Pennsylvania Austria Twin Rocks	Cholera Infantem Mary Prebula
Serverite Jacob J.I. Pollum 11/4/1902 Finland	10/11/1906 3 Yrs., 11 Mos., 7 Da. Soloman Jacob Krumbine	12:00AM Finland Finland Nanty Glo	Morasumes, Caused by trip from Old Country Sophia Ramstreug
Josephine Rossi None Called, Certificate signed by Dr. Pollum 7/25/1906 Italy	10/12/1906 2 Mos., 17 Da. William Rossi Krumbine	 Wehrum Italy Twin Rocks Station	Pneumonia Felicia Malospine
Jno Labius J.A. Comerer 7/16/1906 Austria	10/31/1906 3 Mos., 17 Da. Mike Labius Krumbine	2:00AM Vintondale Austria Wehrum G.S. (Sic)	Pneumonia Mary Risko

Rachael Rager J.I. Pollum 7/22/1904 Johnstown	10/31/1906 2 Yrs., 3 Mos., 9 Da. S.H. Rager Krumbine	11:30AM Cambria Co. Cambria Co. Dunkert (Sic), Belsano	Pneumonia Effie Lefevere
Stillborn Female J.I. Pollum 12/1/1906 Hungary	12/1/1906 Frank Zoraza Krumbine	11:00PM Vintondale Hungary Twin Rocks	 Julia Carsmont
Margaret Brown, Mrs. J.I. Pollum 8/16/1842 England	12/3/1906 64 Yrs., 3 Mos., 16 Da. William Hunter Krumbine	6:45AM England England Ebensburg	Diabetes Mellitus Married Lucy Beuthfield

Death Certificates: 1907

Laura Bracken J.A. Comerer	1/8/1907 Harry Bracken Krumbine	8:00AM Belsano	Chronic Ileo-Colitis
Mary Morin J.A. Comerer 2/6/1907 Hungary	2/6/1907 1 Hour Steve Morin Krumbine	5:00PM Pennsylvania Hungary Wehrum	Premature Mary Patig
Maria Comerer F.C. Jones, Ebensburg 4/7/1907 Pennsylvania	4/8/1907 1 Day J.A. Comerer Krumbine	4:00AM Pennsylvania Pennsylvania Ebensburg	Premature Mary Lynch
Pernell Shreck J.A. Comerer 10/17/1906 Pennsylvania	4/11/1907 5 Mos., 25 Da. J.A. Shreck Krumbine	12:00 Pennsylvania Pennsylvania Auterbine Church	Pneumonia Laura Framton
Geo. Samuel Lewis J.A. Comerer 9/22/1905 Pennsylvania	4/27/1907 1 Yr., 7 Mos., 5 Da. William Lewis O.H. Osman, Ebg.	5:30PM Pennsylvania Pennsylvania Ebensburg	Larygotrachistis Bessie Commigs
Mary Savely J.A. Comerer 2/8/1906 Hungary	5/26/1907 1 Yr., 3 Mos., 18 Da. Joe Savely Krumbine	3:00AM Vintondale Hungary Twin Rocks	Pneumonia Rose Hercik
Ford Rockman, Miner Geo. Yearick Not Given Norway	6/21/1907 50 Yrs. F. Rockman Krumbine	 Norway Norway Blakeley's Cemetery	Dropsy Married Not Known
Stillborn Male J.A. Comerer 7/2/1907 Hungary	7/2/1907 M. Labanitz No Undertaker	12:30AM Vintondale Hungary Wehrum	Seen by Dr. after birth Mary Risko

Unnamed Male J.A. Comerer 7/18/1907 Jackson Twp.	7/18/1907 Vintondale Jno McKay Krumbine	Pennsylvania Near Belsano	Hemorrhage From Naval Pearl Rager
Jno Kovach J.A. Comerer Not Given Hungary	7/30/1907 2 Yrs., 2 Mos. Metro Kovach Krumbine	3:00AM Pennsylvania Hungary Wherum	Measles Anna Kobich
Julia Munkosy J.A. Comerer 6/29/1907 Hungary	8/18/1907 7 Weeks Lewi Murkosy Krumbine	6:00PM Vintondale Hungary Twin Rocks	Obscure, Premature Julia Sheshog
Mary Stankovic, Mrs. J.A. Comerer 10/15/1870	9/7/1907 36 Yrs., 20 Mos., 7 Da. Lizzie Finges (?) Krumbine	10:30AM Hungary Hungary Ebensburg	Dropped Dead Mital Insufficiency Married
Mike Brihara J.A. Comerer Do Not Know Ulsoszstistys	9/15/1907 50 Years Andy Brihara Krumbine	 Hungary Ulso Szstistys Wehrum	Found dead near PRR tracks near Expedit Married Could Not Find Out
Joe Gelles, Miner J.A. Comerer	10/24/1907 29 Years John Galles Krumbine	10:00AM Hungary Hungary Buffington, PA	Accidental Married
Annie Steifel J.A. Comerer 3/27/1825 Germany	11/29/1907 82 Yrs., 8 Mos., 12 Da. Adam Knopp R. Hunter, Irvona	3:00AM Germany Germany Utahville	Cerebral Hemmorhage Widowed Do Not Know
Unnamed Female J.A. Comerer 12/10/1907 Hungary	12/10/1907 One-Half Hour Paul Palscer Krumbine	3:00PM Vintondale Hungary Buffington, PA	Asphyxiated during birth Mary Balogh
William S. McGuire Drayman 6/13/1851 Westmoreland Co.	12/12/1907 56 Yrs., 6 Mos, 30 Da. Wm. McGuire Krumbine	 Indiana Co. Westmorlenad Co. Ebensburg	Found dead in bed Married Jane Jewell

Death Certificates: 1908

Mary Risko J.A. Comerer 11/23/1907 Russia	1/30/1908 2 Mos., 7 Da. Joseph Risko Krumbine	3:30PM Pennsylvania Russia Wehrum G.C.	Do Not Know Elsie Risko

Richard Maddison, Miner Geo. Yearick, Wehrum 12/23/1885 England	5/11/1908 22 Yrs., 4 Mos., 18 Da. John Maddison Krumbine	2:45AM Pennsylvania England Summerhill	Typhoid Fever Married Eliza Wilkison
Anne Hilli Soffia Kivesta J.A. Comerer 5/18/1907 Finland	8/9/1908 1 Yr., 2 Mos., 22 Da. Jacob Kivesta Krumbine	2:00AM Pennsylvania Finland Nanty Glo	Acute Gastro- intestinal Intox. Soffia Ramstram
Stillborn Female George Yearick 9/9/1908 Pennsylvania	9/9/1908 Gilbert Marsh None	8:00AM Pennsylvania Pennsylvania G. Marsh Farm	Elsie Wagner
Peter Frank Loie Massa J.A. Comerer 6/20/1908 Italy	9/19/1908 2 Mos., 30 Da. Azure Massa Krumbine	6:00PM Pennsylvania Italy Twin Rocks	Ileo-Coletis Molyuse Metereza
Andy Doncevic, Miner J.A. Comerer Do Not Know Austria	11/20/1908 34 Yrs. Paul Doncevic Krumbine	11:00AM Austria Austria Twin Rocks	TB of stomach Uncertain diagnosis Married Mary Taritar
Selva Marie Koova (?) J.A. Comerer 12/8/1908 Finland	12/10/1908 2 Days Victor Koova Krumbine	2:00AM Pennsylvania Finland Finlander's, Nanty Glo	Do Not Know Mary Arplud

Death Certificates: 1909

Robert Morris J.A. Comerer	2/6/1909 Samuel Morris Krumbine	11:30AM Coalport	Accidental Drowning Single Mary Adams
Piroska Polsui (?) J.A. Comerer 1/12/1909 Hungary	2/14/1909 1 Mo., 12 Da Paul Polsui Krumbine	6:00AM .Pennsylvania Hungary Ebensburg	Influenza Mary Balogh
Annie Rebinsky J.A. Comerer 1/6/1909 Russia	2/27/1909 1 Mo. 22 Da. Mike Rebinsky Krumbine, Rexis	8:00PM Pennsylvania Russia Wehrum	Influenza Marie Jocholk
Robert Lynch, stable boss J.A. Comerer 8/28/1850	3/11/1909 59 Yrs., 6 Mos., 11 Da Krumbine	1:00AM .Montgomery Co., PA Ebensburg	Apoplecti form Bulbar Paralysis Married

Premature Male J.A. Comerer 3/26/1909 Pennsylvania	5/7/1909 Geo. Bracken Father	Pennsylvania Pennsylvania Belsano	Maggie Rhodes
Ellie Recko Evi Petrilla, Midwife 4/19/1909 Austria	4/19/1909 Joseph Recko Krumbine	11:00PM Vintondale Austria Wehrum	Premature Ellie Recko
Stillborn Male J.A. Comerer 5/7/1909 Austria	5/7/1909 Mike Rosic Father	Pennsylvania Austria Nanty Glo	Asphyxiated Katie Katabal
Unnamed Male No doctor 6/9/1909 Austria	6/9/1909 Joseph Delaney Kumbine, V'dale	Pennsylvania Austria Expedit	Unknown, attended by Midwife Mary Posztor
Jennie Alice Daniels No Doctor Called 5/20/1909 Illinois	8/15/1909 2 Mos., 26 Da. Frank Daniels Krumbine	Vintondale Somerset Co., PA Buffington, PA	Emma Bassinger
Susie Estog J.A. Comerer 9/28/1908 Hungary	8/17/1909 10 Mos., 18 Da. Geo. Estog Krumbine	10:00PM Vintondale Hungary Twin Rocks	Acute Gastro-Intestinal Intox. Susie Vargo
Annie Fedanyer J.A. Comerer & Prideaux 8/10/1906 Russia	8/22/1909 3 Yrs., 13 Da. Mike Fedanyer Krumbine	5:00PM Russia Russia Wehrum	Laryngeal Diptheria Do Not Know
Ilka Toht J.A. Comerer 9/30/1909 Hungary	10/13/1909 13 Days Gabor Toht Krumbine	8:00PM Pennsylvania Hungary Wehrum	Inanition Lizzie Nage

Death Certificates: 1910

Vince Balogh J.A. Comerer 10/13/1909 Austria-Hungary	2/5/1910 3 Mos., 22 Da. Vince Balogh Krumbine	3:00PM Pennsylvania Austria-Hungary Twin Rocks	Unknown Krisstin Kovacz
Mary Ribnicki J.A. Comerer 12/20/1909 Austria-Hungary	2/9/1910 1 Mo., 19 Da. Michael Ribnicki Krumbine	7:30AM Pennsylvania Austria-Hungary Wehrum	Catarrhal Pneumonia Mary Dahok

Joseph Semko J.A. Comerer 12/10/1909 Austria-Hungary	2/12/1910 2 Mos., 2 Da. John Semko Krumbine	8:00AM Pennsylvaina Austria-Hungary Ebensburg	Catarrhal Pneumonia Ester Kossay
Harry Miller, engine hostler J.A. Comerer 6/2/1868 Pennsylvania	2/12/1910 41 Yrs., 8 Mos., 10 Da. _____ Miller Krumbine	6:40AM Pennsylvania Pennsylvania Belsano	Croupous Pneumonia Married Martha Campbell
Irma Gavazzi J.A. Comerer 6/28/1909 Italy	2/16/1910 7 Mos., 19 Da. John Gavazzi Krumbine	3:00PM Pennsylvania Italy Twin Rocks	Unknown Fernina Casseri
Stefka Babicz J.A. Comerer 3/10/1910 Russia	3/16/1910 6 Days Jan Babiecz Krumbine	4:00PM Pennsylvania Russia Twin Rocks	Erysipelas, negligence of Midwife Anges Sabo
Katie Griscisui J.A. Comerer 2/9/1910 Poland	3/22/1910 1 Mo., 13 Da. Joseph Griscisui Krumbine	4:00AM Pennsylvania Poland Wehrum	Influenza Mary Marelok
Bertha May Duncan M. Dustin, Wehrum 5/12/1910 (sic) Pennsylvania	5/12/1910 3 days Geo. Duncan Krumbine	12:00 Noon Pennsylvania Pennsylvania Strongstown	Premature birth Ella Stiles
Metro Romanick J.A. Comerer 5/18/1910 Austria	7/14/1910 Pennsylvania Stanco Romanick Krumbine	11:00AM Austria Wehrum	Gastro-Enteritis Neglect, Mother died 7/2/1910, see Bracken list. Eva Misco
Isabel Madill J.A. Comerer 3/16/1910 Pennsylvania	7/22/1910 4 Mos., 16 Da. William Madill J. Lanighan, N. Glo	4:00PM Pennsylvania Pennsylvania Ebensburg	Gastro-Enteritis Mary Chambers
Francis Endress J.A. Comerer 7/1/1910 Pennsylvania	7/24/1910 23 Days Herman Endress Krumbine	7:00PM Pennsylvania Pennsylvania Buffington	Inanition Nina Gaster
John Sabo, Miner J.A. Comerer Do Not Know Austria	7/30/1910 44 Yrs. Totio Sabo Krumbine	4:00PM Austria Austria Twin Rocks	Fall of rock in mines Married Mary Saleden
David Conrad, Flagman J.A. Comerer 5/12/1883 	8/5/1910 27 Yrs., 2 Mos., 24 Da. Krumbine	2:40AM Pennsylvania Altoona	Run over by his train Married

Michael Kacsank, Miner J.A. Comerer Not Obtainable	8/8/1910 26 Yrs. Krumbine	3:00PM Austria Twin Rocks	Accident by fall of rock Single
Zofie Kozer J.A. Comerer 1/28/1910 Hungary	8/9/1910 6 Mos., 12 Da. George Kozer Krumbine	10:30PM Pennsylvania Hungary Wehrum	Ileo-Colitis Zofi Hollod
Vincenty Luerzyuski M. Austin 7/19/1910 Russia	9/7/1910 1 Mo., 18 Da. Paul _____	6:50AM Pennsylvania Russia	Cholera Infantum, no medical attention Annie Tontalsky
Augustina Corti J.A. Comerer 9/4/1910 Italy	9/20/1910 16 Days Oswaldo Corte Krumbine	11:00AM Pennsylvania Italy Twin Rocks	Inanition Vecillia Reani
John Klein J.A. Comerer 7/26/1910 Austria	9/27/1910 2 Mos. 1 Da. Charley Klein Krumbine	5:00AM Pennsylvania Austria Wehrum	Do Not Know Annie Balogh
Ellie Kozer J.A. Comerer 5/3/1908 Austria	11/4/1910 2 Yrs., 6 Mos., 1 Da. Charley Kozer Krumbine	12:00 Pennsylvania Austria Wehrum	Probable Pneumonia Ellie Balogh
Mike Sobo J.A. Comerer 11/3/1910 Austria	11/5/1910 2 Days George Sobo Krumbine	1:00PM Pennsylvania Austria Wehrum	Acrania Julia Senkla
Bertha Dutento, Mrs. J.A. Comerer 8/28/1866 Pennsylvania	11/10/1910 44 Yrs., 2 Mos., 13 Da. Martin Rowe Krumbine	3:00PM Pennsylvania Germany Tyrone	Aseites, due to Hepatic Insufficiency Married Rebecca Rodgers
Mary Beechey, Mrs. J.A. Comerer 3/4/1831 Wales	11/14/1910 79 Yrs., 8 Mos., 10 Da. Morgans Krumbine	10:00AM Wales Wales Punxsutawney	Found dead, most likely Apoplexy Widowed Do Not Know
Anna Wisinicki J.A. Comerer 11/28/1910 Russia	11/28/1910 6 Hours Vilem Wesinicki Krumbine	7:00PM Pennsylvania Russia Twin Rocks	Premature Mary Andrchtise (?)
Mary Dill, Mrs. J.A. Comerer 9/3/1890 Pennsylvania	12/13/1910 20 Yrs., 3 Mos., 10 Da. Edward Wright J. Lonigan	8:00AM Pennsylvania England Ebensburg	Influenza Married, adopted Sarah Myers

Death Certificates: 1911

Charley Cyrus	1/5/1911	4:00PM	Strangled after castor oil admin. for constipation
J.A. Comerer	6 Mos., 29 Da.	Pennsylvania	
6/7/1910	Albert Cyrus	Austria	Lizzie Skura
Austria	Krumbine	Twin Rocks	
Tony Wojtawics	1/20/1911		Influenza
J.A. Comerer	8 Days	Pennsylvania	
1/12/1911	Andy Wojtawics	Poland	Stella Rabelki
Poland	Krumbine	Twin Rocks, via buggy	
Susanna Lewis, Mrs.	1/26/1911	7:40PM	Cold, exhaustion
J.A. Comerer	80 Years	Wales	Married
Do Not Know	Samuel Townsend	England	Do Not Know
	Krumbine	Brisbane, Clearfield Co.	
John Borycik	2/4/1911	10:00PM	Whooping Cough
J.A. Comerer	2 Mos., 4 Da.	Pennsylvania	
11/30/1910	John Borycik	Austria	Lizzie Roh
Austria	J. Lanigan	Twin Rocks	
Julian Balog	2/10/1911	5:00AM	Unknown
J.A. Comerer	5 Mos., 6 Da.	Pennsylvania	
9/4/1910	Vense Balog	Hungary	Christine Koovcs
Hungary	Krumbine	Twin Rocks	
Joseph Eszenyi, Miner	2/20/1911	1:00PM	Rupture of Ileum and Peritonitis
J.A. Comerer	37 Yrs., 11 Mos., 2 Da.	Hungary	Married
3/18/1873	Joe Eszenyi	Hungary	Lizzie Gulyear
Hungary	Krumbine	Twin Rocks	
Stillborn Male	3/9/1911		Prolapse of cord
J.A. Comerer	Full Term	Pennsylvania	
3/9/1911	John Shustock	Russia	Katie Krivok
Russia	John Lazaresok	Wehrum	
Stillborn Female	4/1/1911		
J.A. Comerer	Full Term	Pennsylvania	
4/1/1911	Stif Pisok	Austria	Mary Kosach
Austria	Krumbine	Twin Rocks	
Dempster Nolan, bartender	4/20/1911	10:00PM	Croupous Pneumonia, DT's
			Married
J.A. Comerer	45 Yrs., 3 Mos., 25 D.	Pennsylvania	
12/25/1865	Not Mentioned		Harriet Fox
Pennsylvania	J. Lonigan	Ebensburg	
Anna Irvin, Mrs.	4/30/1911	6:00AM	Pulmonary TB
J.A. Comerer	20 Years	Ehrenfield, PA	Married
	Patrick McQueeney	Ireland	Mary Adams
England	Krumbine	Wilmore	

Unnamed Stein J.A. Comerer 5/3/1911 Pennsylvania	5/3/1911 3 Hours Thomas F. Stien Krumbine	4:00AM Pennsylvania Pennsylvania Twin Rocks	Miscarriage Mary Oldonan
Mary Pisak J.A. Comerer 2/12/1910 Hungary	5/25/1911 1 Yr., 2 Mos., 23 Da. Steve Pisak Krumbine	4:00AM Vintondale Hungary Twin Rocks	TB, Peritonitis Mary Leosash
Mary (Pate) Marian, Mrs. J.A. Comerer 4/18/1885 Hungary	5/25/1911 26 Yrs., 1 Mo., 6 Da. John Pate Krumbine	5:00AM Hungary Hungary Twin Rocks	Pulmonary TB Married Borbala Goncol
Isabel Brown, Mrs. J.A. Comerer 6/20/1870 Pennsylvania	6/9/1911 40 Yrs., 11 Mos., 20 Da. William Fink Krumbine	7:00AM Centre County Pennsylvania Altoona	Netrine Cancer Married Nancy Albright
Stillborn Female J.A. Comerer 7/3/1911 Russia	7/3/1911 Metro Ballogh Greek Catholic Sexton	Pennsylvania Russia Wehrum	Do Not Know Mary Verbich
Frederick Halls M. Austin 4/8/1911 Russia	7/30/1911 3 Mos. 22 Da. Enick Halls Krumbine	10:00PM Pennsylvania Russia Twin Rocks	Cholera Infantum Mary Lacosky
Mary Babitch M. Austin 1/14/1911 Austria	8/1/1911 7 Mos., 18 Da. George Babitch Krumbine	11:00AM Pennsylvania Austria Buffington Twp.	Cholera Infantum Mary Risko
Stillborn Male J.A. Comerer 8/21/1911 Pennsylvania	8/21/1911 Thomas Altemus Kurmbine	Pennsylvania Pennsylvania Strongstown	Accbuctumeal Force Lilly Bennett
Lilly M. Altemus, Mrs. J.A. Comerer 9/6/1890 Pennsylvania	8/26/1911 20 Yrs., 11 Mos., 20 Da. Charles Bennett Scott Clark, Seward	10:00AM Pennsylvania Pennsylvania Strongstown	Puerperal Eclampsia Married Kate Peddicord
Charley Donge, Miner J.A. Comerer Hungary	9/20/1911 25 Years Mike Donge Krumbine	Hungary Hungary Wehrum	Accident - fall of rock Single Anna Gonge
Mary Griscisin J.A. Comerer 8/16/1911 Russia	11/6/1911 2 Mos., 21 Da. Joseph Griscisin Krumbine	5:00AM Pennsylvania Russia Twin Rocks	Enteritis Mary Marcyak

Lewey Marian	11/15/1911	2:00AM	Disseminated TB inherited from birth
J.A. Comerer	5 Mos., 25 Da.	Pennsylvania	Mother died 5/25/1911
5/20/1911	Steve Marian	Hungary	Mary Pate
Hungary	Krumbine	Twin Rocks	
Walter Treaster, laborer	12/4/1911		Accident - caught in fine shaft in coal washer
J.A. Comerer	35 Yrs., 4 Mos., 10 Da.	Pennsylvania	Married
7/24/1876	William Treaster	Pennsylvania	Jenny Ramsey
Pennsylvania	Krumbine	Strongstown	
William Marlin Hunter	12/7/1911		Found dead, weak from birth
J.A. Comerer	27 Days	Pennsylvania	
10/10/1911	William Hunter	Pennsylvania	Margaret Brooke
England	Krumbine	Gallitzin	
Stillborn Male	12/20/1911	2:30PM	Asphyxia Neonatoum
W.H. Nix, Wehrum		Pennsylvania	
12/20/1911	Daniel Rager	Pennsylvania	Nellie Selders
Pennsylvania	S. Clark, Seward	Blacklick Co., PA	
Thomas Cowan	12/17/1911	12:00AM	Paralysis, Cerebral Hemorrhage
J.A. Comerer	95 Yrs., 1 Mo., 4 Da.	Ireland	Widower
11/23/1816	Joseph Cowan	Ireland	
Ireland	J. Lanigan	Punxsutawney	

Death Certificates: 1912

Joseph Borzek, Laborer	1/10/1912	2:00PM	Pneumonia, Asthma
J.A. Comerer	48 Years	Hungary	Married
1/10/1868	John Borzek	Hungary	Mary Ladeski
Hungary	Krumbine	Twin Rocks	
Anna Seleck	1/12/1912	11:00AM	Burned to death, clothing caught on fire
J.A. Comerer	3 Yrs., 3 Mos., 11 Da.	Pennsylvania	
9/26/1908	Joseph Seleck	Austria	Anna Jacobs
Austria	Krumbine	Wehrum	
Frank Risko, Laborer	1/15/1912	12:30PM	Accident - arm torn off by conveyor belt in coal washer
J.A. Comerer	17 Yrs., 11 Mos.	Hungary	Single
2/16/1894	Frank Risko	Hungary	Anna Hricz
Hungary	Krumbine	Wehrum	
Unnamed Female	1/30/1912	6:00PM	Premature birth
J.A. Comerer		Pennsylvania	
1/30/1912	Joseph Bennett	Pennsylvania	Mary McCreary
Pennsylvania	Charles Bennett	Strongstown	

Christina Gyrogyog J.A. Comerer 1/9/1912 Austria	1/31/1912 22 Days John Gyorgyog Krumbine	10:00AM Pennsylvania Austria Wehrum	Premature Birth anna Ossok
Augusta Pesaci, Miner J.A. Comerer Italy	3/1/1912 26 or 27 Yrs. Henry Pesaci Krumbine	Italy Italy Twin Rocks, via PRR	Accident - rockfall Single Filomeno Manferdelli
Mariano Sobo J.A. Comerer 3/28/1912 Hungary	4/15/1912 8 Days George Sobo Krumbine	5:00PM Pennsylvania Hungary Wehrum	Unknown Julia Sankgerian
Alex Pafko J.A. Comerer 2/26/1912 Hungary	4/19/1912 1 Mo., 24 Da. Andy Pafko Father	1:00PM Pennsylvania Hungary Twin Rocks	Premature Birth Lizzie Csartos
Stillborn Female J.A. Comerer 6/8/1912 Italy	6/8/1912 John Bossoli Father	Vintondale Italy Twin Rocks	 Noka Angelo
Martin Skubike J.A. Comerer 5/10/1912 Russia	7/2/1912 43 Days Martin Skubike Krumbine	Pennsylvania Russia Twin Rocks	Premature Birth Anne Boch
Andy Gaboda J.A. Comerer 12/26/1911 Hungary	8/9/1912 7 Mos., 13 Da. Mike Gaboda Krumbine	Pennsylania Hungary Wehrum	Accute Gastro- intestinal Intox. Helen Fenscak
Anna Ellen Stoyka J.A. Comerer 7/24/1912 Austria-Hungary	8/11/1912 18 Days John Stoyka Krumbine	10:00AM Pennsylvania Austria-Hungary Wehrum	Accute Gastro- Intestinal Intox. Christine Kooves
Eugene Schroyer J.A. Comerer 11/20/1911 Pennsylvania	8/13/1912 8 Mos., 24 Da. Wm. L. Schroyer Krumbine	9:00AM Pennsylvania Pennsylvania Mundays	Accute Gastro- Intestinal Intox. Anna Peale
Mike Pekas J.A. Comerer Mike Pekas Hungary	8/18/1912 1 Month Austria Krumbine	10:00AM Pennsylvania Annie Toat Twin Rocks	Acute Gastro- Intestinal Intox.
Andy Tot J.A. Comerer 1/30/1910 Austria	8/31/1912 1 Yr., 7 Mos. Andy Tot Krumbine	3:00PM Pennsylvania Austria Wehrum	Ileo-Colitis Anny Keysics

John Rxalecstsa	9/1/1912	3:00AM	Ideo-Colitis
J.A. Comerer	1 Year	Pennsylvania	
9/1/1911	Frank Rxalecstra	Hungary	Julian Cresman
Hungary	Krumbine	Twin Rocks	
Stephen Soos	9/3/1912	4:00AM	Measles
J.A. Comerer	1 Mo., 18 Da.	Pennsylvania	
7/14/1912	Joe Soos	Hungary	Freca Sost
Hungary	Krumbine	Twin Rocks	
Anna Koss	9/3/1912	5:00PM	Acute Gastro-Intestinal Intox.
J.A. Comerer	5 Mos., 25 Da.	Pennsylvania	
3/10/1912	Stephen Koss	Low Germany	Anna Volocein
Low Germany	Krumbine	Twin Rocks	
Frank Kerekas	9/4/1912	6:00AM	Acute Gastro-Intestinal Intox.
J.A. Comerer	4 Mos., 18 Da.	Pennsylvania	
4/17/1912	John Kerekas	Hungary	Rosie Dobisis
Hungary	Krumbine	Twin Rocks	
Margaret Schroyer	9/8/1912	3:00AM	Premature - Neglected nursing
J.A. Comerer	1 Day	Pennsylvania	
9/7/1912	Joseph Schroyer	Pennsylvania	Mary Grant
Pennsylvania	Krumbine	Twin Rocks	
Arpad Sobo, Miner	9/21/1912		Crushed by fall of rock in mine
J.A. Comerer	21 Years	Hungary	Single
	Joe Sabo	Hungary	
	Krumbine	Buffington, PA	
Catherine J. Delaney	10/4/1912	1:00PM	Ileo-Colitis
J.A. Comerer	1 Year, 5 Mos., 2 Da.	Pennsylvania	
5/2/1911	Joe Delaney	Hungary	Mary Paster
Hungary	Krumbine	Twin Rocks	
George Kiraly, Miner	10/10/1912		Cholera Morbus
J.A. Comerer	30 Years	Hungary	Married
	Not Known		
	Krumbine	Twin Rocks	
Ivy Kathryn Parsley	11/14/1912	11:00PM	Inanition
W.H. Nix	3 Months	Pennsylvania	
8/15/1912	John Parsley	Rockwood, PA	Maggie Warner
Cresson, PA	Krumbine	Twin Rocks	
Cristena Serlardey	12/12/1912		Pneumonia
W.H. Nix	9 Mos., 10 Da.	Pennsylvania	
3/2/1912	Charley Serlardey	Austria	Christina Risco
Austria	Krumbine	Wehrum	

Name / Informant / Birth / Birthplace	Date / Age / Father / Undertaker	Time / Birthplace / Father's BP / Residence	Cause / Other
Stillborn Male W.H. Nix 12/16/1912 Hungary	12/16/1912 Paul Ignac Krumbine	 Pennsylvania Hungary Twin Rocks	 Victoria Fink

Death Certificates: 1913

Name / Informant / Birth / Birthplace	Date / Age / Father / Undertaker	Time / Birthplace / Father's BP / Residence	Cause / Other
Andy Bernot, deputy coroner A.F. Duncannon Don't Know Hung himself in Vintondale lockup	2/14/1913 About 35 Yrs. Don't Know Krumbine	9:30AM Hungaria Don't Know Twin Rocks	Hanging for suicidal intent, prolonged use of alcohol Don't Know
Bill Manhit A.F. Duncannon, deputy coroner Don't Know Don't Know	2/15/1913 About 30 Yrs. Don't Know Krumbine	5:30AM Don't Know Hungaria Twin Rocks	Crushed under train, trespassing Don't Know
Andy Robert Kempfer J.P. MacFarlane 12/14/1910 Pennsylvania	2/28/1913 2 Yrs., 2 Mos., 14 Da. David H. Kempfer Krumbine	7:00PM Pennsylvania Pennsylvania Buffington	Broncho-Pneumonia Measles Anna Ritta Treaster
Stephison Alupka J.P. MacFarlane 1/4/1913 Russia	3/11/1913 2 Mos., 7 Da. Stephison Alupka Krumbine	1:00AM Russia Wehrum	Convulsions, Broncho-Pneumonia Mary Auseso
Stephen Gribosky J.P. MacFarlane 1/4/1913 Hungary	3/28/1913 Louis Gribosky Krumbine	6:00AM Vintondale Hungary Twin Rocks	Acute Diletation of heart Lizzie Kovacs
Lewis Gribatski J.P. MacFarlane 1/4/1913 Austria	3/30/1913 Lewis Gribatski, Sr. Krumbine	11:00PM Vintondale Austria Twin Rocks	Rachitis Lizze Covatch

Note the two certificates for the same death with the conflicting dates and spellings.

Name / Informant / Birth / Birthplace	Date / Age / Father / Undertaker	Time / Birthplace / Father's BP / Residence	Cause / Other
Frank Barate J.P. MacFarlane 2/18/1889 Hungary	4/23/1913 24 Yrs., 2 Mos., 5 Da. Michael Barate Krumbine	9:00AM Hungary Hungary Buffington	Pulmonary TB Single Kolobocz Eszbo
Wilbur Lawrence Myers J.P. MacFarlane Russia	4/5/1913 1 Mo. Isaac Myers Krumbine	10:00AM Vintondale Russia Altoona	Septic Orchitis, infection following circumcision Rachel Brett

Earl Edward Rager John Grubb, Armagh 4/7/1913 Clearfield Co.	4/7/1913 Daniel Rager W.S. Clark, Seward	2:00AM Jackson Twp. Cambria Co. Buffington	Stillborn Nellie Selders
Augustus Rager John Grubb 4/10/1853 East Wheatfield Twp.	4/14/1913 60 Yrs., 4 Da. Hezekiah Rager W.S. Clark	11:10AM Jackson Twp. Jackson Twp. Mondy Cemetery (sic)	Injured spine in fall Married Mary J. Swartzwalder
Unnamed Male J.P. MacFarlane 5/11/1913 Pennsylvania	5/11/1913 John Steifel J.H. Lanigan, Nanty Glo	 Buffington Twp. Pennsylvania	Premature Annie Houser
George Washington Berkey J.P. MacFarlane 1/1/1838 Pennsylvania	7/23/1913 75 Yrs., 6 Mos., 22 Da. Samuel Berkey Krumbine	4:00PM Pennsylvania Pennsylvania Bethel	Apoplexy, invalid for 10 yrs. Widowed Susan Livengood
Agnes Babecz J.P. MacFarlane 5/22/1913 Gerigieu	7/28/1913 2 Mos., 5 Da John Babecz Krumbine	1:00PM Vintondale Gerigieu Twin Rocks	Cholera Infantum Agnes Sobo
Margaret Irene Lynch J.P. MacFarlane 7/24/1912 Cambria Co.	8/15/1913 1 Yr., 16 Da. Eugene Lynch Krumbine	10:00PM Cambria Co. Pennsylvania Belsano	Diptheria Gritta Strawsbaugh
Goldie Agnes McMullen J.P. MacFarlane 10/24/1911 Pennsylvania	8/21/1913 1 Yr., 9 Mos., 21 Da. Anthony McMullen Krumbine	4:00AM Pennsylvania Pennsylvania Twin Rocks	Gastro-Enteritis Emma Irvin
Helon Ezo J.P. MacFarlane 4/13/1913 Austria	8/31/1913 4 Mos.,18 Da. William Ezo Krumbine	3:00AM Vintondale Austria Belsano	Gastritis, poor feeding Elizabeth Brisko
Jim Telecskametva Dr. Fitzgerald, Coroner 10/29/1913 Hungary	11/30/1913 1 Mo., 1 Da. Jim Telecskametva Krumbine	2:00PM Vintondale Hungary Wehrum	Heart Failure Hafa Valasen
John Jewhas None in attendance 12/14/1913 Hungary	12/14/1913 5 minutes Andy Jewhas John Jewhas	 Vintondale Hungary Wehrum	Congenital Malformation of heart, Premature Mary Parole
Unnamed Female J.P. MacFarlane 12/15/1913 Italy	12/15/1913 Luey Trivilina Krumbine	3:00PM Vintondale Italy Twin Rocks	Stillborn Anna Morett

Unnamed Female None 12/18/1913 Austria	12/18/1913 Mike Balok Mike Balak	 Vintondale Austria Wehrum	Premature - 6 months Lena Balok
Unnamed Male J.P. MacFarlane 12/28/1913 Hungary	12/28/1913 Jos. Sabo Krumbine	9:00AM Vintondale Hungary Buffington	Stillborn Rosa Cassa

Death Certificates: 1914

Nikolof Lata, Miner J.P. MacFarlane Nichus Falva, Hungary Autopsy at request of coroner	1/2/1913 (sic-1914) 30 Years Peter Lata Krumbine	10:00AM Mches Falva, Hungary Nichus Falva, Hungary Johnstown G.C.	Acute Ditatation of heart Married Mary Kosnir
Jon Cozar, Miner J.P. MacFarlane Austria Death Certificate authorized by coroner	1/7/1914 25 Years Michel Kozar Krumbine	 Austria-Hungary Austria Wehrum	Hemorrhage of laceration of neck, not known if accidental Married Anna Evans
Charlie Ochko J.P. MacFarlane 10/15/1911 France	2/17/1914 2 Yrs., 4 Mos., 2 Da. Pete Ochko Krumbine	4:00AM Pennsylvania France Wehrum	Erysipilas of left arm Mary Ener
Anna Holupka A.W. Beatty, Colver 12/13/1913 Russia	2/22/1914 2 Months Steve Holupka Krumbine	 Pennsylvania Russia Wehrum	Broncho-Pneumonia Mary Yarbis
John Morris J.P. MacFarlane 2/23/1914 Pennsylvania	2/23/1914 1 hour Walter Morris Krumbine	7:00AM Pennsylvania Pennsylvania Twin Rocks	Premature birth-5 1/2 mo. Lottie Barry
Sutt A. Calman J.P. MacFarlane 1/16/1914 Austria-Hungary	2/23/1914 1 Mo., 6 Da. Sutta Calman Krumbine	 Pennsylvania Austria-Hungary Ebensburg	Broncho-Pneumonia Lizzie Toth
Luey Frederici, Miner J.P. MacFarlane 10/11/1869 Italy Information was obtained from an Italian passport found in his trunk	3/4/1914 44 Yrs., 6 Mos., 24 Da. Peter Forderici Krumbine	6:30PM Italy Italy Ebensburg	Chest crushed, fall of coal Single Folomeni Zarli
John Celody J.P. MacFarlane Hungary	4/1/1914 8 Mos., 23 Da. Charley Celody Krumbine	2:00AM Vintondale Hungary Wehrum	Peretonitis (Tubercular) Resco Kresteno

Stillborn Male J.P. MacFarlane 4/4/1914 Austria	4/4/1914 Georges Kontnez Father	Vintondale Austria Wehrum	Premature Mary Tetosk
George Slerko J.P. MacFarlane 4/5/1914 Austria	7/23/1914 3 Mos., 18 Da. George Slerko Krumbine	4:00AM Pennsylvania Austria Wehrum	Gastro-Enteritis, Poor feeding Mary Galo
Annie Mehalko J.P. MacFarlane 2/25/1914 Hungary	8/9/1914 5 Mos., 15 Da. John Milhalko Krumbine	7:00PM Pennsylvania Hungary Wehrum	Gastro-Enteritis, Artificial Feeding Mary Strawcher
Daniel F. Thomas J.P. MacFarlane 7/19/1910 Pennsylvania	8/21/1914 4 Yrs., 1 Mo., 1 Da. Daniel A. Thomas Krumbine	4:00AM Pennsylvania Pennsylvania Ebensburg	Gastro-Enteritis Mary A. Davies
Stillborn Female J. Roy St. Clair, Vintondale 8/30/1914 Slav	8/30/1914 Joe Delasko Krumbine	Pennsylvania Slav Twin Rocks	Premature-8 Mos. Ellen Yeana
William Lewis Thomas W.A. Prideaux, Twin Rocks 10/16/1913 Pennsylvania	9/10/1914 10 Mos., 25 Da. Dan Thomas Krumbine	8:00AM Pennsylvania Pennsylvania Ebensburg	Ileo-Colitis Mary Ann Davis
Mary Kruniake J. Roy St. Clair 9/3/1914 Hungary	9/17/1914 14 Days John Kruniake Krumbine	3:30PM Vintondale Hungary Wehrum	Inanition Anna Volosin
John Larance J. Roy St. Clair 5/25/1914 Hungary	9/22/1914 3 Mos., 22 Da. John Larance Krumbine	2:00PM Canada Hungary Twin Rocks	Gastro-Enteritis Mary Redos
Geo. Vaskovich W.A. Evans, Nanty Glo 1/20/1914 Scranton	11/20/1914 Certificate issued by order of coroner Geo. Vaskovich Krumbine	10:30AM Austria-Hungary Wehrum	Stillborn Anna Estok
Mary Kish, Mrs. W.A. Evans, Nanty Glo 11/2/1887 Hungary	11/21/1914 27 Yrs., 19 Da. John Nrailancesek Krumbine	8:00PM Hungary Hungary Twin Rocks	Hypertropic Curolin of liver Married Barbara Iwofcingi (?)
Stillborn Male W.A. Evans 12/17/1914 Austria-Hungary	12/17/1914 Frank Faliczko Krumbine	Vintondale Austria-Hungary Twin Rocks	 Julia Ysenencey

Frank Hodar, Miner W.A. Evans 1869 Hungary	12/21/1914 45 Years John Hodar Krumbine	5:00AM Hungary Hungary Lloyds, Ebensburg	Paralytic Stroke Julie Hodar
Nora Rager Wm. Salisbury, Armagh 2/25/1913 Pennsylvania	12/22/1914 1 Yr., 9 Mos., 22 Da. Martin Rager W. Clark & Son, Seward	 Pennsylvania Pennsylvania Mondays Cemetery (sic)	Infantile Paralysis Harriet Whysong

Death Certificates: 1915

Stillborn Female J.P. MacFarlane 1/8/1915 Pennsylvania	1/8/1915 Arthur Kempher Father	 Vintondale Illinois Buffington Twp.	 Gertrude Hassinger
Robert M. Clyde, motorman J.P. MacFarlane 3/16/1873 Scotland	1/13/1915 41 Yrs., 9 Mos., 28 Da. Robert Clyde Krumbine	7:30AM Hazelton, PA Ireland Ebensburg	Chest crushed in mine accident Married Mary Ceregg
Giovannina Marian J.P. MacFarlane 6/12/1914 Italy	1/17/1915 7 Mos., 5 Da. Samuel Marian Krumbine	4:00PM Pennsylvania Italy Twin Rocks	Convulsions, Broncho-Pneumonia Josephine Pordulla
Lloyd Duncan Thomas W.A. Evans 1/6/1915 Pennsylvania	1/18/1915 12 Days Daniel Thomas Krumbine	10:00PM Pennsylvania Pennsylvania Ebensburg	Edema of Lungs Mary A. Davies
Adam Smydia (Twin) C.A. Fitzgerald, Coroner 1/25/1915 Hungary	1/25/1915 Frank Smydia Krumbine	 Vintondale Hungary Twin Rocks	Stillborn Rose Nagy
Eve Smydia (Twin)	See Information Above		
Lizzie Harvath J.P. MacFarlane 1/10/1915 Austria-Hungary	1/31/1915 21 Days John Harvath Krumbine	1:00AM Vintondale Austria-Hungary Twin Rocks	Bronchitis, Premature Anna Vareb
George Hanwell J.P. MacFarlane 11/2/1914 Wilton, ND	2/6/1915 3 Mos., 4 Da. Thomas Hanwell Krumbine	9:30PM Vintondale Arnot, PA Ebensburg	Bronchitis Josephine Nelson
Andy Suto J.P. MacFarlane 1/1915 Hungary	2/26/1915 1 Month Coleman Suto Krumbine	2:00PM Vintondale Hungary Ebensburg	Broncho-Pneumonia Lizzy Toth

Bill Toth J.P. MacFarlane 9/4/1914 Hungary	3/7/1915 6 Mos., 3 Da. Charles Toth Krumbine	6:00AM Vintondale Hungary Ebensburg	Acute Dilitation of Heart Pista Gerl (?)
John Reniska No medical attention, signed with coroner's permission 2/12/1915 Hungary	4/7/1915 1 Mo., 21 Da. Mike Berniska Krumbine	Curwensville, PA Hungary Wehrum	Inanition Mary Reniska
Andy Batsko, Laborer C.A. Fitzgerald, Coroner 1862	4/14/1915 53 Years Do Not Know Krumbine	9:00PM Austria Wehrum	Neck broken, caught on belt in coal washer Widower
Mary Kosma J.P. MacFarlane 3/25/1914 Hungary	4/29/1915 1 Yr., 1 Mo., 4 Da. Mike Kosma Krumbine	11:00PM Vintondale Hungary Buffington Cem.	Laryngral Diptheria Lovy Bartos
Stillborn Male J.P. MacFarlane 5/1/1915 Austria	5/1/1915 Joseph Rabit Father	Vintondale Austria Wehrum	Premature-6 Mos. Vera Sharney
Ora B. Altimus J.P. MacFarlane 4/10/1887 Pennsylvania	5/2/1915 27 Yrs., 22 Da. N.D. Altimus Geo. Ondreizak, Nanty Glo	1:30AM Pennsylvania Pennsylvania Strongstown	Lobar Pneumonia Single Annie Duncan
John Harnock W.A. Prideaux 3/16/1883 Hungary	5/17/1915 32 yrs., 2 mos. John Harnock Krumbine	2:00PM Hungary Hungary Ebensburg	Pneumonia Married Susie Pop
George Thomas Wray W. Prideaux 12/12/1914 Pennsylvania	5/18/1915 6 Mos., 5 Da. Thomas Wray Krumbine	7:00AM Vintondale Pennsylvania Ebensburg	Pneumonia Mary Shultz
Male - No Name Listed J.P. MacFarlane 1/3/1915	6/8/1915 5 Mos., 5 Da. Peter Todorovich Krumbine	1:00PM Vintondale Austria Wehrum	Gastro-Enteritis Josipa Glogiski
Stillborn Male J.P. MacFarlane 7/21/1915 Hungary	7/21/1915 Joseph Soos Joseph Soos	Vintondale Hungary Vintondale	Premature Teressa Cotar
George Gubosh J.P. MacFarlane 3/27/1915 Austria	7/28/1915 4 Mos., 1 Da. John Gubosh Krumbine	10:00AM Vintondale Austria Wehrum	Gastro-Enteritas Mary Toth

Steve Majerik, Miner	7/30/1915	11:00AM	Killed by fall of stone, head crushed
J.P. MacFarlane	26 Yrs.	Austria	Married
9/9/1889	Unkown	Austria	
Austria	Krumbine	Twin Rocks	

Information was taken from the records of the Vinton Colliery Co.

Peter Balog	8/8/1915	3:00AM	Gastritis, Artificial Feeding
J.P. MacFarlane	1 Mo., 18 Da.	Vintondale	
7/21/1915	Mike Balog	Austria	Anna Kozir
AustriaKrumbineWehrum			
Grace Misner	8/11/1915	6:30PM	Gastro-Enteritis
J.P. MacFarlane	3 Mos., 19 Da.	Vintondale	
4/23/1915	Henry Misner		Annie McMahon
Johnstown	Krumbine	Strongstown	
Metro Kerekes	9/1/1915	8:00PM	Malformation of heart
J.P. MacFarlane	1 Day	Vintondale	
8/31/1915	George Kerekes	Austria-Hungary	Anna Cyifra
Austria-Hungary	Krumbine	Wehrum	
Stillborn Male	9/7/1915	3:00	Premature
J.P. MacFarlane		Vintondale	
9/7/1915	Adam Molnar	Hungary	Ida Klid
Hungary	Krumbine	Twin Rocks	
Edward Jack Little	9/18/1915	2:00PM	Cholera Infantum
W. Prideaux	4 Mos., 24 Da.	Pennsylvania	
4/25/1915	Edward Little	Pennsylvania	Ruth Musser
Pennsylvania	Krumbine	Reynolds, PA	
George Blewitt	9/27/1915	5:00PM	Apoplexy
J.P. MacFarlane	67 Yrs., 4 Mos., 13 Da.	England	Married, Retired
5/14/1848	George Blewitt	England	Sarah Gardner
England	Jeff Evans, Ebens.	Ebensburg	
Frank Such	10/3/1915	2:00PM	Diptheria
J.P. MacFarlane	4 Yrs., 6 Mos., 25 Da.	Virginia	
3/8/1911	George Such	Hungary	Annie Kelliman
Hungary	Krumbine	Twin Rocks	
Annie Hollet	10/17/1915	1:00AM	Congenital, Malformation of Heart
J.P. MacFarlane	7 Days	Vintondale	
10/11/1915	Metro Hollet	Russia	Anny Zoba
Hungary	Krumbine	Wehrum	
Stillborn Male	10/20/1915		Premature-6 mos.
J.P. MacFarlane		Vintondale	
10/20/1915	Steve Borahish	Hungary	Mary Franks
Hungary	Krumbine	Twin Rocks	

Geo. Felde, Miner W. Prideaux 1871 Austria-Hungary	11/1/1915 44 Years Mike Felde Krumbine	2:00PM Austria-Hungary Austria-Hungary Wehrum	Probable Heart Disease Anna Luder
Alvera Marpelli J.P. MacFarlane 7/23/1915 Italy	11/13/1915 3 Mos., 21 Da. Luia Malpelli Krumbine	5:30PM Vintondale Italy Twin Rocks	Acute Gastritis Josephine Canninmap
Stillborn Male J.P. MacFarlane 11/15/1915 Italy	11/15/1915 Guerino Averi 274 Sixth Krumbine	Vintondale Italy Twin Rocks	Premature-8 mos.

Death Certificates: 1916

Joseph Poline, coke yard worker Certificate authorized by Coroner Coke Yard Worker Siberia	1/1/1916 4 (Sic) Ananipa Poline Geo. Ondrezak	10:30AM Siberia Siberia Wehrum	Broken back, abdomen torn open by railway car pressing over body Married Jennie
Hugh Thomas Williams J.P. MacFarlane 5/5/1915 Pennsylvania	1/5/1916 8 Months S.R. Williams Krumbine	1:00AM Vintondale Pennsylvania Mt. Union Cem., Pine Twp.	Broncho-Pneumonia, Chicken Pox Margaret Sides
Joe Glode J.P. MacFarlane 10/15/1915 Hungary	1/23/1916 3 Mos., 8 Da. Andy Glode Krumbine	1:00PM Vintondale Hungary Twin Rocks	Acute Gastritis, Artificial Feeding Lizzie Barat
Ione Hamilton J.P. MacFarlane 2/27/1916 Summerhill	2/27/1916 Three Hours C.A. Hamilton Krumbine	7:00PM Vintondale Indiana Co. Monday P. Cemetery (sic)	Premature-7 1/2 mos. Lettie Bittorf
John Toth M. Freeman, Vintondale 3/2/1916 Austria-Hungary	3/2/1916 Steve Toth Ondriezek	Vintondale Austria-Hungary Twin Rocks	Stillborn Ester Weakski
Thomas B. Wray, blacksmith M. Freeman 8/28/1863 Pine Twp.	3/4/1916 52 Yrs., 7 Mos., 3 Da. Wm. J. Wray Krumbine	Pine Twp., PA Pine Twp. Ebensburg L. Cemetery	Pulmonary TB Married Askin
No Name W. Prideaux, did not attend it, signed by Coroner's permission 3/30/1916 Italy	4/2/1916 3 Days Vincent Alvino Krumbine	7:00AM Vintondale Italy Twin Rocks	Died suddenly premature Jennie Alvino

Lloyd H. Gongloff J.P. MacFarlane 7/27/1913 Pennsylvania	4/5/1916 2 Yrs., 8 Mos., 9 Da. Grant Gongloff Krumbine	11:00PM Vintondale Pennsylvania Ebensburg	Lobar Pneumonia Edith Gardner
Jennie Alvino, Mrs. C. Fitzgerald Do Not Know Italy	4/8/1916 17 Years Fred Alvini Ondriezek	Italy Italy Twin Rocks	Puerperal Septaemi died without medical attention Anne Constable See Above 4/2/1916
Rosa Pindur J.P. MacFarlane Austria	4/10/1916 10 Mos., 19 Da. Mike Pindur Krumbine	5:00PM Pennsylvania Austria Wehrum	Intestinal Obstruction Mary Mancheck
John Molnar, Miner C. Fitzgerald 4/20/1874 Hungary	4/20/1916 42 Yrs., 2 Mos. Krumbine	 Hungary Hungary Twin Rocks	Myocaditis, Alcoholism Married Anna Molnar
Steve Feyeston F. Freeman 6/4/1916 Hungary	6/29/1916 3 Wks., 4 Da. Steve Feyester Krumbine	12:30PM Vintondale Hungry Twin Rocks	Cholera Infantum Anna Rusin
Elona Molnar J.P. MacFarlane 7/1/1916 Hungary	7/4/1916 4 Days Alix Molnar Krumbine	6:00AM Vintondale Hungary Twin Rocks	Volirilus Mary Nagy
George Drabind, Miner H.C. Thomas 3/10/1862 Austria	8/15/1916 54 Yrs., 5 Mos., 5 Da. John Drabink Ondriezek	6:50PM Austria Austria Johnstown Greek Catholic Cemetery	Chronic Myocarditis & Endocarditis Married Do Not Know
Joe Sracic H.C. Thomas 5/6/1916 Hungary	8/24/1916 3 Mos., 18 Da. Frank Sracic Krumbine	3:10AM Vintondale Hungary Twin Rocks	Cholera Infantum Antonia Fiben
Antonio Tote H.C. Thomas 11/21/1915 Austria-Hungary	8/29/1916 9 Mos., 8 Da. Joseph Tote Ondriezek	4:45PM Vintondale Austria-Hungary Holy Name, Ebensburg	Gastro-Enteritis, Improper Feeding Perosha Bodoz
Mary Hagadish C. Fitzgerald 12/1/1914 Hungary	9/15/1916 1 Yr., 9 Mos., 14 Da. Martin Hagadish Krumbine	2:30AM Vintondale Hungary Twin Rocks	Convulsions Anne Govily

Name / Undertaker / Date of Birth / Birthplace	Date of Death / Age / Father / Informant	Time / Place of Death / Father's Birthplace / Place of Burial	Cause of Death / Mother
Helen Marie Swanson J.P. MacFarlane 10/19/1916 Pennsylvania	10/30/1916 1 Day Hilding Swanson Ondriezek	9:00PM Pennsylvania Pennsylvania Strongstown	Patulous Foramen Ovale Alice Frampton
Peter Kerekes J.P. MacFarlane 6/17/1916 Austria-Hungary	11/14/1916 4 Mos., 28 Da. Michael Kerekes Ondriezek	Pennsylvania Austria-Hungary Wehrum	Broncho-Pneumonia
Mike Skubik J.P. MacFarlane Andy Skubik Austria	11/16/1916 1 Yr., 1 Mo., 13 Da. Austria Krumbine	1:30PM Vintondale Elizabeth Ruch Twin Rocks	Broncho-Pneumonia
William Markus C. Fitzgerald 10/4/1914 Hungary	12/5/1916 2 Yrs., 2 Mos., 2 Da. Kalman Markus Krumbine	 Vintondale Hungary Ebensburg	Tubercular Peritontis, died in streetcar in Johnstown Julia Flyis
Nichola Feshovick J.P. MacFarlane 8/8/1915 Russia	12/8/1916 1 Yr., 4 Da. John Feshovick Krumbine	2:00AM Conemaugh Austria-Hungary Conemaugh G.C.C.	Broncho-Pneumonia Annie Yangwak
Esaias Ellen Letitica, Mrs. J.P. MacFarlane 9/2/18? Wales	12/12/1916 50 Yrs., 3 Mos. Only in Vintondale 4 days, Richard Evans Jeff Evans, Ebens.	6:30 Wales duration of illness unknown Wales Blossburg, PA	Diabetes Meditus Widowed LetiticaWesley
Stillborn Female J.P. MacFarlane 12/14/1916 Italy	12/14/1916 Vintondale Fred Alvino Father	2:00AM Italy Twin Rocks	Premature Rachila Dulluca

Death Certificates: 1917

Name / Undertaker / Date of Birth / Birthplace	Date of Death / Age / Father / Informant	Time / Place of Death / Father's Birthplace / Place of Burial	Cause of Death / Mother
Joseph Ostofice J.P. MacFarlane 6/5/1910 Hungary	1/8/1917 6 Yrs., 7 Mos., 3 Da. Grist Ostofice Ondriezek	4:00AM Vintondale Hungary Twin Rocks	Broncho-Pneumonia Mary Irene
Stillborn Female J.P. MacFarlane 1/12/1917 Hungary	1/12/1917 Joseph Sabo Ondriezek	 Vintondale Hungary Twin Rocks	 Rosa Juasa
Geo. Volosin By permission of Deputy Coroner 5/6/1916 AustriaKrumbineWehrum	1/14/1917 8 Mos., 8 Da. Geo. Volosin	6:00AM Vintondale Austria	Acute Pericarditis Annie Levi

Stillborn Female J.P. MacFarlane 2/10/1917 Pennsylvania	2/10/1917 Vintondale Henry Mizner Father	Pennsylvania Buffington Cem.	Annie McMahan
Franceska Mazza J.P. MacFarlane 6/16/1917 Italy	4/13/1917 9 Mos., 28 Da. Caesar Mazza Ondriezek	4:00AM Pennsylvaina Italy Twin Rocks	Broncho-Pneumonia whooping cough Teressa Malpole
Charles Martin J.P. MacFarlane 5/2/1917 Austria-Hungary	5/3/1917 1 Day Charles Martin Krumbine	10:00PM Vintondale Austria-Hungary Ebensburg L.C.	Premature-7 1/2 mos. Julia Lave
Frank Martin J.P. MacFarlane 5/2/1917 Austria-Hungary	5/6/1917 4 Days Charles Martin Krumbine	2:00AM Vintondale Austria-Hungary Ebensburg L.C.	Premature-7 1/2 mos. Julia Lave
Joseph Molnar J.P. MacFarlane 6/23/1917	5/12/1917 10 Mos., 4 Da. Adam Molnar Krumbine	6:00AM Vintondale Austria-Hungary Twin Rocks	Acute Bronchitis Ida Koza
Elmiria Doyka J.P. MacFarlane 2/18/1914	6/29/1917 3 Yrs., 4 Mos., 11 Da. Stanley Doyka Ondriezik	6:00PM Pennsylvania Austria Twin Rocks	Entero-Colitis Vernie Griak
Filomena Spoamblek, Mrs. C. Fitzgerald 5/17/1895	7/26/1917 22 Yrs., 2 Mos., 8 Da. Joseph Kesche Krumbine	Hungary Hungary Portage	Acute Dilitation of Heart Married Filomena Hanucia
Irvin Gerald Bennett W. Prideaux 8/7/1917 Indiana Co.	8/16/1917 9 Days J.D. Bennett Krumbine & Son	Vintondale Indiana Co. Strongstown	Premature May McCreery
Edith Molnar Kim D. Curtis, Vintondale 8/21/1916 Austria-Hungary	9/6/1917 1 Yr., 15 Da. Paul Molnar Krumbine & Son	1:00AM Vintondale Austria-Hungary Ebensburg	Entero Colitis Julia Szerko
Steve Toth J.P. MacFarlane 11/29/1905 Austria-Hungary	10/26/1917 11 Yrs., 10 Mos., 27 Da. Steve Toth Krumbine & Son	12:00PM Austria-Hungary Austria-Hungary Ebensburg	Tubercular Peritonitis Ester Vakshof
Helen Skubic J.P. MacFarlane 11/2/1917 Austria	11/4/1917 3 days Andy Skubic, 47 Plank Krumbine & Son	Vintondale Austria Twin Rocks	Patulous Foramen Ovale Elizabeth Roth

Fronace Laballe, Miner J.P. MacFarlane 10/10.1874 France	11/10/1917 48 Yrs., 1 Mo. Eugene Laballe Krumbine & Son	11:45AM France France Ebensburg	Chronic Endocarditis, TB Married Josephine Perlin
Wm. Kalanca J.P. MacFarlane 9/1/1915 Austria-Hungary	12/16/1917 2 Yrs., 3 Mos., 16 Da. Mike Kalanca Krumbine & Son	7:00PM Twin Rocks Austria-Hungary Twin Rocks	Tubercular Meningitis Elizabeth Kovach
Lizzy Englodi J.P. MacFarlane 9/21/1917 Austria	2/2/1917 2 Mos., 11 Da. Andy Englodi Krumbine & Son	8:00AM Vintondale Austria Twin Rocks	Morasmus, Hereditary Syphilis Lizzie Barata

Death Certificates: 1918

Bertha Tote Balog, Mrs. J.P. MacFarlane 1/1/1867 Hungary	1/3/1918 56 Yrs., 6 Mos., 3 Da. Andy Toth Krumbine & Son	10:00AM Hungary Hungary Ebensburg	Canser (Sic) of Stomach Married Bertha Hegerus
Nick Korich J.P. MacFarlane 10/18/1917 Austria-Hungary	2/3/1918 3 Mos., 16 Da. Geo. Korich Krumbine & Son	2:00AM Vintondale Austria-Hungary Twin Rocks	Pulmonary Edema Mary Kosterjak
Charles E. Berkey J.P. MacFarlane 2/17/1918 Milheim, PA	2/17/1918 13 Hours Wm. P. Berkey Krumbine & Son	Vintondale Bethel, PA Ebensburg, L.C.	Premature Marion Huey
Joseph Sadusky C. Fitzgerald 10/18/1917 Pennsylvania	3/8/1918 4 Mos., 17 Da. George Domanick Krumbine & Son	3:00PM Cleveland Pennsylvania Twin Rocks	Unknown, no medical attention Annie Sidusky
John Kriscolusion J.P. MacFarlane 1/17/1918 Austria	3/16/1918 1 Mo., 29 Da. John Krescolusion Krumbine & Son	5:00AM Vintondale Austria Wehrum	Pulmonary Edema Annie Kerekes
Stephans Page J.P. MacFarlane 8/7/1917 Austria	3/17/1918 7 Mos., 10 Da. Peter Page Krumbine & Son	6:00AM Detroit, Michigan Austria Wehrum	Lobar Pneumonia Josephine Gloyoki
Steve Sous J.P. MacFarlane 3/20/1918 Hungary	3/31/1918 11 Days Joseph Sous Krumbine & Son	3:00PM Vintondale Hungary Ebensburg	Broncho-Pneumonia Freaba Zotak

Mike Boconich, Miner	4/16/1918	9:00AM	Lobar Pneumonia
J.P. MacFarlane	42 Years	Austria-Hungary	
11/1876	Charley Boconich	Austria-Hungary	Do Not Know
	Krumbine & Son	Wehrum	
Stillborn Male	6/30/1918		
J.P. MacFarlane, by permission of Coroner		Vintondale	
6/30/1918	Frank Kizolski	Russian Poland	Magay Majtuzto
Russian Poland	Krumbine & Son	Twin Rocks	
Lizzy Evan	7/2/1918	6:00AM	Measles
J.P. MacFarlane	2 Yrs., 2 Mos., 4 Da.	Vintondale	
4/27/1916	John Evan	Austria-Hungary	Veron Acies
Austria-Hungary	Krumbine & Son	Twin Rocks	
Mike Todorvick	7/16/1918	5:00PM	Pneumonia, Measles
J.P. MacFarlane	2 Years	Vintondale	
7/16/1916	John Todorick	Austria	Josephine Prinosock
Austria	Krumbine & Son	Wehrum	
John Holobin	8/17/1918	4:00AM	Gastro-Enteritis
J.P. MacFarlane		Expedit	
	Harry Holobin	Austria	Milare Slobagen
Austria	Krumbine & Son	Wehrum	
Alax Molnar	8/21/1918	5:00AM	Gastro-Enteritis, Rachitis
J.P. MacFarlane	6 Mos., 17 Da.	Vintondale	
2/4/1917	John Molnar	Austria-Hungary	Mary Chariadia
Austria-Hungary	Krumbine & Son	Twin Rocks	
Charles Risko	8/21/1918	11:30AM	Gastro-Enteritis Malnutrition
J.P. MacFarlane	3 Mos., 21 Da.	Vintondale	
5/1/1918	Mike Risko	Austria	Annie Josie
Austria	Krumbine & Son	Wehrum	
Anna Koponitek	8/24/1918		Ileo-Colitis
W. Evans, Wehrum	1 Yr., 1 Mo., 8 Da.	Allegheny, PA	
7/16/1917	John Kopontek	Austria	Anna Daiok
Austria	Krumbine & Son	Wehrum	
James Colos, Spragger	9/4/1918		Crushed skull, mine accident
J.P. MacFarlane, with Coroner's permission			
	23 Years	Austria-Hungary	Married
	Mitro Hanna Colos	Austria-Hungary	Anna Cifid
Austria-Hungary	Krumbine & Son	Wehrum	
Blanch Clark	9/25/1918	10:00AM	Gastro-Enteritis
J.P. MacFarlane	1 Yr., 11 Mos., 12 Da.	Bakerton, PA	Mal Metritea
10/13/1916	H.J. Clark	England	Ida M. Wilkins
Pennsylvania	Krumbine & Son	Ebensbrg L.L.C.	

Death Certificates: 1919

Frank Barate, Miner	4/23/1919	9:00AM	Pulmonary TB
J.P. MacFarlane	24 Yrs., 2 Mos., 5 Da.	Hungary	
2/18/11889	Michel Barate	Hungary	Kolobcz Eszbo
Hungary	Krumbine & Son	Buffington	

Death Certificates: 1920

Stillborn Female	9/4/1920	11:00AM	Unknown
J.P. MacFarlane		Vintondale	
9/4/1920	Metro Balog	Austria	Mary Verbish
Austria	None Engaged	Wehrum	341 Main Street
No Name Given - Male	9/5/1920	8:00AM	Broncho-Pneumonia
H. Gockley, Vintondale	11 Mos., 13 Da.	Pennsylvania	
9/23/1919	Austin Donaldson	Michigan	Treasa Lantzy
Pennsylvania	Krumbine	Bakerton Cemetery	240 Main Street
John Minchenko	9/13/1920	10:00AM	Gastro-Enteritis
J.P. MacFarlane	7 Mos., 9 Da.	Pennsylvania	
3/4/1920	Daniel Minchenko	Russia	Dolly Somerchiko
Russia	Krumbine	Wehrum	
Murice Robanky	10/28/1920	7:30AM	Patulous Formen Ovale
J.P. MacFarlane	1 Day	Pennsylvania	
10/27/1920	Ignats Robanky	Russia	Anna Kochunsky
Russia	Krumbine	Wehrum	
Stillborn Female	11/28/1920	11:00PM	Mother fell from second story window to ground floor
J.P. MacFarlane		Vintondale	
11/28/1920	John Kritzman	Hungary	Julia Risko
Hungary	Krumbine	Twin Rocks	194 Plank Road
John J. Kempfer, carpenter	12/5/1920	8:00AM	Acute Alcoholism, found dead in bed
By Coroner's permission	52 Yrs., 6 Mos., 19 Da.	Elkart, Indiana	Single
J.P. MacFarlane, by Coroner's permission			
5/16/1868	S.O. Kempfer	N. Carolina	Adda Schoch
Middleburg, PA	Krumbine	Lloyds Cemetery	
John Lazar	12/6/1920	3:30PM	Broncho-Pneumonia
J. Gockley	2 Mos., 21 Da.	Vintondale	
9/27/1920	Paul Luzar	Austria	Sophia Stanslaw
Austria	Krumbine	Wehrum	
Helen Stachy	12/14/1920	1:00AM	Broncho-Pneumonia
J. Gockley		Vintondale	
3/20/1920	Louis Stachy	Hungary	Julia Locadino
Hungary	Krumbine & Son	Twin Rocks R.C.C.	

Death Certificates: 1921

Willard Cook, Laborer	1/3/1921	9:30AM	Endocarditis, Chronic Intestinal Nephritis
Dr. MacFarlane	65 Yrs., 1 Mo., 3 Da.	Pennsylvania	Married
12/1/1855	John R. Cook	Pennsylvania	Lydia Abbott
Pennsylvania	Krumbine		
John Senachko	2/27/1921	2:00AM	Measles
J.P. MacFarlane	9 Months	Pennsylvania	
5/20/1920	John Senchko	Austria	Mary Hollod
Pennsylvania	Krumbine	Wehrum	
Unnamed Female	3/23/1921		Premature, Malformation of Heart & Palate
J.P. MacFarlane		Vintondale	
3/23/1921	Geo. Katza	Austria	Sophia Pelich
Austria	Father	Wehrum	
Eleck Pronko	4/10/1921	8:00PM	Measles
J.P. MacFarlane	11 Mos., 5 Da.	Pennsylvania	Broncho-Pneumonia
5/4/1920	Eleck Pronko	Russia	Julia Mehuncko
Russia	Krumbine	Wehrum	
Ronald Feldman	4/17/1921	1:00AM	Broncho-Pneumonia, Measles
J.P. MacFarlane	2 Yrs., 8 Mos.	Pennsylvania	
9/18/1918	Sam Feldman	Michigan	Joy Cook
Pennsylvania	Krumbine	Ebensburg	
Murtle M. Wray	4/20/1921	11:00PM	Malformation of Pharynx, Premature
J.P. MacFarlane	3 Days	Pennsylvania	
4/18/1921	Frank Wray	Pennsylvania	Zetha Brown
Pennsylvania	Krumbine	Ebensburg	

Appendix D
Death Certificates: Bracken

Name Doctor Date of Birth Place of Birth	Date of Death Age Father's Name Undertaker	Time Place of Birth Place of Birth Place of Burial	Cause Mother's Name
Eva Rumanic, Mrs. Dr. W.A. Prideaux Do Not Know	7/2/1910 28 Yrs. Krumbine	 Austria Wehrum	Probable heart trouble, died suddenly, not seen by doctor Married
John W. Utter Dr. W.A. Prideaux 11/21/1905 West Virginia	9/17/1910 4 Yrs., 9 Mos., 26 Da. Peter Utter G. Empfield	 Pennsylvania Germany Belsano	Typhoid Fever Lillie Sutton
Steve Gregorger W.A. Prideaux 9/28/1881	3/19/1913 31 Yrs., 7 Mos., 9 Da. Vincere Greger Krumbine	 Hungary Hungary Twin Rocks	TB of lungs & throat Married
Andy Cossor Dr. W.A. Prideaux 10/4/1910 Austria	9/19/1914 3 Yrs., 11 Mos., 19 Da. Geo. Cossor J. Krumbine	7:00PM Pennsylvania Austria Wehrum	Larynginal Diptheria Sophia Balast
No Name Given - Male Dr. W.A. Prideaux 11/30/1915 Austria	1/2/1916 1 Yr., 1 Mo., 1 Da. Mike Juba Geo. Ondriezek	3:00AM Pennsylvania Austria Wehrum	Never seemed well after birth Annie Scuba
Stafna C. Kososky Dr. W.A. Prideaux 4/3/1916 Russia	8/17/1916 4 Mos., 14 Da. Steve Kososky Krumbine & Son	 Pennsylvania Russia Twin Rocks	Chronic Ileo-Colitis Alice Melaski
Joseph Supstaling Dr. W.A. Prideaux 1/12/1916 Russia	9/21/1916 8 Mos., 9 Da. Joseph Supstaling Krumbine & Son	10:00PM Pennsylvania Russia Twin Rocks	Chronic Gastric Ileo-Colitis Katie Korda
John Mazzano Dr. W.A. Prideaux 4/9/1916 Italy	9/24/1916 5 Mos., 15 Da. John Mazzano Geo. Ondriezek	11:00PM Pennsylvania Italy Twin Rocks	Ileo-Colitis Mary Whren

Name / Doctor / Date / Place	Death Date / Age / Informant / Undertaker	Time / Place of Death / Birthplace / Burial	Cause of Death / Spouse / Mother
Barbara Boski Dr. W.A. Prideaux 7/9/1907 Italy	8/24/1917 10 Yrs., 1 Mo., 15 Da. Bert Boskio Geo. Ondriezek	7:00PM Pennsylvania Italy Twin Rocks	Acute Heart Trouble Rosa Monithi
Annie Maniski Dr. W.A. Prideaux 4/4/1917 Austria	9/2/1917 4 Mos., 28 Da. John Maniski J. Krumbine	2:00AM Pennsylvania Austria Wehrum	Acute Ileo-Colitis Mary Balog
Barbara Boskie Dr. W.A. Prideaux 7/8/1917	11/9/1917 4 Mos., 1 Da. Brrt Boskie Geo. Ondriezek	7:00PM Pennsylvania Italy Twin Rocks	Pneumonia Mary Rosa
Stillborn - Andy Krichko Dr. W.A. Prideaux 2/6/1918	2/6/1918 George Krichko Father	 Pennsylvania Austria Wehrum	Faulty Position, Long Labor
Sofia Kreatihko Dr. W.A. Prideaux 3/8/1918 Hungary	3/8/1918 1 Yr., 2 Mos., 22 Da. George Kreathko J. Krumbine	1:00PM Wehrum Hungary Wehrum	Not attended by doctor, died within two hours Hellen Sobots
Nick Balazok, Laborer Dr. W.A. Prideaux	4/28/1918 49 Yrs. Metro Balazok Geo. Ondriezek	5:00AM Austria Austria Wehrum	Lobar Pneumonia Married Unknown
Jacob Kososky Dr. W.A. Prideaux 5/27/1918 Poland	9/26/1918 1 Yr., 4 Mos., 13 Da. Steve Kososky J. Krumbine & Son	9:00PM Jerome, PA Poland Twin Rocks	Catanbine Pneumonia Alase Milewsko
Mary Britski Dr. W.A. Prideaux 7/18/1920 Austria	9/6/1920 1 Mo., 19 Da. Mike Britsky Geo. Ondriezek	 Bracken, PA Austria St. Mary's, Nanty Glo	Ileo-Colitis Helen Holochok

Appendix E
Vinton Colliery Company

ANNUAL REPORT
ANNUAL PRODUCTION: VINTON COLLIERY COMPANY*

Year	Mine	Superintendent	Tonnage	Coke Made	Days Worked	Employees
1894	#1	C.R. Claghorn	4,586		43	22
1895	#1	"	53,893		231	58
	#2 (Madeira Hill)	"	3,001		24	20
1896	#1	"	65,445		154	134
	#2 (Madeira Hill)	"	5,071		29	25
1897	#1 & #2	"	105,968		203	112
1898	#1 & #2	"	108,149		238	142
1899	#1 & #2	Henry Douglas	116,178		273	147
	#3 Opened In Aug.	Talmage Bloss, Asst. Supt. C.R. Claghorn Gen. Supt.	18,010	8 Ovens	125	50
1900	#1 & #2	C.R. Claghorn	120,134		256	151
	#3	Henry Douglas, Supt.	60,069	8 Ovens	266	62
1901 Lackawanna C & C	#1	Gen. Supt. Claghorn, Wehrum	107,808		226	119
	#2	W.P. Morgan, Supt.	107,808		226	43
VCC	#3	R.G. Ware	75,467	8 Ovens	271	145
1902	#1 & #2	W.P. Morgan	95,402		214.3	137
						43
LCC	#3 & #4	G.R. Dalamater Wehrum Claghorn, Gen. Supt.	6,478 9,915			69
VCC	#3	R.G. Ware	95,095		234	144
1903 LCC	#1 & #2	J.M. Jones Claghorn, Gen. Supt.	76,742		156	150
VCC	#3	R.G. Ware	100,714		240	130
1904	#1 Leased by VCC	R.G. Ware	10,997		32	125
	#3	"	118,639		249	145

Year	Mine	Supt.	Production		Employees	
1905	#1	Charles Hower	79,475		207	158
	#2		Idle			
	#3		77,375		190	141
	#5		2,131		20	27
1906	#1 & #4	Charles Hower,	104,619		241	197
	#2 & #3	Gen. Supt.	84,185		220	158
	#5		72,332		250	101
	#6		10,491	150 Ovens Being Built	57	22
1907	#1 & #4	"	116,876	25,498	280	189
	#2 & #3	J.I. Thomas,	72,235		209	153
	#5	Supt.	97,366		274	131
	#6		60,223	20,089	198	187
1908	#1 & #4	"	106,348		161	216
	#2		Idle			
	#3		4,345		13	
	#5		55,449		149	123
	#6		62,806	6,127	133	223
1909	#1 & #4	S.K. Smith	98,818		242	176
	#2 & #3	Gen. Supt.	Idle			
	#5		4,570,		16	113
	Abandoned in Feb., to be worked from #6					
	#6		94,468	31,837	265	152
1910	#1 & #4	"	226,295		221	255
	#2		Idle			
	#3		7,896		38	71
	Started in Sept., after idle 3 years					
	#6		197,353	50,421 (120 Ovens Operating)	271	244
1911	#1	T.W. Hamilton	125,933		140	160
	#3	"	58,666		142	78
	#6	"	163,271	52,321	198	198
1912	#1	Sheldon Smillie,	184,218		258	
	#3	Gen. Supt.	76,229		236	530
	#6		179,659	71,979	263	
1913	#1	T.W. Hamilton	159,591		275	201
	#3	"	71,988		271	76
	#6	"	196,522	75,219	278	306
1914	#1	"	253,047		221	
	#3	"	62,475		250	
	#6	"	297,218	62,168	256	891, Total
1915	#1	Otto Hoffman	325,298		261	311
	#3		2,113		34	
	#6		488,295	75,892	307	491

Year	Mine	Operator	Production		Days	
1916	#1	"	319,829		266	219
	#6		426,200	78,464	299	427
1917	#1	"	296,326		303	277
	#6	"	397,237	82,021	304	510
Claghorn	#11	C.R. Claghorn	8,150	New Mines	35	86
	#12	"	5,575		32	87
1918	#1	Otto Hoffman	292,087		288	239
	#6	"	391,271	83,218	293	447
	#11	Abe Abrams	4,131	Idle 8 Mos.	87	25
	#12	"	7,130	Idle last part of year	172	40
	#16	"	42,212		299	70
1919	#1	Otto Hoffman	253,547		280	220
	#6	"	417,815	75,410	287	450
	#16	"	64,783		296	76
1920	#1	"	186,139		266	189
	#6	"	383,081	73,336	285	455
	#16	"	70,741		286	123
1921	#1	"	187,304		231	206
	#6	"	370,308	65,945	236	502
	#16	"	62,014		144	91
1922	#1	"	190,568		274	182
	#6	"	441,331	69,813	287	488
	#16	"	32,756		263	54
	#17	"	2,221		43	19
1923	#1	"	122,832		183	150
	#6	"	354,005	82,729	228	451
	#16	"	15,353		175	28
	#17	"	12,810		160	15
1924	#1	"	20,748	48	140	
	#6	"	252,258	59,294	163	507
	#16					
	#17	"	9,580		63	28
1925	#1	"	Idle			
	#6	"	367,543	77,241	205	470
	#16	"	Idle			
	#17	"	Idle			
1926	#1	"	13,648		47	79
	#6	"	367,627	77,853	224	422
	#16	"	Idle			
	#17	"	Idle			

Year	Mine	Operator	Production		Days	Employees
1927	#1	"	40,854		140	16
	#6	"	295,127	58,013	194	363
	#16	"	Abandoned			
	#17	"	Abandoned			
1928	#1	"	19,118		89	
	#6	"	290,192	66,053	204	426
1929	#1	"	24,080		95	25
	#6	"	347,789	88,207	226	429
1930	#1	Otto Hoffman	15,260			
	#6	and	265,660	69,887		461 Total
	#16	Milton Brandon	Idle			
	#17		Idle			

Source: Annual Report: Pennsylvania Department of Mines. Harrisburg, 1894-1930.

FINANCIAL RECORDS
VINTON LAND COMPANY
BALANCE SHEET - JANUARY 31st, 1925

ASSETS

Real Estate, Vintondale	$ 52,888.00
Lands & Minerals, Vintondale	347,601.36
Real Estate, Claghorn	85,047.69
Lands & Minerals, Claghorn	445,277.36
Claghorn Water Co. Stock	5,000.00
Claghorn Water Co. Advances	5,448.06
Penna. Co. Trustee 5%	7,511.34
Penna. Co. Trustee 6%	14,556.44
Organization	2,200.02
Income Tax	41.00
Cash	2,463.10
Profit & Loss	282,366.68
Profit & Loss 1925	2,558.34
Vinton Colliery Co.	507.16
	$1,253,466.55

LIABILITIES

Capital	$200,000.00
1st Mtge. Bonds 5%	188,000.00
1st Mtge. Bonds 6%	355,000.00
Interest Accrued	15,350.00
V.C. Co. Loan	185,118.18
V.C. Co. Adv. Rylty	27,482.56
V.C. Co. 1st Lease	18,867.03
V.C. Co. 2nd Lease2	14,820.52
Rents, Vintondale	315.00
Royalty, 1st Lease	2,524.94
Royalty, Lease "E"	192.16
Depreciation Reserve - Vdale.	16,710.00
Depreciation Reserve - Clghn.	28,955.16
Income Tax Payable	131.00
	$1,253,466.55

VINTON COLLIERY COMPANY
BALANCE SHEET
DECEMBER 31st, 1925

ASSETS

Lands	$ 91,139.24
Leasehold Rights	500,000.00
Mach. & Equipment	860,813.35
Inventory	9,296.11
Stock in Subsidiaries	202,500.00
Loans to Subsidiaries	225,986.63
Cambria Co. Fair Stock	250.00
Advanced Royalties	290,593.49
Prepaid Insurance	3,339.44
Bonds-Vinton Land Co.	196,675.63
Bonds-Vinton Colliery Co.	98,965.00
Accts. Receivable	52,496.43
Cash	57,060.89
	$2,589,116.21

LIABILITIES

Capital Stock	$ 1,000,000.00
Undivided Profits	329,692.77
1st Mtge. Bonds	256,000.00
Depreciation Reserve	600,275.24
Depletion Reserve	60,448.41
Unclaimed Wages	5,554.51
Medical Fund	6,513.46
Non-Res. Alien Tax	13.15
NY State Non-Res. Income Tax	11.00
Interest Accrued	2,133.34
Bills Payable	275,000.00
Accounts Payable	53,474.33
	$2,589,116.21

VINTON COLLIERY COMPANY
CONDENSED BALANCE SHEET
JANUARY 31st, 1927

ASSETS

Lands	$ 91,139.24
Leasehold	500,000.00
Mchy. & Equipment	818,621.48
Development #17	52,593.80
Warehouse Stock	2,262.75
Stock Coal & Coke	7,630.00
Insurance Advances	2,794.85
Cambria Co. Fair Stock	250.00
B.L. Water Co. Stock	1,000.00
Jackson Water Co. Stock	1,000.00
Water Co. Loans	13,125.47
V.L. Co. Stock	200,000.00
V.L. Co. Loans	183,618.18
V.L. Co. Rylty. Adv.	306,771.33
V.C. Co. 1st Mtge. Bonds	87,440.00
V.L. Co. 1st Mtge. Bonds 6%	70,433.88
V.L. Co. 1st Mtge. Bonds 5%	47,221.74
V. Amusement Co. Stock	500.00
V. Amusement Bills Rec.	18,000.00
W. Griffith Adv. Rylty.	4,476.89
Lacka. C. & C. Co.	9,464.88
Tax Suspense	2,500.00
Cash	53,760.94
Accts. Receivable	<u>69,932.70</u>
	$2,554,538.13

LIABILITIES

Capital	$1,000,000.00
Undivided Profits	432,336.49
1st Mtge. Bonds	243,000.00
Bills Payable	150,000.00
Interest Accrued	3,037.50
Taxes Accrued	2,000.00
Accounts Payable	46,091.35
Unclaimed Wages	5,797.57
Dispensary	7,232.29
Deprcn. Reserve	600,275.24
Depletion Reserve	60,448.41
Jackson Water Co.	9.35
Profit & Loss 1927	<u>4,309.93</u>
	$2,554,538.13

VINTON LAND COMPANY
CONDENSED BALANCE SHEET
JANUARY 31st, 1927

ASSETS

Real Estate V'dale	52,888.00
Lands & Minerals V.	337,316.72
Real Estate Clghn.	85,047.69
Lands & Minerals Co.	445,227.36
Clghn. Water Co. Stock	5,000.00
Clghn. Water Co. Advances	5,748.06
Organization	2,200.02
Penna. Co. Trustee 5%	7,637.90
Penna. Co. Trustee 6%	14,685.20
Accounts Receivable	870.03
Cash	168.32
Profit & Loss	324,825.87
Bond Int. 2,423.34	
Expense 36.35	2,459.69
	$1,284,124.86

LIABILITIES

Capital	200,000.00
1st Mtge. Bonds 5%	176,000.00
1st Mtge. Bonds 6%	338,000.00
V.C. Co. Loan	183,618.18
V.C. Co. Rylty.	306,771.33
Interest Accrued	14,540.00
Deprcn. V'dale	19,354.40
Deprcn. Clghn.	33,207.54
Income Tax Payable	114.90
Rents 315.00	
Royalty 2,183.35	
Interest 20.16	2,518.51
	$1,284,124.86

COKE DATA & COSTS
JANUARY 31st, 1927

Ovens Drawn 1214	Days Worked 24		Men Employed 40
Coal: Fresh Mined 11,163.00	tons net @ $2.00		$22,326.00 Charge
Stock 12/3 1,315.00	tons net @ $2.00		2,630.00 Per
12,478.00	tons net		$24,956.00 Oven
Stock 1/31 1,436.00			8.87
Rewashed 120.00			
Raw 150.00 1,706.00	tons net @ $2.00		$3,412.00
10,772.00	tons net		$21,544.00

Coke: Foundry 6,077.02	tons net	(6989)
Domestic 482.00	tons net	Yield per oven 5.76
Braise 520.09	tons net	Percent of Coal 64.89
7,079.11	tons net	
Stock Decrease 90.00	tons net	
6,989.11	tons net	

Costs:					
	Coal	$21,544.00	@	$3.082 per ton	(6989)
	Washing	1,862.41	@	.266 per ton	
	Coking	6,140.88	@	.879 per ton	
		$29,547.29	@	4.227 per ton	
	Management	177.69	@	.025 per ton	
	Domestic	397.65	@	.057 per ton	
		$30,122.63	@	4.309 per ton	

RESULT OF MONTH'S OPERATION

Sold:	Foundry	6,077.02	tons @	$5.035		$30,599.08
	Domestic	482.00	tons @	3.267		1,574.69
	Braise	520.09	tons @	.50		260.23
		7,079.11	tons			32,434.00
Stock Fdry.						
Increase: 60.00			@	4.75	$285.00	
Stock Domestic						
Decrease: -150.00		90.00	@	4.25	637.50	352.50
		6,989.11	tons			$32,081.50
				Gain Foundry	4,784.40	
				Loss Domestic	2,825.53	1,958.87
				Cost		$30,122.63

SUMMARY FOR MONTH
JANUARY 31st, 1927

	Shipped	House	Washery	Power	Total
Vinton No. 1 V.C. Co		549.05			
Vinton No. 1 Lacka.	6,826.08	38.12			
	6,826.08	587.17			7,414.05
Vinton No. 6 V.C. Co.			610.01	1,227.10	
No. 5	2,426.15		4,142.18		
V.L. Co.	18,805.07				
Lease E				6,410.01	
	21,232.02		11,163.00	1,227.10	33,622.12
Grand Totals	28,058.10	587.17	11,163.00	1,227.10	41,036.17
		Deduct Dec. Stock #1- 100.00			
		Deduct Dec. Stock #6- 310.00			410.00
Shipped & Consumed	41,036.17			Product	40,626.17

Coal Cost			Income		Deductions	
Labor	$62,696.22	-$1.543	Sales	$87,357.39	Interest	$ 954.16
Supplies	14,435.06	-.355	Rents	1,890.16	Fixed Charges	8,074.90
	77,131.28	-1.898	Coke	1,958.87	Idle Expense Claghorn	
			Gain on Sale of Bonds	725.00	Cost of Coal	77,131.28
			Expl. & Supplies	473.01		
			Electric	300.85		
			Disct. & Interest	1,436.70		
			V'dale. Inn.	121.61		
				90,545.27		$86,309.93

Fixed Charges					
Royalty	$3,167.32		Total Income		$90,545.27
Taxes	2,000.00		Total Deductions		86,235.34
Insurance	500.00				
Salaries	2,171.67		Gain for Month		$4,309.93
Misc.	235.91				
	$8,074.90				

Vinton Mines Cost $1.898	Sales Lump.	3,742.06	tons @	2.856
	Stoker	5,212.13	tons @	2.095
	Slack	4,548.19	tons @	2.121
	R/m	4,554.12	tons @	2.685
	Local	587.17	tons @	3.066

Source: Frederic Delano Files, Franklin Delano Roosevelt Library, Hyde Park, NY.

FATAL ACCIDENTS
VINTON COLLIERY COMPANY

1898 October 13
Herman Dishong, American, machine helper, 19, single #2 mine, cauhgt in machine for two hours, leg amputated, but he never gained consciousness.

1900 November 30
Tony Brook, Hungarian, loader, 36, single, #3 mine, crushed by cars.

1901 January 11
John Koran, Slav, miner, 38, married, Vinton #1, fatally injured by fall of coal.

1901 June 25
Clinton Jordan, American, machine runner, 34, married, two children, Lackawanna #1, killed by electrically current charged machine while he was setting jack in the roof.

1908 July 30
James Komitsoe, Slavonian, company man, 35, married, widow and six children, #1, fatally injured while assisting a machinist to change a pinion wheel, a seven inch gear wheel fell on him, died five hours later.

1910 July 30
John Sabo, Hungarian, miner, 44, married, widow and four children, #6, killed by fall of slate at face of his room while loading car.

1910 August 8
Mike Calchery, Slavonia, triprider, 24, single, #6, killed instantly, riding between cars of loaded trip and failed to lower his head when passing through the door, head caught by top of door frame and neck broken.

1910 October 14
Joseph Segarty, Hungarian, miner, 27, married, widow and one child, #1, fatally injured by fall of rock in his room, back broken, died November 19.

1910 December 20
Alex Roza, Hungarian, miner, 33, married, widow and three children, #6, fatally injured by fall of coal, skull fractured while undermining a skip of coal off the rib near the face of his heading, died six hours later.

1911 September 20
Charles Dancho, Slavonian, miner, 23, single, #6, killed instantly by fall of roof at face of room, props not properly set.

1912 January 15
Frank Resco, washery crew, #6, arm torn off by #1 belt, claimed to be applying dressing to belt, but was either riding on belt and carried too far, or had fallen on belt from discharge hopper.

1912 March 1
Agustian Pizzi, Italian, miner, 26, single, #1, killed instantly by roof fall while removing pillar, gone beyond working place for some unknown reason.

1912 September 21
Arpot Sabo, Hungarian, miner, 21, single, #6, killed instantly by roof fall at face.

1914 January 2
Mike Latto, machine helper, #6, dropped over dead, heart trouble, had thought that he had died from electrical shock, new pulmotor machine tried on him.

1914 March 5
Luey Frederici, Italian, miner, #1017, 48, single, #6, killed by fall of draw slate while undermining coal in removing pillars in 7R, old slope, ribs broken and lungs punctured. Accident happened at 5:30; Frederici died at 8 PM. The company got Krumbine to bury him for $50.

1915 January 13
Robert Clyde, American, motorman, 42, married, widow and five children, killed instantly by mine cars, attempted to jump off fast moving cars and fell between them.

1915 April 14
Andy Boca, Slavonian, belt tender, 54, married, #6, body found with foot caught in pully wheel, no witness, outside.

1915 June 1
Metro Popovitch, Slavonian, miner, 29, single, killed by fall of rock at face of room.

1915 June 30
Steve Mayerik, Austrian, miner, 26, married, widow and one child, #1, killed instantly by fall of slate at face of pillar, went in to get tools and caught by fall before he could get out.

1916
Joseph Polimoc, Horwat, laborer, 41, married, widow and four children, #6, instantly killed, runover by railroad car that was being sloped by gravity at slow speed, brake of car on rear and man dropping car saw no one on track when he started it, outside accident.
1918 September 14
Jim Colas, Hungarian, car handler, 28, married, #6, killed instantly by cars on entry.

The above information was obtained from: **Report of the Department of Mines of Pennsylvania, Part II Bituminous, 1894-1918**. Author's note: Detailed records of fatalities and accidents are not recorded after 1918. The sources for the following list of mine fatalities are the **Nanty Glo Journal** and interviews.

1927 March
Joseph Simon killed in accident.
1927 April
Mike Hazy, 39, recently arrived from Hungary, died of blood poisoning sustained in mine accident.
1929 February
Frank Matyus, 49, run over by mine car, began work four days earlier, came from Limburg, Canada, left wife and seven children there, wife unable to have remains shipped north, buried in St. Mary's Cemetery, Nanty Glo.
1929 June 28
Gust Oustfine, 50, killed by rockfall, survived by wife and six children, buried at St. Charles' Cemetery, Twin Rocks.

NON-FATAL ACCIDENTS
VINTON COLLIERY COMPANY, 1898-1918

1898 August 18
Steve Hannan, 30, married, miner, injured on head and body was caught between mine wagons.
1901 August 2
John Shenetski, 22, single, car runner, Hungarian, Vinton #2, four ribs broken by cars.
1903 January 24
Andy Jacobs, 38, married, miner, Slavonian, Lackawanna #2 mine, head and face cut by fall of coal.
1903 February 18
Amos Rissinger, 40, married, machine runner, American, Lackawanna #1 mine, foot injured by being caught between coal and machine.
1903 July 15
Joseph Walker, 19, single, miner, American, Vinton Colliery (#3?), internally injured by mine car, caught between car and side of rib.
1904 July 15
Joseph Walker, 22, single, car runner, American, #3 mine, internally injured by being squeezed between mine cars and door frame.
1904 September 7
T.M. Robertson, 17, single, motorman, American, Lackawanna #1 mine, four toes crushed and amputation necessary. Foot caught in gear of motor.
1904 September 8
John Rego, 28, married, miner, Hungarian, Vinton #1 mine, leg broken, caught between cars and side of heading.
1905 April 24
Mike Volnar, 38, married, runner, Slavonian, Vinton #4, seriously injured by falling off a trestle outside.
1906 January 22
George Helss, 31, married, miner, Slavonian, #1
John Matie, 21, single, miner, Finnish
Mike Oros, 30, married, miner, Hungarian
1906 January 22
John Kovach, 20, single, miner, Hungarian, burned by powder which ignited in hands of John Matie.
1906 January 24
W. Chambers, 45, married, trip rider, Scotch, #3 mine, foot crushed by cars inside mine.

1906 May 4
Mike Romano, 28, single, miner, Slavonian, #1 mine, leg broken by roof fall while taking out props.
1906 August 2
Aubrey Morton, 23, single, motorman, American, #1 mine, back injured by fall of roof.
1906 December 1
Dan Gray, 25, married, company man, American, #3 mine, head and back injured by fall of roof.
1906 December 8
John Benza, 35, single, miner, Slavonian, #6 mine, leg broken by car inside mine.
1907 March 23
Mike Poconick, 21, single, driver, Slavonian, #6 mine, leg broken by cars.
1907 July 9
John Nogonick, 27, married, miner, Slavonian, #6 mine, leg broken by cars.
1907 October 24
John Balogle, 38, married, miner, Hungarian, #1 mine, arm broken by fall of rock at face of his heading, rock had been tested by sounding by foreman.
1907 December 13
John Mahi, 50, married, company man, Finnish, arm broken by being caught between motor and roof.
1908 August 8
Mike Robinski, 29, married, miner, Slavonian, #1 mine, collar bone broken by fall of coal.
1908 August 11
Louis Mash, 27, married, miner, Italian, #5 mine, leg broken by fall of roof.
1908 September 4
Andy Anderson, 25, single, miner, Finnish, #1 mine, leg broken by haulage rope.
1909 July 12
Charles Yukas, 32, married, miner, Hungarian, #6 mine, collar bone broken by fall of coal while undermining at face of his room without the use of sprags.
1909 September 23
Joseph Masnica, 17, single, miner, Slavonian, #1 mine, legs broken by cars in his room.
1909 November 23
Angelo Muska, 20, single, Italian, #1 mine, leg and nose broken by fall of rock at face of his room. He had been told by the assistant foreman to set props.
1910 April 6
Julius Gary, 35, married, miner, Hungarian, #1 mine, bad wound on shoulder, detonating squb that he was carrying behind his ear came in contact with lighted lamp and exploded.
1911 June 2
Paul Palaney, 35, married, miner, Slavonian, #1 mine, ribs broken by cars on heading.
1912 August 7
Will Findley cut about head by washery belt which flew off rolls, returned to work August 12.
1912 September 27
Dave Goughnour and Will Findley both got fingers pinched.
1913 November 11
John Hongo, 15, single, car oiler, Hungarian, flesh wound on leg by cars, outside.
1914 January 5
John Bodgon had leg pinched at #6.
1914 January 16
Miner in #1 had two toes injured at 8 AM by fall of rock.
1914 January 28
Tom McQuerry ran sliver through hand - ugly wound.
1914 January 31
Victor Steiffell, 20, single, machine scraper, American, #1 mine, end of little finger taken off by being caught in sheave wheel of short wall machine at face of room.
1914 February 19
Joseph Raffas, 23, single, runner, Polish, #6 mine, hand lacerated by coupling cars at entry, making a ride on end of empty trip, got chin cut when cars jumped track. All had been warned not to ride the trip. Discharged by Foreman James.
1914 February 20
Andy Boco, washery crew, fell through coal bin into lorry, cut face and broke three ribs. Taken to Memorial by Timmons & Gronland.

1914 March 3
Chalmer Dodson, hand smashed spragging, #6, taking piece of wood pipe with iron on one end from car, pipe 3'10" long, set pipe on end. Motorman stated he had hand between rod and pipe.
1914 April 12
Charles Harasity injured changing steam pipe at tipple, cut nose, Timmons took him to Memorial.
1914 April 28
Joe Dorchick, #1033, had leg broken in #6.
1914 May 13
Paul Rusko, machine runner, #1, had right arm bruised about 10 AM when machine hit sulphur ball - nothing serious.
1914 May 15
George Varge, #1531, on Meisner's crew, had 2nd finger taken off while unloading car of 60 lb. rails at #6.
1914 June 3
John George, loader, #1 was cut above knee by chain machine.
1914 June 4
Miner in #3 had one of his fingers smashed by fall of rock.
1914 June 5
Alex Dorchick, scraper on a puncher in #6, while scraping a small piece of slate fell on his shoulder injuring him slightly, treated by Dr. Nix.
1914 July 2
Wm. Findley bruised left hand putting spring on tables in washery.
1914 July 14
Pete Page had hair singed by fooling with electric wire on coke machine #2 after being warned repeated by Shaffer. Pay stopped until he goes back to work.
1914 July 16
Art Johns sprained wrist while moving armature on hand car. Car jumped track at crossover.
1914 July 17
John Babash, #1082, 10R 2nd Slope Room 11 hit by fall of rock, not serious. Dr. St. Clair says he will be back in 10 days.
1914 July 18
Sat. car spotter, #1, track #1601, got hand bruised trying to stop car with a prop, got hand between ground and prop.
1914 August 26
Miner, Check #324, House 49, had 3rd finger smashed moving machine.
1914 September 24
Bracken sprained ankle hitching neck yoke on mule team.
1914 September 25
Track layer #1 had arm cut with ax laying track.
1914 November 13
Albert Kendall had nail torn off on one finger #1 tipple putting car on track.
1914 December 3
Pete Page got hit with sledge holding chisel on left hand.
1914 December 8
Bob Clyde got hand caught on trolley pole and had small fingers pinched - nothing serious.
1914 December 10
Frank Kish, #570, sandboy at #1 dropped tub of sand on fingers. #563-Thomas Esaias got fingers split in 10 hdg. neither case was serious.
1914 December 18
John Syka, 24, single, runner, Polish, #6 mine, leg broken by cars on entry while trying to board a trip of cars while they were in motion. 2nd Slope, 12R.
1914 December 28
Andy Jussi, house 121, got leg pinched in 2nd Slope, 12R heading, puncher got away from him.
1914 December 31
Victor Steifel, #1, spragger, slipped and bumped head on sand tub. Hole in head not serious.
1915 May 3
Allie Cresswell, 19, single, laborer, American, #6 mine, head lacerated, fell from railroad car, outside.
1915 May 3
Peter Twos, 28, married, miner, Slavonian, #1 mine, leg lacerated by being struck by mine car.

1915 December 14
Welenty Smelera, 40, married, miner, Polish, #1 mine, leg fractured, struck by mine car.
1916 April 28
Oscar Dishong, 26, married, craneman, American, #6 mine, knee severely lacerated and contused, stepping on moving conveyor belt and was thrown off, outside.
1916 November 8
Andy Kovach, 40, married, pick miner, Hungarian, #6 mine, shoulder bone fractured by fall of coal at face of room.
1916 November 21
Aaron Kovach, 47, married, pick miner, Hungarian, #6 mine, knee fractured, caught between cars.
1917 January 11
Gust Ostafin, 40, married, pick miner, Hungarian, #1 mine, knee severely contused by fall of coal at face.
1917 January 15
Peter Kalinoski, 38, married, pick miner, Slavonian, #6 mine, clavicle fractured and scalp lacerated by fall of coal at face of pillar workings.
1917 March 10
Peter Kerekish, 38, married, pick miner, Slavonian, #6 mine, radius fractured by mine car he was lifting on track on entry.
1917 May 15
Joe Vargo, 31, married, pick miner, Hungarian, #1 mine, ulna fractured and forearm contused between car and roof while lifting car onto track on entry.
1917 July 3
Joe Barta, 38, married, pick miner, Hungarian, #1 mine, injured internally while lifting car in room.
1917 July 23
John Ledes, Slavonian, spragger, 25, married, #6, pubic bone fractured by derailed car on entry.
1917 August 6
Mike Kerchinsky, Polish, pick miner, 40, married, #6, femur fractured by fall of coal at face.
1917 August 16
Miner "John", #1080, Steve Roman House, broke three ribs, hit by piece of rock shooting bottom.
1917 October 18
W.W. Preston, American, machine runner, 44, married, #1, thigh fractured between rib and coal cutting machine on entry.
1917 November 2
Herbert Dexter, American, bricklayer, 23, married, #6, leg fractured, struck by locomotive on entry.
1918 January 16
Roman Danksza, Polish, laborer, 46, married, #6, arm fractured, fell on ice near mine entrance.
1918 January 17
John Ochino, Hungarian, pick miner, 32, married, #6, knee injured, caught between cars on entry.
1918 January 22
John Novatanni, Hungarian, pick miner, 19, single, #1, compound fracture of carpal bone, caught between cars on entry.
1918 January 31
John Dobnich, Russian, pick miner, 31, married, #6, cartilage of knee joint dislocated while drilling hole in room.
1918 February 5
Mike Budge, Slavonian, pick miner, 30, single, #6, fourth and fifth metacarpal bones fractured, fell while dragging rails into room.
1918 February 12
Andy Wijtovich, Slavonian, machine runner, 28, married, #6, compound fracture of tibia and fracture of metatarsal bones, struck by coal cutting machine in room.
1918 February 14
Mike Feser, Slavonian, laborer, 45, married, #6, index finger crushed while coupling cars - outside.
1918 March 27
Peter Fettman, Swiss, carpenter, 49, single, #6, spine injured by fall from building - outside.
1918 March 27
John Nazy, Austrian, car handler, 23, single, #6, pelvis fractured and bladder ruptured, caught between locomotive and door frame on narrow side of entry.

1918 April 13
Joseph Booner, Hungarian, pick miner, 36, married, #6, leg cut off, caught in coil of rope he was using to drop cars down entry.

1918 May 10
Peter Kiss, Hungarian, pick miner, 39, #6, clavicle fractured by fall of coal at face of entry.

Sources: Annual Report of the Pennsylvania Department of Mines, 1894-1918. Vinton Colliery Company Mine Superintendent Diaries, 1912, 1913, 1914, and 1917.

1928 February 6
Vincent Carr, 46, fell under a trip of mine cars. He was attempting to board, contusions of abdomen and right arm, taken to Memorial Hospital.

1928 June 16
George Crowell, 65, hurt in mines, fractured clavicle and broken ribs, abrasions and contusions, taken to Memorial.

MINE OPERATIONS
1912, 1913, 1914, 1917, and 1923

MINE DIARY - 1912

January

Even though January 1, 1912 was a holiday, #6 operated because coal was needed for the washery; repairs were made to the grade of #2 slope in #1, and new arches were put into #3 under the Sterling boiler. On January 3, the repair gang was kept busy until 10 PM repairing the Allentown pump. On the next day both the twelve-ton and the sixteen-ton locomotives were brought out of #6 for repair. The crew was kept up all night repairing the journals on the large machine, which were worn 5/8 inches into the casings. Only eight ovens were drawn because of the weather. Due to the inclement weather, the men could not be kept around for cleaning up.

With Russian Christmas on January 7th, the mines had difficulty operating on Saturday, January 6th and Monday, the 8th. Many of the men refused to work. Mr. Delano and Mr. Schwerin arrived on January 6th to inspect the mines and some of the machinery that needed to be repaired. Little coke was drawn during those days due to cold weather and lack of workers. The winds were so high on January 10th that the tipple crew was sent home.

On January 11, all mines were working; production was good. However, due to an accident to the electric brake, the #6 tipple broke down. On the next day, Mr. Hamilton left for Pittsburgh and Erie; #6 lost two hours because the armature of the Goodman locomotive burned out. Pillar cutting using chain machines started in #1 on the 12th. Old man winter blew into the valley with full force on January 13. The temperature dropped to -25 degrees. The car wheels froze, and the mines closed early because of the weather. The coke foreman was advised to give the ovens more air on the 17th.

Continually during 1912 coal production was hampered by the shortage of coal hoppers. Also the inclement weather and low production led to problems at the coke ovens. Personnel problems also filled the time of the supervisors. On January 23rd, Mr. Smillee contacted Mr. Pardoe, #1 foreman, about increasing production in #1. Then on February 1, Mr. Pardoe was given his thirty-day notice. There was not enough coal from #6 on the 25th to charge additional ovens to fill a special order for Canada. Erie City representatives were at the powerhouse to inspect the engines, and Crawford and Cameron were given an order for an Alderich pump. The timbers in the #3 drift collapsed, closing the mine for the day.

February

The Goodman locomotive continued to cause problems in #6. The spare armature burned out on February 1. A set of wheels was borrowed from Big Bend Coal Company in Twin Rocks, but the electricians still were unable to fit out an armature. On February 5th, the office had to summon an armature-winder from Cherry Tree who arrived on the afternoon train. However, even after rewinding, the armature still would not work, keeping #6 idle. The armature was finally repaired in late afternoon on February 7, and #6 went back into operation on the 8th. With the mine back in operation, the washery decided to cause problems. That same day, the #1 belt in the washer broke, cutting off the supply of coal. The drive pulley was changed to thirty inches. Two new five-ton gathering locomotives and two chain cutters were shipped by the Goodman Company of Chicago on February 3 and reached Vintondale on the 9th.

Problems continued to plague #6. On Saturday the 10th, there was a derailment in 6R. The next day, the chain axle broke at the tipple. On the 13th, a delay was caused by a trip off the track and a broken motor. The following day saw a derailment on the head of the slope and a breakdown of the rack rail locomotive which then received new wheels. The set of wheels borrowed from Big Bend was returned on the 16th. State mine inspector, Joseph Williams, inspected #1 on the 20th. Chief electrician Neff left on the same day. An order was placed with American Steel & Wire Company for 10,000 feet of trolley wire.

The weather continued to be bothersome to the company in February. On the 22nd, a severe storm forced #3 and #6 to close down before work started. #1 put in a full day. However, no coke ovens were drawn, and no attempt was made to start up any. #3 experienced two wrecks on the 23rd and a third on the 24th of February. On two different occasions that week, two women and a man were caught stealing coal from the cars at #3. All were fined $5 and costs by the local justice of the peace.

March

There was a shortage of water at the pumping station in late February and early March. Consequently, the washer had to be shut down on Friday, March 2nd and only operated one table the following Monday. Also the mine continued to suffer from a chronic shortage of hoppers. On March 5th, the dinkey track between #1 and #6 was completed, and the #1 power plant was fired up while repairs were made to #6's. #1's only ran two days before having to shut down due to a hot crank pin. Allie Cresswell was temporarily put in charge of the ovens while Maginnis was filling in at the engineering department. Repairs to the #6 powerhouse, which were completed on the 9th, included application of 250 pounds of babbit (a friction reducing alloy of tin, antimony and copper) to bearings, crosshead and box, crank pin box and outboard journal. The exhaust valve seats were rebored, and new valves and stems were installed.

Float ice was blocking the intake pipes at the pumping station on the 10th and 11th. Fifty cars for #6 were ordered from Cherry Tree Iron Works with delivery expected in two weeks. The rebuilt Goodman rock rail arrived March 12. On its first trip the next day, the armature bearings got hot; then clutch broke on the 15th and it had to be brought out for repairs. Otto Hoffman, chief engineer, was sent to Altoona to inquire about rebuilding the twelve-ton Porter engine. The Goodman Company sent its chief draftsman to look at the locomotive on March 22nd.

VCC also received new doorblocks for the ovens at the end of March from the Union Mining Company, a Delano concern near Cumberland. On March 29th, #3 had to shut down due to leaking boilers. The dinkey engine began acting up, and Mr. Hoffman made another trip to Altoona to inquire about repairs. The engine was rebuilt and back in operation in June. Penn Central Power representatives paid business calls to encourage VCC to purchase power from them.

April

A bid for moving machinery from #1 powerhouse to #6's was presented by Mr. Anderson. Penn Central representative Harris made tests on power consumption. The foundations were poured for a new take-up at the #6 tipple, and the new mine cars arrived. The shop foreman of the Maine Belting Company began repairs to the #1 belt in the washery, which had worn badly since it was installed. Several employees went to take mine examiniations. A new pump was installed in #6. Woods of Penn Central came to check the switchboard. #1 powerhouse shut down of April 20th, and #6's carried the load on two units. Personnel changes at the flat were numerous. Jack Drews' last day at #6 powerhouse was on the 10th; an unnamed shop foreman began on the 11th; and Anderson became the master mechanic on the 22nd. A culvert was built under the dinkey track at #1. On April 22nd, there was a collision of two trips in #1, one coming down 3L and one going up 4L. No men were hurt, but both bumpers were badly cracked, and new ones had to be ordered by telegraph. On Monday, April 29th, the cage at #6 tipple broke down, shutting down the mine. Mr. Beil called and made a final proposition on the purchase of power from Penn Central.

May

Beginning in May, the time set for drawing coke was moved back to 3 AM due to the hot weather. The length of time for charging the ovens was set at 96 hours. Later in the summer, the company experimented with charging at 72 and 120 hours. On May 17, #1 was put on a machine-mining basis which led to a decrease in tonnage. The amount of coal used for charging an oven was reduced to 1,800 pounds. Throughout the month, coal for coking was in short supply. An ethnic holiday on the Monday before Decoration Day resulted in a shortage of workers. The ovens were held until the next day to give "the Americans their holiday." Scrap copper was sold for 13 cents a pound.

June

Two loads of scrap iron were shipped to unknown places. Water was running low in both branches of the Blacklick. The washery needed a new gear, which took two days to locate. June 17th, "Otto Hoffman arrives to stir things on flat again." The dinkey hauled coal from #1 for the washery on the 21st; one load derailed because it came down the grade too fast. Vintondale's foremen won fourth place at a first aid meet in Johnstown on the 26th.

July

Mr. Delamater, Wehrum's superintendent, paid a visit on July 2nd to tour the washery. Either Hamilton or Smillee walked him back to Wehrum. The next day was spent in Wehrum with Delamater making tests on coal. Delamater returned again on the evening of the 5th; a foreman's meeting was postponed. July 12th was St. Peter's Day; the company tried to get mantrips going at #1 and #3 but failed. Anderson promised the office an experimental dump car which would reduce the expense at the rock dump. The BR&P failed to bring any rack rails, and the office had to contract Fraser at Dubois concerning the matter. Mr. Gill was hired as a replacement electrician at #6 for Keith. Winn of Heyl & Patterson gave pointers on running the Campbell tables in the washery.

August

August 2nd, "a kind of holiday for the Slavs". Enough men were located to run the coke ovens. Arbogast became chief clerk on the 5th. Two different experts from Heyl and Patterson spent the day in the washer and made suggestions for improvement, especially in the bumping blocks. Coke production in the summer was off due to high sulphur in the washed coal and to the lack of a coke boss. Mr. McDevitt of Keystone National Power paid a visit to observe how powder was being used in #6 and suggested that VCC use his XLF powder. An attempt to locate a coke boss sent Brown, VCC mason, to Greensburg on the 16th. On August 27th, sixty-two coke ovens were drawn, the "biggest day in history of ovens".

Also in August, top brass and company lawyers from Vintondale and Wehrum made several visits to Dilltown to assess stream damage caused by mine water pollution. Both companies were being sued by a grist mill owner in Dilltown. The companies brought in a chemist in September to take water samples. They also made trips to Indiana and Johnstown to consult with lawyers. The case dragged on into December without any indication of a verdict. In the diary, the mill is called the Hess Mill, and the case is referred to as the Brendlinger dam case. Attempts to locate trial records and/or newspaper stories have proven unsuccessful.

September

Mother Nature played havoc with the company in early September. On Labor Day, September 2nd, few miners reported for work, and that night, a thunderstorm washed out part of the dinkey track along Shuman Run. A cloudburst followed the next night. At 3 AM, water was running into the #6 boiler house and into #6. Luckily the water was diverted and did not quench the coke ovens. Inside #6, the water was deep in 4L and almost reached the 8R pump. Due to washouts along the track, PRR service was sporadic. The following week, a four inch pipeline in #3 burst and cut off all air. A bonus system for good tonnage was posted. Galbreath arrived to begin duties as assistant engineer. He stayed until 1915 or 1916 when he moved to Wehrum where he served as superintendent for several years.

The company was looking into selling the scrap from the ammonia plant, which was built in 1906 to use the by-products of coke manufacturing to produce ammonia, coal, tar, etc. However, the ammonia plant was never a successful venture. The two stacks from the #1 powerhouse were dismantled, and engine frame from the powerhouse was moved to #6.

The feasibility of draining #5 mine was studied, and the old company compressors at #5 were removed. Diversion ditches were built around the #5 drift. A "squeeze" (gradual uplifting of the floor or sagging of the roof) was discovered in #3, and inspections were made by the superintendent to see if it could be bypassed. The Sullivan compressor at #3 was sold to Lochery Brothers Coal Company in Windber.

October

Most of the month's events centered around labor problems which are discussed in another chapter. October 26th, C.A. Fetzer took the job of assistant clerk.

November

New electric lines from #6 were strung in town, and the #1 powerhouse was demolished. On Thanksgiving Day, the 28th, all mines and all departments worked until 4 PM.

December

Steps were constructed to #1 mine. Water in #4 was a problem; on the 14th, it had receded one foot. At the washery, a Pennsylvania Crusher arrived on the 9th, and was placed on a foundation and put into operation on December 16th. It started to cause trouble on December 28th. Coke was pulled on Christmas. a new GE motor was installed in #6. Williams of United Gas and Improvement Company of Philadelphia paid a visit to inspect the by-products plant for possible purchase.

Overall, 1912 was a year plagued with production and mechanical problems. Producing top quality coke seemed to allude the company. Management problems also existed by the evidence given in the diary entries; improvements in the mines and the ovens were yet to come.

Mine Diary-1913 Incomplete

In 1913, two overcasts were built which connected #6 with the #5 workings. This helped the drainage problem and improved ventilation in #6. The only mines operating were #1, #3, and #6. The year began with the layoff of two men in #1 and #6 for union activities. January's tonnage was off because of four scheduled weddings on January 18th and the usual Monday hangovers. In March, the same situation arose. The superintendent was "unable to do anything as Slavs believe if the Italians and Hungarians have a holiday, they ought to lay off." Mr. Smillee and family moved from Vintondale on April 26th, and Hamilton moved to the superintendent's house within the next week. Jack Huth was drilling from #6 into #5 and successfully broke through on April 20th. A second water hole from #5 to #6 was drilled on July 30th. the surveying crew was also busy on #3. A union meeting was held at the union hall on October 2 with 18 in attendance. The diary reported that no high officials were present.

In November of 1913, John Burgan's salary was raised to $80 per month, and George Duncan's was raised to $75. On the 24th, the Buffalo, Rochester and Pittsburgh train came in too fast and knocked a car off the tracks. A work train was brought in to put the car back on. The track was reported in good condition. Mr. Hamilton departed Vintondale for the West on December 16th. Mable Davis Updike, office employee from 1912 to 1916, said that Hamilton was not fired, but returned to his home area of Denver because he was not interested in small town coal mining. According to the diary entries, Hamilton was frequently out of town between 1912 and 1914.

Considerable improvements were made in 1914, especially to #1. The workings in #10 plane were reopened after being idle for five years. A 500-foot drift for ventilation was driven to the #10 heading; forty pound rails replaced the light ones. All the mine cars were rebuilt, and the motor barn was extended. In addition to widening #1 slope, new machinery was also installed. An Ajax electric pump, one Goodman shortwall machine, and two ten-ton General Electric locomotives were put in use.

#6 also received extensive improvements, especially to the ventilation system. An electric-powered eight-foot Sirocco fan was installed and was capable of sending 200,000 cubic feet of air per minute into the mine. A new brick and concrete fan house was built, and a brick tube, eleven feet in diameter and one hundred feet long, was constructed to the first left heading, first slope. This opening was two hundred feet west of the drift mouth. A 1,400-foot road for electric haulage was constructed, and thirty-five shelter holes were constructed out of rock along that road. Other improvements included digging a drainage ditch that was 700 feet long. #5 was drained into #6 by using bore holes. The main slope was widened four feet, and sixty pound rails were laid on the main haulage for 2,700 feet.

Outside improvements included constructing a roof over the tracks from the mine mouth to the tipple. New buildings included: weigh-office, mine foreman's office, sand house, and a five hundred ton storage bin. An electric spotter replaced a steam spotter at the tipple. New equipment included 150 mine cars, a fifteen-ton Goodman electric locomotive, two Jeffrey machines, a fifteen-ton General Electric locomotive, four Ajax electric pumps and nine pneumelectric puncher machines.

MINE DIARY - 1914

January

1914 was another of the "diary" years, so more day-to-day events concerning the mines can be offered. On January 1, 1914, not an ethnic holiday, #6 worked all day; #1 worked until 2:45, and #3 worked two hours. On January 2nd and 3rd, production was light due to the death of Mike Latto, who died of a heart attack. Also the crossbar and chain came off the sprocket in #6, adding to the low production.

The new fan for #6 was getting some attention. The General Electric salesman, Mr. Marble, was asked to give a quotation on a 175-horsepower motor for the fan. On January 9, after a two day layoff for "hunk Xmas", #6 was experiencing trouble with the armature in the rack rail. Two cars built for washery refuse with Hockensmith roller bearings were delivered on January 10. A "genuine blizzard" on January 12th did not prevent the mines from working, but the dinkey was unable to run. The next day, a spring broke on it and delayed the delivery of coal to #6.

John Huth staked out the location for the new #6 fan house; a hoist was prepared for use in starting the slope for the new fan. The air course in #6 was double-shifted for the new fan.

On January 20th, #1 lost one hour due to a roof fall 100 feet inside of the slope, a chronic problem in #1. On January 22nd, Mr. Goodman of Dean Pump Company called to repair a defective plunger. A fifteen-ton Goodman motor arrived on the 24th; when it was unloaded the next day in the shop, the motor hangar was broken. It was tried out in #6 on January 27th.

In January, Jack Huth and Herb Daly were sent to #3 hill to find a location for a ditch to handle water runoff. The pump in #3 was unable to keep water down, and 1,000 feet of wooden pipe were ordered to be put on the Cameron pump.

February

On February 6th and 7th, the dinkey crew was laid off because the washery was not washing coal. The new fifteen-ton motor was causing a few problems. Otto Hoffman spent the afternoon of February 9th in #6 observing its operation; on the 11th, the motor was off the track for one hour and thirty-five minutes because the substitute motor man had the brakes on too tight on a curve.

In February, Art Kempfer was using the old #1 powerhouse to build cars. On February 14th, there were fifteen inches of snow at 6 AM. #3 was idle because the sidings were loaded with filled hoppers. The dinkey was snowed in at #1. Henry Misner's crew with Mr. Moore and four others shoveled snow for four days. The snow delayed payday until Monday, the biggest pay in Vintondale to date. The Goodman Company sent its representative, Mr. Smithinger, on February 15th to inspect the fifteen-ton motor. He spent the next two days in #6 working with the motor. Finding no cause for the overheating, he wired for Mr. Pray to come look at it.

March

Mr. Smithinger looked at the motor again on March 1 and still could not find the problem. Mr. Pray arrived on March 3rd on the 4:04. He spent two days working on the motor, but still did not find the cause of the overheating. He promised VCC two new armatures and a full set of "8" fields for the motor.

There was another cave-in in #1 on March 9th. This one was at 10R, 2nd Slope. A double-shift cleanup crew was sent in to clear the passage. The new fan for #6 arrived on March 10th. Two men from the engineering crew were sent to the top of #3 to "prospect" for the Miller seam.

The Goodman motor hauled out successfully on March 16th, but the next day, it burnt the leads off the controller. It was brought out to the shop, and the GE motor was used for hauling out.

All mines, but especially #1, were experiencing a shortage of miners. Fifteen quit from #1 within a few days. From March 15th to 20th, the company was experiencing problems with the pumping station because of ice jams and slush ice in the Blacklick.

The work on the new fan house was proceeding rapidly. The foundation was excavated on March 18th, and the cement was poured on the 21st. Bricks for the fanhouse were unloaded on Sunday the 22nd. Also on the 21st, Mr. Lemmon and Mr. Pray helped Shaffer install the new armatures, but when tested on the 23rd, the new one heated up as much as the old one. Pray and Lemmon remained in Vintondale until March 26th. Coal loaded at #1 on Monday March 30th was for Barrett Manufacturing Company. All mines were working, but were short on men because the previous Saturday was a payday.

April

On April 3rd, Vinton Colliery Company purchased two 5 x 6 inside packed pumps from Mr. Rodden of Harris Pump Company when he made a business call. On April 4th, the suspension bar on the fifteen-ton broke on the first trip. The GE motor hauled it out, and a new bar was installed by 9:30 PM. On April 10, a ten-ton GE motor arrived on the BR&P railroad, but its headlight was broken. On Saturday the 11th, a boxcar of cement arrived along with 200 rolls of red roofing.

Mr. Marble of General Electric made a business call on April 14th and was asked for bid on a fifteen-ton GE motor. On April 15th, the company bought a team of mules, one horse for heavy hauling, and one saddle horse from Joe Shoemaker, who ran the Blacklick Stock Farm near Wehrum. Shoemaker held public livestock sales every several months in Indiana at Moorhead's Livery Barn.

May

The new #6 fan was moved to its foundation on May 8th. There was continued interest in the prospect hole that had been drilled on #3 hill. On Sunday May 10th, a delegation of Galbreath, Misner, Daly, Arbogast and Hoffman visited the site. It was decided to block the hole until the spring rains were over. Brush and trees were removed from the mule pasture, and a fence was built around it. A new air shaft was being surveyed for 10R in #1 by the engineering crew. The work started on May 20th in Mrs. Jones' field on top of Chickaree. In May, the company also decided to get into the potato business. Five acres of the Blair (company) farm were plowed and planted with potatoes.

#1 continued to cause problems. On May 27th, John Burgan and Otto Hoffman inspected #1's loaded cars for sulphur. They spoke to the foreman, Abe Abrams, concerning his sulphur pickers whom they thought were not good workers.

June

On June 6th, there was a serious water main break near the coke ovens. The location was not discovered until the next day at 3 PM. Because of a "Greek Holiday", #3 and #6 were idle on the 8th, and #1 only worked a half day. The #1 tipple crew wrecked two cars. The fifteen-ton motor burned out the resistance on the 15th. Its operator tried to haul out forty-two cars and ran out of sand. The ten-ton was used to haul out.

There was a frost on June 17th. The new #6 fan was tried out on the 18th. Mr. Haskell, a GE representatve and Mr. Shaffer, company electrician, worked on the fan motor on the 21st; a five hour test was conducted on July 1st. In the meantime, work continued throughout June on the new #1 aircourse to 10R. Several trips to this site and to the #3 bore hole were made by Hamilton, Galbreath, Adams, and/or Hoffman. The #1 prospect hole was stopped up by Mr. Daly on June 29th. A house coal tipple was constructed at #1. Galbreath also initiated a project on Shuman Run, starting with clearing underbrush at the resevoir on May 28th. On June 29th and 30th, the washery crew was sent to work at the resevoir.

July

On July 1st, there was another bad break in the ten inch water main at the coke ovens, and Mr. Misner spent the day looking for it. The coke oven crew was sent up to the resevoir site again on the 2nd. The same day, a PRR engine went off the track on its line, came into the VCC yard on the #2 coke track and went off again. The section boss cabled the PRR to take a look at the track. He thought it was the engine and not the track.

July 3rd was a busy day for the company. A cracked flange was found in the main header of the steam line in #6. A new one had to be ordered by express. The potato patch of the Blair farm was plowed by Mr. Goughnour. Mr. Hoffman went to the #3 prospect hole, and the new fan at #6 ran all day. The next day all departments of the Vinton Colliery Company were off, and Vintondale had "one of its very best 4th's."

Henry Misner continued to work on the water line on the 5th; he stopped three large breaks. Shaffer and John Morton worked on the new #6 fan. It ran all day on the 6th and was put into service on the 7th. GE representatives Gibbons and Bryant paid a business call on the 7th and assisted in test on the new fan. They also received an order for a new ten-ton locomotive. The Keystone Powder Company representative also paid a visit and wished to contract for the next year at the 1914 prices.

Throughout July, coke oven foreman Dempsey and engineer Galbreath continued to supervise the coke oven and washery crews who were excavating the new resevoir. A small amount of gas was found in 12R in #6 on July 8th. Finishing touches on the brickwork of the #6 fanhouse were made; a tile drain was installed at the #6 tipple; outside steam lines were covered. Leaks in the ten-inch and the twelve-inch mains that led from #6 tipple to a tank on the hill were plugged.

On July 13th, a crew started timbering at the site of the new #1 air course. Old PRR railroad ties were obtained for use in the new air course. #3 was idle on the 15th, and all rails were being hauled out of it.

Mr. Redden of the Harris Pump Company arrived on July 16th to look at the broken gear on the Harris pump. On the 17th, Hoffman was busy with Mr. James inspecting the air course, water ditches and all pumps in #6. A representative of the Dean Steam Pump Company arrived to inspect the defective plunger on its pump. On Saturday the 18th, Mr. Misner spent the day looking for water main leaks. On the 23rd, Mr. Cameron of the Goodman Company paid a business call and received an order for a chain machine. Delano and Schwerin arrived in town for a week's stay which included the annual director's meeting. They made trips to the #3 prospect hole on the 23rd and 26th. After each visit, the hole was stopped up again.

Work on the new dam on Shuman Run progressed; valves and pipes were procured. Water was in short supply during the summer of 1914, and there seemed to be a rash of water main breaks. On July 25th, there were two breaks in the twelve-inch main where it crossed the Blacklick.

On the 25th, Mr. Darling of the Natrona Salt Manufacturing Company called, and the VCC shipped him its barrel of sulphur balls. Sunday the 26th was clean-up day at the #6 powerhouse. In addition to a good general cleaning, the red ash in all four boilers was flushed out.

A mule was often viewed as more valuable than a miner and harder to replace. On July 30th, a mule was killed in #1 in 1L, 10 heading. The driver was fired; he had been warned by the foreman not to treat the animal in the manner which led to its death.

August

On August 3rd, the Goodman fifteen-ton motor went off the track in 5R in #6. The next day, the band on the armature let go, and the cores and fields were ruined. The motor got so hot that it melted the sodder.

During the first week of August, sand was hauled from #1 to the resevoir site, and the concrete forms were constructed. On the 13th, washed coal was shipped to Lackawanna Steel in Buffalo. The Wehrum washer had burned on the 8th, and their washer at #3 had not been used for ten years. Eventually, the Lackawanna #3 mine was drained, and the coal was brought out through #3 and cleaned there.

Delano and Schwerin returned on the 18th; and while they were in town, an order was placed with the Hockensmith Company for fifty new cars for #6. Also because miners in #1 were loading slack coal, Jack Staub was assigned to watch the men loading in the new slope in #1. Two men were discharged for loading dirty coal.

A new rock dump trestle to the PRR bridge was started on the flat, and the PRR engineers were surveying near the #1 tipple for more room for empties. A carload of sulphur balls was sorted and cleaned, and the surveillance for dirty coal continued at #1. A foreman's meeting concerning dirty coal was held on August 25th. Even Harry Hampson, who had resigned earlier as the weighmaster at #1, was put to work inspecting coal.

Mr. Lemmon of Goodman Manufacturing made a professional call to demonstrate his company's shortwall machine. On August 26th, there was a twelve car wreck in #1, 2nd Slope at 11 AM; the hitching broke. In addition, a trip came off the track in the old dip at 8 AM. #6's Goodman motor burnt up again on the 28th; the insulation burnt to a powder because of heat caused by daily work.

Water shortage problems continued to plague the company in August. The concrete work on the first half of the new dam was completed, but the dam on Bracken Run was dry. Additional carloads of sand and cement were unloaded and sent to the resevoir.

September

On September 3rd, the washery loaded an additional twelve cars for Lackawanna. Abe Abrams fired one miner on September 4th and two more on the 7th for loading dirty coal. Nine more were called into the office concerning the problem. To help ease the problems with the motors, Otto Hoffman went to Wehrum to borrow one thirteen-ton motor and one six and one half-ton motor until the new GE motor arrived. A new plan for a double shift at the washery was being devised by Galbreath and Hamilton. At a foreman's meeting on the 7th, Hamilton discussed dirty coal and changing the motors, which were not delivered by PRR as scheduled. The Wehrum motors arrived on the 9th, but the six and one half-ton had a broken axle. Mr. Chase of Wehrum sent an axle and wheels to Vintondale by a team of horses. Mr. Pray of Goodman arrived the same day to install the new motor in the fifteen-ton. The thirteen-ton motor was to be repaired on the 10th, and the fifteen-ton Goodman burnt up an armature on the last trip. The thirteen-ton Jeffrey was put into service at 6 PM at #6.

On the 11th, the thirteen-ton went off the track and caused a one hour delay. Mr. Pray was still in town assisting in the installation of a new motor in the fifteen-ton. He also spent a lot of time discussing electric haulage with Hoffman and left on the 16th. During that week a Jeffrey shortwall machine was put into operation, and the Vinton Colliery Company began to pay rent to Wehrum for their machines. Two miners from #6 were also discharged that week for loading dirty coal. A miner with either a sense of humor or a major grievance sent a parcel post package to the Vinton Colliery Company office. Inside were sulphur balls, slate and a note signed, "From Some One".

On the 18th of September, twenty-five of the new mine cars for #6 were unloaded; the new fifteen-ton GE motor arrived on the 26th. At #3, all machine mining ceased on the 26th, and the company was getting it ready for pick mining. There were eight miners, one tippleman, one foreman and two mules left to work in #3. The new ten and fifteen-ton GE motors were put into operation on September 29th, and the fifteen ton was off the track on the drift mouth the next day.

October

In October, a decision was made to build a five hundred ton coal bin at the washery. Work began on the 21st. A roof was built over the #6 drift, and a runaround for empty mine cars was constructed there. A new set of tires was put on the GE ten-ton locomotive; it had received a new motor in April. A chute was built at the washery to load washed Lackawanna coal.

The potato crop at the Blair farm was harvested on October 20 and 21st, but the yield was not good. Several company representatives arrived in Vintondale during October. GE sent Mr. Barry to "look over" the fifteen-ton motor. The Temple Company of Chicago sent Mr. Mayer to inspect VCC's cars for the American Aluminum Company of Pittsburgh. A contract for explosives was signed with McAbee Powder and Oil Company. A representative of Carnegie Steel also called.

During the night of October 20th, a PRR crew pushed a box car of sand and two empties off the track at #1. The next day a work train was in to put them back on track. Company teams began hauling rails and ties from #3; a fence was constructed around the new air course at #1 and one around the #1 fan house in the mule pasture.

In #6, a pump in 5R was moved to 1st Slope in 9R. Work was progressing as fast as the timber arrived for the coal bin and the runaround. Company blacksmiths were busy welding rods for the coal bin. The #3 hoist was brought to #6 and placed under the tipple to spot cars. A fourth small Ajax pump was installed in one of the headings at #6.

November

The last of five Penn Electric punchers went into service in early November; these had been purchased in the spring. Company representatives visited Vintondale frequently during November. On the 2nd, VCC arranged for a Mr. Whiteside to demonstrate his caps and powder. Mr. Forr paid an inspection visit on the 4th. The #1 boiler in the powerhouse was inspected by Mr. Shephard of Hartford Boiler Company.

PRR sent in a steam shovel on the 4th of November to remove part of the bank bordering the track at #1. Problems with the motors continued to plague the Vinton Colliery Company. A wire was sent to Mr. Marble of GE requesting that he pay a call. The shaft on the fifteen-ton Goodman worked out of the armature; the part was sent to Chicago for repair. The armature on the six-ton Jeffrey was the next to go; a set of coils was procured from Wehrum.

A mule at #3 hurt its foot, so the mine was closed for two days to let the men and mule rest. There was a wreck in the old dip in #1, and one hundred tons of coal were lost. Some bad coal was discovered in 3L of the new slope in #1. The other twenty-five new mine cars for #6 arrived and were unloaded on the 17th.

Another company representative who called in November was Mr. Reed of Hyler and Patterson manufacturers of machinery for the washer. Explosives ordered from the McAber Powder Company were delivered by automobile. Demonstrations of these explosives were held at the office on the 17th and at #6 on the 18th. Comment made in the diary about the powder was, "Their powder done good work."

Two days were spent looking for a water leak at the #3 tipple. Two breaks were found in the four inch pipe. All carpenters were put to work on the coal bin. Abrams and Hoffman went into #6 on the 20th to look at the face of the new 7L Slope which had a 7% grade. Some gas was found in 2nd Slope, 12th Heading on the 23rd. That same day, #6 started working an eight hour day, but switched back to nine hours on the 29th. The fan in #6 was increased to 100,000 cubic feet per minute on the 29th.

Another powder demonstration was arranged with Mr. Jonas of the Dupont Powder Company. GE sent Mr. McWilliams to install a ventilation system in the GE fifteen-ton motor. However, this ventilation machinery did not cool the motor as much as hoped. The temperature on the back motor was 95 degrees at noon.

December

On December 4th, GE sent Marble and Larson to look at the fifteen ton motor. The voltage of the powerhouse was raised to 2,650V to help improve the performance of the GE motor. On December 9th, #6 loaded 1,472 tons at the rate of about 128 tons per hour; 1,157 cars were dumped. Thirty-six trips were made with thirty-two cars in a trip. On the 10th, the tonnage increased to 1,502. The resistance on the Jeffrey thirteen-ton grounded on the 11th; a team was sent to Wehrum to bring back a new one. Two carloads of coke oven brick and a car of seventy pound rails were unloaded.

On Monday December 14th, there were fifteen inches of snow on the ground. As fast as Misner's crew could clear the tracks, they filled up again. The carpenters on the coal bin were let go early because of the wind and snow. On the 16th, the temperature droppd to -14 degrees making it difficult to move cars at the tipple. Two days were spent by the steam pipe fitter thawing pipes at the Vintondale Inn. Because of a belt breaking in the washery, the crew had to hurry to catch up with a shipment of Lackawanna coal. On the 19th, the men quit work on the coal bin at noon because of rain.

A gear in the Bradford Breaker in the washer was stripped. Blewitt was sent to Pittsburgh for a new one. Mr. Goughnour was hauling house coal for the company and was thirty orders behind on the 23rd, but was able to catch up on Christmas Eve. The last major job of December was setting up a track for Lackawanna's washed coal. Misner and his crew were in charge of the job, and a second-hand switch was ordered from the Buffalo, Rochester and Pittsburgh Railroad.

MINE DIARY - 1917

January

Sunday, January 21 saw twelve inches of snow, accompanied by rain. The yard crew was brought out to dig ditches to carry the water away from the tipple. The next morning, there were eighteen inches of water on the tracks at the tipple. Car shortages plagued VCC at the end of January. Three calls each were made to PRR and BR&P offices on the 26th. On Sunday the 28th, a crew was installing the elevator in the new #1 coal bin. A tube in the #4 boiler at the powerhouse let go at 3 PM. Problems with the Hess Brothers Construction Company, who were building the coal bin and new houses, is dealt with in the town improvements section.

February

On February 3, #1 was idle due to lack of hoppers, and many miners did not come out to work because of the -15 degree temperature. On the 5th, the temperature remained at -15. The company sent to the coke crew's houses to get them out to draw coke. The verticle shaft broke on the #2 coke track, and the repairs were finished at 3 AM. Both branches of the Blacklick were jammed with ice.

A test run on the new coal bin was tried on Tuesday, February 6th. Fifty tons were loaded into it, and the results were satisfactory. The representative of the Dupont Power Company paid a business call on the 12th. That same day, Hoffman spent the afternoon in #1 with foreman Abrams visiting all areas, starting at 5L and the substation. Two days later, he spent the day in 1st and 2nd Slopes in #6 with foreman James. A first aid meet was held in the council rooms.

On the 16th, Mr. Faye, contractor for the #6 shaft, paid a call and requested a $4,000 advance on the estimate for the shaft. Eight miners from Wyoming were hired, easing the labor shortage. Mr. Shilling of the Association Insurance Company made a two-day inspection of #6-inside and out. A third dinkey was unloaded on the 19th. The horsepower of the #6 fan was increased to 225, producing 138,000 cubic feet per minute. The motor in #6 was also "running warm."

On February 20th, the Eire Engine Company sent Mr. Wodsides to inspect VCC's four engines. Another first aid meeting was held on the 22nd. On Saturday the 24th, fourteen cars for LC&C were loaded at #6. By 3:45, all empties in the yard were filled. Mr. Vogel of Hely & Patterson called on March 2 to work on the coal bin conveyor.

Mine foreman James resigned on February 28th with the effective date as March 15th.

March

There was another derailment on the PRR tracks on Monday, March 3rd. An engine switching two empties from #1 to #6 ran off the track because of a broken rail. A PRR wreck crew was called in. That same day, the BR&P was to deliver twenty steel coal cars, but arrived at 11:00 with twelve coke racks instead.

On Saturday, March 10, $10 worth of brass and copper was stolen from the #1 scrap bin. The next day, Hoffman examined the washery crane and discovered both legs buckled. He spent the evening in the office examining the crane blueprints.

Delano, Schwerin and Claghorn arrived on March 12th and spent five days in town surveying company property, studying mine maps, making trips to Claghorn and the #1 airshaft near Rummel's Run. Claghorn spent the 16th inside #6 and Friday morning at #1. On Friday the 16th, Delano, Schwerin, Huth, and Hoffman walked to the #1 airshaft. (Presumably via the pole line.)

A new set of wheels was installed on #6's ten-ton motor. At 4 AM on March 22, the top valve on the #2 boiler blew, breaking the casting and scalding Joe Millar. A new pulley was installed on the #1 fan.

April

On April 1, the old dip in #1 continued to be a problem; three hundred feet of six-inch wooden pipe burst at 8 AM. Mr. Herbster of the Hockensmith Wheel and Mine Car Company informed the office that he was sending VCC ten sets of car iron grates to replace the one that broke.

After a fifty-mile-per-hour wind on the 5th, the washer roof was in bad shape. The dinkey hauled coal to the washery until midnight. The high winds continued the next day, which was "Roman" Good Friday. The mines were short of men. Abrams, Olson and Crocker spent the day in #6 checking ventilation; Crocker closed 13L because of gas. On Easter Monday (the 9th), both mines closed early because of a shortage of miners. Mr. Burgan traveled to Cresson to examine coal car distribution in PRR's Allegheny Division. Mr. Crocker also returned to complete his inspection of #6, which passed except #13 and #14 2nd Slope. Abrams and Olson spent the day with Crocker in #6 on ventilation.

On the 15th, Orthodox Easter, the following repairs were made: new brushes on #6 fan; new pinions on both GE 15 ton motors; repaired #3 boiler walls. On Easter Monday, forty miners reported at #1-68 tons of production and sixty-five showed up at #6-478 tons. A ten percent bonus for all employees was granted. The oven crew drew coke as usual, starting at 1 AM and finishing at 7:20 AM.

Saturday, April 12, Mr. McDinell of Ridgway Engine Company arrived to install a piston, but the threads on the new rod were too small to fit the piston. A new GE ten-ton motor arrived for #1 on the 25th.

May

Abe Abrams, #1 foreman, was promoted to general mine foreman. On May 7th, the 13L, 2nd Slope, closed because of gas, was reopened. The next day 13R, 2nd Slope was filled with water, idling fourteen men; it had been raining intermittently for four days. Mr. Pray of Goodman paid a call on May 10th. Coal cars remained scarce, and the washed coal was high in sulphur, causing Hoffman to spend part of a day in the washery. He also discussed changing some of the wires with Abrams and Shaffer.

Saturday, May 12th was mishap day. The suspension bar on the thirty-ton motor broke; a car came off the track in the cage; and a sprocket broke on the spotter. The next day, Mr. Meisner raised the rock track. A new blow-off pipe was installed on the #1 boiler. Hoffman Bros. completed the hole for the new shaft for Faye. Huth and assistant mine foreman Millar broke through into the air shaft. Two weeks later, Faye had to drill through very hard rock.

On Monday, May 21st, a PRR inspector arrived to check #6 for a car rating. Because of a defective brake, a car at #1 ran out and derailed. On Wednesday, five cars broke away from the 10 heading of #1.

June

In June, the Scaife Erecting Company engineer arrived to build a filtration plant. A cloudburst hit Vintondale at 2:30 PM on Thursday, June 7th and lasted three hours, washing out streets. The drift at #1 was blocked, and there was fifteen inches of water at #6's tipple. Work in #6 had to be stopped several times because of water. Even with all pumps working at capacity, mining in the 13R heading was stopped again on Friday. The next day, an extra pump was placed in 15R 2nd Slope because it was full of water. Miners were to be put on other jobs until the following Monday, but the water had still not receded.

Scrapped machinery from the #3 powerhouse was sold to Seifon Brothers who also loaded a PRR car with scrap from the ammonia house. VCC sold them some scrap rails for $150. Mr. Meisner unloaded a Goodman flame-proof chain machine. On June 12th, a rockfall broke the trolley wire in 11R in #6.

The pumps in #6 continued to be strained to capacity, and Hoffman spent the morning of the 19th in second slope with Abrams. For an unspecific reason, Hoffman sent boxes of cigars to the #6 motor dispatcher, motormen and spraggers.

On June 22, hoppers were again scarce at both mines; the washery was full; and the BR&P was late. Repairs for Sunday the 24th were the #1 coke machine, the steam filters on the filtration plant, and new tubes in #2 and #3 boilers. The filtration plant was put into operation on the 25th, and the first filtered water filled the boilers the next day. Because of the sulphur water, the boiler tubes for the previous ten days had to be renewed. Also on the 25th, 13R, 2nd Slope had to closed because of gas. A fireboss was assigned there, and the gas dissipated by 3 PM.

July

On July 1st, the maintenance schedule included: raising the track on the rock dump, rewheeling the ten-ton motor, and installing a new shaft in the #6 spotter. The next day saw a shortage of miner's-"pay day drunks and Italian Fourth of July in the old country." The mines remained "short" the rest of the week because of the holidays. On the 8th, two sets of steel timbers were placed in #1's drift. Mr. Cook of Cambria Steel paid a business call pertaining to an order for rails.

#6's shop completed repairs to the Goodman chain machine, and Mr. Misner unloaded a carload of sand for both mines. On Sunday the 15th, repairs were made at the washery and the tipple. Mr. Gahler, powerhouse plant engineer, was learning how to operate the filtration plant. The following Sunday, exhaust valves on the #4 direct current generator were blown out; side walls on the #2 boiler were repaired; the red ash was washed out. A double shift was working on the leg of the washery crane. It was completed at 6:55 PM the next day. Mr. Williams, BR&P inspector, paid a call concerning overloading and underloading of cars. On July 25th, gas was again discovered by Abrams in 13L 2nd Slope. Hoffman spent the entire day in #6 inspecting all working places.

VCC was leasing a 700 acre tract for #1 from Lackawanna. Galbreath, now superintendent of Wehrum, made several visits to Vintondale and to #1 in particular to check on the leased workings.

On July 27th, the #6 scales were adjusted; the boring on the #4 DC generator engine exhaust valves was out. The rock haulage system was down for one hour the next day. On Sunday, a new slack chute was installed in the #6 tipple. The #1 boiler was cleaned, and its side walls were repaired. As usual, the mines were short of men the following Monday. Due to a lack of cars, the washery was full of washed coal. On Tuesday the 31st, Hoffman visited all working places in #1 with Abe Abrams.

August

Cars remained scarce all week, and #1 was idle for that reason on Saturday, August 4th. In #6, there was a wreck five hundred feet from the drift mouth; the tipple crew reported it as a fire to Hoffman. A small fire was caused by the trolley wire on the car iron. Thirty minutes were needed to clear the ten wrecked cars.

Lack of hoppers continued to be a problem. Bins at both mines were full. On August 10th, the leg on the washery crane dropped into the pit when the clutch slipped. Two hours were spent retrieving it. The armature on the motor of the washery table was changed. Additional steel timbers were erected in the #1 drift, which was prone to cave-ins. (Those who lived in Vintondale in the 1950's should remember the two large deep sinkholes, caused by the collapse of the #1 drift, between Brandon's property and the Green Grass. For those unfamiliar with Vintondale, the Green Grass was an open grassy area behind the houses on Chickaree Hill. It may have served as Vintonale's first ball field.)

On Saturday August 11th, #1 loaded 10,467 tons into twenty-nine hoppers for use as fuel for the PRR. #6 was not as fortunate; it loaded out at 12:40 and had to wait until 2:15 for cars. Only four arrived; the washer remained full.

On Monday the 13th, a severe electrical storm at 2 PM put the #6 pumps out of commission. They were repaired and working at 7 AM. The boom on the #1 coke machine broke at 3 AM, and repairs took two days.

Mr. VanCleve and Mr. Shultz of the Wilson Snyder Pump Company spent several days in Vintondale inspecting #6's pumps. On the 16th, VCC purchased a centrifugal pump for the pumping station. The next day, the twelve inch main was tapped for Mr. Shultz so that he could get the correct data from the Worthington pumps which belonged to the Jackson Water Company. A capacity test was made the next day by Shultz, VanCleve and Slutter.

On Saturday, August 18th, the motor on the Bradford Breaker burned out and was repaired on Sunday. The #4 DE engine in the powerhouse was also fixed, as was the #3 locomotive boiler. Shultz returned on the 21st to make tests on the Jackson Water pumps. The Crighton Associated Insurance Company inspector arrived to look over the #6 filtration plant.

The draw bar on the thirty-ton motor broke causing #6 to close down at 2:45 on the 23rd. Hoffman spent most of the following day at the flat supervising the repair jobs at the washer and the motor. The washery chain broke on Saturday, but caused no delay at the ovens. Sunday repair jobs included putting a new cradle on the #2 coke machine, repairing #6 mine cars, and clearing the #4 boiler.

There was a small rockfall in 9R, Third Slope which was cleaned up by 8 AM. Mr. Devers of Scranton Electric paid a visit to "look over our power situation." The meaning of the "situation" was never explained in the entries. Slutter spent the day showing Devers around. Huth, Dempsey, Butch and Roschea drove to Kingport to look at coke ovens. Berkey took the cracked Bonnett cap of the #4 DC engine to Johnstown to be welded. On Wednesday the 29th, Mr. Devers tested the #4 generator and spent the next day in the powerhouse with Shaffer. The coke braize (or breeze) chain broke requiring two hours for repair. (Braize are the small particles of coke and ash which are left behind when coke is drawn.) On Friday, the mines were short because of the Wednesday payday. Hoffman spent the morning in 2nd Slope with Abrams. The chain on the washery crane broke at 5 AM and was repaired at 7 AM.

September

The mines continued to be short-handed on Saturday the 1st. The miners wanted a rest, and the coke and yard workers found their excuse by attending the wedding of John Roschea, coke drawer. That day Abrams and Hoffman drove to Patton in the morning to hire an electrician. Monday, September 3rd was Labor Day, but the mines operated as usual. However, a full shift of miners did not turn out, and some even took a second day off. A second representative of Scranton Electric Company, Mr. Rae, arrived on the 4th to inspect the power plant and the inside lines.

Hoffman spent 3-5 PM on the 5th discussing chain machines with Mr. Askin of Jeffrey. Mr. Cuthbertson arrived to look over the #6 powerhouse with Mr. Slutter and Mr. Rae. Having completed their inspections, the two men left on the 8:14 the next morning. At 8 PM on the 6th, Mr. Crawford of the Goodman Manufacturing Company called to discuss delivery of the chain machinery.

Saturday, September 8th was an Italian and Hungarian holiday. #1 had 75 miners report and only worked until noon. Hoffman spent the morning at #3 rock dump and #6 flat. The new set of fields in the #2 DC generator was working well. Scheduled repairs on Sunday the 9th included: a new piston installed on the #1 AC engine in #3, new brushings, and rivets in the crane chain. Cameron and Dickinson of Goodman paid a business call and set delivery date of a month for the new machine. There was a leak at the bore hole in #6; the water discharged into the mine.

There was a frost on Tuesday, September 11th, which was also a Russian holiday. Consequently, the mines were short again. Hoffman sent John Morton to the Berwind-White mines in Windber to inspect Goodman chain machines. Morton was accompanied by Mildon of the Goodman Company.

On Wednesday, the BR&P knocked a car off the track at the washery, but no damage was done. Schwerin wired from New York instructing Hoffman to purchase the Windber chain machines. Mr. Mildon visited #1 and inspected VCC's chain machines and reported them in good shape. The next day, two Berwind trucks delivered a chain machine; the second machine arrived on the 15th. Mildon changed the voltage on the machines from 500V to 250V.

Because the coal seam ran out in #1's 2nd R10 heading, Abrams and Hoffman inspected the surface above that heading and estimated that it would be at least 100 feet before they could pick up the seam again.

September 16th was a Sunday, and the usual repairs were lined up. The #2 Ridgway engine needed a new piston, and some minor repairs were made at the washer. Inspector Murphy went over the #1 and #2 Sterling and #2 dinkey engines and gave Hoffman a good verbal report.

Mr. Simpson of Goodman reduced the voltage of three chain machines from 500v to 250v on the 20th. State Mine Inspector Crocker conducted a two day inspection of #1. Due to men volunteering or being drafted for the army, VCC was short-handed and actively recruiting miners. Former employees John Klem and George Cosser, who left two months previously because of dissatisfaction with #1, returned to VCC employment. That same day, Madigan brought in twenty men on the 7:40; four were for Vintondale, and the rest were destined for Claghorn. Officer Timmons housed them in the town hall for the night.

On Sunday, September 23rd, Murphy inspected the #3 and #4 boilers and #3 dinkey. The oven crew hauled daubing mud for the winter. The next day, the door shaft of a wooden BR&P car broke after it was loaded with coal; a patching job was done right away. BR&P president Noonan and other officials made a quick stop in Vintondale on an inspection trip of its lines. Their car inspector Williams spent several hours in town.

The compensation inspector, Mr. Shilling, arrived on the evening train and spent the next day in #1 with Abe Abrams. Hoffman also talked via telephone to Schwerin about purchasing additional chain machines from Berwind-White. On the 27th, Shilling made an inspection of #6. Holt was off work on the 29th trying to locate a fireboss. Mr. Yoeman of Conemaugh was hired as the #6 electrician.

October

Schwerin and sons, Joseph and Clarence, Jr. arrived on October 2nd and spent two days at Claghorn. On the 4th, Schwerin, Williard, Huth, Claghorn and Hoffman spent most of the afternoon in #1 at the 2R 10 Heading and the evening in #6. While Hoffman spent the next two days at the flat, Mr. Schwerin and sons spent one day at Claghorn and one at Cambria Steel in Johnstown.

On Sunday the 7th, the repair crew worked at the #6 tipple and the washery; the dinkey cars were in the shop. Four additional miners from West Virginia arrived on the 7:40. On the 8th, a former employee, Joe Schrayer, returned from Robindale. Goodman representative Dickinson paid a business call and received an order for six chain machines. A new set of wheels was installed on the ten-ton motor in the #6 shop. Madigan brought in three laborers on the 10th. The ram motor on the #1 coke machine burnt up. The #2 machine finished drawing coke at 7 PM. The new armature was to arrive the next day by noon.

On Thursday the 11th, the tipples loaded run of mine for most of the day, keeping the dinkey busy all night hauling coal from the #1 bin to the washer. Other than the usual payday absentees, things were quiet at the mines until the 17th when the fifteen-ton Goodman stripped a pinion at 9 AM and the ten-ton burned out an armature. A mine car went through the cage at the tipple at 11:30 and held up the mine for one hour. "#6 was just a series of mishaps all day."

When the roof of 2R 10 Heading at #1 caved in at noon, Huth was sent to the surface and Olsen to the inside to investigate. The ovens started the next day at 3 AM to get the coke brieze moving early; it started at 11 AM. A new coke brieze was being equipped with machinery and put into operation on the 23rd. Henry Misner unloaded a car of sand for both mines and a car of props for #6.

October 24th was recorded as mild weather in the diary, but there were ten inches of snow on the ground by midnight. The yard crew had to spend the day shoveling snow. After midnight, the snow changed to rain. Blacksmiths Jenks and Wray were recruited to empty the #1 coal bin and dump it at the washery so that #1 might have a full day on the 26th. Both mines ran short of hoppers for the next two days.

On Sunday the 28th, repair crews were scurrying all over the flats, and Hoffman made three trips to check on them. Electricians from both mines moved half of the haulage and/or power cable for their mines; the remainder was to be shifted the following Sunday. The rock chute on #6's tipple was repaired. Surface crews ballasted the tracks on the flat. The #1 day shift was digging a short ditch in 10 Heading 2L to do away with a pump estimated cost was $100. The #3 boiler was inspected internally and judged in good condition. A tool shanty for the surface crew was started on the 29th.

November

In the beginning of November, a ten percent wage increase was posted, but many of the monthly employees were unhappy with the size of the increase. Slutter inspected the pump station and substation on the 3rd and found them in satisfactory condition. Misner and a crew from PRR raised the track at #6 on Sunday the 4th. Dempsey put new sprockets and shaft in the foot of the #1 elevator.

Richard Rager reported on the 9th that the creek on his property above 2R, 10 Heading was collapsing. Huth was sent to investigate, and Hoffman himself went the next day. Hoffman ordered Rager to put a fence around the area. Slutter inspected the #6 substation and showed the attendant how to use the power factor and improve voltage. The next day, he examined the #1 substation and gave the same instructions to its attendant.

Sunday, November 11th was another busy repair day. The PRR section crew raised the main tracks at #6. Coke oven crews hauled the winter's supply of daubing mud. The sidewalls of #2 boiler was repaired, and the #2 dinkey boiler was cleaned.

The carpenters began roofing the washery on Monday. There was a foreman's meeting in the evening. The next day, Huth and some laborers were sent to the surface 2R, 10 Heading at #1 to install tile drains because water was leaking into the brick wall of the air shaft. Two men were assigned to cut ties on VCC land. The company had been unsuccessful in attracting any area farmers to cut ties. Mr. Pierce of the Dictaphone Company paid a call and demonstrated how to operate the machine to the office personnel. He departed the next day on the 12:38 PM. The 14th was a busy day on the flat. Hoffman inspected the washery roof and found parts of it in bad shape. Slutter and Art Kempfer, boss carpenter, began work on a conveyor from the washer to the powerhouse bunker. Meisner unloaded a ten-ton GE motor for #6, which was earmarked for later use at Claghorn. Hoffman also sent the #6 foreman Holt a box of cigars and congratulations for the high tonnage. (And for the miners?)

On the 15th, Hoffman spent time on the flats both morning and afternoon and also found time to check on the men cutting ties. The next day, he and Abrams spent the morning in #1-10 Heading in the old #4 workings - 8 and 9 Heading. Cleanup work and drawing some room and pillars were taking place. The dinkey crew hauled coal to the washery from the #1 bin until 4:30. A prop and tie contract was signed with Mr. Fitzpatrick. Repairs on Sunday the 18th included having: the mason repair all the Smith forges in the #6 shop, the rock cars at #6 shop fixed, and the planking on the #1 tipple weigh office renewed.

On November 19th, at a foreman's meeting, all foremen and policemen were instructed to get the men out to work. The next day, Huth and part of Meisner's crew were sent again to the surface of 2R, 10 Heading to lay 24" tile pipe across the surface. The 21st was a "Russian holiday," and #1 was short fifty men while #6's shortage was 75. The company raised the price paid for props to a nickel each and to fifteen cents for a 5"x5"x5' tie. Mr. Kirkpatrick called again to discuss a tie and prop contract.

Hoffman spent the morning of November 22nd at the flat inspecting work being done on the Jones Stocker and the washery table. A boom was being built for the coke machine, and a new roof was installed on the #1 tipple.

On Saturday, November 24, the #1 substation "goosed itself" when someone pulled the switch at the powerhouse because there was too big a load at 8 AM. By the time Slutter got to it, Steve Christian had reversed it. School boys were cleaning up brick at the #3 powerhouse and clearing brush on both pole lines to the bore holes. Hoffman inspected their work at both places. The washery crew installed new chutes from the rewash table on Sunday. The carpentry crew put new sheeting and paper on the #2 belt runway from #6 to the washery. The oil in the oil switches was renewed in the #6 powerhouse and #6 substation.

Monday the 26th, the Dictophone representative checked the machines in the office and found them in "good order." A foreman's meeting was held at 7 PM. Hoffman had a busy day on the 27th. Besides visiting the flats in the morning and afternoon, he also checked on the tie cutters and brush clearers on #3 hill. He put Dunmeyer, Hoffman and Cook on hauling house coal in addition to the two regular teams. The old coke brieze was torn down, and work was begun converting the conveyor from the washery to the powerhouse to handle fuel. The next day, there was a wreck in 5R, 2nd Slope in #1. The mule teams began hauling ties from #1 hill to #6. Because of this, four miners quit. Their objections to this unfortunately were not recorded.

December

On December 1st, Slutter traveled to Johnstown to pick up an electrician and an armature winder. There were only two men working on the repair crew at the #6 tipple that Sunday. The #4 boiler was also cleaned that day. Because of payday and religious holidays, the mines were short of men from the third to the fifth of December. On the 3rd, the attendant at the #1 substation burnt up the middle bearing at 7:30 AM. Many of the miners refused to wait for the repairs to be made. A foreman's meeting was held at 7:30 PM, and a GE representative arrived on the 3rd to inspect the motors. On the 5th, the dump shaft at #1 broke at 8 AM and took one hour to fix. The carpenters and machinists continued to work on the conveyor to the powerhouse.

Hoffman took Abrams, Holt, Olsen and Wagner to get an explosives license for the VCC. Location of the procurement was not revealed, but was most likely Ebensburg because they departed on the 4:04 and returned on the 7:50. Work in the powerhouse included putting engines and fan for the stockers on their foundations. Crews were also working on the conveyor from the washery to the powerhouse.

Saturday the 8th was a Roman Catholic holy day. The company posted notices in English and Hungarian requesting that the men come out for work. Both mines were able to fill the mantrips. However, Officer Timmons checked in at the 7:30 mass and reported to Hoffman that there were only two miners and one cokeman among the congregation. Besides, the holy day, production was down due to one of the worst blizzards in years. Misner's crew was busy all day clearing tracks, and the dinkey struggled to haul coal to the washery.

Repairs completed on Sunday included the sidewalls of the #3 boiler and #6's ten-ton motor's brakes and resistance. The cold, windy weather continued for several days, slowing down all operations. The dinkey was at #6 spotting empties and taking out loads. The ovens were late in finishing because of the slow movement of cars. Hoffman spent most of the day at #1 tipple spotting cars and then conducted a foreman's meeting at 7 PM.

By Tuesday, conditions had not improved, and extra men and the dinkey were needed at #6 to get the cars around at the tipple. A carload of sand was unloaded. Mr. Faye was trying to thaw out the #6 air shaft. Delano and Schwerin arrived on the 4:04 and spent the evening studying the mine maps. The dinkey was kept busy all night hauling coal from the washery to #1.

Perhaps because the top boss was in town, both mines had few absentees. 200 men showed up at #1 and loaded 1,016 tons, and 225 men at #6 mined 1,081 tons. Delano, Schwerin and Hoffman spent the day in Claghorn to close some coal contracts. That evening they inspected #6 with Slutter, Abrams and Holt.

Both mines loaded out early on the 13th and 14th. The surface crew and all spare hands, including station agent Bill Clarkson and policemen Butala and Timmons, were shoveling snow, starting at 3 AM. Hoffman spent the day in the yards trying to keep the cars moving. Payroll did not arrive on time on Saturday the 15th; it had been shipped from New York the previous Thursday. At 4 PM, there was a wreck involving the #3 dinkey above the bridge. It could not hold five rock cars and cut them off. The cars jumped the switch at the bridge.

On Sunday the 16th, the surface crew unloaded a car of sand at #6; others were shoveling snow at the tracks. The machine runners repaired the coke drawing machines. Hoffman and son Kenneth visited #6 airshaft in the afternoon. He had to find a boiler which could provide a steam jet to thaw out the shaft.

The payroll finally arrived on Monday the 17th at 11:05. (one hour late) Payout began at 4 PM. Mr. Corner of Niles-Bernent and Pond Company, manufactures of punchers and shears, paid a business call. A foreman's meeting was held at 7 PM, and Hoffman sent out Christmas cigars. On the 18th, both mines loaded out at noon. For a change, there were few "payday drunks". Huth and Reed plugged a leak in the #6 air shaft. However, manpower dropped on the 19th, a "Russian holiday". #6 only had 150 miners, and #1 was idle for lack of cars. Work on the conveyor belt connecting the powerhouse and washery continued, and excavation started on the Jones Stocker blower fan. Mr. Crocker, state mine inspector, was in to look over #6.

On the 19th, there was a small cave-in on the road above #1's fan. Lack of cars caused #1 to be idle again on the 21st and 22nd. Corkcer continued examining #6, accompanied by Abrams. Coke was drawn on Sunday; the side walls of the #6 boiler were repaired. The electrical crew repaired the breakers on the switchboard.

On Christmas Eve, both mines worked with a full contingent of miners. Mr. Misner spent the day unloading two cars of rails, and the carpenters worked on the conveyor. Hoffman sent John Morton to Pittsburgh to inspect punchers and shears. All departments were idle on Christmas. The next day, the mines operated with about one half the usual numbers. On the 27th, #1 had only two cars and loaded out at 11 AM. #6 managed a full shift and loaded run of the mine all day. Misner unloaded a car of sand and worked on converting the air duct for the Jones Stoker. On the 28th, due to a lack of cars, #1 was idle. Hoffman spent the morning in 2nd Slope with Abrams and Holt. Saturday the 27th saw #1 idle again. Four men were put on the tipple dumping coal into the large bin and breaking lumps. The pay also arrived from New York.

Sunday the 30th was a busy day for the repair crews. Carpenters put in chutes to run the rewash into large bins. The sidewalls were rebuilt in #3 boiler, and Misner continued to excavate for the Jones Stoker air duct.

The last day of 1917 was a hectic one. Hoffman convinced the #1 miners that coal hoppers would be arriving at 8 AM. So the miners returned home to change into work clothes and get their tools. 150 miners reported for work; at #6, 208 miners showed up, an excellent number for the Monday after a payday. Mr. Shilling of the Associated Insurance Company called; he and Hoffman discussed the new standards. The #2 DC engine was acting up; upon investigation, the meter was way off; the engine was doing 900 amps.

Perhaps someone planned to usher in the New Year's with a big band. Mr. Buck found two cases of dynamite in the woods above Rexis. Timmons and a teamster retrieved them and stored them at the company magazine. Timmons also notified the Indiana County Sheriff's office.

MINE DIARY - 1923

The 1923 diary follows much the same pattern as that of 1917. Hoffman recorded his daily route from home to the office via #1 and #6. On the few occasions when Hoffman was out of town, such as his mother's death in May, Arbogast made the diary entries. Fear of a strike on April 1st required Hoffman and Arbogast and mounted police to be present for the start of the mantrips for the rest of the year. The following is a summary of the mundane events happening at the mines in 1923.

January

The first week and one half of the entries were missing, having been recorded in the 1922 book. First entry of note for the mines was the visit of the Eastern Car representative on January 11th; he paid a business call concerning using steel cars for #1's dinkey. Sunday, January 14th had its usual repairs scheduled. The mason was building side walls for the #3 boiler; a rockfall in #6's 18L 2nd Slope at 8 PM was cleared by George Clark and Frampton. Penrod, the state insurance rating inspector, was in #1 on the 15th and 16th. The following Sunday, one man keyed up boxes on the new dinkey; two worked on putting sheet iron on the #1 coke machine; and a crew was laying 60# rails in 1st Main in #6. On Thursday January 25th, the chronic hopper shortage sent all the men home except tracklayers and those working in low coal in #1.

February

On Saturday February 3rd, the tower on the #1 coke machine buckled. After patching it, coke was drawn at noon. The shop was to build a tower and install it the next week. Patterson of Home Scale Company arrived to repair the #6 scale. Because of a very bad roof, the pump and rails were pulled from 18N, 2nd Main; the stumps were left in so that one or four could be pulled later. The Heyl and Patterson representative met with Hoffman on the 5th to discuss tables of load sizes of coal. The same day, Hoffman met with Peffer of Penn Central Power company to discuss questions of the powerhouse purchasing power. The same day, the Link Belt representative called and departed with an order for a chain drive machine, one which was $251 cheaper than a Jeffrey. State inspector Crocker was in #6 from the 26th to the 28th. On the 26th, an empty motor run by John Hazy ran into a mantrip; there were no injuries, but both motors were damaged. Hazy had been given instructions to go to the right and off the heading, but went on by and hit the mantrip. The next day, #1 was shut down for lack of cars, so one of its motors was used in #6 while the two damaged motors were in the shop for repair.

March

On March 1st, the surface crews were busy unloading a car of beams and a car of sand for #1. The representative for Penn Central Power Company read meters in the powerhouse on the 6th. The #1 dinkey, loaded with rock cars, got too fast a start and jumped the track near #8 houses on Dinkey Street. Later that month, it went off the track again near the machine shop and hit a BR&P car, causing a leak in the dinkey's boiler.

April

Dinkey accidents continued into April. On the 10th at 3 PM, the dinkey and some coke cars sideswiped, turning over 3 rails. The trucks were off the tracks, taking 115 minutes to repair. The rails were replaced by 6:00PM. Gas was discovered in 2nd Slope on Saturday, the 14th, and the miners had to be sent home. By 2 PM it had dissipated some. On Wednesday April 18th, Hoffman went into #6 with foreman Clark and visited all of 2nd and 3rd Main and the face of 2nd Slope. Conditions were bad at 4. Hoffman was out by noon. However, the 1st Slope motor went off the tracks, and three trips were lost. On the 17th, the suspension bar on the fifteen-ton motor broke; a shortage of men closed #1 at 10 AM. Because of bad conditions in #6's 2nd Slope, the miners refused to work on the 28th until Hoffman made some undisclosed "satisfactory arrangements."

May

On May 1st, there were additional problems in #1. Two hours were lost when the motor's brake rigging dropped, and the trip ran away, tying up the track. Clark, Huth, and Hoffman looked at the gas feeder in 14R on the 3rd and decided to lay 2" pipe from the sealed stopping valve to the airshaft to drain the gas. Thomas and Abrams made arrangements with Hoffman to lay off a car dropper and a car oiler. There would only be one man on each job from then on. Hoffman also drove to Johnstown to discuss the exams for assistant and for fireboss. (Was he asking for special consideration?) On the 14th, Crocker was in to see the 14R gas feeder with Clark and Hoffman. At noon, Crocker began an inspection of #1. At 3 PM, he notified the office that he would be prosecuting a miner for shooting the solid. The business callers of the day were Morse and Chief Engineer MacFarlane of the Borden Company.

On May 8th, Hoffman went to the depot, in a steady rain, to meet Mr. Wilson of the New Jersey Link Company to show him the ovens. At 10:00, Hoffman drove him to Graceton; upon return, Wilson took a train to Uniontown. Hoffman found everything "OK" in his absence. To keep the yard cleaner and save labor, the company started making ash slurry to dump on the rock dump rather than the previous practice of making piles along the track. Two miners were laid off on the 16th for loading a large sulphur ball. On the 17th, both mines were closed, but the ovens worked. At 8 AM, Clark, Huth, Hoffman and Gongloff made a motor inspection of #1, visiting 2nd Slope 7W and 1st Main. The mine was in good shape. In the afternoon the same group, plus Dempsey and Abrams, went into #6, visiting the 14R and everything below it in 2nd Slope. Conditions at the face of 2nd Slope were bad. On the 22nd, the current hitching broke at #1 and a trip ran down the 2nd Slope and wrecked thirteen cars, delaying the tipple three hours. The tipple crew ate lunch early at 11:30, and the first trip after the wreck came out at 12:05. Finkbenin and Fulmer of New Jersey Link Company came to look at the coke machine. At noon, Hoffman received a message from New York to close #1's 10 Heading workings for the time being; farmers who mined that heading laid off each summer. The dinkey ran off the track again on the 26th, taking one hour to get it back on the tracks.

June

Early on June 4th, Hoffman had to go to #6 because he received word that the miners refused to go in because yardage had been cut in 2nd Slope. After Hoffman talked to the men, they went to work and "all was well". The #1 miner who was going to be fined for shooting solid was called into the office with no mention of any repercussions from Hoffman. On Saturday the 9th, Hoffman paid a visit to the rock dump, which was being moved back by the surface crew. Washery refuse was also being sent to the dump. The mines closed early on the 18th because not enough men showed up after payday. New York sent a message to close the mines on Saturday the 23rd. A thunderstorm that day hit the #1 fan and burnt off the lead, caused water damage to the streets and blew down large trees.

July

On July 4th, the AC board in the powerhouse was hit by lightning at 11:30 AM. Three insulators and a fifty volt potential transformer were hit and took three hours to repair. The next day, the mines were short because of the Fourth of July revelries. Goodman Representative Maloney entered 3rd Slope of #6 to see if conditions were right for using a top cutting machine. The surface crew put two lengths of cast iron pipe under the dinkey track at the rock dump on the 11th.

August

John Jones, State Compensation Inspector, arrived on the 31st and started his inspection on the 3rd; he spoke at an Inspection Bureau meeting on August 6th. All mines were idle on the 10th due to the funeral of President Warren G. Harding. The next event of any note at the flat happened on August 25th; the second armature in two days blew up on the 5 & 6 pump in the old dip in #1. Boiler inspector Murphy made a good report on the two Stearlings and the #1 locomotive.

September

On September 6th there was a rockfall in 1st Left, new slope in #1; wires were down. Cleanup took one hour. Broundee of the Ridgway Dynamo and Engine Company arrived to examine the #1 engine. The next day, the ten-ton locomotive armature blew up. On the 8th, #1 closed at 11 AM due to a "Roman Holiday", and at quitting time, the fields of two fifteen-ton motors burned out, sending them to the shop. The dinkey ran away with rock cars again on the 8th, but no damage was done. The engineer ran up the grade at the rock dump and stopped. One wheel was flattened, but repairable. On the 25th, the ice wagon's team ran away and damaged Lybarger's car. Hoffman went into #6 at 8 AM on the 28th with Clark and went through all of the 2nd Slope and 3rd Main looking at variants. The face of the slope was in firm shape.

October

In October, in addition to tending to Claghorn, the movie theatre and the bank, Hoffman also met with two men on the 9th who wanted to start a drug store in Vintondale. His reaction was not recorded, but probably was not very positive. The same day, Assistant Superintendent Evans and Chief of Police Manion of Revloc paid a visit. Also car inspector MacArthur was in #6; the next day, he was in #6 with VCC's safety engineer. Crocker was back on the 16th to inspect #6. Visitors on the 18th were Attorneys Little and Jones and Philip Caufield. Hoffman met Crocker in the office on the 22nd, and then went in #6 and visited all of 2nd and 3rd Slopes. Dyer of Ridgway Engines worked on the #1 Ridgway valves on the 30th.

November

McNelles of Cresson paid a visit on November 13th; he was in charge of the Taylor ovens. On Monday, November 19th, the usual payday drunks were off, and #6 had a rough day with three motors down at different times. A fire was reported by Clark on the 28th in 1st Main, 18L Heading in #6. It was discovered to be gob burning between rooms 9 and 10. This heading had been abandoned because of a dip and lack of a pump for it. The fire was quenched by 3 PM by moving the gob and using a Pyrine extinguisher. According to Russ Dodson, there was an "amazing amount" of methane escaping out the shafts and drifts of #6. There were several small explosions, but never any large ones. The higher the oxygen level in a mine, the greater the chance of an explosion. Spontaneous combustion fires were also very numerous in #6. The gob, or slack coal and other refuse, would smolder. These fires were frequently in abandoned headings and often detected first at the air shafts. If the smoldering refuse was shoveled into a mine car, it often burned out the bottom of the car. The usual method of extinguishing such a fire was to move the gob.

December

In December, the rock dump was moved back to allow dumping in a new place, and the washery washed a bin of coal. A car of lumber was unloaded for the new picking table. Mr. Hoffman visited the new Penn Central power plant in Saxton. On Saturday the 8th, the mines were idle, but the ovens worked. Carpenters put several braces on the #6 coal bin. Some bolts were made for the sand plant. Hoffman spent the afternoon of the 9th on his weekly letter to New York and the evening visiting miners at home. A full crew of carpenters worked on the new picking table for #6. Both mines were idle on the 11th and 12th. The surface crew worked on track repairs while the carpenters worked on the picking table. #1 lost one hour on the 13th when the arch in the boiler dropped. It was -10 degrees on the 14th, and both mines were idle that day and the next. Coke was drawn as usual; the carpenters continued working on the picking table; and Hoffman met the miners while they received their pay.

A leak in the 4" wooden pipe at the rock dump at noon on the 18th resulted in the repair crew being called out. Hoffman spent the afternoon with Blewitt discussing the inventory of repair stock such as rails and timber. Repair crews were out on the 21st repairing the GE motor, and a carpenter was working on the picking table. Hoffman visited the flats in the morning and afternoon, met all the trains, and visited the miners at home in the evening. Coke was drawn as usual on the day after Christmas; the mines were idle; the carpenters worked on the picking table; and the erecting engineer for the Link Belt Company arrived to assist with the #6 picking table. That Friday, both carpenters and machine shop crews worked on the table; both mines were idle. On Sunday the 30th, a rainy day, the carpenters and machine shop crew worked on the tipple. Hoffman was called to the flat when five studs in the #1 Ridgway broke. Part of the electrical load was transferred to Penn Central and repairs started immediately.

HESHBON LANDS

These tracts in Brush Valley, East Wheatfield, West Wheatfield and Buffington Townships were sold on January 13, 1917 to Vinton Land Company by Lackawanna Coal and Coke Company. The deed book references are included for some of the tracts.

1. A.V. Barker coal/surface, 681a 63p., BV Twp.
   July 28, 1902, ICDB 74B, p. 375
   sold by Barker to LC&C, excepting railroad right of way granted by George Dias and D. Harbison Tomb on two tracts, 16 a. March 12, 1912, EW and BV Twps.
2. A.V. Barker coal and surface, 65a. 39p. BV Twp.
3. A.V. Barker coal and surface, 96a. 35p. BV Twp.
4. A.V. Barker coal and surface, 5a. 122p. BV Twp.
5. A.V. Barker coal and surface, 28a. 88p. BV Twp.
6. A.V. Barker coal and surface, 54a. 139p. EW TWP.
7. A.V. Barker coal and surface, 35a. 153p. EW Twp.
8. A.V. Barker coal and surface, 74a. 138p. Buffington
9. A.V. Barker coal and surface, 81.15a. EW and WW TWP.
   timber rights reserved to Hugh Mack.
10. A.V. Barker coal 100a. 66p. BV Twp.
11. A.V. Barker coal 52a. 121p. BV Twp.
12. A.V. Barker coal 67.18a. & 1.82a. EW Twp.
13. A.V. Barker coal 154.77a. EW Twp.
14. A.V. Barker coal 84a. 66p. Bv Twp.
15. A.V. Barker coal 27a. 103p. BV Twp.
16. A.V. Barker coal 123a. 48p. BV Twp.
17. A.V. Barker coal 87a. 41p. BV Twp.
18. A.V. Barker coal 60a. 104p. Buffington
19. A.V. Barker coal 100a. 150p. Buffington
20. A.V. Barker coal 111a. 5p. BV Twp.
21. A.V. Barker coal 223.42a. WW Twp.
22. A.V. Barker coal 34a. 156p. BV Twp.
23. A.V. Barker coal 35a. 51p. BV Twp.
24. A.V. Barker coal 93a. 2p. Buffington
25. A.V. Barker coal 84.45a. BV Twp.
26. A.V. Barker coal 105.71a. EW Twp.
27. A.V. Barker coal 109a. 114p. BV Twp.
28. N.L Wagner coal and surface 107.65a. WW Twp.
    same tract conveyed by Wagner to LC&C
    August 19, 1902, ICDB 74A, p. 232.

---

29. William B. Overdorf coal 29.51a. BV Twp.
    subject to LC&C paying Overdorf for all surfaces needed
    October 30, 1902, ICDB 77B, p. 98

---

30. C.C. Fisher coal 69a. 73p. BV Twp.
    subject to LC&C paying a reasonable price for surface.
    November 8, 1902, ICDB 76B, p. 237

---

31. Martha McKee coal and surface 27.46a. BV Twp.
    November 7, 1902, ICDB 78A, p. 278

---

32. Joseph Campbell coal 110.88a. BV Twp.
    November 5, 1902, ICDB 77B, p. 101.

---

33. Samuel McHugh coal 80.94a. WW Twp.
    surface agreement November 7, 1902 ICDB 76B.

34.	Daniel Hendricks coal surface agreement, November 24, 1902, ICDB 76B, p. 292.	122.31a.	BV Twp.
35.	John Wagner coal surface agreement, November 19, 1902, ICDB 76B, p. 536.	133.98a.	BV Twp.
36.	Alvin T. McNutt coal surface agreement, January 17, 1903, ICDB 76B, p. 546.	102.40a.	BV Twp.
37.	Elizabeth Fenton coal surface agreement, December 22, 1902, ICDB 121, p. 78.	41.60a.	BV Twp.
38.	Peter Garman coal surface agreement, November 17, 1902, ICDB 76B, p. 252.	10.75a.	WW Twp.
39.	Edward Clouse coal surface agreement, November 20, 1902, ICDB 77B, p. 168.	10.75a.	WW Twp.
40.	I.V. Lower coal	69.81a. and 10.13a.	BV Twp.
41.	George W. Dias coal excepting two acres under his buildings surface agreement, December 9, 1902 ICDB B, p. 62.	187.51a.	BV Twp.
42.	Mary Robinson coal and surface excepting small piece of ground for cemetery December 9, 1902, ICDB 78A, p. 446.	81.85a.	Bv Twp.
43.	Edwin Edgar Dias coal/surface excepting 3 acres right of way granted to E&BL RR. March 3, 1903, ICDB 81B, p. 432.	56.49a.	BV Twp.
44.	Clarence Claghorn coal surface agreement, October 31, 1902, ICDB 76B, p. 233.	223.08a.	BV Twp.
45.	C.L. Campbell coal/surface October 3, 1902, ICDB 77B, p. 96.	22.86a. and 34.14a.	BV Twp.
46.	Christopher D. Campbell coal surface agreement, September 15, 1902, ICDB 77B, p. 85.	96.31a.	WW Twp.
47.	Hugh S. Mack coal surface agreement, September 27, 1902, ICDB 77B, p. 90.	109.30a	WW Twp.
48.	Angeline Mack coal surface agreement, September 17, 1902, ICDB 77B, p. 87.	158.16a.	WW Twp.
49.	Jacob Mack coal surface agreement, October 1, 1902, ICDB 77B, p. 93.	133.79a.	WW Twp.
50.	Alexander Morgan surface January 20, 1903, ICDB 77A, p. 397.	7.42a	BV Twp.
51.	Alexander Morgan coal includes all coal under parcel #50 surface agreement, January 20, 1903, ICDB 77A, p. 394	115.17a.	BV Twp.

52.	Catherine Mikesell et al coal	95.33a.		WW Twp.

surface agreement, November 1, 1902, ICDB 77B, p. 167.
one third of above parcel, undivided 2/3 conveyed by J.S. Walbeck, Guardian of Ralph Mack and Mrs. Myrtle Alsop, minor children of the late Thomas Mack, DB 77B, p. 170.

53.	William S. Adams coal	16.47a.		BV Twp.

surface agreement, December 16, 1902, ICDB 78B, p. 124

54.	D.I. Cunningham coal	213.92a.		BV Twp.

surface agreement, December 16, 1902, ICDB 78B, p 426

55.	David McCormick coal	21.34a.		BV Twp.

excepting four lots in Heshbon sold to A.E. Wagner, W.B. Wagner, James Dias, and David Cunningham. Surface agreement, December 29, 1902, ICDB 76B, p. 422.

56.	Hugh Mack coal/surface	35a.		WW Twp.

January 17, 1903, ICDB 76B, p. 544.

57.	Amanda Adams coal	100.75a.		BV Twp.

surface agreement, December 27, 1902, ICDB 78B, p. 126
reserving UP church lot and two acres/house.

58.	Samuel Dias coal/surface	54a.		WW Twp.

January 7, 1903, ICDB 79A, p. 140.

59.	Cyrus Wakefield coal/surface	21.08a.		BV Twp.

November 3, 1902, ICDB 78A, p. 14.

60.	Robinson Heirs coal/surface	110.41a.		BV Twp.

December 9, 1902, ICDB 78B, p. 73
excepting surface sold to Mary Craig, 59.08a. by LC&C,
May 17, 1904, Vol. 84, p. 500.

61.	Alexander Morgan surface	7.29a.		BV Twp.

same mining rights as #51, March, 1903, ICDB 82A, p. 272.

62.	Hugh Mack surface	68.29a.		WW Twp.

same coal rights as #13, January 17, 1903, ICDB 76B, p. 539.

LANDS SUBJECT TO THE 1916 LEASE

All of the following coal tracts are located in Buffington Township, Indiana County.

No. 1	(Reed and Griffith),	299 a.	95 p.
No. 2	(Reed and Griffith)	143 a.	68 p.
No. 3	(Reed and Griffith)	13 a.	68 p.
No. 4	(Reed and Griffith)	213 a.	10 p.

sold by George Roberts to J.L. Mitchell, January 6, 1903, known as the Roberts Tract

The following tracts are located in Buffington Township and/or Blacklick Township, Cambria County.

No. 5	(Reed and Griffith)	35 a.	114 p.

sold by Mattie Lydick to J.L. Mitchell, November 2, 1903, known as the Lydick Tract.

No. 6	(Reed and Griffith)	668 a.	49 p.

sold by Evan Morgan to Mitchell, February 16th, 1903; F.B. Williams to Griffith and Reed August 20th, 1903; James B. Graham to Reed, January 30, 1909; James B. Graham to Griffith and Reed, November 29, 1909; Evan Morgan and W.S. Donnelly to Reed and Alvin Evans, February 21, 1902. Also heirs of Alvan Evans to Webster Griffith, February, 1911 and deed from J.L. Mitchell, February 28, 1907.

The above six tracts are designated as such in a deed from Griffith and Reed to Delano and Burr, July 20, 1911, Cambria County Deed Book Vol. 234, p. 406 and Indiana County Deed Book Vol. 124, p. 239.

No. 7 Blacklick Twp. 2 a. 100 p., sold by William George to Delano and Burr, July 22, 1911, known as the George Tract.

No. 8 Buffington Township 45.66 acres, sold by Thomas Davis to Burr and Delano, February 11, 1914, subject to a mortgage of $3,000 due in five years with a 5% interest annually, known as the Davis Tract.

No. 9 Buffington Twp. 75.69 acres, sold by George Kerr to Burr and Delano, April 27, 1916, subject to a mortgage of $4266.24, known as the Kerr Tract.

No. 10 Buffington Twp. 98 a. 74 p., sold by Henderson and William Bracken to Delano and Burr, May 4, 1916, subject to a $6,319,36 mortgage, known as the Bracken Tract.

No. 11 Buffington Twp. 166.683 acres, sold by Thomas Duncan to Delano and Burr, May 15, 1916, subject to mortgage of $8,000.79.

No. 12 All lots in borough of Vintondale marked on town plan as Lots 2, 4, 6, 8 ,10, and 12 in Block T; Lot 2, Block F; Lots 6, 8, and 10 Block AD; Lot 2 and 30/50 of Lot 4, Block M; 1/2 of Lot 2 in Block D; Lot 7 in Block J; and Lot 10 in Block L.

No. 13 4.136 acres lying along Shuman's Run Valley road in Vintondale with all improvements erected on lots including sixty-two two-story houses.

Tracts 1-11 were granted by Burr and Delano to the Vinton Land Company on July 29, 1916. Tracts 8-11 were subject to mortgages totaling $21,586.39. Tracts 12 and 13 with sixty-two houses erected on them were conveyed to Vinton Land in July, 1916 by Delano. All of the above lands were subject to a lease with the Vinton Colliery Company.

Appendix F
Lackawanna Coal & Coke Company

ANNUAL REPORT:
LACKAWANNA COAL AND COKE COMPANY, 1902-1922

Year	Mine	Superintendent	Production		
1902	#3	Claghorn, Gen. Supt.	6,478		43
	#4	G.R. Delameter, Supt.	9,915		69
1903	#3	Claghorn, Gen. Supt.	15,651	204	87
	#4	John Reed, Supt.	27,974	227	127
1904	#3	Claghorn, Supt. &	38,199	208	
	#4	Gen. Supt.	48,729	245	
1905	#3	C.M. Blanchard	Idle		
	#4		35,510	75	162
1906	#3	Ward N. Johnson	Idle		
	#4		179,744	240	334
1907	#3	"	Idle		
	#4		262,378	255	364
1908	#3	"	Idle		
	#4		61,136	95	153
1909	#3	"	Idle		
	#4		116,409	153	191
1910	#4	"	193,221	245	260
1911	#4	H.J. Mehan	259,144	300	377
1912	#4	"	294,573	305	397
1913	#4	W.A. Luce, Asst. Gen. Man., Pgh.	341,222	288	425
1914	#4	F.B. Dunbar, Luce	189,989	123	407
1915	#4	"	102,738	246	164
1916	#3	"	259,264	303	288
	#4		Idle, All Coal Taken Through #3		
1917	#3	R.E. Galbreath	316,784	307	383
	#4		Abandoned		
1918	#3		380,526	306	381
	#4	George Lindsey	Abandoned		
1919	#3	"	351,219	262	382

1920	#3		385,015	299	362
1921	#3		220,430	151	409
1922	NO STATISTICS GIVEN UNDER LACKAWANNA COAL AND COKE				

ANNUAL REPORT: BETHLEHEM MINES CORPORATION, 1923-90

1923	#78	Lindsey	423,625	283	489
1924	#78		376,625	238	462
1925	#78 #79		345,839 Idle	235	288
1926	#78 #79		406,867 Idle	303	436
1927	#78 #79	R.E. Abrams	352,962 Idle	204	461
1928	#78		346,133	190	431
1929	#78		148,928	99	354

1930　#78 and #79
Temporarily abandoned (for 60 years)

MINE FATALITIES: LACKAWANNA COAL AND COKE COMPANY

1902	July 23:	Ernest Monroe, Negro 21, Killed instantly while sinking slope for #4. Stone fell on him while he was striking on a drill, tried at 7:30 AM and at 11:00 AM to bring down stone. It fell at 1:30 and brushed the side of a fellow worker.
	October 11:	Mikeal Cowash, 50, married, five children, Hungarian, miner and Charles Remish, 30, married Hungarian, miner, killed instantly at bottom of hoisting shaft of #4. Five men employed at taking down rock, hole drilled preparatory to blasting, much water coming in, rock sounded solid, a prop was taken out by victims, while one of men was preparing a blast, a rock fell.
	November 14:	Paul Hertell, Slav, 28, married, killed in #3 by blast of dynamite, charging a hole in bottom. In putting charge to back of hole, he used an iron tamping drill, exploding charge, claimed to be experienced in rock work, held miner's certificate from anthracite.
1903	August 4:	Simon Boldi, Hungarian, miner, 42, married, #4, killed instantly by roof fall, slate was not properly propped.
	December 11:	Emerie Schmidt, Hungarian, timberman, 44, married, 4 children, fatally injured by fall of roof while putting up timber on main slope, roof fell in, causing internal injuries, died on the 13th, was using all reasonable precautions at the time.
1904	April 28:	William Diehl, 24, single, working on top of coal crusher, fell into hopper at 4 PM, crushed to a pulp, took four hours to find multilated remains. Bert Noel, single, from Munster. Rescued before reaching crushing wheels, was probably fatally injured. (Indiana County Gazette, May 4, 1904.)
	May 11:	John Vantroga, Austrian, miner, 40, married, three children. George Shippley, Austrian, machine runner, 20, single. Andrew Drubant, Austrian, machine scraper, 23, single. All killed by an explosion of firedamp in #3.
1905		Lackawanna Mines closed most of the year.
1906	March 30:	George McLosh, Hungarian, miner, 32, married, 4 children, fatally injured by car inside of mine. Handling car on heading when partner let another car from face down the heading which struck McLosh. Accident occurred through misunderstandings between McLosh and partner.
1907	May 28:	Lewis Young, American, machine runner, killed instantly by fall of roof while under cutting coal, props set within six feet of face, slip in roof ran parallel with face, tailing in over face of coal.
1909	June 23:	1. Lovely Louis, Italian, miner, 22, married, 2 children. 2. Ernest Barrochi, Italian, miner, 41, single. 3. Dominick Lilton, Italian, miner, 21, single. 4. Tony Batista, Italian, miner, 20, single. 5. Tony Totena, Italian, miner, 22, single. 6. Charles Foldy, Slavonian, miner, 32, married, 4 children. 7. A.D. Raymer, American, pumpman, 31, married, 1 child. 8. George Kowash, Slavonian, trackman, 23, single. 9. Simon Rominski, Russian, miner, 36, single. 10. Steve Base, Polish, miner, 35, single. 11. Kosti Sevick, Lithuanian, miner, 31, single. 12. George Lenn, Lithuanian, miner, 34, married, 3 children 13. Joe Meniott, Italian, miner, 25, single. 14. Mike Lilton, Italian, miner, 23, single. 15. Alex Shaftock, Slavonian, miner, 46, 2 children 16. Charles Georda, Italian, miner, 22, married, 1 child 17. Charles Loray, Italian, miner, 20, single.
	June 26:	18. Frank Delegram, Italian, rockman, 28, single.
	June 29:	19. William Burns, American, electrician, 37, married, 5 children.
	June 30:	20. Patrick F. Burns, American, trackman, 32, married, 3 children.
	July 2:	21. Clarence Huey, American, runner, 37, married, 2 children.
1911	September 2:	Mike Sesta, Slavonian, spragger, 24, single, instantly killed while trying to get on some loaded wagons that he and motorman has been gathering in pillar workings.

1913	February 21:	James Woods, American, machine runner, 38, married, 4 children, killed by being caught by revolving threadbar on heading and thrown against rib of place, crushing his skull.
1914	June 29:	Wasel Honnatick, Austrian, miner, 45, married, 4 children, fatally injured by being squeezed between rib and side of wagon while he was leaning on side of passing wagon close to his room parting on heading, died three days later.
1917	February 17:	John Bockovick, Austrian, pick miner, married, no children, killed instantly by fall of coal at face of room, coal fell over sprag, improperly placed.
1918	September 9:	George Balko, Austrian, pick miner, 34, married, 2 children, fatally injured by cars on entry, died September 19.
	October 2:	Frank Testa, Italian, pick miner, 28, married, 2 children, instantly killed by fall of slate at face of pillar workings.

MINE ACCIDENTS: LACKAWANNA COAL AND COKE COMPANY

1905:	December 20:	John Orwas (Orris), Hungarian, miner, 42, married, #4, leg broken, necessitating amputation below knee. Empty cars ran to the face of his place.
1907	October 3:	Steve Butch, Italian, miner, 22, single, right hand blown off at wrist with dynamite firing two shots at same time. One exploded, thought both had gone off, other went off when he returned.
1909	June 23:	1. Chris Frazier, Canadian, trackman, 60, married
		2. A. L. Johnson, American, mine foreman, 33, married
		3. Joe Orwart, Polish, miner, 29, single.
		4. Pat Betest, Italian, miner, 19, single.
		5. John Tobin, American, machine boss, 28, single.
		6. John Kessler, American, miner, 34, married.
		7. Fred Thomas, English, assistant foreman, 31, married.
		8. Tom Batesta, Italian, miner, 53, single.
	Above burned by dust explosion.	
		9. Sam Koncha, Italian, miner, 26, married.
		10. Louis Koncha, Italian, miner, 29, single.
		11. Tony Martin, Italian, miner, 35, married.
		12. Nick Spelli, Italian, miner, 44, married.
	Above overcome by afterdamp following dust explosion.	
1911	April 21:	George Balkal, Slavonian, miner, 21, single, foot broken by fall of rock that he was trying to pull down.
	July 12:	John Ponboly, Italian, miner, 21, single, one finger amputated by being caught in gear wheel of hoisting machine.
1912	April 12:	Tony Fatchey, Italian, miner, 23, single, back injured by fall of slate while cutting coal with a machine at face of heading.
	July 23:	Lucas Rich, Polish, miner, 35, single, face and eyes injured by explosion of a shot in room. He thought it misfired and went back to relight it, exploded just when he reached point of hole.
	September 26:	Morgan Powell, Welsh, pumper, 55, married, wrist broken by falling down and striking his arm against rail on main heading.
1913	June 2:	Tony Mash, Italian, miner, 26, single, leg broken by fall of slate at face of room while undermining coal.
	August 14:	John Stoaik, Russian, miner, 26, single, burned about head and face by explosion of powder.
	September 4:	Andy Korgan, Polish, miner, 26, married, leg broken in two places by being struck by wagons that he has neglected to sprag.
1914	March 3:	Joseph Melenski, Lithuanian, scraper, 42, married, leg broken, attempted to step on a moving mining machine on heading and fell and machine caught his leg.
	August 5:	Peter Buchoski, Slavonian, motorman, 22, single, leg broken, motor came in collision with another motor on heading
1915		John M. Smith, Scotch, spragger, 28, married, squeezed about hips by mine car on entry.

1916	February 17:	John Kalis, Slavonian, motorman, 23, single, scalp wounds, thrown against side of entry in jumping off runaway trip.
	May 9:	Andy Prebela, Austrian, pump repairman, 45, married, #3, hand lacerated while fixing packing on a pump while it was running. His hand was caught between plunger and packing gland.
	June 24:	Mike Lako, Slavonian, pick miner, 36, single, #4, thumb and fingers smashed, while removing block from under car, wheel passed over his hand.
	July 25:	Joseph Martin, Hungarian, brakeman, 26, married, #4, foot crushed by locomotive on entry.
	August 14:	William Bell, Scotch, machine runner, 31, married, #4, end of backbone injured, fell on rail in room.
	December 5:	Peter Hotchen, Hungarian, pick miner, 39, married, #3, clavicle fractured by fall of coal at face of room.
	December 16:	George Dibble, American, spragger, 18, single, #3, foot contused, caught between derailed motor and car on entry.
1917	January 30:	Barto Bolitza, Slavonian, pick miner, 40, married, #3, back severely bruised by fall of slate at face of room.
	February 28:	August Biangasto, Italian, machine runner, 30, single, hand lacerated and contused, caught in gears of mining machine on entry.
	July 17:	Tony Gordono, Italian, pick miner, 43, married, clavicle fractured, head and shoulder contused, thrown from wagon - outside.
1918	July 22:	Nicola Ripola, Italian, bratticeman, 48, married, leg cut off and scalp lacerated by locomotive on entry.

LACKAWANNA COAL & COKE COMPANY
Daily Time Report - Washery Outside, Maarch 9, 1923
NOTE: This report must be handed into the office before 6 p.m. each day.

Check No.	Name	Hours Worked	Description Of Work
	R. Stinson	9	Foreman
701	M. Urbanski	10	Table repair
702	J. McKeel	10	Crewman
703	Geo. McGinnis	9	Machine Picks
704	H. Brown	9	Laboratory
705	A. McKeel	9	Table repair
706	W. Stutzman		Off
708	J. McKeel	10	Crewman
709	J. Reynolds	10	Recovery Repair
710	J. Craig	10	Hoisting Engr.
712	C. McGinnis	10	Separators
713	A. Handlon	10	Dinkey Engr.
714	E.R. Helsel	4	Shipper
715	C. Zizok	10	?
716	J. Churilla	10	Tableman
717	M. Churilla	10	Elevators
718	R. Tulis	10	Refuse Dump Recovering Car Wreck
719	F. McCloskey	12	9x11 Breaker-Jacket Screen Repair
720	M. Brown		Off
721	J. Risko		Off
722	J. Dodd	9	Car Dropper
723	R. Wagner	14	Fireman Nights
724	D. Stutzman	9	Recovering Refuse Car Wreck
725	Geo. Marlit	10	Washed Coal Bins
726	Geo. Lesko	10	Tableman
727	Geo. Lesko	10	Pushing Coal
728	J. McGinnis	9	Oilman
729	P. Foyer	9	Recovering Refuse Car Wreck
730	H. Davis		Off
731	J. Krenotch	12	9x11 Breaker Jacket Screen Repair
732	W. Yohnut	9	Repairing ?
733	P. Hoffman	10	Dinkey Brakeman
734	L. McCloskey	10	Fireman (Day)
735	S. Patterson		Off
736	P. Maholley	18	10 Hours Refuse Dump 8 hrs. Wed Night Pushing Coal
737	J. Kinesak	12	Jacket Screen Repair
738	H. Condi	10	Clean Up Coal and Refuse Dump
740	W. Dill		Off
741	A. Yohnert	10	Recovering Refuse Car Wreck

Source: Wehrum Reunion Committee Collection

BETHLEHEM MINES CORPORATION: POLICE DEPARTMENT REPORTS, 1926-1929

JOHNSTOWN DIVISION: Headquarters: Number of Officers: 4 Wehrum, .68 sq. mi, 1650 population; Slickville, .29 sq. mi.

Patrolmen: Wehrum, 1928

E.Z. Adams, seniority 11/1/25; H.E. Fegard, seniority 11/1/25;

1929: Adams and Fegard resigned 5/28/29, Patrolmen Fuchick and Thompson were their replacements. Superintendent of Police, O.R. Check, and Captain of Police, L.W. Kranz, were also on duty along with Patrolmen Galbreath and Bryant due to large amount of money in the bank for withdrawls. They remained until June 5; Bryant was on the job until June 12. M.J. Dando, was reassigned from Ellsworth July 6, 1929, seniority 7/1/29. He was removed on January 16, 1930 and reassigned to Slickville. On January 15, 1933, he was laid off following the closing of the Slickville mine in July, 1932.

Items to be included in the daily report:

Safety devices, tresspass signs, safety and danger signs, powder magazines, conditions of stables, sanitary and welfare conditions and observe for vice celebrations and any such conditions as might interest plant superintendent will be reported to superintendent by the superintendent of police.

Duties to be observed:

Agitation, assemblages, arrests, abuse of safety devices, abuse of machinery and property, boards with nails protruding, boards off fences, brass, copper, etc., unprotected and exposed to theft.

Arrests	1926	1927	1928	1928	1930
assault and battery	2	2	1		
misdemeanors			9		
felonies		1	1		
disorderly conduct	21	21	1		
larceny	2	4	1		
motor vehicle violation	1	2			
malicious mischief		3			
resisting arrest	1				
surety of the peace	1				
tresspass	2	4			
violation of liquor laws	1	2			
drunk and disorderly			3		
defrauding boarding house keeper			3		
employees			5		
non-employees			5		
convictions			7		
discharge			1		
fines collected			$38.00		
costs			$13.80		
store bills collected			35.00	$102.26	
fraudulent checks			1		
violation of school law	0		1		
Exclusions:					
labor agents	3				
peddlers	159		240	278	
union agitators			45	2	
undesirables			315	120	
Special investigations	3	3			
Patrol dances, picnics, athletic events		3			
Fights and disturbances quelled	6	119	302	67	
First aid and conveyed to hospital		3			
Eviction notices	2	48	27	13	
Assisted at fires	3	5	5	8	
Employees not working, checked up				1260388	
Complaints investigated			84	54	
Broken doors and windows reported	57	18	106		

Calls for ambulance and physician		101		12244
Defective equipment reported	7	2		1
Employees escorted-labor	7			
Arrivals seeking employment, referred to super.	275	648		237
Lights left burning-turned out	127	212		240
Labor assemblages and meetings patroled		9		
Leases checked	18	51	42	28
Rents collected		$106	$204	
Lost and missing persons found	2	2		1
Mad and/or maimed animals killed	3	6		1721
Necessary lights reported out	204	116		117
Settling disputes between boarders and bosses	4			
Fire Hazards and dangerous places reported to the superintendent	2437			
Unsanitary conditions reported	2	22	3	6
Liquor destroyed-Wehrum and Slickville stills	3			
Gallons of moonshine	42		104	34
quarts of wine	172			
Miles traveled				
Mounted	2859	1464		
Dismounted	8359	5760		
Vehicle	843	1440		
Trains, street cars, buses met	720	718		
Persons rescued	1			
Quarantine card posted	11	59		21
Stealing coal and wood-apprehensions		31		319
Union literature destroyed	342	52		21
Visitors refused admission without passes	405	402		202
Visitors admitted with passes	603	264		288

Stolen property of employees recovered:

1928: $82.00, 1 bicycle, 1 watch, 3 pairs of shoes, 2 suits, 1 box tools, 1 ladder, 1 trunk, 1 sweeper, 1 electric heater, 5 wall brackets, 1 spool Bell wire.

1929: 1 tire, 1 robe, 1 garden hose, 1 wheelbarrow, 1 hatchet, 1 boiler, 1 porch swing, 1 dinner pail, 2 pair pants, 1 window shade, 1 shirt, 2 pairs shoes, 1 suit clothes, 2 trunks, 2 suitcases, $25 cash.

Stolen Property of company recovered:

1928: 2 torches, 1 safety lamp, 8 picks, 1 roll copper wire, 11 pit wheels, 18 electric sockets, 5 window panes.

1929: $25.43, cigars, cigarettes, 186 pounds copper wire, 2 gallons paint, 1 pack lampblack, 6 pick handles, 2 watches.

Source: Hellman Library University of Pittsburgh Industrial Archives

APPENDIX G
GRACETON COAL AND COKE COMPANY: ANNUAL REPORT

Year	Mine	Superintendent	Tonnage	Coke Made	Days Worked	Employees
1921	#2 & #3 202 ovens, 152 operating		210,703	31,703	229 230	80 220
1922	#2 & #3 202 ovens operating		200,635	50,718	289 290	89 209
1923	#2 & #3 195 ovens operating		187,264	71,045	237 237	81 222
1924	#2 & #3 199 ovens operating		161,205	66,510	38 231	84 249
1925	#2 & #3 198 ovens operating	M.F. Brandon	Idle 176,395	71,119	244	264
1926	#2 #3 199 ovens operating	M.F. Brandon	1,827 190,507	75,511	34 265	19 271
1927	#2 #3 134 ovens operating	M.F. Brandon	24,262 169,453	60,401	254 254	30 233
1928	#2 #3 134 ovens operating	M.F. Brandon	33,445 123,355	42,168	182 182	36 227
1929	#2 & #3 110 ovens operating	M.F. Brandon	1125,742	33,132	147 147	36 184
1930	#2 #3 70 ovens operating	Homer Robertson	28,943 184,653	21,264	240 240	40 250
1931	#2 #3	"	25,718 129,701	10,217	147 147	38 210
1932	#2 #3	"	17,286 103,481	3,913	134 134	27 203

1933	#2	Homer Robertson	7,163	None	50	29
	#3		115,675	Listed	135	210
1934	#3		158,047	5,502	187	245
1935	#3		182,435	4,311	214	262
1936	#3		56,169	1,618	69	213

Went into receivership, 5/29/1936, Became Coal Mining Company of Graceton.

APPENDIX H
COMMERCIAL COAL MINING COMPANY

ANNUAL REPORT

Year	Mine	Superintendent	Tonnage	Days Worked	Employees
1904	#4	New mine, began shipments of coal in December			
1905					
1906					
1907	#4	William Smith, General Superintendent, Expedit Herman Oldham, Superintendent	171,871	272	198
1908	#4	William Smith Herbert Oldham	145,245	213	234
1909	#4	William Smith Herbert Oldham	106,230	200	215
1910	#4	William Smith John Hammond	129,604	219	174
1911	#4	William Smith	15,495	26	132
1912	#4	"	87,427	197	170
1913	#4	"	Combined with commercials #3		
1914	#4	"	"		
1915	#4	"	73,302	147	102
1916	#4	"	73,829	131	142
1917	#4	"	87,830	144	142
1918					

1919-1924, Production Listed With Commercial #3.
1925-1928, #3 Listed as Idle, No mention of #4
1929, Commercial #3 no longer listed in the Department of Mines records.

MINE FATALITIES: COMMERCIAL COAL COMPANY: BRACKEN

1905	June 8:	Joseph Lovish, Hungarian, miner, killed instantly by coal falling on him, began work early in morning on the 8th, engaged in completing undermining across the room. Most of the mining done on previous day, and two shots had been fired which failed to bring down coal. 7 AM coal fell on both men, killing Lovish and seriously injuring partner, Frank Nodge. No attempt to comply with rule 14, to protect themselves by setting sprags.
1908	February 22:	Joe Vassanilles, Italian, miner, 34, married, three children, instantly killed by fall of coal, had been loosened by shot, did not fall, set sprag under it after blasting when he went under it to mine farther in, coal fell over the sprag.
1910	January 26:	Joseph Bokar, Polish, miner, 47, married, two children, fatally injured by fall of draw slate while taking out pillar in his room.
1915	April 21:	Joseph Caprillo, Italian, motorman, 28, married, four children, killed by fall of slate while traveling on entry.

MINE ACCIDENTS: COMMERCIAL COAL COMPANY: BRACKEN

1907	January 4:	Julian Feddock, Slavonian, miner, 23, single, head bruised by fall of coal, failed to set sprag.
	January 4:	James Lower, American, car coupler, 16, single, foot crushed by cars outside mine.
	August 8:	Mike Buncle, Slavonian, miner, 32, single, foot sprained by fall of draw slate.
	August 9:	John Roguo, Slavonian, miner, 45, single, leg broken by fall of coal.
	September 28:	Mike Tintosh, Slavonian, driver, 35, married, right leg hurt by being caught between two cars.
1909	August 2:	John Burnot, Slavonian, tippleman, 29, single, leg broken by cars - outside.
1910	August 20:	Simon Bardosky, Russian, miner, 19, single, collar bone broken by fall of coal while undermining at face of room.
	October 18:	Joseph Valina, Italian, miner, 32, single, spine injured by fall of coal while undermining with a loose end at face of his room.
	December 22:	Albert Pidolski, Austrian, miner, 49, married, collarbone broken by fall of coal while undermining at face of his room.
1912	May 9:	Mike Choyka, Slavonian, miner, 35, married, hand and fingers bruised by cars on headings.
	September 10:	Ed Butterbaugh, American, driver, 21, married, compound fracture of leg by cars on heading.
1913	January 31:	Albert Podolski, Slavonian, miner, 52, married, pelvis bone broken by fall of roof while pushing car out of room. Roof fell on his back.
	May 2:	Samuel Keith, company man, 35, married, jawbone broken by cars while riding in empty trip down slope.
	November 8:	Frank Gavasky, Slavonian, laborer, compound fracture of one leg and other broken, struck by rope on entry.
1914	November 23:	Carlo Anton, Italian, miner, 38, married, leg broken by fall of drawn slate while cleaning up a fall in room.
1915	January 18:	George Bellock, Slavonian, loader, 42, married, leg broken by fall of slate at face of room.
	April 9:	Steve Grantz, Slavonian, brakeman, 24, single, contusion of knee and foot, gear case became loose, throwing end of locomotive off track, caught between locomotive and side of entry.
1916	August 30:	Valent Pingsaw, Slavonian, pick miner, 23, single, hand and eye burned, went back to a shot that he supposed had misfired.
	September 19:	Joe Wotonsky, Russian, pick miner, 22, single, contusion of limbs and back, sprained by fall of coal at face of entry.
1917	July 12:	Lewis Kovash, Hungarian, machine runner, 35, married, compound fracture of leg, caught between cars on entry.
1918	August 19:	Premo Decarlo, Italian, driver, 20, single, sternal end of clavicle fractured, slipped and fell while removing harness from mule - outside.

APPENDIX I
MINE EXAMINATIONS

1907: Mine Foreman, First Grade
Robert Jones, Vintondale
Mine Foreman, Second Grade
John Tobin, Wehrum
A.L. Johnston, Bernice (Son of Wehrum Supt. Ward Johnson)
Arthur Davis, Vintondale

1908: Mine Foreman, Second Grade
S.N. Hewlett, Wehrum

1909: Mine Foreman, Second Grade
Thomas Hogarth, Wehrum

1910: Mine Foreman, First Grade
Thomas Hogarth, Wehrum
William Young, Wehrum
Fireboss
Thomas Hogarth

1911: Mine Foreman, First Grade
James Parks, Wehrum

1912: Mine Foreman, First Grade
Charles Adamson, Wehrum
Mine Foreman, Second Grade
John Maginnis, Vintondale
Fire Boss
John Maginnis, Vintondale
John H. Huth, Vintondale

1913: Mine Foreman, Second Grade
John Daly, Vintondale
Joseph Hotter, Wehrum
Assistant Mine Foreman
Thomas Kavanaugh, Vintondale
Fire Boss
Jesse James Miller

1914: Mine Foreman
William Madill, Wehrum

1915: Assistant Mine Foreman
Jesse James Miller

1917: Mine Foreman, First Grade
Milton Brandon, Homer City
Gordon Francis, Wehrum
Assistant Mine Foreman
Charles Wilton, Wehrum

1918: Mine Foreman, First Grade
William Francis Powell, Wehrum
Joseph Woodward, Wehrum
Alfred Andrew Engell, Bracken

1922:   Mine Foreman
         Robert Magee, Vintondale
         Tom Madigan, Wehrum
         Tom Buckingham, Wehrum
         W.H. McCracken, Vintondale
         George Cooke, Vintondale
         J.M. Waldren, Wehrum
         W.G. Gongloff, Vintondale
         W.E. Hunter, Vintondale
         Tom Whinnie, Vintondale

APPENDIX J
PROSECUTIONS FOR VIOLATIONS OF MINING LAWS

1909	December 23:	Andy Mudge, Hungarian miner pleaded guilty to charge of passing a danger board in Vinton Colliery Company's #6 mine before fire boss had returned to make report. Bound over in $200 for his appearance at court.
1913	January 22:	Peter Rebinsky, miner, Vinton Colliery Company's #6 mine, pleaded guilty before court at Ebensburg to passing a danger board. Ninety days in jail.
	February 14:	Mike Babich, pleaded guilty before Ebensburg Court to passing danger board in Vinton Colliery Company's #6 mine, fined $60 and costs. Charley Dyock - same charge.
	December 31:	Alexander Rube charged with failure to tamp full length of hole and using short fuse at Lackawanna Coal and Coke's #4 mine. Pleaded guilty, fined $100 and costs and thirty days in jail. Joseph Warmus - same charge. Lewis Busick - same charge, warrant issued, but defendant so far has avoided arrest.
1914	August 25:	M.J. Mulvehill, mine foreman, Lackawanna Coal and Coke, charged with failure to have a clear space of two and one half feet between side of car and rib on certain headings, pleaded guilty on August 31, costs and $25 fine.
1917	July 9:	William Blacklock, pumpman in Vinton Colliery Company's #6 mine, charged with taking smoker's articles into a section of mine worked exclusively with locked safety lamps, pleaded guilty, fine of $10 and costs or three months in jail.
1918	January 2:	Frank Sape, miner, Commercial Coal Company's #4 mine, charged with taking matches in section worked exclusively with locked safety lamps, pleaded guilty. $10 fine and costs. Jack Balick - same charge.

*Source: **Annual Reports**, Pennsylvania Department of Mines.*

APPENDIX K
COAL MINERS' POCKETBOOK

Principles, Rules, Formulas and Tables

THE BUYERS' MANUAL
1. Link-Belt Company: Philadelphia, Chicago, Indianapolis

"We design as well as build complete machinery for the handling and preparation of coal at the mine. Submit your problems to us for solution. We make no charge for service, layouts or estimates. Catalogs on request."
Link-Belt Machinery for the handling and preparation of coal at the mine includes: tipples, car hauls, retarding conveyors, coal washeries, coal driers, etc." (continues for two paragraphs)
2. Sirocco (trademark) Multi-Blade Mine Ventilating Fans American Blower Company, Detroit, Michigan
3. Goodman Coal Mining Machines, Many types, breast machines, straight-face machines, longwall machines, direct or alternating current, compressed air. Electric Mine Locomotives, "A Locomotive for Every Mine Service"
Goodman Manufacturing: Chicago, IL. NY PGH, Cincinnati Charleston, WVA, Birmingham, St. Louis, Denver, Seattle

AUTHOR'S NOTE: These manufacturers sold their products to the Vinton Colliery Company as can be seen in the mine superintendent diaries.

Source: Coal Miners' Pocketbook 11th edition, Revised, Enlarged, and Entirely Reset. McGraw-Hill Book Company, Inc., 239 West 39th Street, New York, 1916.

APPENDIX L
EVICTIONS

Vinton Colliery Company

Vinton Colliery Vs:	House Number or Address
1906: Cambria County Continuance Docket, Book 55, June Term	
Joe Statinchock	House #9
1906: December Term	
William Gardner	House #25
1909: Continuance Docket, Book 58, June Term	
Angelo Betelli	House #119
John Gillingham	House #145 and #135
Shandor Nagy	House #158
John Upside	House #129
Edward Brown	House #2
Frank Leaninatti	House #151
Mike Cefro	House #63
John Chirkola	House #66
David Duncan	House #94
Samuel Duncan	House #3
Frank Fankovish	House #96
Mike Frasfolution	House #68
Joe Risko	House #117
John Robosky	House #153
John Valosin	House #69
Adam Budvesel	House #154
William Chambers	House #31
Samuel Rager	House #106
David Watkins	House #16
George Watkins	House #30
Adam Yonsie	House #161
Charles McCloskey	House #15
Andy Kochlany	House #113
Alex Nodge	House #114
Edward Penrose	House #88
Joe Tomanek	House #59
Louis Walpelo	House #152, Sixth Street
2/3/1910	
Frank Harraghe	
2/3/1911	
Irvin Williams	House #122
8/17/1911	
W.R. Patterson	House #123
10/17/1916: Continuance Docket, Book 66, pp. 663-670, 674-677	
Joe Bianetta	#294 Chickaree
Tony Celini	House #40
Domenico Coletti	#154 Sixth
Peter de Bagnno	House #38
Louis Glass	#95 Barker
Tony Martinelli	#140 Fourth Street
Louis Nemes	#161 1/2 Sixth Street
10/18/1916	
Peter Krechknosky	#281 Sixth Street
Lamberto Larse	#158 Sixth Street
Charles Rubis	#152 Sixth Street

John Rubis	#165 Plank Road
Frank Coballi	#321 Chickaree
John Kerekish	#111 Barker
Cesar Mazza	#327 Chickaree
Joe Toth	#35 Plank Road

10/20/1916

George Balko	#118 Barker
Ernest Beraturi	#250 Plank Road
John Bosso	#136
Mike Jelor	#109 Barker
George Lukas	#133 Fourth Street
Louis Malpeli	#284 Fourth Street
John Rocosi	#37 Plank Road
Pete Rocosi	#250 Plank Road
Edward Regis	#60
Frank Slavic	#135 Fourth Street

1924: Continuance Docket, Book 74, pp. 290-457, Amicable Action of Eviction.

3/19/24

Edward Berkey	#94 1/2 Main Street
Edwin Carlson	#92 Main Street
Willard Carlson	#25
Ralph Carlson	#84
Ivan Clouser	#2
W.R. Hunter	#207, E.
Charles Lynch	#207, B.
J.F. McGuire	#21
John Peterson	#89
Dewey Rairigh	#165
Lawrence Simmons	#321
Otto Yahnke	#125

3/28/1924

Balazi Bajur	#84
Robert Conlin	#93
W.J. Carlson	#351
Joe Gasko	#111
Louis Griboczky	#72
John George	#297
Joe George	#348
Richard Eisais	#95

4/5/1924

John Durczak	#349
Joe Beich	#135
Paul Jekovich	#134
Andrew Novitak	
Mike Doncevic	#140
Steve Vargo	#301

4/23/1924

Gust Degand	#332 Main Street
Dan Evans	#105 Main Street
John Kerchinski	#283 Fourth Street
George Kurtenecz	#167 Plank Road
John Morton	#143 Third Street
W.H. McCracken, Jr.	#335 Narrow Gauge Street
Vincenty Pastula	#69 Goat Hill

4/30/1924

W.H. Brubaker	#149
George Banyaz	#36
George Draga	#287

Oscar Dishong	#3
Milton Gongloff	#108
Grant Gongloff	#106
Joseph Lewis	#340
Chesoili Lukoicnino	#304
Mrs. M.B. Mitchell	#122
Colis Mazioruiti	#330
Adam Matyovie	#257
Thomas Maly	#159
William Mazey	#124
Bevfalon Orban	#276
Johan Pipsak	#278
Petar Page	#155
Joe Palko	
Mike Spontak	#163
J. Sadowski	#128
Gjuro Susnjar	#151
George Reed	#168
Josef Tsltonski	#274
Joe Tasnady	#6
Tom Wiesznar	#295
Charles Zuba	#63

5/2/1924

George H. Cooke	#337

7/2/1924

John Sucry

VINTON COLLIERY COMPANY
CLAGHORN, WEST WHEATFIELD TOWNSHIP

Ejection Docket #2 Indiana County, June 1922

James George	House 32 Main Street
Andrew Swanson	House 12 Main 1Street
Charles Gust	House 14 Main Street
Frank Zerin	House 46 Main Street
Mike Morin	House & Lot 13 Main Street
Joe Stepko	House & Lot 23 2nd Street
Steve Fitz	House & Lot 20 2nd Street
John Ronges	House & Lot 42 Main Street
Mike Pavalko	House & Lot 17 2nd Street
Mike Murray	House & Lot 84 At #16 Main

GRACETON COAL AND COKE COMPANY, GRACETON

Ejection Docket #2, Indiana County, June 1922

John Rush	House & Lot 53
John Stills	House & Lot 99
John Kolessar	House & Lot 100
Max Olshofsky	House & Lot 71
John Pavlik	House & Lot 22
Michael Smandra	House & Lot 95
John Frank	House & Lot 103
Frank Olshofsky	House & Lot 59
Frank Pavlik	House & Lot 57
Andy Zedik	House & Lot 34
Dan Ohar	House & Lot 47
Mike Ohar	House & Lot 65
Gusti Rura	House & Lot 51
Tony Morini	House & Lot 40

Mike Duffalo	House & Lot Shanty B
John Herbik	House & Lot 48
Den Brody	House & Lot 52
John Kostak	House & Lot 45
Louis Spiaggi	House & Lot 87
George Stelmach	House & Lot 50
John Vresilovic	House & Lot 97
John Pilrvic	House & Lot 91
John Kundla	House & Lot Shanty A
Steve Knapik	House & Lot 46
John Kaletta	House & Lot 80
John Scrociak	House & Lot 49
William Uswa	House & Lot 89
John Stoklosa	House & Lot 38
Steve Smolis	House & Lot 67
Joe Hercheck	House & Lot 90
Andy Fertal	House & Lot 24
Ben Smerk	House & Lot 62
Charles Gutt	House & Lot 44
Joseph Konkoly	House & Lot 44
September, 1922	
Steve Luchis	House & Lot 131
Filip Siska	House & Lot 63
Max Veslovsky	House & Lot 70
John Varga	House & Lot 101
George Bjalko	House & Lot 94
Vincent Rabatin	House & Lot Shanty E
Steve Kuzna	
June, 1924	
M.M. Templeton	House & Lot 81

LACKAWANNA COAL AND COKE COMPANY

September, 1922

Name	House and Lot
Tony Gastega	31
Peter DeBagno	215
Premo Delegram	3
John Chelosky	1
John Chelosky, Jr.	5
Enrico Palermini	32
Charles Kozar	22
Joe Sanzo	24
Jacob Mirus	140
John Ponzng	266
Joe Gigliotti	162
Mike Lemick	258
Charles Olessio	208
Arthur Brutout	141
Joe Martin	18
John Chamberlain	146
Sam Roberts	173
Peter Donorcy	223
Frank Dicenzo	43
Peter Cowan	230
Chas Deyok	210
Frank Pondoli	243
George Babish	180
John Kmecak	177

Appendix M
Receipts from the 1909 Vintondale Strike

PAYORS: John King and Richard Grego, Organizers for the United Miner Workers

DATE	PAYEE	AMOUNT	REASON
May 10	Joe Tomonik	4.00	Rent for May
May 11	Albert Chyrus	4.00	Rent for May
May 12	Mike Farkas	80.00	Grocery Order
	S.R. Williams	100.00	Subject to orders
	David Nevy	70.00	For merchanidse
May 13	Jon Risko	8.50	House Rent
May 19	S.R. Williams	123.00	Subject to orders
	Mike Farkas	116.66	Grocery order
	David Nevy	50.00	
	Mike Farkas	6.00	Grocery order
	John Evans (?)	10.00	Death (?) of child
May 26	John Wozniak	20.00	For Merchandise
	Horvath & Co.	10.00	For Merchandise
	J.H. Krumbine	50.00	To be taken out in store goods
	Mike Farkas	50.00	Groceries
	S.R. Williams	125.00	Subject to orders
	David Nevy	40.00	Merchandise
May 28	S.R. Williams	1.60	Moving
	Mike Farkas	3.50	Hauling
May 29	David Nevy	.50	(?) house goods
May 31	L.H. Davis/Vinton Lumber Company	40.00	2 leases of J. Carney due May 5, 3 leases of Dominick Jelotti due May 29, 5 houses one month in advance
June 1	H.E. Miller	6.50	Rent for house in Rexis for James McCloskey for June
	B.M. Daughenbaugh	8.87	Fine & costs on case of Robert Bianchintis
June 2	W.R. Treaster	2.50	Moving
	George Kerr	6.00	Moving
June 3	S.R. Williams	100.00	Subject to orders
	Mike Farkas	70.00	Grocery order
	J.H. Krumbine	50.00	To be taken out in store goods
	John Wozniak	14.00	
	Horvath & Co.	7.00	Merchandise
	David Nevy	70.00	Merchandise
June 4	Vinton Lumber Co.	63.00	Rent as per list
	Jos. Dominak	4.00	Rent for June
	Emma Shaffer	10.00	Rent for Storeroom June 5-July 5, 1909
	S.R. Williams	6.00	Rent for June, House 190
June 7	John Risko X his mark	10.00	House rent in advance from June 5

Date	Name	Amount	Description
June 8	Jon Silik	4.00	Rent June to July
	Albert Cyrus	4.00	House for June
June 9	S.S. Kinkead	24.63	Costs: J.I. Thomas
		29.03	Costs: W.J. McGough
		11.95	Ex Bill: W.J. McGough
June 10	S.R. Williams	100.00	Subject to orders
	Mike Farkas	50.00	Grocery order
	David Nevy	100.00	Groceries
	Horvath & Co.	9.50	Order
	Mike Farkas	4.00	Grocery order
	S.S. Kinkead, Clk by DeLancey	25.48	Coweth (?) vs. Shendor Kuihas (?)
	Dr. Comerer	25.00	Excising bullet & reducing fracture
No Date	J.H. Krumbine	40.00	To be taken out in store goods
June 16	For Andy Jacobs	7.00	Rent of house in Vintondale
	Frank (X) Korhahik		1 month beginning June 4
June 18	S.R. Williams	119.00	Subject to orders
	Mike Farkas	80.00	Grocery order
	John Wozniak	20.00	
	Horvath & Co.	5.00	Merchandise
	David Nevy	133.00	Groceries
	J.H. Krumbine	50.00	To be bought out in goods
	Mike Farkas	2.00	Groceries
June 25	David Nevy	100.00	Sub. to order
	S.R. Williams	113.50	Subject to orders
	George Kerr	2.50	Moving
June 28	Lorenzo Cavalli	6.30	Freight charges of household goods from Vintondale to Van Horner
June 29	Doughty & Graham	2.44	Pads & books
June 30	H.E. Miller	6.50	House in Rexis for July
July 1	David Nevy	50.00	Per Merchandise
	J.H. Krumbine	20.00	For merchandise
	John Wozniak	10.00	
	Mike Farkas	20.00	Grocery order
	S.R. Williams	117.50	Subject to orders
	Horvath & Co.	6.00	For order
	Horvath & Co.	3.50	Rent of (?)
July 2	Dr. Prideaux Twin Rocks	5.00	Services rendered the man shot in Vintondale
July 5	Emma Shaffer	10.00	Small storeroom in Advance
July 6	Jon Silik	4.00	Rent in advance for July
July 7	John Risko	10.00	Rent July 5-Aug. 5
July 8	Frank Farkvec	7.00	House Rent for July in advance
	Geo. Tomaneck	4.00	House Rent for July in advance
	Vinton Lumber Co.	117.65	Rent as per list
July 9	S.R. Williams	6.00	Rent for House 190 for july

July 14	David Nevy	50.50	For merchandise
	J.H. Krumbine	8.00	For merchandise
	John Wozniak	15.00	For merchandise
	Mike Farkas	22.00	Grocery order
	S.R. Williams	118.00	Subject to orders
	Horvath & Co.	8.50	Towing (?)
	S.R. Williams	4.00	Water (?) for May & June for Joseph
July 22	J.H. Krumbine	23.50	For union orders issued July 14
	John Wozniak	17.00	Merchandise
	Mike Farkas	23.00	Grocery order
	S.R. Williams	87.00	Subject to orders
	Horvath & Co.	2.00	Towing (?)
	David Nevy	33.00	Groceries
No Date	S.S. Kinkead, Clk. by DeLancy	35.08	Costs in case of Coweith vs John Storick
No Date	George Kett	2.50	Moving

Source: District 2, United Miner Workers of America Papers, 1909 Strike Receipts, Box 8, Folders 345, 346 and 347, Special Collections, Stapleton Library, Indiana University of Pennsylvania, Indiana, PA.

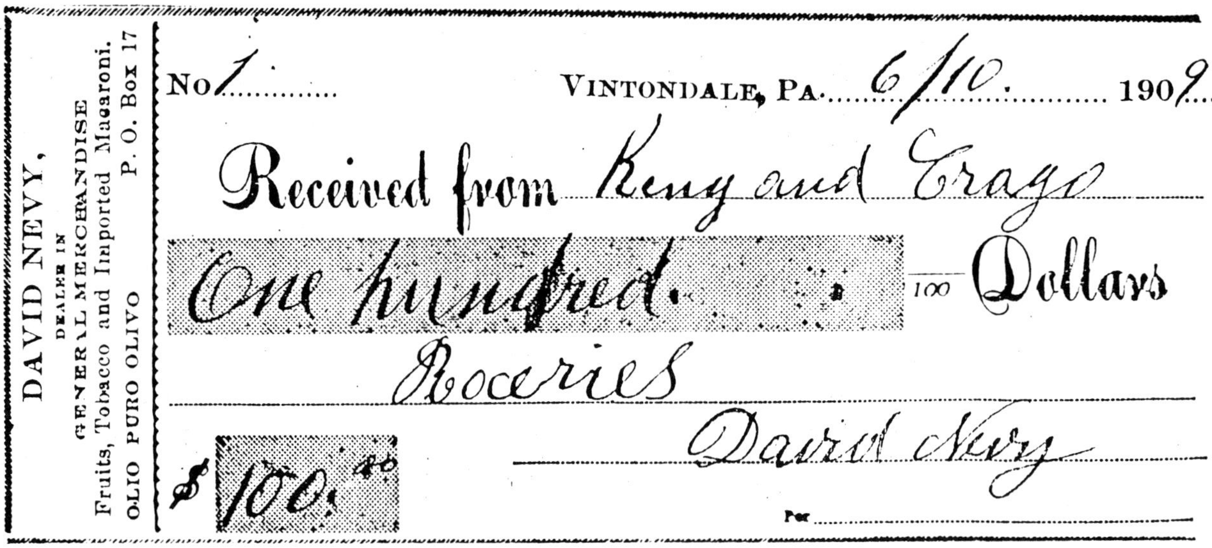

No 1   VINTONDALE, PA. 6/10. 1909

Received from Keny and Crago

One hundred ———————— 100 Dollars

Groceries

David Nevy

$ 100.⁴⁰

DAVID NEVY,
DEALER IN
GENERAL MERCHANDISE
Fruits, Tobacco and Imported Macaroni.
OLIO PURO OLIVO   P. O. Box 17

Per _____

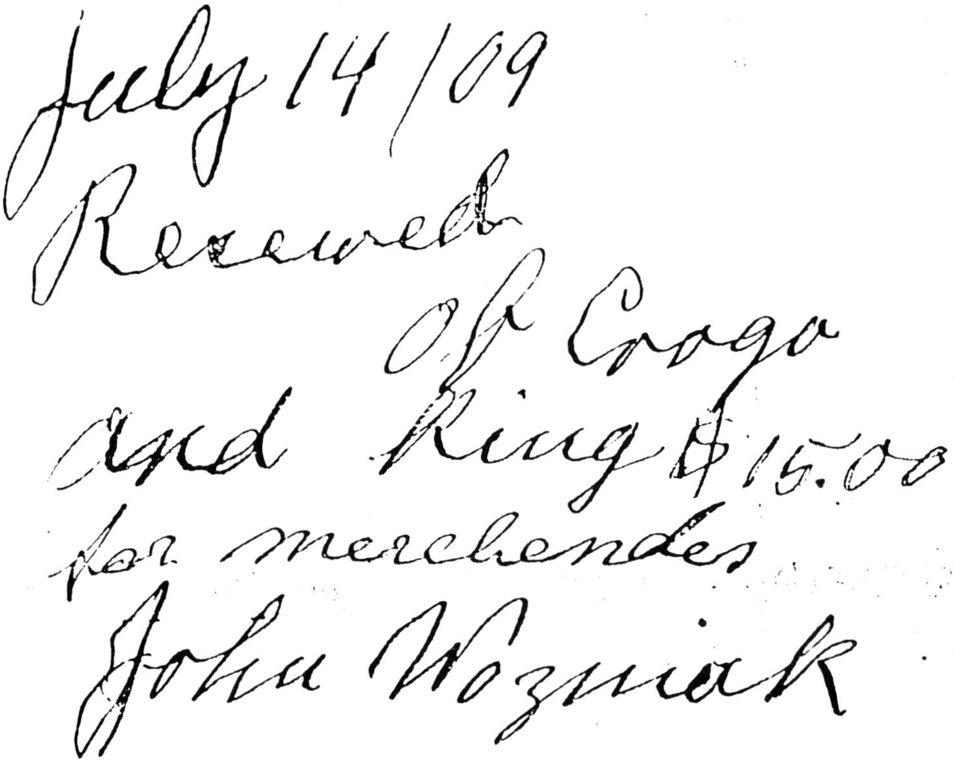

July 14/09

Resewed

of Crago
and King $15.00
for merchendes

John Wozniak

LEF, PRESIDENT   JNO. W. WRIGLEY, SEC'Y AND TREAS.   GEO. H. GEARHART, SUPT.

VINTON LUMBER CO. Limited
MANUFACTURERS OF
ROUGH AND DRESSED LUMBER
BILLS CUT TO ORDER

TELEGRAPH OFFICE   WESTERN UNION, JOHNSTOWN   POSTAL, VINTONDALE   HEMLOCK & HARDWOODS   BELL TELEPHONE

REXIS, PA., June 4/09.

Paid to.

Date	Name	House #	Amount	Total
July 2nd.	Duncan & Chambers,	House #22.	$15.00	
5th	John Evans,	18	8.00	
5th	Charles McCloskey,	15	8.00	
3rd	Walter Rager,	16	8.00	
3rd	John Lup,	3	8.00	
A/C	E. E. Thompson,	17	8.00	
5th	Edw. Brown,	1	8.00	$63.00

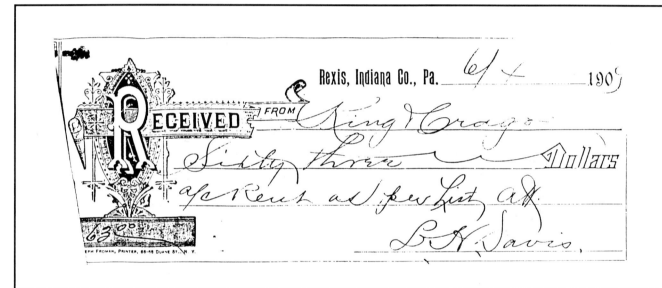

Rexis, Indiana Co., Pa. 6/4 1909

Received from King & Crago Sixty three Dollars a/c Rent as per List att.

L. H. Davis

$63.00

July 5th 1909

Received of King & Crago Ten Dollars for rent for small store room in advance.

$10.00

Emma B. Shaffer

DOUTY & GRAHAM
...DEALERS IN...
FRESH AND SALT MEATS, EGGS, BUTTER AND CHEESE.

VINTONDALE, PA. *June 29 1909*

Received of King Crago $2.00 for pads books

Douty Graham

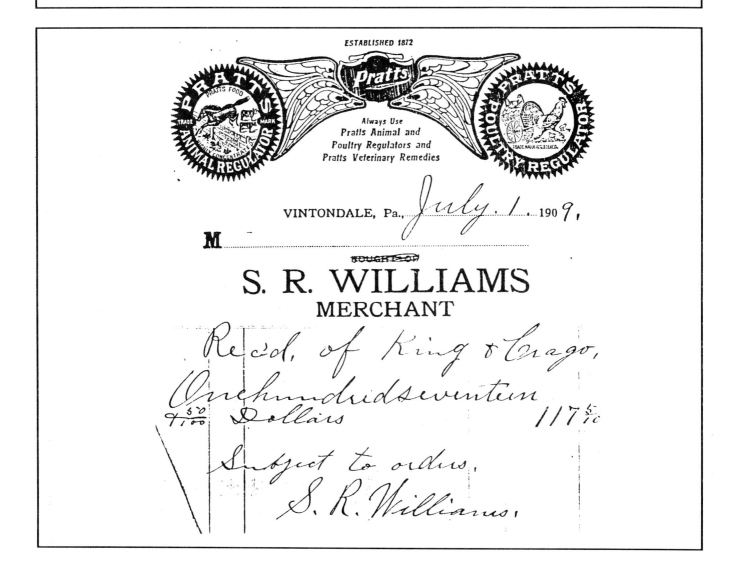

June 7 /09

Receive from John Kink. $10.00
Ten dollars for house rent from
June 5th in advance

John Resko X his mark

WILLIAM A. PRIDEAUX, M. D.

Received from June 13 /09   King Crige 19—

seventy   00 Dollars

Grocers order

$70.00   Mike Parker

STATEMENT.

DR. J. ALVIN COMERER,
VINTONDALE, PA.

June 10, 1909

To M. N. M. W. of A.
for care of J. K. King & Cragg

No. 77,703.

For Medical Services		
(visits)		
" Surgical Services	25	00
" Obstetrical "		
For excising bullet		
& reducing fracture	"	"
of right radius, for	"	"

Received Payment,

J. A. Comerer, M.D.

Expedit, Pa. July 7, 1909

Richard Cragg Dr. to W. J. Prideaux $5.00 for services rendered the man shot in Vintondale

Rec'd Payment five dollars

W. J. Prideaux

S. R. WILLIAMS,
...Dealer in...
General Merchandise, Dry Goods, Groceries,
FLOUR AND FEED, ETC.

Vintondale, Pa., May 28, 1909,

Rec'd of King & Crago,
for moving household goods
One + 60/100 Dollars $1 60/100
                           S. R. Williams,

VINTONDALE, Pa., July 9, 1909,
M
BOUGHT OF
S. R. WILLIAMS
MERCHANT

Rec'd of King & Crago
Six Dollars, Rent
for House No. 190
for Month of July, 1909.
"$6 xx"
S. R. Williams.

July 14 — 1909.
Received from King and Crago
Eight — — — Dollars
                           100
For Merchandise.
$ 8.00   John H. Krumbine

SOURCE: United Mine Workers of America District 2 Files, Special Collections, Stapleton Library Indiana University of Pennsylvania, Indiana, PA.

Appendix N
Fires in Vintondale

1900	May	Serious forest fires on north branch of the Blacklick Creek.
1903	May	"Ugly" forest fires raging close to town limits. Vinton Lumber Company mill threatened. Altoona fire department brought on special train to help fight the flames.
1914	February 2	Small fire at #9 house, extinguished with chemicals by firemen. Roof burned off #33 & #34 house at 7:30 PM.
1914	March 5	Mrs. Tot's barnm on Plank Road burned to the ground at 7 PM. Firemen were able to save the neighbor's barn, which was only two feet away.
1914	July 17	Flue fire at 1 PM in house #11, owned by Carl Salada.
1914	August 6	Sam Brett's chimney struck by lightning, igniting a small blaze, extinguished with buckets of water.
1914	August 8	Wehrum washer burned. Vintondale sent its hose and twelve men to assist.
1914	September 15	#6 hill caught on fire from a spark off the 12:38 train; Meisner and crew sent out to fight it.
1914	September 16	Meisner and crew still fighting forest fire.
1914	October 21	Another fire on #6, Meisner and crew went to prevent it from spreading to oak tract.
1914	November 12	Mr. Graff of PRR refused to pay bill for fighting #6 fire, claimed it was a C&I fire.
1914	November 27	Forest fire at 7 PM on #1 hill, started by kids or hunters. Burned 20 acres before being extinguished by Dempsey, Cresswell, Meisner, Huth, Gordon, Swanson, Hoffman and Morgert.
1914	November 28	Huth went over burned area of #1 hill to make sure the fire was out.
1917	April 17	Small forest fire behind upper row of houses on Chickaree.
1917	May 3	Fire department to boiler house, corner burning, caught on fire from a spark from the dinkey.
1917	September 8	Fire department called out to extinguish a mattress fire at John Hitler house on Plank Road.
1923	May 28	Three railroad men, six of Meisner's surface crew and Boy Scouts fighting forest fires behind #6.
1923	May 29	Crews out fighting forest fire, was not extinguished properly the first time.
1923	May 30	Ludwig, District Forester called. Crew of four men out all night on forest fire. Back at 7 AM.
1923	December	Old tipple in Bracken reported on fire December 7 about 9:00 PM, two laborers, three foremen and 1 policeman sent, returned about 11:00 to report old sand house and part of tipple burned. Kids built fire on dinkey on Sunday, December 9th, police were sent after them.
1926	May	Forest fire on Chickaree, fought by Scouts and Nicola Construction.
1930	January 1	The Hungarian Reformed Church was a total loss as a result of a fire which began at 10 PM.
1930	October 19	Four room brick schoolhouse built in 1917 destroyed.

Appendix O
Privately Owned Lots in Vintondale, Sold by the Blacklick Land and Improvement Company, 1894-1900

Lot Number (s)	Block Number	Purchaser	Date
9 and 11	I	George Blewitt	1894
2 and 4	P	Vinton Colliery	1894
8	AD	Jacob Bennett	1894
13 and 15	B	Valentine Barker	1894
2 and 4	AE	Blair Shaffer	1894
4	K	Minnie Bamond	1895
1	I	Benjamin Bennett	1895
8 and 10	U	Theodore Bechtel	1895
4 and 6	G	Theodore Bechtel	1895
1	J	Warren Delano	1895
4 and 6	I	Warren Delano	1895
2	V	Warren Delano	1895
2	K	J.B. Rager	1895
1	B	Francis Farabaugh	1895
2	D	Henry Taylor	1895
3, 5, and 7	I	Dr. Abner Griffith	1895
16	B	A.J. Miller	1895
2	AD	Emma Shaffer	1895
10 and 12	K	Annie Lyle	1895
12	I	Isaac Michaels	1895
11, 13, and 15	C	Dr. Abner Griffith	1896
2	L	Nicholas Altimus	1896
12	AB	Thomas Morris	1896
3	C	Clarence Claghorn	1896
4	AD	Merritt Shaffer	1896
2	AB	M.B. Shultz	1896
15	D	First Baptist Church	1897
1	K	S.J. Ruffner	1897
1	E	Jackson Township Schools	1897
1	B	C.S. Invilliers	1898
2	C	Mary Davis	1898
5	K	Matilda Wray	1899
4, 6, 8, 10, 12, and 14	B	Warren Delano	1900
4, 6, 8, 10, 12, 14, and 16	C	Warren Delano	1900
4 and 6	J	Warren Delano	1900
4	AA	Warren Delano	1900
10, 12, 14, and 16	A	Augustine Vinton Barker	1900
2	A	Augustine Vinton Barker	1900
6	K	Warren Delano	1900
3, 5, 7, 9, and 11	A	Warren Delano	1900
11	B	Barker Brothers	1900
2	I	George Berkey	1900
4	AE	E. Findley	1900

Appendix P
Tax Assessor's Records
CLAGHORN

Name	Occupation	Years on Tax List
Abrams, A.	superintendent	1919, 1920
Alerson, Alexander	miner	1922, 1923
Alerson, Mrs. Alexander	housewife	1923
Anderson, Carl	miner-1918, mine foreman-1919, miner-1920, 1921, 1 cow-value $20 miner-1923, 1924, laborer-1925	
Anderson, Mrs. Joe (Mary)	housewife	1923, 1924
Anderson, Peter	miner	1918, 1919, 1920, 1921
Anderson, Verner	miner, non-citizen	1921
Andricko, George	miner	1918
Backman, John	miner	1918
Baldo, Pete	miner	1922
Banosky, Joseph	miner	1919
Baran, John	miner	1918
Baran, Pete (Barany, Barren)	miner, non-citizen	1921, 1923, 1924
Baran, Mrs. Pete	housewife	1923
Barns, D.E.	miner	1918
Barr, Edd.	miner	1918
Barr, Frank	miner	1918
Barr, Mike	miner	1918
Barr, William	miner	1918
Bastante, Joe	miner	1923
Bastante, Mrs. Joe	housewife	1923
Batchick, Micke	miner, non-citizen	1921
Beckhind, Charley	miner	1919
Beller, Gasper	miner	1923
Benson, Ed	miner	1919, 1920
Beresh, Mike (Berich-1923)	miner, non-citizen	1921, 1922, 1923, 1924
Beresh, Mrs. Mike	housewife	1923
Berg, Andrew	miner, non-citizen	1920, 1921
Berg, Axel	miner, non-citizen	1921
Bergstron, Venner	miner	1920
Biro, Andy	miner	1924
Borczeck, Egnatius	trackman	1919
Borzilla, A.	miner	1918
Brown, Chas.	miner	1923, 1924
Brown, Mrs. Chas.	housewife	1923, 1924
Buck, N.L.	chief clerk	1923, 1924
Buck, Mrs. N.L.	housewife	1923, 1924
Burger, William	miner	1921
Burger, Dave	laborer-1919, miner-1921	
Bush, Andrew	miner	1920
Campbell, Lester	miner	1924
Campbell, Mrs. Lester	housewife	1924
Campbell, Merle	miner	1923
Campbell, Mrs. Merle	housewife	1923
Carchesio, Sylvester	miner	1923
Carlson, Andrew	miner	1921
Carlson, Edwin	miner	1920
Carlson, Emil	miner	1920, 1921

Carlson, J.B.	miner	1918, 1919, 1921
Carlson, J.C.	miner	1920
Carlson, John E.	miner	1920
Carlson, Oscar	miner	1921, 1922, 1923
Carlson, Mrs. Oscar	housewife	1923
Chachio, Mike	miner	1923
Cachio, Mrs. Mike	housewife	1923
Chapple, Peter	miner	1918
Charlesworth, W.C.	miner	1921
Cherry, A.B.	blacksmith	1923
Cherry, Stanly	miner	1924
Cherry, Toney	miner	1924
Chestock, Charles	miner	1919
Cindrich, John	miner	1918
Clevenger, John	miner	1923
Clevenger, Mrs. John	housewife	1923
Closky, Joe	miner	1920
Cocolic, Andy	miner	1918
Conrad, Frank	miner	1918
Cresswell, A.F.	miner-1922, mine superintendent-1923, mine foreman-1924	
Cresswell, Mrs. Allie	housewife	1923
Cupsick, John	miner	1919, 2 horses, 2 cows
Curtise, D.R.	doctor	1918
Dabris, Sam	miner	1918
Dalberg, John	miner	1920
Danielson, C.B.	miner, non-citizen	1920, 1921, 1 cow
Dauch, John	miner	1918
Davis, Chas.	miner	1918
Davis, Frank	miner	1918
Davis, H.B.	miner	1920, 1921
Dayack, Chas.	miner	1918
Deok, Lewis (Luey Dock, Deak)	miner	1921, 1922, 1923, 1924
Derst, Joe	miner, non-citizen	1921
Direnaldo, Joe	miner	1923
Dombrich, Mike	miner	1918
Dominik, Mike	miner	1918
Dominik, Mrs. Angelo	housewife	1923, 1924
Donelop, J.E.	miner	1922
Donahue, Pat	miner	1923
Droleric, Thomas	miner	1922
Drolice, Lewis	miner	1922
Drusake, Mike	miner	1918
Eaton, Fred	miner	1918
Erickson, John	miner	1921
Erickson, John W.	miner, non-citizen	1921
Erickson, Walter	miner	1920
Erricson, John	weighmaster	1923
Erricson, Mrs. John	housewife	1923
Erwin, James	miner	1919
Evans, B.L.	chief clerk-1919, clerk-1920, miner-1921	
Felico, Chas.	miner	1918
Fitz, Steve - see eviction list	miner	1922
Fivecoats, Elmer	miner-1922, mine foreman-1923	
Fivecoats, Mrs. Elmer	housewife	1923
Foster, Ed	miner	1920
Forth, Hugo	miner	1921, 1922
Fox, Daniel	miner	1923

Fox, Mrs. Daniel	housewife	1923
Frankson, Emil	miner	1919
Fred, Andrew	miner-1918, 1920, 1921, blacksmith-1919	
Frederick, C.F.	miner	1918
Fritz, John	miner	1922
Gabrilelson, John	cutter	1919
Gaki, Geo.	miner	1918
Galiski, Blarch (?)	miner	1918
George, Arthur	miner	1920, 1921, 1922
George, James - see eviction list	miner	1922
George, S.W.	miner	1924
Gezich, Fred	laborer	1922
Gispor, Andy	miner	1924
Glagolis, John	miner	1924
Gromley, Samuel (Grumley-1923)	motorman-1919, miner-1920, 1922, 1923	
Gromley, Mrs. Samuel	housewife	1923
Gunda, John	miner	1921
Gunnile, Gunnar (Gunnell)	miner	1920, 1921
Gust, Chas. - see eviction list	miner	1920, 1921, 1922
Guestfasaan, Ed	miner	1918
Gustfasaan, Sam	miner	1918
Hacobich, Pete	miner	1922
Hagen, Mat (Hagan)	miner	1919, 1920
Hagins, John (Hagains)	miner-1918, motorman-1919	
Halasburdo, Jacob	miner	1918
Halasko, Chas.	miner	1918
Haldin, Gust	miner, non-citizen	1921
Harish, Jos.	miner	1922
Harshaw, Alex	miner	1922
Hart, August	miner	1922
Harvatt, George	miner	1919
Harvies, Alex	miner	1921
Hazi, John	miner	1919
Henderson, Carlo	miner, non-citizen	1920, 1921
Henderson, John	miner	1920
Hendrickson, Charley	miner, non-citizen	1918, 1921
Henry, Clyde	miner	1922
Holinquist, Harry	miner, non-citizen	1918, 1921
Holochock, Joe	miner	1918
Holope, Alex	miner	1922
Holun, Alex (Holm)	miner	1919, 1920
Hoover, Clair	miner	1920
Hudson, Joseph	miner	1923
Hudson, Mrs. Joseph	housewife	1923
Icker, John	miner	1921
Ingres, Conrad	miner	1921
Isaacson, Erick	miner	1921
Jackson, Matthew	miner	1921
Jaska, Mike	miner	1919,. 1922
Jensen, Jay	police	1919, 1922
Jensen, Mrs. Jay	housewife	1923
Johston, Algot	miner	1919
Johnes, Thomas	laborer	1922
Johnson, Andrew (Johston)	miner	1919, 1920, 1921
Johnson, Ed (Johston)	miner	1919, 1920, 1921
Johnson, Ed. J.	miner	1920, 1921, 1 cow
Johnson, Ralph	miner	1921
Johnson, Waldmar	miner-1918, 1920, 1921, 1922, motorman-1919	

Name	Occupation	Year
Jones, Thomas (same as Johnes?)	clerk	1923
Kallskata, Alfred	miner	1919
Kaonovich, Mike	miner-1918, blacksmith-1919	
Karolchak, Chas.	miner	1920, 1 cow
Kasonovich, Mike	miner-1918, blacksmith-1919	
Keller, Clinton	miner	1923
Keller, Mrs. Clinton	housewife	1923
Keyser, Jacob	miner	1918
Kosa, Charles	miner	1924
Kosa, Mrs. Charles	miner	1924
Landelin, Ed	miner	1918
Landrus, George	miner	1920
Lantz, Chas. Sr.	miner, non-citizen	1921
Lantz, Chas. Jr.	miner	1918, 1920, 1921
Larsen, S.C.	miner	1920, 1921
Larson, Joseph	miner	1918
Larson, Raynar (Rognar)	miner	1918,. 1920
Lawrence, Frank	miner	1923
Lawrence, Mrs. Frank	housewife	1923
Lent, Metro, Jr.	miner	1919
Ling, Ervin	miner	1923
Ling, Mrs. Ervin	housewife	1923
Lohr, Charles	carpenter	1919
Lomont, Charlie	miner	1923
Lomont, Mrs. Charlie	housewife	1923
Ludanick, Frank	miner	1918
Lutheran, John	miner	1918
Machiniak, Mike	miner	1918
Madison, Earnest	miner	1923
Madison, Mrs. Earnest	housewife	1923
Malm, Victor	miner, non-citizen	1921
Manse, John	miner	1918
Masolich, Mike (Heshbon)	miner	1918
Mattson, Mike (Matson-1922)	miner	1921, 1922
Mattson, Earnest	miner	1923
Mattson, Mrs. Earnest	housewife	1923
McClusky, Wm. (McCusky)	miner	1920, 1921, 1 horse
McMasters, Chester	watchman	1923
McMasters, Mrs. Chester	housewife	1923
McNutt, Robert	laborer	1918
Mezei, John	miner	1923
Mezei, Mrs. John	housewife	1923
Michaelson, Edwin	miner	1921
Miller, Dallas	miner-1921, laborer-1922	
Miller, Merit	miner	1924
Miller, Mrs. Merit	housewife	1924
Miller, Milton	miner	1923
Miller, Mrs. Milton	housewife	1923
Moberg, Augustes	trackman-1919, miner-1920, 1921, 1922, 1 cow	
Molney, J.A.	miner	1918
Moran, Mike - on eviction list?	miner	1921, 1922
Moris, Mike (same as above?)	miner	1923
Moyers, Ernest	miner	1922
Moyes, E.J. (same as above?)	miner	1918
Nasman, Oscer	miner	1920
Nasman, Walter	miner	1920
Nesman, Alex	miner	1922
Nucyn, Mike	miner	1918

Name	Occupation	Year
Ogyn, Steve	miner	1924
Ogyn, Mrs. Steve	housewife	1924
Onesky, Jacob	driver	1919
Orelli, Sam	miner	1924
Orelli, Clarentina	housewife	1924
Oravzi, Frank	miner	1923
Pant, Alfred	motorman	1919
Parsons, Chas.	miner-1918, teamster-1919	
Parsons, E.C.	teamster	1919
Papst, Rudolph	miner	1918
Paul, Lewis (Louis)	miner	1922, 1923
Pavalka, Mike (Pavaka) - see eviction list	miner	1918, 1919, 1921, 1922, 1 cow
		1922, 1 cow
Pavalko, Alfred (same as above?)	miner	1920, 1 cow
Peace, Thomas	miner	1921, 1922
Peterson, Martin	laborer-1919, miner-1921, 1922, 1 cow	
Petriskey, John	miner	1918
Petrof, Ralph	miner	1918
Petuzzi, J.B.	miner	1918
Pierce, Thomas	miner	1920
Popodines, John (listed twice)	miner	1918
Potolonis, Geo.	miner	1918
Pott, J.N.	miner	1918
Preston, Edward	miner	1918
Preston, W.W.	miner	1918
Prindle, J.R.	miner	1921
Pusposki, Louis	miner	1918
Rafty, John	miner	1918
Raney, Wm.	miner	1922
Reilly, F.S.	doctor	1921, 1923, 1924
Reilly, Mrs. F.S.	housewife	1923, 1924
Ressler, Earl	miner	1921
Restler, Laurence (Ressler)	miner	1919, 1920
Ringler, M.J.	miner	1918
Ripino, Peitro	miner	1918
Rodas, Evert	miner	1919, 1920
Rominycio, Mytro	miner	1918
Ronges, Dodar (Rodgers, Dodar)	miner-on eviction list?	1921, 1922
Ronn, Winna	miner	1922
Rorve (?), Walter	laborer	1919
Rornosko, Mike	miner	1918
Roush, J.A.	miner-1918, electrician-1919	
Rubis, Geo.	miner	1918
Rubish, John	miner	1918
Rushnock, Steve	miner	1919
Sadistit, John	miner	1918
Sager, H.J.	laborer	1921
Sand, Emil	laborer	1920, 1921
Sandbak, Gust	miner	1920
Sandelin, Ed	miner	1920
Sasso, Dornick	miner	1918
Savich, Geo.	miner	1918
Saxberg, Varner (Verner)	miner	1919, 1920
Searle, G.R.	miner	1922
Senberg, Gotfred	miner	1920
Shaffer, C.D.	laborer	1922
Shepard, John	laborer	1922

Sherish, John	miner	1918
Shook, J.H.	miner	1918
Simon, George	miner-1922, laborer-1923, 1924	
Simon, Mrs. George	housewife	1923
Smith, Frank	miner	1918
Smith, Charles	miner	1923
Smith, Mrs. Charles	housewife	1923
Smith, John	miner	1923
Smith, Mrs. John	housewife	1923
Solesky, Mike	miner	1920
Sovich, Wm.	miner	1920
Staff, John	miner	1922
Stahl, Wm.	miner	1918
Starr, Ernest	laborer	1921
Starr, Wm.	laborer	1921
Steblack, Andy	miner	1918
Steele, Wm.	electrician	1923
Steele, Mrs. Wm.	housewife	1923
Stenman, Carl	miner-1919, 1920, laborer-1921	
Stepic, Mike	miner	1919
Sterko, Joe - on eviction list?	miner	1922
Sternberg, Richard	laborer	1921
Stiles, B.D.	miner	1918
Storm, Edwin	laborer	1921
Storjin, Peter	scraper	1919
Stroud, Wm. J. (Stroad)	trackman-1919, miner-1920, laborer-1921	
Stufico, Paul	miner	1918
Stupack, John	miner	1918
Sturnberg, Gotfred (same as Senberg above?)	miner	1921, 1922
Suckel, George	laborer	1921
Sula, Tom	miner	1923, 1924
Sula, Tom, Jr.	miner	1924
Sula, Mrs. Tom, Jr.	housewife	1924
Surnick, Frank	miner	1924
Swanson, Andrew - see evictions	miner	1921, 1922
Swanson, Swan	miner-1918, electrician-1919, laborer-1921	
Swanson, Tom	miner	1921, 1922
Sylian, Geo.	miner	1922
Tomkin, R.D.	miner	1918
Treka, Joe	miner	1923, 1924
Treka, Mrs. Joe	housewife	1924
Unger, George	miner	1918
Utberg, Adolf	miner	1919, 1920, 1921
Vallenius, Verner	miner	1919, 1920, 1921
Vallenius, Mrs. Verner	housewife	1923
Vasko, Mike	miner	1918
Wagner, Carl	laborer	1923
Wagner, Mrs. Carl	housewife	1923
Wagner, Cecil	laborer	1918
Wagner, M.K.	miner	1920
Wagonis, Mike	miner	1918
Watson, William	miner	1920
Weaver, Henry	laborer	1919
Wetzel, Roy A.	blacksmith	1923, 1924
Wetzel, Mrs. Roy A.	housewife	1923, 1924
Williams, John	miner	1923
Williams, Mrs. John	housewife	1923

Name	Occupation	Year
Williams, Thomas	miner	1923
Williams, Mrs. Thomas	housewife	1924
Wickman, Edward	miner	1921
Yarchuk, Ignatz	miner	1924
Yarchuk, Mrs. Ignatz	housewife	1924
Younkin, Chas.	miner	1921
Zovich, Wm.	miner	1918

AUTHOR'S NOTES:
The above names have been deciphered from the West Wheatfield tax assessors' books, 1917-1925. Names which the author was unable to interpret are followed by a question mark in parenthesis.

From 1917-1920, the county tax for a miner was $1.00; a laborer was $.50; a superintendent was $2.00. In 1922, the rate paid by a miner rose to $1.20, and a laborer paid $.60. The housewives were not listed until 1923; their tax was $.15. In 1918, 1919, and 1920, the assessor listed separately almost all the employees in the back of the assessment book. The occupations were usually listed as miners, rather than specific occupations. Also each assessor had a different spelling for the town. Examples include Claigehorne and Claighorn.

REXIS

The following names were taken from the Indiana County Tax Assessor's Books for the years 1901 to 1911. Names were listed according to post office address. The Rexis address disappeared from the list in 1910 and residents were then listed as having a Vintondale mailing address.

Name	Post Office	Occupation	Years on List
A.J. Aldms	Rexis	Blacksmith, House & Lot	1904
D. Dow Altemus		Single	1908
D.H. Altemus	V'dale	Laborer	1908
L.D. Altemus	Rexis	Laborer	1905, 1906, 1907
Nelson Altemus	V'dale	Laborer-1904, Carpenter-1905, 1906, 1907	
J.N. Antes	Rexis	Laborer	1906, 1907
Fred Arndt	Rexis	Laborer, married	1908
Thomas Baker	Rexis	Laborer, married	1908
A.A. Barry	Rexis	Miner	1906
Wm. Barry	Rexis	Laborer-1903, Inspector-1904, Yard Foreman-1905, 1906, 1907	
Wm. Barndollar	Rexis	Mill Foreman-1905, Filer-1906, 1907	
J.E. Barnhill	Rexis	Agent, 1 cow	1909
John W. Baum	Rexis	Laborer	1903, 1904
Edgar Beard		Laborer	1911
Harry Beam	Rexis	Laborer	1904
Porter Beers	Rexis	Laborer	1908
Thomas Bennett		Fireman	1911
John Blakely, Jr.	V'dale	Merchant-1 horse, 1 cow	1903
David Bracken	Rexis	Laborer	1901, 1911
Harve Bracken	Rexis	Laborer	1904
Wm. Brandall	Rexis	Foreman	1904
Harve Bratten (Bracken?)		Laborer	1905
Henry Breth	Rexis	Engineer-1903, 1904, 1905, 1906, Laborer-1907, 1 cow	
Saml & Jacob Brett	Rexis	Vacant lot	1907, 1908, 1909, 1911
Thomas Buckingham	V'dale	Laborer	1904
	Rexis	Laborer	1905
Thomas Buckingham	V'dale	Laborer	1904
	Rexis	Laborer	1905
Y.A. Burkhart	Rexis	Laborer	1904
Z.C. Burns	Rexis	Engineer	1903
Warren Buterbaugh	Rexis	Laborer	1903
Orlie Butler	Rexis	Carpenter, single, 1 horse	1907

Name	Location	Occupation	Years
Harry Byrne	Rexis	Laborer	1905
Marlin (Morland)	Cameron	Carpenter, house & lot	1905, 1906, 1907, 1908 1909, 1910, 1911
Peter Cannon (?)	Rexis	Mine foreman	1904
John Causer	V'dale	Laborer	1904
Vernon Clauser		Laborer	1908, 1909, 1910, 1911
A.B. Conrad	Rexis	Carpenter, single	1905, 1906, 1907
Joseph F. Conrad	V'dale	Prentesh (sic)	1903
John Crossley	Rexis	Laborer-1904, Contractor-1905, 1906, 1907	
S.S. Curry	Rexis	Laborer	1903
E.S. Davis	Rexis	Laborer	1909
James K. Davis		Laborer	1911
Wm. Davis	Rexis	Sawyer, 1 cow	1906
John Dearmey	Rexis	Laborer	1904
Edward Dehass	Rexis	Laborer	1905
J.B. Dick	V'dale	Dairyman, married 3 horses, 13 cows	1903, 1904, 1905, 1906 1907, 1908
Laurance Dodson	Rexis	Laborer-1904, 1905, 1906, 1907, 1909, 1910, 1911	
	V'dale	Laborer-1908, 1 cow	
Loshel Dodson*	Rexis	Laborer	1907
Lynwood Dorr	Rexis	Bookkeeper-1903, 1905, Clerk-1904	
William Downs*	Rexis	Sawyer, 1 cow	1903, 1904, 1905, 1907
G.W. Duncan	Rexis	Laborer, 1 horse, 1 cow 1905	
Samuel Duncan	Rexis	Farmer-1903, Laborer-1906	
	V'dale	Laborer, married, 1 horse, 1 cow, 1907, 1908	
Webster Duncan	Rexis	Laborer	1907, 1908
John Dynmyre			1911
T.E. Dunn		Supt-B&Y RR	1911
Herman English	Rexis	Laborer	1904, 1905
Berk Fagans	Rexis	Laborer, vacant lot	1904
James Fagans	Rexis	Laborer, H&L	1904, 1905
Edward Findley*	Rexis	Supervisor-1905, 1906 Laborer-1909 2 horses, 1 cow	
	V'dale	Baker-1908, 1909, 1910, 1911, house & lot	
Wm. Gallow	V'dale	Laborer	1904
Edward W. George	V'dale	Miller	1905, 1906
	Rexis	Farmer-1907, 1908, Laborer-1909, 2 horses, 1 cow	
Frank George	Rexis	Mixed-1906, Contractor-1908, 4 horses, 1 cow Laborer-1904, 1905, 8 acres cleared	
	V'dale	Laborer, H&L	1903
James Gibson	V'dale	Laborer	1904
C.B. Gill	Rexis	Laborer, house & lot	1903, 1904, 1905, 1906, 1907, 1909, 1910, 1911
Frederick Gill	Rexis	Engineer	1905
Edward Gilligan	Rexis	Laborer	1904, 1905
Harry Good		Laborer	1911
Evan Griffith	V'dale	Laborer	1904
Archie Hadden	Rexis	Laborer	1906, 1907
J.R. Hancock	Rexis	Laborer	1909
John Hankasin (Hankersen)	Rexis	Laborer, single	1906, 1907, 1908
David Henderleiter	Rexis	Laborer	1906
Joseph Herr	Rexis	Laborer	1905
George W. Hoffman	Rexis	Laborer	1906
Milton Hoffman	V'dale	Laborer	1909
John Hovel (?)	Rexis	Teamster, married	1908
Charles Hullihan	Rexis	Miner, married	1908
R.C. Hunter	Rexis	Laborer, 1 cow	1905, 1906
William Hunter	Rexis	Laborer, single	1906,. 1907

Name	Location	Occupation	Year(s)
George Johnston	Rexis	Laborer	1904
Lloyd Johnston	Rexis	Laborer	1906
? Kilpatrick	Rexis	Insurance agent	1907
Charles Kims	Rexis	Railroader, married	1908
John H. Krumbine	Rexis	Painter & undertaker	1905, 1906, 1907, 1908
G.A. Kunkle	Rexis	Electrician, married	1908
John A. Lewis	Rexis	Laborer	1908, 1909
Edward Lines	Rexis	Laborer	1906, 1907
Eli Lines (Lyons?)	Rexis	Blacksmith-1904, Foreman-1905, Laborer-1906, 1907	
H. Ling	Rexis	Liveryman	1907
John Lloyd		Laborer	1911
James Lockett	Rexis	Laborer	1905, 1906
A.D. Lowlas (?)	Rexis	Carpenter	1910
Harvey Luke	Rexis	Laborer	1906,. 1907
Frank Lunday	Rexis	Laborer	1903
John Lundhing (?)		Miner	1911
Arthur Lunt	Rexis	Laborer	1903
F. Lyons	Rexis	Bookkeeper	1903
Wm. J. Lybarger	Rexis	RR-1908, Laborer-1909, Boss-1910, 1911	
John Mack		Laborer	1911
James Mahan	Rexis	Laborer	1906, 1907, 1908
Edward Manuel		Laborer	1909
Jackson Mardes	Rexis-1906 Strongstown-1907	Sawmill and cider press	
J.R. Marshall	V'dale	Machanist (sic)	1904
Thomas Marthers	Rexis	Laborer	1907
B.L. McCanullty		Laborer	1911
Bernard McClusky (McCloskey)			1908, 1909, 1910
Frank McClusky		Single	1907, 1909, 1910
F. McCornuch	V'dale	Engineer	1904
A. Frank McGuire*	Rexis	Fireman-1903, 1906, Laborer-1904, 1905, 1907	
Joe Miller		Laborer	1911
Samuel Misener	Rexis	Laborer	1909
Harriet Misner	V'dale	Underage, 1 cow	1903, 1904
John Misner*	V'dale	Laborer, H&L	1903, 1904, 1905, 1906 1907, 1911, 1912 Single-1909
Michael Misner*	Rexis	Laborer, house & lot	1906, 1907, 1908, 1909, 1910
William Misner	Rexis	Laborer	1903, 1904, 1905
Henry Misoner	Rexis	Laborer	1903
E. Mood	V'dale	Machanist (sic)	1904
J.F. Morgan	Rexis	Fireman	1905
Harry Mullen	Rexis	Foreman-1906, Trackboss-1907	
Andrew Nicewonger	Rexis	Laborer-1904, 1907, Foreman-1906	
Grant Orner	Rexis	1 horse, 1 cow	1903
William Orner	Rexis	1 horse, 2 cows	1903
John Overlitner		Trainman	1912
L.P. Phesendend	Rexis	Laborer	1903
Samil Philips	V'dale	Laborer	1904
Thomas Possatt	Rexis	Laborer	1908
Mike Powell	Rexis	Laborer, 1 cow	1907
Kada Princheck		Laborer	1912
Martin Rager		Conductor-C&I	1912
Robert Rairigh	V'dale	Watchman	1903, 1904
Amos Reninger	Rexis	Laborer	1907
George Rhodes	Rexis	Laborer	1903

Geo. B. Rodgers		Farmer-1903, 1904, 1905, 1906, 1907, 2 horses, 5 cows Married-1908, 6 horses, 2 cows Laborer-1909, 1910, 1 cow	
Charlie Ross	Rexis	Laborer	1903
Irvin Scott	Rexis	Laborer, 1 cow	1906, 1907
O.J. Scott		Laborer	1904
Blair Shaffer Heirs	Rexis	17 a. timber	1904, 1905, 1906, 1907, 1908, 1909, 1910, 1911, 1912
V.W. Shaffer		Laborer	1906
Shaw & Co.	Rexis	1 horse	1903
D. Shirey	Rexis	Laborer	1903
H. Shorey	Rexis	Laborer	1903
O.A. Shorey	Rexis	Shipping clerk	1903
J.A. Shreck	Rexis	Blacksmith, 1 cow	1908, 1909, 1911, 1912
Andie Shultz		Laborer	1903
Harvey Shultz	V'dale	H&L	1903
John Smart	Rexis	Laborer, married	1909
J.P. Stephel	Rexis	Miner	1912
J.H. Steiffele	Rexis	Laborer, 1 horse	1910, 1912
H.H. Stephens	Rexis	?	1905
Frank Stewart	Rexis	Laborer	1906
W.W. Stewart		Laborer	1912
James H. Straw		Laborer-1907, 1908, 1909, Railroader-1910	
Willard C. Straw	Rexis	Laborer	1907, 1908, 1910
E.E. Thompson		Carpenter-1910, Laborer-1911	
John W. Townsend	Rexis	Laborer	1905, 1907
Welse B. Townsend	Rexis	Single	1905, 1906, 1907
John L. Wells	Rexis	Engineer-1903, 1905, Laborer-1904	
Clarence Wetze	Rexis	Laborer	1903
W. Wetzel	Rexis	Lumber inspector	1903
Edward Wheeling	Rexis	Laborer	1907
Harry Whelland	Rexis	Laborer	1903
George Williams	Rexis	Laborer-1904, 1905, Foreman-1906, 1907	
John Wise	Rexis	Laborer, married	1906, 1908
John Wood*	Rexis	Laborer	1904, 1905, 1906, 1907
J.G. Yocum	Rexis	Carpenter	1904
A.M. Zimmerman	Rexis	Fireman	1904
A. Zulick	Rexis	Miner	1904
Chas Zwick	Rexis	Laborer	1904

denotes known employees of the Vinton Lumber Company.

WEHRUM
1903-1912

Name	Occupation	Years on List
Wm. Abbotts	Blacksmith	1903, 1904
J. Abrams	Miner	1904
Wm. Abrams	Miner	1904
Abe Abrems	Laborer	1903, 1904
Wm. Abrens	Laborer	1903
E.F. Adams	Painter	1903, 1904
Samuel Adams	Laborer	1906
Charles Admansen	Pumper-1910, Laborer-1911, 1912-Laborer	
Wm. Adrich	Laborer	1903
Charles Anderson	Bartender-1907, Electrician-1910	
Albert Algot	Laborer	1905
Adam Antell	Laborer	1912
Frank Artley	Bricklayer	1904
Jacob Artley	Bricklayer	1904
A. Ashton	Teamster	1904
Mike Asper	Merchant	1912, 1 horse
C.H. Atherholt	Brickmaker	1903
P.S. Atherholt	Foreman	1903
M. Austin	Doctor	1911
John E. Barnhart (Barnhill?)	Laborer	1903, 1908
Thomas Barnes	Miner	1912
James Barr	Machinist	1904
John Beals (Beale)	Engineer-1903, 1904, 1905, 1906, 1907, 1908, Master mechanic-1910, Foreman-1911, Clerk-1912	
Joseph Beel	Laborer	1912
John Biedling	Laborer	1904
G.F. Bill	Electrician	1912
Fred Bitterf	Laborer	1905
Andrew Black	Laborer	1912
E.J. Blackley	School teacher	1910
C.M. Blanchard	Superintendent	1906
Geo. S. Blewitt	Foreman-1903, Clerk-1904	
Thomas Bony (?)	Miner	1912
W.H. Booth	Miner foreman	1903
P.F. Bowers	Miner	1906
John Boyl	Bartender	1904
William Bracken	?	1903
Hugh Brady	Bartender, single	1908
H.H. Brennan	Foreman	1904
Thomas Brennan	Laborer	1903
Joseph Breulick	Miner	1911
Christ Brixner-house & lot bank building & lot-1906	Single-1903, Hotelman, landlord-1905, 1906, 1907, 1908, 1909, 1910, 1912	
Elmer Brolsh	Laborer	1904
J.H. Brown	Laborer	1912
Thos. Buckingham	Laborer	1907, 1908, 1909, 1910, 1911, 1912
William G. Buckingham	Tending pump-1903, Engineer-1905, 1906, 1907, 1910, 1911, 1912, Track layer-1908, Laborer-1903, 1909	
S.Y. Buchanan	Carpenter	1904
T.A. Buggy	Engineer-1903, Miner-1904	
L.P. Byers	Engineer	1904
Chas. Byroads	Plumber	1904

Name	Occupation	Years
Burkhart and Link (Changed from Link & Burkhart in 1905)	Livery Stable	1904, 1905, 1906, 1907, 1908, 1909, 1910, 1912
B.F. Burkhart	Liveryman	1904, 1905, 1906
Patrick Burns	Laborer (Died in 1909 explosion)	1907, 1908, 1909
William Burns	Laborer-1907, 1909, Miner-1908 (Died in 1909 explosion)	
D.M. Buterbaugh	Merchant, married, 2 horses	1906, 1908
John Byerly	Laborer	1903
Wm. Cairnes	Laborer-1903, 1904, Pumpman-1905	
Robert Cairns	Laborer	1903, 1904
S. Cairns	Miner	1904
Daniel Campbell	Laborer	1912
Frank Campbell	Laborer	1912
W.P. Canghern	Chief engineer	1904
Charles Carlin	Laborer	1907
D.H. Carlin	Laborer	1907
David Carlin	Laborer	1907
Elmer Carlin	Laborer	1907
H.R. Carney	?	1911
Frank Carron	Civil engineer, single	1905, 1906, 1907
Daniel Carter	Pumpman	1905
L. Darwin Cassat (Cassett)	Laborer, single	1906, 1909, 1910, 1911
John Caudile	Laborer	1907
Louis Cichina	Porter	1911
Archy Collins	Miner-1906, 1907, Engineer-1909, Pumper-1910, Laborer-1911, Foreman-1912	
Cal Collins	Laborer	1903
Chas. Collins	Laborer	1903
W.E. Colton	Foreman	1905
G. Conrado (?)	Hotelkeeper	1910, 1911, 1912
H.L. Cook	Unemployed	1905
Frederick Cornman	Retired	1903, 1906, 1907, 1908, 1909, 1910
Michal Core	Engineer	1903
James Cortes	Track foreman	1903
H.E. Cover	Sup.	1904
Emanuel Coy	Laborer	1904
Jesse Craig	Laborer	1911, 1912
John Crandle	Miner	1908
James Craven	Laborer	1903
John C. Croft	Laborer-1903, Foreman-1904	
A.G. Croft	Laborer-1903, Carpenter-1904	
John Crumb	Laborer-1903, 1905, 1906, 1907, Engineer-1904	
G.Y. Dale	Carpenter	1903, 1904
G.R. Dalemator	Supt.-1903, Foreman-1904	
D.J. Daugherty	Laborer	1903, 1904
Jno Daugherty	Laborer	1904
M. Daugherty	Brickmaker	1903
Benjamin Davis	Laborer-1903, 1907, Engineer-1904, 1905, 1906	
David Davis	Laborer	1907
Isaac Davis	Laborer	1903
Wm. Davis	Laborer	1903
Thos. A. Davison	Motorman	1904
W.G. Deck	Laborer, V'dale	903
Samuel Degsozi	Laborer	1910
Walter DePew	Laborer	1907
P.W. Dieter	Clerk	1906

Name	Occupation	Year
H.M. Dill	Laborer	1912
John Dodd	Fireboss	1912
James Donahugh	Laborer	1912
Pat Doyle	Laborer	1903
Andy Dratter (Dratten, Dratton)	Laborer	1912
John Dratter	Laborer	1912
Michael Dratter	Laborer	1912
William Dull	Blacksmith	1904
B.F. Duncan	Laborer	1903
John Duncan	Teamster	1910, 1911, 1912
William Duncan	Laborer	1911, 1912
Bruce Dunlap	Laborer	1903
George Dyer (Gyer?)	Laborer	1907, 1909, 1912
Andrew Eastlake	Laborer-1903, 1904, 1906, 1907, Miner-1909	
J.C. Edwards	Miner-1903, Engineer-1904	
John Elder	Laborer	1906, 1907
Edwin Emanue	Laborer	1912
John Emmaus (?)	Laborer	1912
Frank Emigh	Laborer	1904
H.A. Endress	Butcher	1905
H.D. English	Laborer	1903
Jacob Enos	Foreman	1910
D. Evans	Miner	1904
John Evans	Laborer	1903, 1904, 1905
Ed Farrell	Laborer	1903
Mike Famy (?)	Laborer	1912
David Felmly	?	1908
J.B. Felmly	?, 1 horse, 1 cow	1908
Albert Felton	Carpenter	1905
Chas. Filday	Miner	1904
Geo. Filday	Miner	1904
Saml. Flickinger	Butcher	1903, 1904
Christopher Frazier	Track layer-1904, Disabled-1910, 1911, 1912 Laborer-1903, 1908, 1909, 1912	
A.B. Frean	Laborer	1903
F.L. Freeman	Blacksmith	1904
Oscar Freeman	Blacksmith-1903, 1904, Laborer-1905	
Albert B. Friar	Laborer	1904
John Galer (Gailer)	Laborer-1906, 1907, 1910, Electrician-1905	
H.L. Gardnes	Laborer	1903
Harry Gardners	Clerk	1906
Charles Gaster	Single-1905, Laborer-1906	
John Gaster	Single	1907
William Gaster	Contractor-1907, Laborer-1903, 1904, 1905, 1906, 1908, 1909, 1910, 1911, 1912	
John M. Gaylor (same as Galer?)	Miner-1908, Laborer-1911, 1912	
John Gibson	Teamster	1903
Ben Gile	Dr., House & lot	1903, 1904
T.H. Gillman	Track foreman	1912
J.G. Gilligan	Carpenter	1903
David Gittens	Carpenter	1903
A.J. Given	Miner	1904
E.P. Goodwin (Goodson)	Manager-1907, Clerk-1908, 1909, 1910, 1911, 1912	
B.S. Goss	Carpenter	1903
E.P. Goss	Teamster	1903
H.F. Gorsuch	Blacksmith	1903
L.G. Gorsuch	Clerk	1903
Jas. Graner (?)	Laborer	1912

Name	Occupation	Year
Chas. Greenan	Laborer	1904
Burton Griffith	Barber	1903
Edward Griffith	Laborer	1903
J.M. Grow	Carpenter	1903
Gillen Grown (?)	Laborer	1912
Harry Gurns	Laborer	1912
George Gyer	Laborer	1912
J. Hallton	Laborer	1903
D.W. Hamond	Fireman	1904
Frank Hand	Blacksmith	1903
Charles Hanghey	Laborer	1904
Edward Hannah	Miner	1904, 1905
Christ Hansan	Carpenter	1907
L.A. Harding	Foreman	1903, 1904
C.J. Harigan	Laborer	1912
James Hatter	Laborer	1912
Albert Haywood	Miner foreman	1903
Sampson Hewlett	Mining engineer	1908, 1909, 1910, 1911, 1912
Wm. Hipp	Laborer	1904
Chas. Hoffman	Laborer	1904
William Hogan	Laborer	1903
Frank Hogarth	Laborer	1904, 1905
Thomas Hogarth	Boss-1908, Laborer-1904, 1905, 1909, Mine foreman-1910, 1911	
A.D. Hoke	Store clerk	1903, 1904
J. Holmes	Laborer	1903
Jos. Holoda	Laborer	1912
Mike Holt	Laborer	1912
Thomas Hook	Laborer	1907
Elb Hoover	Telegrapher	1908
Martin Hopkins	Laborer	1904
D. Horwatt	Miner	1904
A. Huey	Watchman	1904
Clarence E. Huey (Hughey)	Teamster-1903, Laborer-1904, 1905, 1907, 1909 Stable boss-1908 (Died in 1909 explosion)	
Julious Hufmann	Barber	1904
Charles Hulihan	Miner	1910, 1911, 1912
P.S. Humphries	Carpenter	1904
James Irwin	Laborer	1903
David G. Isaacs	Bricklayer	1903, 1904
Reese James	Laborer	1905
C.E. Jarvice	Engineer	1904
Charles Jarvis	Laborer	1906
D. Jefferson	Laborer	1903
S.E. Jenks	Engineer, married	1908
Ambrose Jenkins	Fireman-1905, Laborer-1912	
Cyrus Jenkins	Laborer	1906, 1909
Joseph Jenkins	Fireman	1904
C.F. Jennings	Laborer	1903
Matthew Jensen	Fireman-1905, Laborer-1906, 1907, 1909, 1910	
L. Johnes	Laborer	1903
Thomas Johnes	Laborer	1903
William Johnes	Laborer	1903
Frank Johns	Railroad	1911
Lewis Johns	Electrician	1908
A.L. Johnson	Mine Foreman-1908, 1909, Chemist-1910	
Earl Johnson	Electrician	1910, 1911
Frank Johnson	Section laborer	1912
George Johnson (Johnston)	F. boiler-1910, 1911, Laborer-1909, 1907, 1912	

Name	Occupation	Years
Chas. Johnston	Laborer	1903
Ed Johnston	Laborer	1903
S.M. Johnston	Laborer	1905
William Johnston	Mine foreman	1906, 1907
John Jones	?	1904
L.W. Jones	Electrician	1907
Wm. Jones	Bricklayer	1904
William J. Jones	Pumpman-1906, Laborer-1905, 1907	
B.F. Jordan	Fireman	1904
Geo. E. Jury	Clerk	1907
C.A. Kalriter	PRR brakeman	1912
Geo. Kating	Plumber	1904
Jno Keating	Plumber	1903
C.K. Keough	Carpenter	1912
F.R. Kern	Machinist	1904
E.W. Kerr	Laborer	1904
L.W. Kerr	Engineer	1905
Charles Kesker	Laborer	1912
John Kessler (Kesker)	Laborer-1907, 1908, 1909, Carpenter-1912	
Thomas Kiley	Laborer	1906, 1907
Allison King	Laborer	1905, 1906, 1907, 1908
Anderson King	Laborer	1903, 1904
Charles Kirker	Yard boss	1912
Fred Kuchenbrod	Laborer	1912
Gus K. Kuchenbrod	Laborer	1911, 1912
Harry Kuchenbrod	Laborer-1907, 1908, 1909, 1910, 1911, 1912	
Sylvester A. Kuhn	PRR agent-1906, 1907, 1908, 1909, 1910, 1911, 1912	
Charles Lang	Laborer	1905
John A. Lang	RR foreman	1903, 1904
Walton Langstruth	Archtecture (sic)	1904
Harry Laughney	Laborer	1907
J. Lawannas	Miner	1904
A.D. Lawer	Laborer	1904
Charles L. Lee	Pipefitter-1904, Laborer-1905	
John Lee	Laborer	1904
Charles Lilly	Lineman	1903
Charles B. Ling	Carpenter	1903, 1904, 1905
Mrs. Sadie Ling	Widow, 2 horses, 4 cows	1905
W.H. Lingenfelter	Clerk	1903, 1904
W.H. Littleman	Laborer	1912
Edward Lloyd	Mine Foreman	1912
Chas. Long	Machinist	1904
E.B. Long	Laborer, married	1908
James Long	Laborer	1907
John Long	Foreman-1906, 1908, Track boss, 1907, married	
Arthur Warren Lunt	Bartender-1904, Carpenter-1905	
H. Lunt	Engineer	1904
Robert Lusides (?)	Miner	1912
Wm. Lybergar	Foreman	1903, 1904
Don Lydick	Laborer	1903
James Lydick	Carpenter	1903
Ely Lyons	Blacksmith, married	1908
James Mabry	Laborer	1906
Albert Mack	Married	1908
Gilbert Mack	Laborer	1907
J.M. Mack	Laborer	1907
James Mack	Laborer	1907, 1908
Jesse Mack	Laborer	1905

Name	Occupation	Years
Warren Madox (Maddocks)	Electrician, cripple	1904, 1905, 1906, 1907
J.C. Maddox	Retired	1910
Peter Mahan	Laborer-1903, 1904, 1905, 1907	
W.D. Maizy	Laborer	1911, 1912
L.W. Mardis	Laborer, 2 horses, 1 cow	1905
J.L. Marietta	Laborer	1905
Gilbert Marsh	Laborer	1903, 1904
Smith Marsh	Carpenter	1904
John Martin	Laborer	1910, 1912
Mike-Tony Massimalli	Foreman	1906
Sam Massimalli	Mason	1912
Mike Masy	Laborer	1912
R. Matterson	Laborer	1903
Andy Mazey	Laborer	1912
B.D. McClemons	Laborer	1904
David McCord	Laborer	1904
Patrick McDonald	Clerk	1912
John McFersun	Fireman-1906, Laborer-1907, 1908, 1909, 1910, 1911, 1912	
C. McGee	Laborer	1903
Bernard McGuire	Laborer	1907, 1908
Swethurs McKercher	Foreman	1907
E.W. McKissel	Laborer	1907
John McLaughlin	Laborer	1907
Marlin McMahan	Invalid	1903
J. E. McMillen	Laborer-1903, Carpenter-1904	
Thos. McQuillan	Miner	1911
James McQuillen	Laborer	1911
James Mehan, Sr.	Miner	1912
James Mehan, Jr.	Miner	1912
A. Meredith	Laborer	1903
Charles Messenger	Laborer	1907
Charles Miller	Miner	1908
J.J. Miller	Miner	1907
Archie Miles	Laborer	1907
A. Miner	Laborer	1903
C. Walter Morris	Laborer-1907, Meatcutter-1908	
J.F. Morgan	Laborer	1904
James Morrison	Laborer	1903
Aubrey Morton	Helper	1909, 1910, 1911, 1912
John Morton	Blacksmith	1909, 1910, 1911, 1912
John Nasal	Laborer	1912
John Naylor	Fireman	1905
Ephrian Neely	Carpenter	1903, 1904
Argot Nelson	Laborer	1906
W.H. Nervis (Nivrus?)	Methodist-Episcopal minister	1909
W.K. Nix	Doctor	1912
Henry Nokes	Laborer-1907, 1909, 1910, Miner-1911, 1912	
Luther Nokes	Laborer-1907, 1908, Singleman-1910, Miner-1911, 1912	
Robert Oakley	Miner	1911, 1912
C.P. Oakman	Engineer	1904
Wm. S. Orner	Laborer, 1 cow	1906
D. Grant Orner	Laborer	1904, 1905
James T. Parker	Laborer	1912
J.T. Parks (same as above?)	Fireboss	1911
John Parsley	Laborer-1903, Motorman-1904	
Joe Pastor	Miner	1904
Reed Patterson	Laborer	1903, 1904, 1905, 1906, 1907
Hugh Peddicord	Laborer	1912

Name	Occupation	Year
James Pedicord	Laborer	1911, 1912
George Pendred	Pumper-1904, 1906, Laborer-1903, 1907, Engineer-1905, 1908, 1909, 1910, 1911	
Thomas Phipps	Laborer	1906
Elmer Piper	Laborer	1912
Henry Potter	Miner, single	1908
William Potter	Miner, married	1908
Morton Powel	Laborer	1912
Thomas Powell	Laborer	1903, 1904
Samuel Quinzler	Fireman	1908, 1909
D.F. Rager	Teamster	1911
J.L. Rager	Laborer, married	1908
S.H. Rager	Laborer	1904
R.W. Rairigh	Laborer	1905, 1906
A.D. Raymer	Laborer-1907, Miner-1908, married (Died in 1909 explosion)	
F.W. Ream	Sup, mill	1904
John Reed	Mine foreman	1904, 1905
	46 acres cleared, 12 timber from C. Hoffman	
Ebenezer Remey	Engineer	1906, 1907
John Resh	Barber	1912
Crist Reuther	Engineer	1903, 1904
Sylvester Rich	Laborer	1907
George Richards	Laborer-1903, 1905, Mechanist (sic)-1904, 1906	
James Riley	Laborer	1912
Samuel Roberts	Laborer	1912
Grant Robison	Laborer	1907
Fred Rockman	Miner-1904, Laborer-1905	
Gus Rodeen	Blacksmith, married	1907, 1908
J. Roger	Teamster	1904
Richard Rosemergy	Laborer	1904
George Rymer	Meatcutter	1907, 1908
Alexander Saler	Laborer	1910
J.B. Saler	Watchman	1911
Earl Sanders	Laborer	1912
Jno Sarman	Laborer	1911, 1912
Frederick Sauers	Laborer	1905
L.B. Saxton	Hotelkeeper	1903
H. Saylor	Mechanic	1903
James W. Scanlan	Mechanic-1903, Foreman-1904, Laborer-1905	
Geo. Schoola (?)	Laborer	1912
G.E. Schrader	Foreman-PRR	1910
M.B. Scott	Store manager-1903, Sup. store-1904	
Wm. Sell	Laborer	1904
Andrew Sheesley	Laborer, married	1908
G. Shelton	Laborer	1903
Wm. Sherwood	Laborer	1904
Michael Shields	Miner	1905
David Shilling	Single	1908
Peter Simmers	Laborer	1903
George Simpson	Motorman	1912
R. Skiles	Fireman	1904
Samuel Skitchall	Laborer	1905
John Slather	Laborer	1907
William Slatter	Laborer, married	1908
John Smart	Miner	1910
E.M. Smith	Clerk-1903, Police-1904	
H.G. Smith	Laborer	1903
Thomas Smith	Foreman	1903

Name	Occupation	Year
Thomas Snyder	Motor foreman	1912
Lloyd Specht	Bricklayer	1903
Karl Spicher	Laborer	1912
Scott Steel	Engineer	1904
Wm. Steel	Engineer	1904
Wm. Steel	Laborer, 1 horse	1904
James Stephens	Laborer	1907
Jas Stephenson	Laborer	1907
Sam Stiles	Laborer	1907
Anthony Stilip	Laborer	1904
Dave Stutzman	Carpenter-1904, 1905, Laborer-1906, 1907, 1908, 1912 2 cows	
John Stutzman	Laborer	1907, 1910, 1911, 1912
Jno Swatsunek	Laborer	1912
W.E. Swatecentruver (?)	Mine foreman	1903
Alex Taylor	Laborer	1908, 1909, 1911
Edward Thirlwell	Laborer	1903, 1904
Benjamin Thomas	Laborer	1907
Fred Thomas	Miner-1906, Laborer-1907, Assistant Superintendent-1909	
J.C. Thomas	Fireboss	1912
J.W. Thomas	Clerk	1903
Walter Thomas	Pipeman	1903
L.E. Thurston	Laborer	1909, 1910, 1911
Bert Tilinghan	Laborer	1907
A.D. Tillingham	Laborer	1910, 1911
A.B. Tillinghost	Laborer	1903, 1904
John L. Tobin	Mine foreman-1907, Machine boss-1908, Laborer-1909, 1910	
Thomas Tobin	Laborer	1909
Brown Treaster	Miner	1911
William Trotter	Miner-1906, Police-1907	
C.R. Tyler	Laborer	1904, 1905
Henry Van Horn	Laborer	1903, 1904, 1905
Charles Wallon (Wallen)	Miner-1908, Laborer-1911, 1912	
J. Walton	Laborer	1907
W.E. Warren	Minister	1908
J. Washington	Laborer	1903
Thomas Watson	Miner	1910
Amos Weakland	Master carpenter	1910
Arthur Weaver	Motorman	1912
Richard Weaver	Laborer	1912
Martin Welsh	Carpenter	1912
Peter Welsh	Laborer	1904
A.J. Wetmore	Laborer-1903, Carpenter-1904, 1905	
James Willan	Laborer	1903
B.L. Williams	Carpenter	1903
Ben Williams	Carpenter	1903
David B. Williams	Laborer	1905
Frank Willians	Carpenter	1903
Albert Wilson	Laborer	1907
Charles Wilson	Laborer	1912
Edward Wilson	Laborer	1907
James Wilson	Laborer	1905, 1906, 1907, 1908, 1909
John Wilson	Laborer, married	1908
William Wilson	Laborer	1903
William Wilt	Laborer	1903
James Woltz	Clerk	1906
Benj. F. Wood	Clerk	1906, 1907
Jas. Woods	Laborer	1912

Wm. Worthington	Laborer	1912
William Wosley	Engineer-1907, Track layer-1908, Motorman-1911	
Jas. Wray	Plumber	1904
J. Wright	Laborer	1903
Geo. Yearick	Doctor	1907, 1908, 1909, 1910
D.W. Yost	Railroader-1909, Laborer-1910, 1912, Repairman-1911	
Louis Young	Laborer	1907
William Young	Fireboss	1910
Steve Zania	Laborer	1912
Laird Zimmerman	Fireman-1904, Engineer-1906, 1907, Wayboss-1908, Laborer-1909, 1910, 1911, Engineman-1912	

FOREIGN MINERS: WEHRUM
Buffington Township, 1903
Foreign List: Wehrum, taken from a carbon copy of a typed list furnished to assessor

Check Number	Name
705	E. Kish
708	Joe Bargy
710	Chas Covash
711	Jos Coptic
715	John Sanju
718	Martin Garber
721	Sol Bartalon
722	John Barron
723	L. Kish
724	Joe Vilas
725	John Gansu
726	Sam Pent
728	John Nemish
730	John Rosh
731	John Bogn
733	J. Orosy
735	A. Perlock
736	J. Foseycash
735 (737?)	Louis Collar
738	Geo Drago
741	Chas Stoyka
742	Mike Pacula
743	Mike Cipero
745	John Malody
746	Joe Pastor
747	Chas Gaborder
750	Dick Covash
751	Jos. Covash
752	John Negra
753	Mike Kodi
754	Mike Forkash
756	Jos Flags
757	Thos Phillips
761	John Rennicks
762	Peter Parzika
770	Steve Novack
775	Frank Kosack
781	D. Horwatt
782	A. Zulic
792	John Copsic
793	John Bortush

796	**John Molner**
798	**John Garber**

No. 3 Colliery

801	**Mike Barkech**
802	**Geo Lutes**
804	**Chas Bukas**
805	**Mike Fedoc**
812	John Stanko
813	John Gibon
814	John Kovach
816	Jos. Rubis
818	Mike Bogdan
819	Mike Barkish
823	Mike Yack
824	Andy Franc
825	Frank Nemesh
830	Mike Jerme
831	A. Veskie
833	Steve Hosack
834	John Stoki
835	Nick Snavish
838	Chas Piden
842	Geo Polock
844	Chas Georgi
845	John Balko
846	Geo Raphenus
847	John Panco
848	Jos. Choppel
849	Mike Names
850	Chas Luck
852	Jos Kosack
853	Mike Hardy
854	Jos. Kess
855	John Vallo
856	Mike Yanno
857	C. Feddecki
860	John Robola
862	Geo Bogdan
864	Andy Sherbo
865	Chas Toriski
866	Jos. Hallet
867	John Kovautch
868	Geo Panick
870	John Yanos
871	Mike Vallo
872	Andrew Vallo
873	Mike Sepher
875	Andrew Brunocki
876	A. Chanko
877	John Zays
878	Jas. Sepher
879	John Kovaci
880	Frank Stoki
881	Mike Kertz
882	Paul Lokauch
893	Chas Stoker

934	Andy Kish
935	Chas Lava
936	John Macask
937	John Detora
938	Chas. Kevish
942	John Boroasky
958	Gray Desaina
961	Frank Julit
962	Thos Caneda
937 (967?)	Frank Danch
977	J. Choykas
979	J. Harak
981	M. Ondy
983	Steve Balog
985	W. Pakance
986	J. Kerchok
988	J. Stoyka
990	M. Olak
991	J. Botak
1029	P. Foyt
1030	M. Fudor
1062	G. Mandoki
1063	S. Faraca
1066	J. Deoteo
1068	J. Kish
1070	P. Roka
1073	M. Stevison
1074	V. Kevies
1075	C. Shebo
1078	E. Kosko
1081	J. Roka
1086	A. Fyes
1104	P. Regola
1105	John Delgram
1108	Frank Lennett
1109	Joe Lennett
1110	Frank Belgram (Delgram?)
1113	F. Santre
1115	Joe Cambi
1116	C. Osfolm
1117	S. Caso
1118	D. Cordia
1127	F. Fawcutt
1138	Sylvan Marlough
1147	J. Bennett
1202	John Feyen
1203	John Copching
1204	M. Dohonish
1205	John Myers
1206	Geo. Ratzo
1207	Sam Baldi
1208	Mike Powell
1211	P. Stoyka
1212	John Revee
1213	Mike Putzki
1214	C. Sarwady
1215	Mike Frantz
1216	C. Dohonish

1217	G. Kodi
1219	Louis Ples
1259	Joe Sepra
1260	John Monoch
1261	Andy Batza
1262	Velot Mertac
1263	Velot Desaina
1266	P. Lillinger
1268	Mike Kosak
1270	M. Zula
1271	Steve Demian
1273	Frank Teson
1277	Frank Drago
1282	V. Fayce
1285	Chas Kosack
1303	J. Bogdan
1304	J. Straker
1306	C. Stocker
1308	C. Bogdan
1309	C. Wargo
1310	M. Danch
1312	C. Holoka
1314	M. Holoka
1315	J.T. Risko
1317	G. Kretch
1318	J. Risko
1323	A. Rammko
1327	J. Danch
1329	J. Karnuta
1331	J. Tilcot
1518	L. Chicken
1519	A. Laffarda
1522	Andrew Hericus
1523	Andy Washko
1527	M. Modestina
1528	Louis Hoosak
1395	M. Mussamillo
1396	M. Ponso
1397	E. Mazzei
1399	R. Balardo
1401	Carlo Tomasi
1409	C. Nicola
1411	C. Angelo
1413	P. Lassano Pelegrime

Buffington Township, 1906
Foreign List: Wehrum

Name	House Number
Adam Antel	195
Steve Amastic	176
Andy Boynce	194
Steve Bolgan	196
Andy Bolgan	196
Frank Bolgan	196
Gigtislli Bruno	11
Frank Cemanaski (?)	27
Louis Chick	13

John Colomphia	193
Frank Dugan	89
Lowman (?) Dugan	89
Dair (?) Enedy	194
Garrew Enegoits	194
Frank Fedeli	53
Rome Fedeli	53
Mike Gousuch (?)	140
Julen Garew	195
John Hady	194
Mike Litim	89
John Lavy	194
Eginots Lavy	194
Charley Lavy	194
Joseph Leo (?)	15
Frank Lilligraud	13
John Mida	198
John Matana	26
Joseph Mattie	172
George Mechre	193
Andy Nelishe	195
Sam Pitua	198
Joseph Pollen	30
John Risko	20
Paule Rosko	176
John Reseck	194
Martie Risko	52
Andy Stoyka	52
John Stoyka	
Mike Sephra	19
Martin Toyel	194
Peter Tote	176
Mike Tote	172
John Tote	172
Charlie Toynish	172
Nick Tole (?)	11
Joseph Tole (?)	11
Charlie Walton	42
George Wodash	213
Fred Wyte	89
Mike Powell	#3 mine
John Palaski	174
Yenish Mitish	Shanty in Rexis

See town plan for location of house numbers.

Buffington Township, 1912
Foreign List: Wehrum, pp. 79-94

Laborer's tax rate: co. tax-$.50, poor tax-$.07, building tax-$.04

P. 79
Charles Covash
Mike Buchnock
C.T. Tsauess (Travers?)
Steve Risco
Andy Andesko
Geo. Fsienchack (?)

P. 80
Andy Guss
Andy Masinio
Metro (?) Pasico
Vinten (?) Michna
Mike Sasika
Steve Saseski

P. 81
Steve Pamorah (?)
Steve Mar (?)
Steve Satah
Toney Paseueli (?)
Frank Basley (?)
Andy Galzra (?)

P. 82
Joe Pischeney
Peter Petroskey
Jos. Gergay
Mike Balko
Jno Lehnack
Chaso Lash (?)

P. 83
Steve Wosco
Steve Recko
Chas Poschamok
Gino Okits (?)
Mike Groeick (?)
Chas. Prodent

P. 84
Nick Seshen
Luey Saprot (?)
James Geung
Jno Rossmore
Jno chreuse
Tony Schasosin

P. 85
Lemme Consiza (Consigli?)
Joe Frencheck
Joe Daszza
Ellis Caghas
Jas Sabo
Renni Freader (?)

Jos. Degrazie (?)
Chas. Kinkarpe (?)
Charles Sefer
Mike Sebra
Charles Chips
Geo. Robis

Geo. Polick (?)
Mike Dayak
Chas. Lookishina
Jeo Jesney (?)
Geo Zolday
Andy Sabo

Steve Nocklah
Carmel Sovty
Mike Masdas
Mike Csin
Joe Barnos

Frank Scalbos
Mike Galestsko
Dereteslo Restu (?)
Iristina Denuhe (?)
Mike Selinski
Steve Elas (?)

Ralsh Batest
Metro Helod
Metro Relvda (?)
Chas. Karnich
Jno Laney
James Danna

Carl Batist
Vincent Pilot
John Tut (?)
Chas. Saekel
Frank Arnetta
Stephen Osick

John Betosi
Jos Rallie
Steph Schuck
Geo. Chipp
Jno Sabo
Jos Beswick

Joe Leona
Mike Masdzey
Charles Risco
Andy Sincheck
George Kudy

Mike Kush
Jos. Batista
Louey Unsereck
Jno Budnah
August DeMatella

Paul Keten
Mike Paszeroe
Alex Phillipi
Peter Fanssi
Jno Situra

Geo. Osack
Mike Sanno
Peter Paegole
Chas. Gresco
Frank Bulgara

Geo. Regillin
Geo Franco
Mike Dayack
Jno Daik
Steve Rajsie

Joe Robby
Lesi Bovyek (?)
Domy Dilegram
Milas Rajsie
Toney Bosboney (?)

Toney Deroka
Wm. Fatislak
Mike Kestich
Jn. Zaker (?)
Angelo Jacob

P. 86
Thomas Bewditch
Joe Pesino
Gaston Lasko
Jno Vesko
Joe Parchichnar
Ernest Masgyra
P. 87
Charles Passam
Andrew Walla
Joe Kavash
Geo Roszge
Peter Stablick
Joe Heska
P. 88
Jeo (?) M. Delegram
Mike Kisko
Geo Balko
Frank Fedock
Geo Masino
Joe Sabo
P. 89
Steve Beckah
Jno Gresko
Jno Leesah
Geo. Gesekah
Jno Rossum
Felix Demsick (?)
P. 90
Joe Bentilla
Geo. Passwell
Andy Pelko
Tony Besdanelo
Frank Remasie
Angelo Aterium
P. 91
Andy Dudash
Thomas Spinek
Joe Benny
Zesgro (?) Testa (?)
Duneneck Wise (?)
Ernest Penock
P. 92
Salvania Angelo
Patsy Leeaspel (?)
Louie Phillip
Wornen (?) Pastelak
Jno Resksot
Martin Felino

Geo Toten (?)
Jilus (?) Gani
Meston Miner (?)
Joe Melenski
Peter Gnesane
Jno Juntre

Steph Segsta
Charles Stecko
Ignos Geskah
Baset Heustin
Peter Haschehan
Mike Putchack

Frank Capreno
Toney Breay
Joe Levopah (?)
Chas. Haydo
Steve Sabo
Peter Woskeck

Frank Csonyolina (?)
John Barsah (?)
Jno Keleuse
Jno Kannik
Andrew Poloski
Adam Antimere

Angelo Askangelo
Frank Danalo
Philip Panemar
Chas. Povlis
William Meskay
Mike Jeney

Burt Radeck
Jno Chulack
Joe angelo
Nick Senerah
Geo Risko
Tonie Pennu (?)

Pau Mevle (?)
Masko Cranalze
Butch Mondo
Henry Conalyale
Jno Brechkesko
Donenick Angelo

Jno Stoyka
Andy Satab (?)
Gabor Harras
Charles Usick
Peter Hellod

Toney Jasnatona
Charles Alesco
Charlie Ballock
Elik Rabilla
Anthony Zalinski

Joe Popen
Stanley Stonga (?)
Joe Benkoe
Frank Rolko
Jno Chelek

Nicholas Slesges (?)
Andy Barsate
Caynona Datchen
Thos. Sekeskey
Nick Woskarick

Anaelo (?) Misek
Chas. Bessen
Mike Messhesco (?)
Joe Mazalesa
Angelo Banamala (?)

Jacob Raipjack (?)
Peter Cheso
Charles Retsko
Chas. Chanel
Andy Bero

John Simbrush
Jano Mazzesini
Mike Kestah
Martin Tar
Charles Bordo

P. 93

Andy Balko	Charles Fedick	Gasgar Sandes
Sam Stauty (?)	Joe Frank	Jno Balkno
Roger Mash	Alec Shevock	Geo Belko
Tony Angelo	Geo Stephanek	Charles Tetteysko
Angelo Metoskol	Flesly Sesolis	Mike Balya
Steve Chuhey	Mike Osak	

P. 94

Frank Sokolok	Frank Rischko	Mike Shunak
Frank Nickelet	Pete Mikusig	

Parenthesis indicate that the author encountered difficulty in deciphering the tax assessor's handwriting.

Appendix Q
Obituaries

1920's
   Annie Farkas, died January 8, 1923
   Lottie Barry Morris, May, 1923
   John Cresswell, died September 19, 1923
   Henry Berman, died April 13, 1926, father of Mrs. Julius Levison
   J.H. Krumbine, died June, 1926
   Helen Rosocha, 5, died January, 1927, fatally scalded when she fell into a tub of hot water
   Mrs. Martha Krumbine, died June, 1927

1930's
   William J. Lybarger, 55, died April, 1931
   James Francis McCloskey, killed at #6, November 12, 1931
   Catherine Nevy, 83, died July, 1934
   Evan Davis, 85, died December, 1935, ran boarding house in Vintondale
   William Evans, died December 30, 1935, janitor of Vintondale Schools
   Rebecca Lynch, 79, died January, 1936
   Herb Daly, April 11, 1884-May 12, 1936
   Dr. James P. MacFarlane, August 28, 1878-September 4, 1937
   Margaret Sides Williams, August 28, 1870-August 15, 1938
   Christean Bracken Goughnour, 81, died July, 1939

1940-1943
   John W. Daly, 68, died June 19, 1941
   Annabelle Burke Cresswell, October 31, 1867-April 6, 1941
   Walter Morris, October 6, 1880-October, 1941
   Ida Cook, February 23, 1869-October 15, 1941
   Sam Feldman, died November, 1941
   Mike Koraca, March 3, 1889-November, 1941, boarded with Eli Abramovich
   Joe Bennett, 1879-December, 1941
   Ada Michelbacher, died January, 1942
   Robert Henry Magee, died February 24, 1942
   Baloz Bagu, September 29, 1889-October, 1942
   Erwin Hakenen, June 16, 1905-November, 1942
   Sam Brett, died January, 1943
   Mike Farkas, April 25, 1943-November 15, 1978
   Anna Jacobs Sileck, 52, died June 6, 1943
   Samuel R. Williams, 76, died July 4, 1943

1947-1949
   Katherine Makepeace, January 20, 1893-January 4, 1947
   J.A. Goughnour, 88, died March 1948
   Guerino Averi, 60, died October 24, 1948
   Gust Nagy, died May, 1949, boarded with John Morey, Sr.
   Joseph Selak, died May, 1949, boarded with John Rabel
   Mike Vereb, 58, died July 4, 1949, invalid for 31 years due to spinal injuries received at #1 mine
   Mary Hallas, 1892-1949
   Arthur Pisaneschi, 1928-1949

1950-1951
   John Brozina, 86, died April 4, 1950 semi-invalid for 10 years after mine accident
   Harry Hampson, September 14, 1856-October 18, 1950
   Paul Saffranka, died September, 1950, boarded with Mike Beres
   Blair Shaffer, Jr., 66, died May 12, 1951
   Mike Budzer, 65, died September, 1951, services held at the Strizak home

1952-1953
    John Sarnosky, 55, died May 17, 1952, boarded with Frank Oravec
    Thomas Cloyd Hoffman, August 22, 1901-September 22, 1952
    Mrs. John Roberts, July 17, 1866-April 7, 1953, she and John Roberts celebrated their 65th wedding anniversary on January 25, 1953
    Anna Chekan, October 14, 1884-February 19, 1953
    Michael Felczan, died November 1953, boarded with Holupkas

1954-1955
    Milso Kosanovich, died May 10, 1954
    Mike Beres, 71, died January, 1955
    Mike Nevida, 73, died August 16, 1955 as a result of a fractured skull due to a fall. Funeral held at home of Pete Lazich
    John Roberts, January 7, 1864-September 15, 1955
    Philip Gnilakevich, September 27, 1921-September 27, 1955, found dead behind the wheel of his car, death due to a heart attack

1956-1958
    Joseph Sileck, March 20, 1884-August 16, 1956
    Josephine Cassol, October 21, 1900-August 17, 1956
    John "Blues" Lonetti, May 26, 1911-February 22, 1957
    Raymond "Biff" Cresswell, died June 23, 1958
    Andrew Wojtowicz, November 15, 1886-December 26, 1958

1959
    Lee Cresswell, May 10, 1903-January 1, 1959
    Meri "Nummie" Morey, May 24, 1878-March 4, 1959
    A.F. "Allie" Cresswell, October 4, 1894-May 1, 1959
    Anna Koscho, November 2, 1878-May 10, 1959
    Ivan Dotts, September 18, 1921-May 14, 1959
    Elizabeth Antal, September 19, 1879-June 20, 1959

1960
    Olin A. Hagens, Sr., October 18, 1910-January 15, 1960
    Thomas Kasper, March 1, 1916-January 22, 1960
    William T. Steele, July 31, 1878-May 10, 1959
    Father Charles Gallagher, born January 3, 1890, ordained June 29, 1918, died June 28, 1960
    Charles Gongloff, November 4, 1897-July 8, 1960
    Mary Nevy Avalli, October 18, 1886-August 4, 1960
    Annie Kuhar, September 22, 1899-October 31, 1960

1961
    Joan Nagy, August 29, 1895-February 22, 1961
    Nick Malenich, Sr., October 14, 1906-July 27, 1961
    Frank Pioli, April 12, 1890-October 31, 1961
    Sam Salvador Pandal, March 17, 1887-December 6, 1961
    John Firko, October 16, 1886-December 11, 1961

1962
    John "Cedor" Hatnovich, January 13, 1894-March 5, 1962
    John J. Huth, November 12, 1888-April 7, 1962
    Helen M. Palovich, August 17, 1918-May 2, 1962
    Mary Harasty, April 12, 1889-July 10, 1962
    Mary Ann Thomas, October 20, 1881-September 21, 1962
    Mike Grosik, Sr., August 15, 1883-November 30, 1962

1963
    Roy Moore, August 23, 1888-January 21, 1963
    Hilda Swope, died February 21, 1963, first grade teacher in Vintondale schools
    Harry "Red" Clark, Sr., March 4, 1903-June 5, 1963
    Mary Gresko Pytash, September 30, 1916-July 22, 1963
    Beatrice Hunter, April 5, 1891-September 11, 1963

1964
- Elizabeth Tourous, September 26, 1877-March 22, 1964
- Daniel A. Thomas, August 15, 1882-April 5, 1964
- Pete Tourous, Sr., November 21, 1884-May 22, 1964
- Rita G. Murphy, December 23, 1942-June 23, 1964
- Charles Crouse, August 17, 1882-September 29, 1964

1965
- Stephen Morey, April 22, 1900-March 17, 1965
- Thomas Madigan, August 1, 1891-June 29, 1965
- Mary Lynch Clarkson, May 26, 1880-August 28, 1965
- Helen Medvik, June 19, 1887-October 15, 1965
- Nicholas Oblackovich, November 2, 1888-August 31, 1965
- Gertrude McPherson, May 6, 1876-November 12, 1965
- Ella Shaffer Daly, February 14, 1888-December 14, 1965

1966
- Peter Toth, October 26, 1889-January 7, 1966
- Helen Balog, September 30, 1892-January 11, 1966
- Pete Lazich, died April 29, 1966
- Frank Monar, Sr., October 13, 1880-July 12, 1966
- Andrew Jacobs, Sr., March 11, 1899-September 1, 1966
- Frank Wray, November 15, 1890-December 31, 1966

1967
- Filomenia Colangelo, April 19, 1894-May 4, 1967
- James Asti, May 18, 1896-June, 1967

1968
- Mary R. Nevy, 1885-1968
- Walter Hunter, June 5, 1887-January 29, 1968
- Karla Jane "Kippy" Toth, February 8, 1966-April 15, 1968
- John W. "Jack" Butala, August 28, 1893-April 24, 1968
- Vincent Joseph Hasen, July 17, 1898-August 13, 1968
- Mary E. Shaffer, November 2, 1896-September 14, 1968
- Julia Markus, November 10, 1887-September 15, 1968
- Mary (Bosci) Buchey, December 5, 1888-October 20, 1968
- Ralph Jansura, March 13, 1918-November 3, 1968
- John Payni, February 27, 1887-December 5, 1968

1969
- Frank Sebulsky, January 1, 1898-January 25, 1969
- Pietro (Pete) Perandi, August 20, 1887-February 15, 1969
- Pete Tomko, April 15, 1887-February 26, 1969
- Father John T. Callan, born March 15, 1903, ordained July 1, 1928, died March 25, 1969
- Steve Kovach, October 17, 1915-July 30, 1969
- Paul McCloskey, August 27, 1969
- Andrew Gresko, October 6, 1954-September 21, 1969
- Charles Shestak, January 2, 1893-December 20, 1969

1970
- Andy Sabo, October 24, 1885-January 28, 1970
- Steve Glowa, Sr., March 13, 1900-March 17, 1970
- Barbara Balko, August 19, 1889-April 1, 1970
- Italina Nevy, December 25, 1888-June 11, 1970
- Joseph Giazzon, July 28, 1897-June 22, 1970
- Metro Karol, November 15, 1894-July 25, 1970
- Mary Stahusky, July 9, 1909-August 3, 1970
- Kalman Antol, Sr., February 4, 1890-August 7, 1970
- Andrew J. Chekan, March 3, 1907-October 24, 1970
- George Vaskovich, Jr., July 28, 1920-December 6, 1970

1971
- Charles Robert Lynch, August 22, 1891-January 26, 1971
- Victor Spongross, October 18, 1912-January 27, 1971
- Andrew G. Anderson, November 16, 1905-March 13, 1971
- Alma Hoffman, March 25, 1902-March 20, 1971
- Andrew Skubik, May 27, 1891-April 8, 1971
- Andrew S. Molnar, April 7, 1894-September 8, 1971
- John Oravec, July 12, 1906-December 4, 1971
- Clarence E. Shaffer, February 25, 1900-December 17, 1971

1972
- Anna Kopenits, November 26, 1886-February 5, 1972
- Theresa Farkas Boykin, December 6, 1929-April 15, 1972
- Theresa Kerekes, died April 28, 1972
- Frances Sebulsky Oravec, August 20, 1918-September 17, 1972
- Gertrude Beistel, January 20, 1906-October 20, 1972

1973
- Homer S. "Jack" Shaffer, September 22, 1893-April 7, 1973
- Mable Hampson Huth, May 2, 1888-June 29, 1973
- Carl Michelbacher, March 11, 1910-August 13, 1973
- Mike Shamko, Sr., January 12, 1909-October 17, 1973

1974
- Grace Ure, November 26, 1890-January 3, 1974
- Michael Monyak, March 21, 1913-April 10, 1974
- Anna M. "Goldie" Grosik, January 23, 1911-April 11, 1974
- Marie Cresswell Mihalik, May 12, 1900-April 23, 1974
- John Monyak, Sr., February 21, 1914-May 5, 1974
- Louis Dusza, February 29, 1920-August 23, 1974
- Alex Oro, December 6, 1888-December 1, 1974
- Michael Velisko, Sr., November 15, 1896-December 9, 1974

1975
- Cuthbert "Cup" Hayes, July 15, 1904-May 17, 1975 (Milkman for Alwines' Diary)
- Helen Oro, October 14, 1895-July 14, 1975
- Roy Roberts, April 3, 1902-September 26, 1975
- Stella "Baba" Wojtowicz, November 1, 1891-October 3, 1975
- Thomas Whinnie, October 9, 1891-October 25, 1975
- Charles Melhalko, October 15, 1894-October 4, 1975
- Victor Viga, October 28, 1919-November 7, 1975

1976
- Martha Morey, February 15, 1914-January 3, 1976
- George Hozik, November 23, 1913-January 31, 1976
- Bruno Bianucci, March 17, 1897-April 24, 1976
- Naydean Rosner Leleck, November 4, 1929-June 15, 1976
- Joseph Pioli, April 2, 1882-July 16, 1976
- Merle Pisaneschi, July 2, 1930-September 3, 1976
- Catherine Gongloff Johnston, September 29, 1906-September 5, 1976

1977
- Joseph Kerekes, March 31, 1895-April 11, 1977
- Charles "Fishy" McPherson, October 19, 1911-June 23, 1977
- George Berish, January 27, 1914-October 23, 1977
- Joseph Oravec, December 6, 1911-October 24, 1977

1978
- Michael Gasser, September 29, 1894-January 22, 1978
- Anna Brozina Butala, March 1, 1901-January 26, 1978
- Arnold Colangelo, August 28, 1926-May 8, 1978
- Martha Toth, April 28, 1921-May 23, 1978

Paul Shandor, Sr., August 15, 1914-June 26, 1978
Michael Hozik, died October 19, 1978
Anna Yelenosky, December 21, 1896-November 13, 1978
Oscar "Pete" Wray, April 9, 1912-November 23, 1978

1979

Joseph Grosik, died September 4, 1979
Charles Verba, January 9, 1913-September 9, 1979
Zetha Mae Wray, July 14, 1896-December 25, 1979
David Pesci, July 4, 1896-December 25, 1979

1980

Monabel Colbert Hill, January 4, 1921-March 2, 1980
George Telesko, died March 24, 1980
Charles Nemish, died December 10, 1980
Anna Nemish, died December 10, 1980
William Colangelo, February 23, 1925-December 20, 1980
James B. Wray, November 22, 1900-December 25, 1980

1981

Helen Glowa, May 11, 1908-June 3, 1981
Helen Dusza Farkas Firko, August 10, 1894-October 18, 1981

1982

Sadie M. Clark, January 26, 1897-January 1, 1982
Mary Brozina, September 2, 1899-April 9, 1982
Charles Harasty, Jr., August 16, 1922-April 18, 1982
William Toth, Sr., April 16, 1908-April 27, 1982
Christine Huth Beresford, November 28, 1920-August 14, 1982
Hugh Pisaneschi, May 13, 1896-August 19, 1982
Joseph Lutsko, June 28, 1908-October 12, 1982
John T. "Cappy" Smith, March 6, 1923-October 13, 1982
Stephen Kish, March 18, 1910-October 27, 1982
Edmond J. Lucas, May 7, 1935-November 2, 1982
Freda Morey, December 12, 1902-November 23, 1982

1983

George O'Hara, April 23, 1915-February 19, 1983
Anna Kosta, died April 13, 1983
Walter Lindrose, Sr., March 14, 1906-May 14, 1983
Frances Wojtowicz Pluchinsky, December 26, 1911-May 23, 1983
Joseph "Bay" Shedlock, March 5, 1915-May 24, 1983
Steve Kosta, August 14, 1915-July 3, 1983
Agnes Beltz, died July 13, 1983
Clara Abramovich, died August 11, 1983
Michael Mastovich, January 24, 1918-August 16, 1983, music teacher at NGV schools
Mary Monyak Drabbant, September 4, 1905-August 23, 1983
Edna Shultz Misner, February 29, 1892-September 27, 1983
Steve "Pishta" Hedgedus, March 12, 1928-October 19, 1983
James Virgil "Verdy" Dempsey, October 26, 1914-October 21, 1983
William Lindrose, August 20, 1901-December 8, 1983

1984

Minnie Clouser Regetz, October 16, 1896-January 10, 1984
Nancy Gnilkevich, died February 4, 1984
Alfred Pioli, April 11, 1919-February 21, 1984
Eula "Lou" Hampson Burr, September 7, 1885-April 1, 1984
Lena Cresswell, September 4, 1913-April 28, 1984
John Borzcik, December 26, 1925-May 11, 1984
Helen Simchak Oravec, December 15, 1907-May 22, 1984
Grace Samitore, died May 22, 1984
Anna Lutsko, January 6, 1912-June 15, 1984

Dorothea Goebert, died July 1, 1984
Joe Silagyi, April 28, 1896-July 24, 1984, Vintondale's last surviving boarder, lived at John and Lizzie Rabels
Peter Faish, March 6, 1914-August 5, 1984
William Grant Gongloff, December 9, 1887-August 12, 1984
Mearl Edmiston, Sr., July 4, 1895-September 12, 1984
Mary (Kish) McConnell Dyson, February 13, 1908-September 28, 1984
Ruth Daugherty, October, 1905-November 18, 1984, third grade teacher, Vintondale Schools
Mary Mehalko, December 6, 1899-December 14, 1984

1985

Martha McFeaters Mack, July 15, 1909-January 29, 1985
Goldie Moore Stophel, February 11, 1896-February 1, 1985
Sarah Zack Edmiston, May 1, 1898-February 3, 1985
Nick Sakal, December 5, 1894-February 11, 1985
Mary Beres, May 14, 1914-February 16, 1985
Mike Bugal, March 16, 1920-March 13, 1985
David "Peg" Mottin, March 20, 1924-March 13, 1985
Willard Donaldson, June 21, 1921-March 20, 1985
David Hagens, July 30, 1941-April 5, 1985
Walter Doyka, June 8, 1917-May 2, 1985
Sadie Lewis, November 5, 1911-May 10, 1985
Clara DeBona, August 17, 1921-May 30, 1985
Frank Mehalko, November 1, 1891-September 8, 1985
Stanley Kovach, Sr., February 14, 1925-September 21, 1985
Simonne (Mrs. George) Gresko, December 2, 1925-September 22, 1985
Mike Gorsik, Jr., October 2, 1910-December 28, 1985

1986

Mary Touris Hrapchak, January 19, 1914-January 1, 1986
Jennie Schreckengast-Beistel, January 1, 1910-January 10, 1986
Thelma Uncapher Fessler, July 8, 1930-February 7, 1986
Helen Farkas Hardisty, December 23, 1923-February 24, 1986
Lester Jansura, June 14, 1919-March 4, 1986
Dorothy Whinnie Emery, November 22, 1917-March 23, 1986
Gustie Pesci, November 22, 1906-March 30, 1986
Sara Bagu Cresswell, died March 30, 1986
Palmina Nevy Pioli, April 25, 1896-April 3, 1986
Alex Yobbagy, January 25, 1915-April 10, 1986
Albert "Ab" Rager, March 29, 1902-April 11, 1986
Mike Mihalik, October 10, 1888-April 28, 1986
Irene Barate, March 28, 1916-May 13, 1986
Anthony Miranda, Novmber 3, 1930-June 25, 1986
Claire "Tood" Sebulsky, October 24, 1918-July, 1986
John "Jack" Daly, October 30, 1927-July 28, 1986
Sarah Mahan Reed, June 4, 1915-October 2, 1986

1987

Betty Daly Schweppe, January 25, 1921-January 5, 1987
Linda Rosner Hayes, February 21, 1948-January 6, 1987
Paul "Igor" Shandor, January 29, 1954-March 8, 1987
Gino Simoncini, July 14, 1902-July 16, 1987
Joanna "Jennie" Simoncini, September 22, 1906-July 16, 1987
Blanche Ruffner Good, March 15, 1899-July 27, 1987
Harry E. Ling, November 26, 1887-July 29, 1987
Edward Lutsko, October 8, 1934-August 7, 1987
Mary Stenko Butala, November 6, 1892-August 16, 1987
Julia Brozina Ure, November 29, 1903-November 4, 1987
James David McCloskey, August 6, 1930-December 7, 1987
Douglas Kuhar, December 13, 1958-Decmber 21, 1987
Robert Carland, died December 25, 1987

1988
- Ann Bisko Sugyik, May 4, 1918-January 11, 1988
- Theresa Dobias Kerekesh, June 29, 1889-January 15, 1988
- Robert F. Donaldson, November 17, 1926-January 31, 1988
- Martha Kovach Lezanic, died July 17, 1988
- Roy J. Michaels, July 7, 1924-February 9, 1988
- Mary Byich Mihalic, September 25, 1927-March 5, 1988
- Marlin L. Bowser, May 26, 1919-March 8, 1988
- Ethel Beres, January 28, 1889-March 16, 1988
- Anna Dancha Yanko, June 5, 1909-March 22, 1988
- Anita Asti, July 5, 1901-May 2, 1988
- Edith Karosi Simmons, March 6, 1906-May 23, 1988
- Wade Frantz, February 24, 1918-May 25, 1988
- Esther Molnar Gulakowski, August 8, 1922-May 26, 1988
- Josephine Pisaneschi, September 11, 1903-August 7, 1988
- Elizabth Ruffner Robison, July 24, 1897-October 1, 1988
- Carl Dishong, died October 12, 1988
- Ivy A. Feldman, died December 8, 1988

1989
- Mary Eleanor Cooke, June 3, 1914-January 13, 1989
- Jay C. Michaels, July 26, 1889-January 15, 1989
- Anna Rose Lucas, March 18, 1909-January 25, 1989
- Joseph Szabo, January 19, 1919-February 23, 1989 (Born in Wehrum)
- Charles Blair, February 26, 1921-March 8, 1989
- Stephen A. Pytash, August 13, 1915-May 10, 1989
- Joseph Shestak, March 12, 1924-May 15, 1989
- Alfred "Tubby" Pisaneschi, 1931-May 24, 1989
- Leonard Yahnke, February 11, 1919-June 8, 1989
- Stephen DeLosh, June 20, 1943-June 13, 1989
- Mary Smith Cramer, March 5, 1912-June 19, 1989
- Joseph Barotina, May 9, 1914-June 19, 1989
- Joseph Dodson, April 11, 1901-June 22, 1989
- Kathleen Balog Gaboda, August 30, 1900-July 10, 1989
- Daniel Merritts, May 5, 1931-August 3, 1989
- Hazel Misner Fleming, February 9, 1907-August 23, 1989
- Susan "Sis" Hagens, April 5, 1912-November 16, 1989
- Harry "Hap" Hess, October 12, 1913-December 1, 1989
- Dora Lazich, July 9, 1906-December 31, 1989

1990
- Jennie Colbert Rager Cramer, May 12, 1896-January 14, 1990
- Alvin Miller, April 28, 1915-January 29, 1990
- John Smidga, September 6, 1912-February 26, 1990
- Ella Gill Lantzy, June 1, 1897-March 6, 1990
- Angeline Gasser Gerley, May 18, 1907-April 4, 1990
- Sarah Simmons Beltz-Hagens, October 28, 1918-April 19, 1990
- George Miller Molnar, March 22, 1910-March 12, 1990
- Catherine Morris Lybarger Uzo, 1906-May, 1990
- Shirley Misner Rubish, January 30, 1935-May, 1990
- Emma Ruth McCloskey, April 26, 1915-June 24, 1990
- Isabelle Wojtowicz Tackett, December 23, 1918-July 28, 1990
- Walter Nemesh, August 11, 1931-July 30, 1990
- Anna Trynovich Banko, April 22, 1914-August 2, 1990
- John Kangur, February 2, 1910-September 8, 1990
- Helen Ure Monar, died October 17, 1990
- Agnes Huth Dusza, February 6, 1913-October 17, 1990
- Bert Szabo, April 18, 1922-November 6, 1990 (Born in Wehrum)
- Hazel Cramer Dill, September 21, 1906-December 27, 1990

1991

Elizabeth Farkas Belk, October 12, 1921-January 3, 1991
Ann Birch Rogalski, March 17, 1908-January 4, 1991
Lewis Fessler, May 13, 1916-January 27, 1991
John Bosley, Sr., Septmber 3, 1942-February 1, 1991
Nino Barchella, July 24, 1911-February 6, 1991

WEHRUM

Virginia Cruciotti Averi, died December 24, 1923
J.E. "Ed" McMullen, died February 7, 1939 in Wehrum
Jesse Craig, died May 18, 1945
Mrs. John Dutko, died 1948
Mary Frazier, 88, died May 12, 1951
Ralph Cresswell, died March 11, 1955
Pearl Craig, died May 12, 1963
Joseph Shook, died March 17, 1964
A.E. Brickley, died June 16, 1965
Mrs. Burns, died November 10, 1965, age 87
W.C. Graham, died January 9, 1968
Barbara Goodwin Lawson, died 1968
Bessie Decker Cassat, March 22, 1893-May 12, 1969
Harry Kime, died September, 1969
Harry Kuchenbrod, November 30, 1883-April 15, 1970
Charles Johns, died November, 1970
Katie Kuchenbrod, November 30, 1883-April 15, 1970
Twilight Gaster, died 1971
Mary Fyock Davis, died March, 1971
Melvin Davis, died July, 1971
George Lindsay, October 12, 1879-October 10, 1972, Wehrum superintendent in the early 1920's
Della George Griffith, died January 1, 1980
Arthur Buterbaugh, died January 4, 1980
Stephen R. Risko, August 12, 1905-July 19, 1980
Jackson Blattenberger, November 16, 1911-July 21, 1980
Dorothy Handlon Ridenour, August 1, 1906-January 10, 1983
Alice Rose, February 4, 1919-December 11, 1983
John "Shorty" Leck, October 26, 1925-January 30, 1985
Reverend Clarence M. Bennett, June 23, 1901-February 12, 1985
Esther Mack Salyards, January 1, 1919-February 16, 1985
Edward F. Buckingham, June 21, 1904-September 2, 1985
Mary Ann Stoyka Mahalko, July 26, 1918-September 10, 1985
John "Jay" Churilla, July 6, 1923-November 8, 1985
Helen Sheftic Grieco, March 26, 1914-April 1, 1986
Emmet Averi, March 1, 1914-April 12, 1986
Emma Huey Stinson, July 18, 1902-July 31, 1986
Bertalan "Bert" Toth, January 22, 1905-December 29, 1987
Anna Perbola Longazel, July 28, 1896-January 10, 1988
Hulda Duncan Wagner, May 14, 1909-February 1, 1989, teacher at Wehrum school
Mary Wizniak DelPratte, February 28, 1916-February 20, 1989
Marco DeRubis, Sr., June 27, 1915-May 17, 1989
John Bokash, June 27, 1928-August 18, 1989
Ralph W. Beard, July 30, 1927-January 31, 1990
Harry C. Graham, April 9, 1897-February 12, 1990

CLAGHORN

LaRue Foster Bowers, July 9, 1919-September 3, 1988

GRACETON

Albert E. Oswalt, November 17, 1912-August 10, 1987, postmaster of Graceton for forty years, former employee of Graceton Coal and Coke.

John Varescak, December 30, 1917-August 15, 1990

OTHER NOTABLES

Webster Griffith, June 5, 1860-April 2, 1928
Honorable Augustine Vinton Barker, June 20, 1849-August 20, 1928
Deck Lane, Ebensburg photographer, died October, 1930
Honorable Samuel Lemon Reed, March 13, 1864-October 13, 1934
Clair Bearer, August 23, 1891-June 24, 1986
Honorable George W. Griffith, December 12, 1894-June 26, 1988
Malcolm Cowley, August 24, 1898-March 28, 1989
Randolph Myers, April 9, 1889-August 30, 1989
Dr. Thomas Dugan, 1921-October, 1990

Appendix R
Wehrum Cemetery
TRANSLATION OF TOMBSTONES

Translation was done by Mrs. Helen Balog Oravec in May, 1985 through an academic excellence grant awarded to Denise Weber by the Indiana Area School District. This cemetery is owned by Sts. Peter and Paul Russian Orthodox Church of Vintondale.

Row 1 is at the back of the cemetery, and the translated stones will read from left (south) to right (north).

ROW 1:
1. Father
   Michael Babich
   10/24/1886
   3/28/1927
2. Stefan Risko
   1/9/1907
   1/19/1907
   Memory Eternal
3. Michael Hritz
   Died June 10, 1911
   Memory Eternal
4. Here Rsts With God
   Charles Dancha (Donge)
   Died 9/20/1911 in Vintondale
   Lived on This Earth 24 Years
   Born in Cumalovo
5. Born October 26, 1908
   Died January 13, 1912
   Anna Sileck
6. John Gorgi
   Born 10/1880 in City of Uhly
   County of Maramaros, Austria-Hungary
   Died 11/30/1912, 32 Years Old
   Survived By Wife Mary and Children
   Mary, John and Helen
7. Christina ?
   Daughter
   1912
8. Andrei Toten
   Born 2/8/1911
   Died 9/1/1912
9. Christine Celidin
   Born 3/22/1912
   Died 11/1/1912
10. Unable to Decipher
    Inscription Worn Off

ROW 2:
1. Here Rests
   Mary Lach
   Died 5/3/1925
   Born in Malovo, County of Maramaros
2. Michael Miklus
   5/21/1907
   Age 28 Years

3. "Rock of Ages"
   Theodore Laska (?)
   Born 3/1876
   Died 10/10/1908, Age 32 Years
   Born in Dravhove, County of Maramaros
   Eternal Memories
4. Here Lies
   Peter Todoravic
   1/3/1915
   3/9/1917
5. Here Lies
   Joseph Poline
   Born 1/19/1874
   Died 1/1/1916
6. Anna Orcak
   Died 3/22/1916
   Eight Years Old

ROW 3:
1. Peter Stoyka
   1920-1923
2. Ilya Rusanuk
   30 Stotnicov, Celecha, County of Maramaros
   Died 1/25/1917
   Age 22 Years
3. Here Lies
   John Bokoch
   Born 1889
   Died 2/4/1917
   Born in Village of Uglia, County of Maramaros
4. Here Lies a Servant of God
   John, son of John Dajak (Daok)
   31 Years Old
   Died 4/16/1918
   Peace on His Ashes (Body?) (Soul?)
   Born in Cumalovo, Maramaros
5. Here Lies a Servant of God
   John, Son of Charles Dovbnec
   30 Years Old
   Died 4/23/1918
   Peace on His Ashes
   Born in Cumalovo, Maramaros
6. Here Lies Body of Servant of God
   George Balko
   Born in Village of Velikelucki, Austria
   Died 9/10/1918
   35 Years Old From Birth
   Married Two Years to Anna
   Eternal Memories

ROW 4:
1. Here Lies
   Theodore Ckba (Skiba)
   Born in Lisko, Galacia in 1888
   Died January 31, 1917
   Memory Eternal to Him

2. Michael Pachnich
   40 Years From Birth
   Born in Pracedo, Maramaros
   Died 4/18/1918
   Peace to the Soul
3. Here Lies a Young Man (Son)
   Alexander Dronko
   Born 3/4/1920
   Died 4/10/1921
4. Here Lies
   John Leleck
   Born in Veleklucki, Beleischa County, Carpathian Russia
   Died 3/4/1920, Age 26 Years
   This Monument Was Erected to Her Loving Husband By His Wife Sophia
   Peace to Your Body (Soul?)
   (This gravesite was originally surrounded by a cast iron fence.)

ROW 5:
1. Here Lies
   George Holod
   Born 11/16/1873 in Veleklucki
   Died 12/8/1917 in Black Lick, PA
   Eternal Memories
2. Belieed to Be a Dedication Marker for the Entire Cemetery
   10/1/1906 (A Church Holiday)
   George Holod, Curator
   Village of Nagy Luscka
   Posz Chresztna, Wehrum, PA
   Wife Sophie
   Kurator, Cz.
   Rear Side of Monument
   Alex Szewczyk
   1864 F.M. Katarina Sz W.
   S. Pat Pokrowy
   Kurator, Okt 1
3. Here Lies
   George (Frank) Risko
   Son of Charles
   Born in Maramaros
   17 Years Old
   Died 1/15/1912
   Survived by Mother Helen Hritz
4. Here Les George Varga
   Son of Charles
   Born in Veleklucki, Austria-Hungary
   Died 5/12/1918
6. Mother
   Palach
   1885-1922

ROW 6:
1. Unable to Decipher Name
   9/27/1923
   9/30/1923
2. Daughter
   Helen Rosocha
   8/15/1921
   8/30/1927
   Gone But Not Forgotten

3. Here Lies
   Anna Risko
   Daughter of Michael Drahova
   Born 3/30/1904
   Died 12/12/1918
   Eternal Memories, Blessed Rest
4. Here Lies
   Anna Elnecki
   6/30/1918
   3/17/1922
5. Nikolai Chulak
   ?/27/1917
   Died 9/8/1921
   Seed of Michael Chulak

WEHRUM CEMETERY LIST

The following names are compiled from death certificate registers dated 1906-1921 found in the John Huth attic in 1982. Question marks designate the author's inability to interpret the handwriting. Death certificates are from residents of Jackson and Blacklick Townships only. Actual Wehrum residents are not listed here. Any name given without an address was a Vintondale resident. These are not complete records for those years.

Name	Age	Dath of Death
Jno Labius	3 1/2 months	10/31/1906
Mary Marin	1 day	2/6/1907
Stillborn Labanitz (no undertaker)		7/2/1907
Jno Kovach	2 years	7/30/1907
Mike Brihara	50 years	9/15/1907
Found dead near RR tracks near Expedit		
Mary Risko	2 months	1/30/1908
Annie Popovich (Twin Rocks)	23 1/2 months	10/7/1908
Ellie Recko (premature)	1 day	4/19/1909
Chas. Poken (Twin Rocks)	16 days	4/20/1909
Annie Fedanyer (Twin Rocks)	3 years	8/22/1909
Ilka Toht	13 days	10/13/1909
Mary Ribnicki	1 month	2/9/1910
George Mucca (mine accident?)	42 years	3/19/1910
Katie Griseisui	41 days	3/22/1910
Eva Rumanic	28 years	7/2/1910
Metro Romanick (child of above)	2 months	7/14/1910
Zofi Kozer	6 months	8/9/1910
John Klein	2 months, 1 day	9/27/1910
Ellie Kozer	2 years, 6 months, 1 day	11/4/1910
Mike Sobo	2 days	11/5/1910
Mary Yotsock (Bracken)	11 months, 7 days	12/2/1910
Stillborn Male Shristock		3/9/1911
Stillborn Ballogh		7/3/1911
Charley Donge (mine accident)	25 years	9/20/1911*
Anna Seleck (burn victim)	3 years, 3 months, 17 days	1/12/1912
Frank Risko (mine accident)	17 years, 11 months	1/15/1912*
John Jordy, (Colver, mine accident)	31 years	1/30/1912
Cristena Gyorgyog	22 days	1/31/1912
Mariano Sobo	7 days	4/5/1912
Ellen Stoyka	18 days	8/11/1912
Anna Stoyka	19 days	8/13/1912
Andy Gaboda	9 months	8/9/1912
Andy Tot	1 year, 7 months	8/31/1912*
Christena Serlardey	9 months	12/12/1912

Name	Age	Date
John Jewhas (premature)		12/14/1912
Stepheson Alupka	2 months	3/11/1913
Jim Telecskametva	1 month, 1 day	11/30/1913
Unnamed Female Bolok (premature-6 months)		12/18/1913
Joe Cozar	25 years	1/7/1914
Charlie Ochko	2 years, 4 months, 2 days	2/17/1914
Anna Holopka (Colver)	2 months	2/22/1914
John Celody	8 months, 23 days	4/1/1914
Unnamed Male Kontnez (premature - stillborn)		4/5/1914
George Slerko	3 months, 18 days	7/23/1914
Annie Melhalko	5 months, 15 days	8/9/1914
Mary Kruniake	14 days	9/7/1914
Andy Cossor (Bracken)	3 years, 11 months	9/19/1914
George Vaskovich (stillborn)		11/20/1914
Andy Batsko (mine accident)	53 years	4/14/1915
John Reneska (Curwensville)	1 month, 21 days	4/7/1915
John Sotcharok (mine accident)	36 years	4/19/1915
Not Named Rabit (stillborn)		5/1/1915
George Haluska (Twin Rocks)	7 months	5/20/1915
Metro Popovash	49 years	6/2/2925
Baby Todorovich	5 months, 5 days	6/8/1915
George Gubosh	4 months, 1 day	7/28/1915
Peter Balog	1 month, 18 days	8/5/1915
Metro Kerekes	1 day	9/1/1915
Anny Hollet	7 days	10/17/1915
Geo Felde	44 years	11/1/1915
Joseph Poline (coke yard accident)	42 years	1/1/1916*
Unnamed Male Juba (Bracken)	1 month, 1 day	1/2/1916
Frank Pepok (Twin Rocks)	1 year, 1 month, 2 days	4/7/1916
Rosa Pindur	10 months, 19 days	4/10/1916
Annie Kraszienen (Twin Rocks)	4 1/2 months	8/16/1916
Peter Kerekes	4 months, 28 days	11/14/1916
Geo Volosin	8 months, 8 days	1/14/1917
Frank Skiba (Twin Rocks)	25 years	1/31/1917*
Charles Zuba (Twin Rocks)	1 month, 19 days	2/10/1917
Daniel Korin (Twin Rocks)	62 years	8/6/1917
Annie Maniski (Bracken)	4 months, 28 days	9/2/1917
Annie Kiasilims (Twin Rocks)	2 months, 20 days	9/3/1917
Andy Krichko (stillborn)		2/6/1918
Sophia Kreatihko	1 year, 2 months	3/8/1918
John Kriscolusion	1 month, 29 days	8/16/1918
Stephans Page	7 months, 10 days	3/17/1918
Mike Boconich	42 years	4/16/1918
Nick Balazok (Bracken?)	42 years, approx.	4/28/1918
Mike Todorvick	2 years	7/16/1918
Charles Risko	3 months, 21 days	8/21/1918
John Holobin	3 days	8/17/1918
Anna Kopontek	1 year, 1 month, 1 day	8/24/1918
Pete Jlcki (Twin Rocks)	1 year, 1 month, 13 days	8/28/1918
Charles Popovich (Twin Rocks)	8 1/2 months	9/8/1918
James Colos (mine accident)	23 years	9/14/1918
Stillborn Balog		9/4/1920
John Minchenko	7 months	9/13/1920
Mike Simsko	1 month	10/22/1920
Murice Robanky	1 day	10/28/1920
John Luzar	2 1/2 months	12/6/1920

John Senachko	9 months	2/27/1921
Eleck Pronko	11 months	4/10/1921*

The asterisk indicates that there is a surviving cemetery marker.

RUSSIAN CHURCH DEEDS
STS. PETER AND PAUL RUSSIAN ORTHODOX GREEK CATHOLIC CHURCH

Indiana County Deed Book 100, p. 149, October 8, 1906
R.W. Mack, Buffington Township to Archbishop Tilchon of the Russian Orthodox Church. $200 for .45 acres in Buffington Township.

Cambria County Deed Book 208, p. 260, September 24, 1907
Vinton Colliery Company to Archbishop Platon. $1.00 for Lot 11, Block J on Third Street.

Cambria County Deed Book 231, p. 45, February 2, 1911
Joaniky Kraskoff, single, to Congregation of the Russian Orthodox Church of Vintondale. $300.00 for Lot 9, Block J on Third Street. Previously conveyed by Vinton Colliery to Kraskoff on September 24, 1907 for $300.00.

Cambria County Deed Book 267, p. 637, February 1, 1915
Platon Rozdestvensky, formerly Archbishop, to Alexander Nemolousky, Bishop of Russian Orthodox Church of North America. $1.00 for Lot 11, Block J.

Cambria County Equity Docket, Volume 3, p. 121
Bill in Equity.

Cambria County Deed Book 354, p. 87, October 24, 1922
Vinton Colliery Company to Russian Orthodox Church. 2.14 acre lot in Jackson Township on south side of Goughenour's road.

Indiana County Power of Attorney, Volume 9, p. 148
October 16, 1923. Eudokim Merchesky, Metropolitan of Russia and former Archbishop of diocese of North America to John S. Kedrovsky, grants his power of attorney to John S. Kedrovsky, Metropolitan of All America and Archbishop of North America for all properties used for the purposes of the Russian Orthodox Greek Catholic Church. The power of attorney was to apply to each parcel standing in his name as bishop or archbishop.

Appendix S
A.V. Barker Property Sales in Vintondale

1. Barker to W.S. Cook, Lot 10, Block A, October 2, 1906, $1,200. Alcoholic restrictions not included.
2. Barker to Pete Begi, lot next to Alex Jacobs, Plank Road area, June 27, 1906, $90. Sale of alcholic beverages prohibited.
3. Barker to Annie Rishko, Plank Road near lot sold to Mike Petrilla, March 11, 1907, $525. Sale of alcoholic beverages prohibited.
4. Barker to Makoi Gernicki, next to land of Eva Jacobs, August 22, 1906, $150. Sale of alcoholic beverages prohibited.
5. Barker to John Gernicki, next to lot of Makoi Gernicki, August 22, 1906, $150. Sale of alcoholic beverages prohibited.
6. Barker to Mike Petrilla, Plank Road borders Annie Rishko and Annie Cilip, August 3, 1907, $675. Sale of alcoholic beverages prohibited.
7. Barker to John Rishko, Plank Road, borders Annie and Mary Poloney, September 30, 1907. Sale of alcoholic beverages prohibited.
8. Barker to Annie Culp, Plank Road, borders Mike Petrilla, June 27, 1906, $100. Assigned by Annie and Joe Cilip to Eva Jacobs for $1.00. Sale of alcoholic beverages prohibited.
9. Barker to Agnes Szucs, house 59, near Albert Cyrus and Annie Farkas, January 24, 1907, $525. Sale of alcoholic beverages prohibited.
10. Barker to Steve Woyinagi, bordering public road, February 23, 1907, $250. Sale of alcoholic beverages prohibited.
11. Barker to Paul Paloney, borders John Rishko, June 17, 1909, $250. Sale of alcoholic beverages prohibited.
12. Barker to John Babinseck, public road, borders Annie Farkas, September 2, 1907, $600. Sale of alcoholic beverages prohibited.
13. Barker to Paul Graham, Lot 14, Block A, borders S.S. Ruffner on east, August 10, 1914, $800. Alcoholic restrictions not included.
14. Barker to Antonio Monza, Cresson, borders Albert Cyrus, VCC and Paul Paloney, October 15, 1913, $450.
15. Barker to Mrs. Minerva Cameron, Lot 2, Block D, March 13, 1915, $2,700.
16. Barker, widower, to Andy and John Jacobs, to take place of deed from Barker to Alex Jacobs, April 18, 1906, which was destroyed before recording in Cambria County, $1.00.

Bibliographies
WORKS CITED
Primary Sources

INTERVIEWS

Anderson, Joseph, Vintondale. Personal interview.
15 February 1990.

Anderson, Mr. and Mrs. John, Vintondale. Personal interview.
1 September 1990.

Anderson, Mrs. Ruth Roberts, fomerly of Chicago, now of Johnstown. With Cora Roberts. Personal interview.
6 July 1981.

Ashman, Dr. Philip, Johnstown. Personal inteview.
3 April 1982.

Averi, Emmet, Vintondale Homecoming. Personal interview.
September 1982.

Babich, Charles and Michael, Vintondale Homecoming. Personal interview.
31 August 1986.

Barclay, John, Clymer. Personal interview.
3 September 1990.

Bearer, Clair, Ebensburg. Personal interview.
22 June 1982.

Bennett, Rev. Clarence, Indiana. Personal interview.
1 July 1983.

Beres, Mrs. Esther, Vintondale. Personal interview. Translated by Stephen Dusza.
1 November 1981.

Biondo, John, Vintondale. Personal interview.
28 December 1981.

Biondo, Mrs. Elizabeth Simon and Mrs. Mary Simon Kerekish, Vintondale. Personal interview.
15 February 1990.

Bracken, George. Vintondale R.D. #1. Personal interview.
9 February 1991.

Brandon, Dr. Boyd, Reynoldsville. Personal interview.
11 July 1983.

Burr, Mrs. Eula Hampson, Roebling, NJ. Personal interview.
8 December 1981.

Butala, Mrs. Mary Strugala, Ebensburg. Personal interview.
1 November 1981.

Callan, Rev. Thomas, T.O.R., Ehrenfeld. Personal interview.
September 1986.

Chick, Mrs. Catherine Hanlon, Blairsville. Personal interview.
17 August 1983.

Dancha, Mrs. Ann Petrilla, Vintondale. Personal interview.
19 June 1982.

DiFrancesco, Samuel, Johnstown. Personal inteview.
29 September 1981.

Dill, Waton and Hazel Cramer Dill, Vintondale, R.D. Personal interview.
14 September 1990.

Dodson, Joseph, Claysburg. Personal interview.
18 September 1982.

Dodson, Russell. Vintondale. Personal interviews.
9 November 1980, 4 January 1981.

Drabbant, Mrs. Mary Monyak, Vintondale. Personal interview.
27 February 1982.

Dusza, Mrs. Agnes and Mary Ellen Pytash. Personal interview.
13 October 1982.

Dusza, Stephen, Vintondale. Personal interviews.
    29 September 1981, 15 January 1984, 7 June 1990.

Esaias, Mrs. Lovell Mitchell, Johnstown. Personal interview.
    14 November 1982.

Ezo, William, Vintondale Homecoming. Personal interview.
    31 August 1985.

Firko, Mrs. Helen Horvath Dusza Farkas, Vintondale. Personal interview.
    25 October 1980.

Frazier, Clarence, Wehrum Reunion. Personal interview.
    8 August 1983.

Fritchman, Vernon, Sr., Indiana. Telephone interview.
    4 August 1983.

Galesh, mrs. Anna Kovac and Eva Palanes, Central City. Personal interview.
    5 July 1984.

Garvin, Mrs. Rhoda Bennett, Indiana. Personal interview.
    5 October 1990.

George, Mrs. Sara Williams, Johnstown. Personal interview.
    5 September 1981.

Glowa, John, Weirton, West Virginia. Personal interview.
    3 September 1988.

Goughnour, William, Clyde. Personal interview.
    3 January 1988.

Graham, Harry, Indiana. Personal interview.
    25 August 1985.

Griffith, Judge George, Ebensburg. Personal interview.
    28 June 1982.

Hubner, Gabriel and Rose Larish Hubner, Vintondale. Personal interview.
    13 February 1982.

Hunter, Merle and Winifred, Rexis. Personal interview.
    14 april 1990.

Kangur, John, Vintondale. Personal interview.
    20 August 1983.

Kerekish, Mrs. Theresa, Vintondale. Personal interview.
    28 December 1981.

Kirker, Miss Thelma, Johnstown. Personal interview.
    10 July 1985.

Kish, Stephen and Mary Kerekish Kish, Vintondale. Personal interviews.
    27 and 28 December 1981.

Kranich, Mrs. Miriam Brett. Johnstown. Personal interview.
    3 August 1982.

Lantzy, Mrs. Ella Gill, Rexis. Personal interviews.
    7 January 1983, 11 February 1990, 25 February 1990.

- - -. Interview with Winifred Hunter
    5 May 1987.

Lazich, Mrs. Dora, Vintondale. Personal interviews.
    20 March 1982, 25 November 1983.

Lesak, George, Ebensburg. Personal interview.
    26 August 1989.

Ligush, John, Vintondale Homecoming. Personal interview.
    6 September 1987.

Ling, Harry, Vintondale Homecoming. Personal interview.
    5 September 1987.

Longazel, Mrs. Ann, Ebensburg. Personal interview.
    25 April 1986.

MacFarlane, James, Butler. Personal interview.
    13 March 1982.

Martich, John, Vintondale Homecoming. Personal interview.
    3 September 1989.

McConnell, Miss Olive, Beaver Dams, NY. Personal interview.
    20 June 1981.
McDevitt, Theresa, Indiana. Personal interview.
    1 September 1990.
McGee, Joseph, Vintondale. Personal interview.
    27 June 1982.
McGuire, Charles, Cresson. Personal interview.
    2 November 1980.
McMullen, William, Johnstown. Personal interview.
    14 November 1982.
McNaulty, Sarah Jones, Indiana. Personal interview.
    5 August 1987.
Mehalko, Frank, Vintondale. Personal interview.
    September 1982.
Metz, Lance, Easton. Personal interviews.
    12 June 1985, 19 October 1989.
Mihalik, Michael, Vintondale. Personal interview.
    2 November 1980.
Miller, Mrs. Charles (Ruth), Mahony City. Personal interview.
    11 June 1985.
Miller, Mrs. Hazel Zimmerman, Indiana. Personal interview.
    29 December 1984.
Miller, Jay, Vintondale Homecoming. Personal interview.
    1 September 1990.
Misner, Wilbur, Rexis. Personal interview.
    8 May 1990.
Morey, Julius, Vintondale. Personal interview.
    29 May 1982.
Mower, Charles, Vintondale. Personal interview.
    16 November 1980.
Myers, Randolph, Ebensburg. Personal interview.
    13 May 1984.
Neumeyer, Mrs. Helen Bretzin, Indiana. Personal interview.
    25 May 1990.
Oblackovich, Stephen and Hazel McCracken Oblackovich, Vintondale. Personal interview.
    13 February 1982.
Oravec, Mrs. Helen Balog, Vintondale. Personal interview.
    24 April 1982.
Orris, Michael and John, Wehrum Reunion. Personal interviews.
    8 August 1982.
Pioli, Mrs. Paulina Nevy, Vintondale. Personal interview.
    21 October 1981.
Pisaneschi, Hugh and Josephine Bossolo Pisaneschi, Vintondale. Personal interviews.
    9 November 1980.
Ploaich, Mrs. Helen Shestak Kasper, Vintondale. Personal interview.
    9 January 1983.
Pytash, Mrs. Anna Sileck, Vintondale. Personal interview.
    23 January 1982.
Pytash, George, Vintondale Homecoming. Personal interviews.
    6 September 1987, 1 September 1990.
Rabel, Mrs. Elizabeth Morey, Rexis. Personal interviews.
    9 January 1983, 17 March 1984.
Roberts, Mrs. Cora Bracken, Vintondale. Personal interview.
    25 October 1980.
Roberts, Joseph, Cresson. Personal interview.
    25 August 1982.
Rogalski, August and Anna Birch Rogalski, Vintondale. Personal interviews.
    1 November 1981.

Rok, Mrs. Dorothy Milazzo, Johnstown. Personal interview.
> 24 October 1990.

Rosenzweig, Aaron, Pittsburgh. Telephone interview.
> 13 August 1990.

Rosenzweig, Donald, Pittsburgh. Telephone interview.
> 13 August 1990.

Sandor, Mrs. Ida Bagu, Vintondale. Personal interview.
> 1982.

Schwerin, Clarence III, New York City. Telephone interview.
> 19 November 1981.

- - -. Personal interview.
> 7 December 1981.

Serene, Margaret, Indiana. Telephone interview.
> 9 June 1990.

Sholter, Sara Galer, Weickert. Personal interview.
> 4 August 1985.

Shook, Thomas, Johnstown. Personal interview.
> 19 August 1985.

Silverstone, Seymour, Johnstown. Personal interview.
> 9 October 1990.

Simoncini, Mrs. Jennie Grassi, Vintondale. Personal interview.
> 21 February 1982.

Smay, mrs. Mary Kreashko Averi, Vintondale. Personal interview.
> 15 September 1990.

Smith, Mrs. Pauline Bostick, Indiana. Personal interview.
> 7 October 1990.

Stinson, Mrs. Emma Huey, Altoona. Personal interview.
> 6 July 1984.

Toth, Baloz, Vintondale Homecoming. Personal interview.
> 3 September 1989.

Updike, Mrs. Mabel Davis, Ebensburg. Personal interviews.
> 28 June 1982, 23 August 1982, October 1982.

Ure, Louis and Aileen Huth Michelbacher Ure, Vintondale. Personal interview.
> 15 February 1990.

Uzo, Mrs. Katherine Morris Lybarger, Vintondale Homecoming Personal interview.
> 6 September 1987.

Williams, Lloyd, Vintondale. Personal interviews.
> 2 August 1981, 27 June 1982, 11 August 1985.

Wray, Mrs. Catherine Mahan, Johnstown. Personal interview.
> 26 June 1982.

Wray, James, Johnstown. Personal interview.
> 28 November 1980.

Young, Mrs. Martha Abrams, Johnstown. Personal interview.
> 13 March 1983.

LETTERS

Adams, Frederick Paris. Letters to the author.
> 23 March 1983, 10 June 1986.

Averi, Emmett, Lilly. Letters to the author.
> 14 September 1983, 18 June 1984.

Bearer, Clair, Ebensburg. Letters to the author.
> 23 July 1982, 4 August 1982.

Britsky, Mrs. Goldie Hubner. Letters to the author.
> 7 January 1983, 30 January 1983, 15 April 1983, and 30 June 1989.

Burgan, John. Letter to the Cambria County Historical Society.
> 15 July 1954.

Burr, Miss Deborah, Roebling, NJ. Letter to the author.
> 15 January 1981.

Caudill, Harry M., Whitesburg, Kentucky. Letter to the author.
    21 October 1985, died November 1990.
Claghorn, Rev. William L., Philadelphia, Letters to the author.
    21 October 1985, November 1990.
Cowley, Malcolm Sherman, Connecticut. Letters to the author.
    29 July 1982, Undated 1982, 17 December 1982, 28 August 1982, 19 May 1985, 11 October 1985, 13 February 1987.
- - -. Letter to Mrs. Iona "Eddie" Thompson.
    17 November 1967.
Delano, Warren. Letters to Charles Hower.
    2 November 1905, 6 November 1905, 8 November 1905, 10 November 1905, and 8 June 1906.
Donner, Mrs. Corinne Morris. Letters to the author.
    24 June 1989, 13 November 1989.
Downs, J.W. President of Lackawanna Steel Company. Letter to Lynn Adams, Superintendent of the Pennsylvania State Police.
    27 March 1922. State Police Files, Pennsylvania State Archives, Harrisburg.
Esaias, Mrs. Lovell Mitchell, Johnstown. Letter to the author.
    2 November 1983.
Griffith, George W. Letter to Milton Brandon.
    26 November 1931.
Griffith, Thomas, Executor of the Estate of Webster Griffith. Letter to C.M. Schwerin, President of the Vinton Colliery Company
    April 1933.
Gronland, Nicholas. Letters to the author.
    18 November 1982, 2 March 1983, 14 April 1983, and 13 September 1983.
Hoffman, Otto. Letter to Webster Griffith.
    18 February 1926.
Juba, George. Letters to the author.
    5 June 1985, 28 February 1990.
Long, John W. Letters to the author.
    22 June 1989, 12 September 1989.
McConnell, Miss Olive, Beaver Dams, NY. Letter to the author.
    17 March 1981.
McMullen, William. Letters to the author.
    10 November 1982, 17 September 1983, 6 June 1984.
Miller, Mrs. Charles (Ruth). Letter to the author.
    10 July 1985.
Miller, Mrs. Dorothy Misner. Letter to the author.
    19 July 1984.
President of Lackawanna Coal and Coke Company. (Luce?). Letter to Lynn Adams, Superintendent of State Police.
    27 March 1922. State Police Files, Pennsylvania State Archives, Harrisburg.
President of the Vinton Colliery Company (Delano?). Letter to J.C. McGinnis, Treasurer of the Cambridge Bituminous Coal Company, Frackville, PA.
    2 May 1906.
Schwerin, C.M. Letter to Lynn G. Adams, Superintendent of State Police.
    25 March 1922. State Police Files, Pennsylvania State Archives, Harrisburg.
- - -. Letter to M.F. Brandon.
    12 May 1933.
Sebulsky, Mrs. Claire (Todd), Anchorage. Letter to the author.
    14 January 1981.
Sternagle, Mrs. Beryl, Hollidaysburg. Letters to the author.
    19 June 1990, 8 July 1990.
Stinson, Mrs. Emma Huey. Letters to the author.
    7 June 1984, 10 July 1984.
Tucker, Mrs. Frank, Corapolis. Letter to the author.
    31 July 1983.

Wagner, Alton, Cheverly, Maryland. Letter to the author.
> 26 October 1990.

Young, Mrs. Martha Abrams. Letters to the author.
> 7 January 1983, 15 January 1983.

MANUSCRIPT SOURCES

Cambria County Historical Society

Jencks, J.H. **History of the Cambria & Indiana County Railroad Company**, n.p., n.d.

FRANKLIN DELANO ROOSEVELT PRESIDENTIAL LIBRARY, Hyde Park, New York

Delano Family Papers

Frederic Delano Papers

Sara Delano Roosevelt Papers

Warren Delano Papers

Photograph Collection

Warren Delano, A Tribute Written by Frederic Delano in September, 1920 following the death of his brother. n.p.

Pluchinsky, Frances Wojtowicz. "Memories of Vintondale". n.p., n.d.

State Line and Sullivan County Railroad. Letterhead. Courtesy of Nedra Snyder, Sullivan County Historical Society.

UNITED MINE WORKERS OF AMERICA, District 2 Files, Special Collections, Stapleton Library, Indiana University of Pennsylvania.

Baldwin, Roger. Letter to John Murphy.
> 2 May 1922.

- - -. Telegrams to John Brophy.
> 6 May 1922, 10 October 1922.

Brophy, John. Letter to Roger Baldwin.
> 10 May 1922, 21 October 1922.

- - -. Telegram to Roger Baldwin.
> 1 November 1922.

- - -. Telegram to P.T. Fagan, Pittsburgh.
> 31 August 1922.

- - -. Telegram to John L. Lewis, Philadelphia.
> 31 August 1922.

Dias, Charles at Wehrum in April 1922. Statement to T.H.W., G.W.F., and J.P.G. 3/20/23 at Heshbon, PA.

Fagan, P.T. Letter to John Brophy.
> 1 September 1922.

Injunction: William Welsh vs. Vinon Colliery Comapny, 1922, Box 31.

Injunction: John brophy et al. vs. Vinton Co.lliery Comapny, Setpember Term 1922, Box 31.

J.J. Kinter. Letters to John Brophy.
> 19 June 1922, 26 June 1922.

Milner, Lucille, Field Secretary for ACLU. Letter to John Brophy.
> 6 November 1922.

Strike Receipts, 1909. Box 8, Folder 347.

UNIVERSITY OF PENNSYLVANIA, Phiadelphia. University Archives. Alumni File of Clarence Claghorn.

UNIVERSITY OF PITTSBURGH. Hillman Library. Industrial Archives. Bethlehem Mines Corporation, Police Department Reports, 1926-1929.

VINTON COLLIERY COMPANY: EXTANT RECORDS

Agreement: Delano, Warren and Albert Fletzer on Rental of Vintondale Inn.
> 7 March 1912.

Agreement: Stauft, D.B., Scottdale and Vinton Colliery Company, use of "Stauft Watering Device" for cooling coke.
> 4 November 1909.

Lease: Vinton Land Company to Vinton Colliery Company. 31 July 1916. Philadelphia: Allen, Lane & Scott.

Mine Superintendent Diaries. 1912, 1913 (incomplete), 1914, 1917, and 1923.

Mortgage: Vinton Land Company to The Pennsylvania Company for Insurances on Lives and Granting Annunities. 1 August 1916. Philadelphia: Allen, Lane & Scott.

Vinton Colliery Company Mine Scrip, 1940.

Walter Scranton's Letter Book, January-June 1902. Archives. Hugh Moore Canal Museum and Park, Easton, PA.

ACCOUNT, MINUTE AND REGISTER BOOKS

Automobile Account Book of John L. Huth, 1916-1917.

Baptist Church, miscellaneous years, 1896-1930.

Borough of Vintondale Minutes, 1921-1930.

Hungarian Reformed Church, 1916-1930, located at the Hungarian Reformed Church Archives at the Bethlehen Home in in Ligonier, translated by Julianna Chaszar.

Immaculate Conception Church Collections Records, 1915. Found among accounts of John Huth.

Vintondale Hunting Club. Courtesy of Lloyd Williams.

Vintondale Inn: Registers, 1911 and 1922. Found in the Huth attic.

ST. FIDELIS ROMAN CATHOLIC CHURCH, Wehrum.

St. Fidelis Parish File, Diocese of Pittsburgh, 1923.

"Mission Churches" **History of St. Bernard Parish**, Indiana, PA. Diocese of Greensburg.

St. Bernard Parish File, Diocese of Greensburg.

COUNTY, STATE, AND FEDERAL RECORDS

DEED BOOKS

Blair County: All deeds, mortgages, agreements, wills, etc., dealing with Lot Irwin oand his heirs.

Cambria County: Vintondale. All deeds, charters, mortgages, and leases related to Eliza Furnace, Blacklick Land and Improvement, Vinton Colliery Company, Warren Delano, A.V. Barker, Clarence Claghorn, Samuel Lemon Reed, Vinton Land Company, Vintondale Supply Company, Vintondale Inn, Vintondale Amusement Company, Cambridge Bituminous Coal Company, Lackawanna Coal and Coke Company, Webster Griffith, Jacob and Samuel Brett, Vinton Lumber Company, Vintondale State Bank, Mike Farkas and other miscellaneous deeds relating to Vintondale until 1930.

Clearfield County: All deeds and wills related directly to A.W. Lee and the Vinton Lumber Company.

Indiana County: Claghorn. All deeds, charters, mortgages, etc. related to Lackawanna Coal and Coke Company, Lackawanna Iron and Steel Company, Vinton Colliery Company, Vinton Land Company, Claghorn Water Company, Augustine Vinton Barker, Clarence Claghorn, and James Mitchell.

Indiana County: Rexis. All deeds, charters, mortgages, etc. related to Vinton Lumber Company, Cambria and Indiana Railroad, Ebensburg and Blacklick Railroad, Blacklick and Yellow Creek Railroad, pennsylvania Railroad, and Blair Shaffer.

Indiana County: Wehrum. All deeds, charters, etc. related to Lackawanna Coal and Coke Company, Lackawanna Iron and Steel Company, Lackawanna Steel Company, Bethlehem Steel Corporation, Bethlehem Mines Corporation, Wehrum National Bank, Wehrum Bank Association, Blacklick Inn, Wehrum Band Association, and James Mitchell.

County of Philadelphia. All deeds, charters, wills, etc. related to Theodore Bechtel, James Raymond Claghorn, Clarence Claghorn, John Krause, Abraham Rex, and George Rex.

EJECTION DOCKETS

Cambria County, 1896-1924

Indiana County, 1922

ROAD DOCKETS

Cambria County

Indiana County Road Docket 8, 1917.

Indiana County Bridge Book.

TAX ASSESSOR RECORDS

Cambria County, Jackson township, 1848-1860

Cambria County, Vintondale, 1925

Indiana County, Claghorn, 1917-1925

Indiana County, Rexis, 1902-1912

Indiana County, Wehrum, 1902-1912

STATE RECORDS

Patent and Warrant Books, Pennsylvania State Archives, Harrisburg.

State Police Files, Pennsylvania State Archives, Harrisburg.

Birth and Death Certificates, 1906-1921. Blacklick Valley. Found in the Huth attic.

FEDERAL RECORDS
The United States Census, Philadelphia, 1850, 1860, 1870, and 1880.
The United States Census, Jackson Township Cambria County, 1840.

PUBLISHED PRIMARY SOURCES
Newspapers
Free Press 25 March 1911, 1 April 1911, 29 April 1911, 20 May 1911, 12 December 1912.
Indiana County Gazette, miscellaneous issues, 1891-1897.
Indiana Evening Gazette, miscellaneous issues, 1922-1924.
Indiana Progress, miscellaneous issues, 1899-1906.
Indiana Times, miscellaneous issues, 1890-1930.
Indiana Weekly Messenger, miscellaneous issues, 1903.
Johnstown Democrat, miscellaneous issues, 1922, 1924, 1930.
Johnstown Tribune, March and April, 1924.
Johnstown Weekly Tribune, miscellaneous issues, 1907, 1915.
McCormick, John. "Vintondale, Prosperous Town at Forks of Blacklick". Johnstown Weekly Tribune. 28 June 1907.
Nanty Glo Journal. 1921-1930, all issues.
New York Times. March 11, 1909, September 10, 1920, April-July, 1922, November 3, 1940, June 6, 1956, July 24, 1966.
Philadelphia Inquirer. August 27, 1884, March 6, 1891, September 11, 1906.

ANNUAL STATE REPORTS
Report of the Department of Mines, Commonwealth of Pennsylvania.
Annual Reports of Mine Inspectors: Anthracite, 1880-1903.
Annual Reports of Mine Inspectors: Bituminous 1894-1930.
Pennsylvania Department of Public Instruction. **Report of the Superintendent of Public Instruction**. Harrisburg: J.H. Kuhn, Printer of the Commonwealth, 1916-1919.
"Clarence Schwerin".
Twenty-fifth Anniversary of the Class of 1901. Columbia School of Mines, 1925.

BOOKS
Atlas of Indiana Co. Pennsylvania. From actual Surveys by and under the Direction of F.W. Beers. New York: F.W. Beers & Co., 1871. Reprinted by Walsworth Don Mills, 1982.
Blankenhorn, Heber. **Strike for Union**. Published for the Bureau for Industrial Research. New York: H.W. Wilson, 1924. Reprinted by Arno Press and the New York Times, 1969.
Brophy, John. **A Miner's Life**. an autobiography edited and supplemented by John O.P. Hall. Madison: University of Wisconsin Press, 1964.
Caldwell, J.A. **Illustrated Historical Combination Atlas of Cambria County, Penna**. Philadelphia: Atlas Publishing Company, 1890.
Caudill, Harry. **A Darkness at Dawn. Appalachian Kentucky and the Future**. Lexington: University Press of kentucky, 1976.
- - -. **Night Comes To the Cumberlands. A Biography of a Depressed Area**. Boston: Little, Brown & Company, 1963.
- - -. **Theirs Be The Power. The Moguls of Eastern Kentucky**. Bloomington: University of Illinois, 1983.
Cowley, Malcolm. **And I worked at the writer's trade: chapters of literary history, 1918-1978**. New York: Viking Press, 1978.
- - -. **Blue Juniata: Collected Poems**. New York: Viking Press, (1968).
- - -. **Exile's Return: a literary odyssey of the 1920s**. New York: Viking Press, 1951.
Grisak, Michael J., compiler. **The Grisak Family History**, Volume I, based on the autobiography of Joseph Grisak, 1878-1950.
Gospell, James, Ed. **Gospell's City Directory**. Phiadelphia, 1870.
Hays, Arthur Garfield. **City Lawyer: The Autobiography of a Law Practice**. New York: Simon & Shuster, 1942.
Josephson, Emanuel M. **The Range Death of Franklin D. Roosevelt. History of the Roosevelt-Delano Dynasty America's Royal Family**. New York: Chedney Press, New & Revised Edition, 1959.
Platt, William G. Pennsylvania **Second Geological Survey**, 1877. Harrisburg: 1878.
McElroy's Philadelphia Directory, 1850, 1860.

MAGAZINE ARTICLES

Hapgood, Powers. "Coal Strike Continues." **The New Republic.** 4 October 1922, 32: 147.

---. "Two Days from a Diary". **The Survey** 22 March 1922, pp. 1033-34.

Blankenhorn, Heber. "After West Virginia-Somerset." The Survey 13 May 1922, pp. 238-240.

Hayes, Arthur Garfield. "Flag in the Coal Mines". The New Republic 14 June 1922, pp. 74-75.

Journal of the United Mine Workers of America. Columns on the Vintondale strike and the Wehrum mine explosion. 21 January 1909, 1 April 1909, 13 May 1909, 20 May 1909, 27 May 1909, 1 July 1909.

Secondary Sources

Published Books

Barnes, Andrew Wallace, ed. **History of the Philadelphia Stock Exchange, Banks, and Baking Interests.** Philadelphia: Cornelius Baker, Inc., 1911.

Belders, Daniels. **Pennsylvania: Birthplace of Banking in America.** Pennsylvania Bankers Association, 1976.

Bender, Frederick Moore. **Coal Age Empire: Pennsylvania Coal and Its Utilization to 1860.** Harrisburg: Pennsylvania Historical and Museum Commission, 1974.

Biographical and Portrait Cyclopedia of Cambria County, Pennsylvania, Comprising About Five Hundred Sketches of the Prominent and Representative Citizens of the County. Philadelphia: Union Publishing Company, 1896.

Biographical Encyclopedias of Pennsylvania. Philadelphia: Galaxy Publishing Company, 1874.

Bodnar, John. **The Ethnic Experience in Pennsylvania.** Lewisburg: Bucknell University Press, 1973.

Burgess, George and Miles Kennedy. **Centennial History of the Pennsylvania Railroad.** Philadelphia: Pennsylvania Railroad Company, 1949.

Burr, Charles Todd. **General History of the Burr Family with a Genealogical Record from 1193-1891.** 2nd edition. New York: Knickerbocker Press, 1891.

Cambria County Sesquicentennial Book, 1804-1954.

Casson, Herbert N. **The Romance of Steel. The Story of a Thousand Millionaires.** New York: A.S. Barnes & Company, 1907.

Claghorn, William Crumby, compiler. **The Barony of Cleghorne AD. 1203 Lanarkshire, Scotland to the Family of Claghorn AD 1912. United States of America.** (120 copies printed for Claghorn Family and their connections and friends), #5. Philadelphia: Lyon & Armor, Printers.

Delano, Daniel. **FDR and the Delano Influence.** Pittsburg, 1946.

Gable, John. **History of Cambria County Pennsylvania.** Two Volumes. Topeka: Historical Publishing Company, 1926.

Hayes, Calvin. **History of the Blairsville Presbytery 1830-1930,** N.P., N.D.

History of Indiana County, Penn'a. 1745-1880. Newark, Ohio: J.A. Caldwell Company. Reprinted Walsworth Publishing Company, 1981.

History of Mercer County, Penna. Its Past & Present. Chicago: Brown, Runk & Co., 1888.

Jalkanen, Ralph J., ed. **The Finns in North America: A Social Symposium.** Hancock, Michigan: University of Michigan Press, 1969.

Kay, John L. and Chester M. Smith, Jr. **Pennsylvania Postal History.** Lawrence, MA: Quarterman Publicatins, Inc. 1976.

Keystone Coal Catalog, combined with Coal Field Directory for the year 1920. Pittsburgh: Keystone Consolidated Publishing Company, 1920.

Kleeman, Rita Halla. **Gracious Lady: The Life of Sara Delano Roosevelt.** New York: D. Appleton-Century Company, 1935.

Kline, Benjamin, Jr. "Wild Catting" on the Mountain. Book No. 2 in the series **Logging Railroad Era of Lumbering in Pennsylvania.** Privately Printed, 1970.

Leary, Thomas E. and Elizabeth C. Sholes. **From Fire to Rust: Business, Technology and Work at the Lackawanna Steel Plant, 1899-1893.** Buffalo: Buffalo and Erie County Historical Society, 1987.

Lee, J. Marvin. **Centre County-The County in Which We Live.** (State College?), 1965, revised 1974.

Maynard, D.S. **Industries and Institutions of Centre County.** Richie & Maynard, publishers, 1877.

Moody, John. **Moody's Analyses of Investments & Security Rating Books. Railroad Investments, 14th Year, 1923.** New York: Moody's Investors Service.

Moser, H.O. **A History of Delano, Pennsylvania 1861-1931.** No Publisher or Date.

Pietrok, Paul. **The Buffalo, Rochester & Pittsburgh Railway.** Privately Printed, 1979.

Reynolds, Patrick. **History and Mystery of Pennsylvania.** Volume V. **Pennsylvania Profiles.** Willow Street, Pensylvania: Red Rose Studios, 1981.

Roberts, Peter. **The New Immigration. A Study of Industrial and Social Life of Southeastern Europeans in America**. New York: MacMillan Company, 1912. Reprinted by Arno Press and the New York Times, 1970.

Roosevelt, Elliot and James Brough. **The Untold Story: The Roosevelts of Hyde Park**. New York: G.P. Putnam's Sons, 1973.

Schotter, H.W. **Growth & Development of the Pennsylvania Rialroad Company, 1946-1926**. Philadelphia: Press of Allen Lane & Scott, 1927.

Sharp, Myron. **A Guide to Old Stone Blast Furnaces in Western Pennsylvania**. Philadelphia: Allen Lane & Scott, 1927.

Souders, D.A., Superintendent of Immigration of the Reformed Church in the United States. **The Magyars in America**. New York: George H. Doran Company, 1922.

Steeholm, Clara and Hardy. **The House at Hyde Park**. New York: Viking Press, 1950.

Stephenson, Clarence. **Buena Vista Furnace**. Marion Center, PA: Mahoning Mimeograph and Pamphlet Service, 1968.

- - -. **Indiana County 175th Anniversary History**. Four Volumes. Indiana: A. G. Halldin Publishing Company, 1978.

Stewart, J.T. **Indiana County Pennsylvania. Her People, Past & Present**. Two Volumes. Chicago: J.H. Beers & Company, 1913.

Storey, Henry Wilson. **History of Cambria County. With Genealogical Memoirs**. Three Volumes. New York: Lewis Publishing Company, 1907.

Swetnam, George and Helene Smith. **A Guide to Historic Western Pennsylvania**. Pittsburgh: University of Pittsburgh Press, 1976.

Ward, Geoffrey. **Before the Trumpet. Young Franklin Roosevelt 1882-1905**. New York: Harper & Row, Publishers.

Wilson, William Bender. **History of the PRR**. Two volumes. Philadelphia: Henry T. Coates & Company, 1895.

Yearly, C.R., Jr. **Enterprise and Anthracite: Economics and Democracy in Schuylkill County, 1820-1875**. Baltimore: The Johns Hopkins Press, 1961.

MAGAZINE ARTICLES

Claghorn, Clarence R., "Notes on the Bernice Anthracite Coal-basin, Sullivan County, PA. **Transactions**. American Institute of Mining Engineers, New York Meeting. Feb. 1889: 606-616.

- - -. "A Modified Longwall System." Notes on the Method Employed at the Vintondale Mine of the Vinton Colliery Company. **Mines and Minerals**. August 1901: pp. 16-18.

Cunningham, Howard. "Ghost Towns of Indiana County." Verticle Files. Notes from meeting of Historical and Genealogical Society of Indiana County.
 11 April 1958.

Fratrick, William. "Forgotten Towns of Indiana County." **Pennsylvania History**. January 1960.

"Henry Wehrum". Obituary. **Iron Age**.
 29 November 1906.

Roddy, Dennis B. "Cowley legacy a generation of U.S. writers." **The Pittsburgh Press**. 2 April 1989: B1.

Sosinski, Stephan, "2 get chair in county train robbery." Cambria County Magazine. 2 March 1977: 7.

Weber, Denise Dusza. "The Wehrum Mine Explosion." **Indiana County Heritage** 9.1 (1984): 20-26.

LETTERS

Franklin Delano Roosevelt Library, Hyde Park, NY. Letter to Gloria Berringer, Historical and Genealogical Society of Indiana County.
 25 October 1979.

Richard Rosenzweig. Letter to author.
 11 August 1990.

Seeber, Frances, Supervisory Archivist, Franklin Roosevelt Library. Letter tro the author.
 24 February 1982.

INTERVIEWS

Coveny, Fr. James. Holy Family Parish, Colver. Telephone interview.
 28 November 1990.

Gift, Kathryn. Snyder County Historical Society. Letter to author.
 6 November 1989.

Kirk, Mrs. Ressie Rairigh. Telephone interview.
 5 October 1990.

Aaron Rosenzweig, Pittsburgh. Telephone interview.
    16 August 1990.
Don Rosenzweig, Pittsburgh. Telephone interview.
    16 August 1990.
Richard Rosenzweig. Letter to author.
    11 August 1990.

ACKNOWLEDGEMENTS

The author is extremely grateful to many people for helping with this book from its inception to its printing. My biggest thanks goes to my husband, Tom, and my children, Heidi, Gretchen, and Mike for putting up with all the mess for the past ten years. Guess what! The mess is not going anywhere until Volume II is done. To my parents: Steve and Agnes Dusza, thanks for all the input and Mum, thanks for hanging around long enough to know it was definitely in the final stages. To my sister Dee for taking the "after" photos of Wehrum and Vintondale and to Fr. Don for helping with the deed searches and for his research on the Catholic churches in the valley.

To Eileen Cooper for getting me hooked on local coal mining history. To Tood Sebulsky and Boyd Brandon for their special support. To the Indiana Area School District for two sabbatical leaves during which a large part of the research was conducted, and for several Academic Excellence Grants, one of which sponsored the translation of the tombstones in the Wehrum Cemetery. To the Pennsylvania Historical and Museum Commission for a Matching Research Grant in 1985, which financed the trip to the Roosevelt Library in Hyde Park. Also to the PHMC for granting permission to use copies of the 1906 glass negatives which were given to them in 1984. To the University of Pittsburgh Ethnic Heritage Studies Center and Joe Makarewicz for the grant to copy and translate the records of the Hungarian Reformed Church. A very special thanks to Juliana Chaszar, at the time a junior in high school, for translating these records.

To the late Joe Dodson for loaning his copies of the mine diaries. To Pauline Jones and children John, Paula and Helen and the late Helen Kimak for allowing me to use the 1923 mine diary, a most important resource. To the late Iona "Eddie" Thompson of Ebensburg for all her letters and contacts, and especially her friendship. It was my privilege to have met her. To the late Malcolm Cowley, thanks for all your advice and for being so gracious when I dragged you all through Cambria County. It was a real honor to have met you. To Lloyd Williams, Pauline Smith, Mabel Updike, Thelma Kirker, Gloria Risko, and Hazel Dill: thank you for all the rare photographs and background information.

To Betty Nedrich, Bill Martin and the late Bob Hutchinson for putting up with me for weeks on end while I was gleaning back issues of the **Nanty Glo Journal** for information. Special thanks to Bill Martin for the publicity. To the Johnstown **Tribune-Democrat** and the Greensburg **Tribune-Review**, your continued news coverage of this project is much appreciated.

A special thanks to Aileen Michelbacher Ure for the family photographs and those of the company store and for all the family history. To Joe Pluchinsky for allowing me to print Frances' marvelous "Memories". To all those who agreed to be interviewed and/or loaned me photographs, the book could not have been possible without all of you.